Psychoactive Drugs

Contemporary Neuroscience

Psychoactive Drugs

Tolerance and Sensitization

Edited by

A. J. Goudie

University of Liverpool, Liverpool, England

and

M. W. Emmett-Oglesby

Texas College of Osteopathic Medicine, Fort Worth, Texas

Humana Press • **Clifton, New Jersey**

Library of Congress Cataloging-in-Publication Data

Main entry under title:

Psychoactive drugs : tolerance and sensitization / edited by A. J.
Goudie and M. W. Emmett-Oglesby.
    p. cm. — (Contemporary neuroscience)
    Includes bibliographies and index.
    ISBN 0-89603-148-9
    1. Psychotropic drugs. 2. Drug tolerance. I. Goudie, A. J.
(Andrew J.) II. Emmett-Oglesby, M. W. (Michael W.) III. Series.
    [DNLM: 1. Antidepressive Agents—pharmacology. 2. Behavior—drug
effects. 3. Drug Tolerance. 4. Psychotropic Drugs—pharmacology.
QV 77 P9694]
RM315.P745 1989
615'.788—dc20
DNLM/DLC
for Library of Congress   89-11094
                                                  CIP

© 1989 The Humana Press Inc.
Crescent Manor
PO Box 2148
Clifton, NJ 07015

Printed in the United States of America

Contents

v

Behavioral and Pharmacological History as Determinants of Tolerance- and Sensitization-Like Phenomena in Drug Action

James E. Barrett, John R. Glowa, and Michael A. Nader

Tolerance to Drug Discriminative Stimuli
Alice M. Young and Christine A. Sannerud

Part 2:   Molecular Mechanisms

Dispositional Mechanisms in Drug Tolerance
and Sensitization
A. D. Lê and Jatinder M. Khanna

Dopamine Receptor Changes During Chronic Drug Administration

Andrew J. Greenshaw, Glen B. Baker, and Thomas B. Wishart

Sensitization to the Actions of Antidepressant Drugs
Paul Willner

Adaptation in Neuronal Calcium Channels as a Common Basis for Physical Dependence on Central Depressant Drugs
John M. Littleton and Hilary J. Little

Part 3: Summary Chapters

Behavioral Tolerance and Sensitization
Definitions and Explanations
Derek E. Blackman

Drug Tolerance and Sensitization
A Pharmacological Overview
Harold Kalant

Contributors

GLEN B. BAKER • *Department of Psychiatry, University of Alberta, Edmonton, Alberta, Canada*

JAMES E. BARRETT • *Department of Psychiatry, Uniformed Services University of the Health Sciences, Bethesda, Maryland*

DEREK E. BLACKMAN • *University of Wales, College of Cardiff, U.K.*

MICHAEL W. EMMETT-OGLESBY • *Department of Pharmacology, Texas College of Osteopathic Medicine, Forth Worth, Texas*

JOHN R. GLOWA • *Biological Psychiatry Branch, National Institute of Mental Health, Bethesda, Maryland*

ANDREW J. GOUDIE • *Department of Psychology, University of Liverpool, Liverpool, U.K.*

ANDREW J. GREENSHAW • *Department of Psychiatry, University of Alberta, Edmonton, Alberta, Canada*

HAROLD KALANT • *Department of Pharmacology, University of Toronto, Toronto, Ontario, Canada*

JATINDER M. KHANNA • *Addiction Research Foundation of Ontario and Department of Pharmacology, University of Toronto, Toronto, Ontario, Canada*

A. D. LÊ • *Addiction Research Foundation of Ontario, Toronto, Ontario, Canada*

HILARY J. LITTLE • *Department of Pharmacology, The Medical School, Bristol U.K.*

JOHN M. LITTLETON • *Department of Pharmacology, Kings College, London, U.K.*

MICHAEL A. NADER • *Department of Psychiatry, Uniformed Services University of the Health Science, Bethesda, Maryland*

CHRISTINE A. SANNERUD • *Division of Behavioral Biology, Johns Hopkins University Medical School, Baltimore, Maryland*

SHEPARD SIEGEL • *Department of Psychology, McMaster University, Hamilton, Ontario, Canada*

PAUL WILLNER • *Psychology Department, City of London Polytechnic, London, U.K.*

THOMAS B. WISHART • *Department of Psychology, University of Saskatchewan, Saskatoon, Saskatchewan, Canada*

DAVID L. WOLGIN • *Institute for the Study of Alcohol and Drug Dependence, Department of Psychology, Florida Atlantic University, Boca Raton, Florida*

ALICE M. YOUNG • *Department of Psychology, Wayne State University, Detroit, Michigan*

Preface

Research into the processes of tolerance and sensitization has escalated at a substantial rate in recent years, presumably because of the fundamental importance of understanding the long-term, as opposed simply to the initial, acute effects of drugs. The rapid growth of such research in recent years is documented clearly by the editors in the introductory chapter to this text.

However, despite the fact that there is a very large amount of literature concerned with the effects of long-term drug treatment, there is, to the best of our knowledge, no published text that has ever attempted to integrate some of the many diverse findings that have been made in this area. Basic research has uncovered a number of different mechanisms by which tolerance and sensitization to drugs can develop. Such mechanisms are of very different types, involving psychological behavioral, metabolic, neuronal, and subcellular processes. Because of the complexity of *each* of these different types of mechanisms, with few exceptions, individual researchers usually tend, understandably, to concentrate on their own specific areas of expertise, paying relatively little attention to relevant research occurring in other areas. Consequently, they neglect or simply ignore the important question of the *relative* importance of the specific mechanism that they are studying, and the related question of the possible interrelationships that may exist between different mechanisms for the production of tolerance and sensitization.

The editors of this book believe that major advances in tolerance/sensitization research will probably only come about as a result of significant interdisciplinary collaboration. We have consequently assembled definitive reviews from acknowledged authorities of specific areas of tolerance/sensitization research, our review authors having very different scientific backgrounds. Thus, we have compiled in one text reviews dealing with important issues in tolerance/sensitization research, in the hope that such a compilation will encourage interdisciplinary collaboration between researchers into tolerance and sensitization.

The initial chapter of the book (by the two editors) consists of a brief historical overview of relevant research and an introduction to later chapters. In the next two chapters, Wolgin and Siegel review the large body of evidence that has accumulated in recent years showing that learning processes (both Instrumental and Pavlovian) can be critically involved in tolerance development. In the following chapter, Barrett and colleagues review provocative behavioral experiments that show clearly that both tolerance- and sensitization-like phenomena can be produced by nonpharmacological factors, such as the past behavioral history of experimental animals. Such experiments provide a substantial challenge to researchers attempting to account for *all* facets of tolerance or sensitization in purely reductionist terms. In the fifth chapter, Young and Sannerud provide a definitive review of the historically controversial question of whether or not tolerance develops in animals to the actions of drugs as discriminative stimuli—such actions of drugs being widely believed to be related to ''subjective'' responses to drugs in humans, which are believed to be important factors in drug abuse and dependence.

Subsequent chapters are concerned with specific mechanisms involved in tolerance/sensitization development that involve more reductionist or ''molecular'' processes. Lê and Khanna provide an up-to-date, detailed review of the role of the historically well-documented dispositional mechanisms involved in tolerance and sensitization to a number of drugs. Greenshaw and colleagues subsequently discuss tolerance and sensitization from the viewpoint of receptor mechanisms. In order to do this, they concentrate on changes in dopamine receptors that accompany chronic drug administration. In the following chapter, Willner also considers the role of receptor mechanisms in detail when discussing the therapeutically important question of the mechanisms involved in sensitization to antidepressants. Finally, Littleton and Little describe recent provocative studies suggesting that important insights into tolerance and dependence may be gleaned from the study of the physiology and pharmacology of the cellular calcium channels that regulate neuronal excitability.

The various review chapters presented in this book consequently survey a diverse range of mechanisms involved in tolerance and sensitization, from complex cognitive processes involved in learning to adaptive responses at the level of receptors and calcium channels. Thus, the articles found here demonstrate the diversity of research approaches in this area and, we consequently believe, highlight the importance of compiling an interdisciplinary text in this area. The need for such a text is further illustrated in the final section of the book, in which we are fortunate to be able to include excellent, critical overviews—by an eminent behavioral scientist, Derek Blackman, and an equally eminent pharmacologist, Harold Kalant—of the review material in the previous chapters. These summary chapters provide insightful critiques of the previous chapters and raise some fundamental issues for tolerance and sensitization researchers, including questions such as: "What *exactly* is tolerance," "How, if at all, do different mechanisms for the production of tolerance interact," and "What is the optimal scientific strategy for the analysis of the processes involved in tolerance and sensitization?"

We believe that this interdisciplinary text, which is both thought-provoking and definitive in its reviews of specific research areas, will be of value to researchers in a number of disciplines (pharmacology, psychiatry, psychology, and physiology) who are interested in tolerance and sensitization. The text should also be useful to graduate students taking advanced courses in these disciplines. Selected review chapters should also be helpful in advanced undergraduate teaching.

We thank all of our contributing authors for their excellent review chapters and prompt attention to editorial requests; the staff at Humana Press for tolerating our (not infrequent) correspondence and, above all, our wives, Jackie and Gloria, for their failures to sensitize to our potentially irksome behaviors while editing this text.

Andrew J. Goudie
Michael W. Emmett-Oglesby

Tolerance and Sensitization: Overview

Andrew J. Goudie and Michael W. Emmett-Oglesby

The theories evolved by the different investigators in explanation of these findings [of morphine tolerance] are both numerous and varied—*DuMez, 1919*

None of the various mechanisms described for the development of tolerance to alcohol may be accepted as a complete explanation for the phenomenon observed—*Bogen, 1936*

The investigation of acquired tolerance to alcohol is a most difficult problem, since a variety of factors, both physiological and psychological may be involved in its explanation—*Newman and Card, 1937a*

It should be apparent from the historical quotations cited above, all of which are at least 50 years old, that for many years tolerance to the actions of drugs has been thought of as being mediated by a number of different mechanisms. Indeed, it can be argued that the term tolerance refers simply to *all* of the various possible mechanisms by which effects of drugs can be reduced during long-term drug treatment (Kalant et al., 1971). The chapters in this book deal with a number of different mechanisms that are currently thought to be involved in tolerance and sensitization processes. All too often, it appears that work on tolerance and sensitization occurs within a relatively restricted frame of reference, and studies are often confined to, at most, one or two mechanisms that may be involved in the development of tolerance or sensitization. This text attempts to take an interdisciplinary approach to tolerance and sensitization, in the hope that it will encourage individuals working on different mechanisms for these phenomena to consider the possible relationships and interactions among the various mechanisms involved.

Opiates are in many ways the prototypical drugs that show a substantial degree of tolerance. There is extensive historical evidence that the medical and nonmedical use of various forms of opiates (opium eating, ingestion of laudanum, and so on) was commonplace in the 19th century in both the U.S.A. (*see* Courtwright, 1982) and U.K. (*see* Berridge and Edwards, 1981). Indeed, there were no legal restraints on the use of opiates in either country in the 19th century (Siegel, 1986). It therefore seems highly likely that knowledge about drug tolerance was relatively widespread in the last century. Given the substantial tolerance that develops to many drugs and the evidence that a variety of psychoactive agents have been used by humans since prehistoric times (Kramer and Merlin, 1983), it is also likely that the basic phenomenon of tolerance has actually been well known since prehistoric times.

Tolerance to the effects of drugs (with its presumed relationship to dependence) has been a well-known phenomenon in scientific circles for many years. For example, when writing a monograph for the Scottish Temperance League in 1855 favoring the virtues of total abstention from alcohol, W. D. Carpenter, Professor of Medical Jurisprudence and Examiner in Physiology at University College London, provided the following graphic description of tolerance to opiate drugs and to alcohol:

No sane man questions that opium is a poison . . . yet the system may be so habituated to its effects (as it may become to those of other narcotics), that doses which would prove immediately fatal to those unaccustomed to them, are taken as necessaries of life by such as have bought themselves into dependence upon these substances, and may be affirmed to be a part of their daily diet. And so it is with regard to Alcohol;. . . .

Siegel (1983) has described vividly how such ideas about tolerance were sufficiently well known in the 19th century for them to be incorporated into the writings of the Victorian novelist Wilkie Collins as a part of the plot of the novel *The Moonstone*, suggesting that knowledge of drug tolerance must have been relatively commonplace at that time.

The experimental investigation of mechanisms involved in the processes of tolerance and sensitization is, of course, a more recent development. The greater part of this research has been into mechanisms involved in tolerance rather than sensitization, although in recent years there has been considerable interest in the processes involved in sensitization, with particular respect to stimulant (Kilbey and Sannerud, 1985) and antidepressant drugs (*see* Willner, this text). However, sufficient experimental research had been conducted into mechanisms involved in opiate tolerance by the early years of this century for A. G. DuMez, a U.S. Public Health Service Official, to write a review in 1919 for the *Journal of the American Medical Association* on experimental investigations of mechanisms involved in tolerance and withdrawal to morphine.

In current usage, tolerance is usually operationally defined as a shift to the right in the dose–response curve that occurs with chronic drug treatment; sensitization is correspondingly defined as a shift to the left in the dose–response curve. Such definitions carry no implications concerning the mechanisms that may mediate these events. An early empirical demonstration of tolerance in a behavioral study involving such a shift in dose–effect curves was provided by the study of Newman and Card (1937b) on tolerance to the actions of alcohol in dogs. Having compiled a rating scale to measure the "degree of drunkenness" induced in dogs by alcohol, these authors habituated dogs to alcohol for 13 mo (by allowing the dogs to ingest alcohol orally). Figure 1 shows the "degree of drunkenness" at various blood alcohol concentrations induced by a test dose of alcohol in two groups of dogs: a control group that was alcohol naive, and a group of animals that was habituated to alcohol

Degree of drunkenness

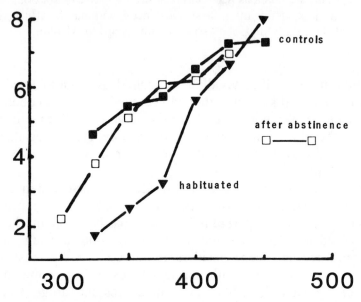

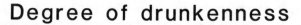

Mgm alcohol per 100 cc blood

Fig. 1. Degree of "drunkenness" at various blood alcohol concentrations induced by a test dose of alcohol in two groups of dogs. The control group was drug naive; a second group was habituated to alcohol for 13 mo. The habituated group was also studied after 7 mo of abstention from alcohol.

for 13 mo. The test dose of alcohol was also studied in the drug-habituated animals when they had abstained from alcohol for some 7 mo.

These data show clearly how habituation to alcohol produced a shift to the right in the dose–effect curve (i.e., tolerance) and how such tolerance was lost after abstinence from alcohol. Such data resemble many current studies in which tolerance is exhibited as a reversible shift in the dose–effect curve.

If tolerance/sensitization are defined as shifts in the dose–effect curve, it follows that research should attempt to define the proportion of the shift in the curve that results from the action of specific mechanisms that induce tolerance/sensitization. Furthermore, it is important to attempt to know how the relative significance of various mechanisms that produce tolerance/sensitization may be modified by basic pharmacological variables, on the one hand (such as the drug studied, drug dose, frequency of dosing, duration of dosing, route of dosing, and so on) and on the other hand, by behavioral variables (such as the specific behaviors studied and the environmental and behavioral contexts within which they are studied). Such goals for tolerance/sensitization research are clearly difficult to achieve, and they may appear unduly ambitious. However, unless basic research is conducted with these goals in mind, how are we to relate the results of fundamental research to the real world and establish the "ecological validity" of the various mechanisms for the induction of tolerance/sensitization that are studied in the experimental laboratory? For example, in a major review of tolerance to nonopiate drugs, Kalant et al. (1971) discussed the question of the significance of the tolerance that was achieved by increased drug metabolism following chronic drug administration. They noted that, for a number of sedative agents, including ethanol, the magnitude of metabolic tolerance observed was relatively small (i.e., it only involved relatively small increases in rates of drug metabolism). As a consequence of this finding, it follows that in studies with large drug doses given intraperitoneally (when the drug reaches the brain relatively rapidly), metabolic mechanisms will have relatively little influence on tolerance, particularly if behavioral testing takes place soon after injection. For small drug doses, the significance of metabolic tolerance will be greater than for large doses, particularly for drugs administered orally. Even with large doses, the relative significance of metabolic tolerance will be greatest if behavioral testing takes place at relatively long times after drug injection, since metabolic tolerance should result in a shortening of the duration of the drug effect.

It should be apparent from the discussion above that the relative significance of a particular mechanism for the development of tolerance/sensitization may depend critically upon basic features of the dosing regime employed and the behavioral test paradigm utilized. The more general conclusion that the significance of a particular mechanism for the development of tolerance/sensitization may vary substantially with basic pharmacological and behavioral parameters would seem to have widespread generality (*see also* Baker and Tiffany, 1985; Goudie and Demellweek, 1986). Investigators involved in research into a particular mechanism for the develop-

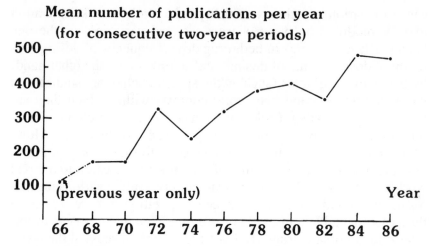

Fig. 2. Mean number of publications/yr that were indexed under the term "Drug tolerance" in Index Medicus for consecutive two-yr periods over the years from 1968–1986.

ment of tolerance/sensitization need to keep this basic point in mind in their studies, because there is little point in analyzing a specific mechanism for tolerance/sensitization development in detail if the mechanism analyzed has little significance beyond artificially constrained laboratory situations. It is probable that only an interdisciplinary approach to the study of tolerance/sensitization will allow the significance of various mechanisms to be evaluated in terms of their relevance to the real world.

Figure 2 shows the results of an analysis of the progress of research into tolerance during the period 1966–1986. The data shown are the mean number of publications/yr, for each consecutive 2-yr period, which were indexed under the phrase "drug tolerance" in *Index Medicus*. Not all of the papers indexed in such a way may have been designed to investigate tolerance as their primary aim; nonetheless, the rate of publication of literature related to tolerance is considerable, at present amounting to some 500 papers/yr. Furthermore, the rate of publication in this area is increasing in an almost linear manner, such that if current trends continue, the rate within 10 yr will approach some 20 papers/wk! A potentially worrying aspect of such a volume of information is that it is quite possible for basic findings to be established and then get "lost" in the experimental literature. Khanna et al. (1982) have shown, in an excellent review, how basic ideas about the nature of ethanol tolerance have frequently been put forward many years ago, only to disappear; "reappearing" as "novel" discoveries some decades later. For example, recent years have seen an upsurge of interest in the idea that tolerance may, under some circumstances, be associated with the acquisition of instrumentally learned coping strategies that allow organisms to cope with behaviorally disruptive effects of drugs (*see* reviews by Wolgin and by Barrett et al. in this text). The hypothesis that tolerance may involve instrumental learning has in recent years been verified empirically, with important early contributions by Schuster et al. (1966) and Chen (1968); interestingly, however, this idea was put forward in the 1930s by Newman and Card (1937a), who suggested that: "psychomotor compensatory mechanisms" were probably involved in ethanol tolerance and that: "The habituated individual probably learns to compensate for his lack of coordination . . . and so does not exhibit the same degree of abnormality as the abstainer."

The researcher interested in this area thus faces two substantial problems: understanding the multiple mechanisms accounting for tolerance/sensitization and their relationship to one another, and coping with the large literature base concerning each of these mechanisms. By drawing together experts to review some of the basic mechanisms in tolerance/sensitization research, we hope to encourage interdisciplinary collaboration, and thus to address these problems. The alternative to such collaboration would seem to be that more and more research will be conducted in greater detail into a constantly expanding number of mechanisms for the development of tolerance/sensitization; research workers will have their hands full attempting simply to keep up to date with research into any one mechanism for tolerance/sensitization development. In such circumstances, very little thought will be given to the overall picture that is (or is not) emerging. It is instructive to note that, in their studies in the early part of this century into ethanol tolerance, Newman and Card (1937a), having recognized that tolerance can be induced by a number of different mechanisms, concluded:

We feel that only by a segregation of these various possible factors [involved in the development of tolerance], with investigation of each of them individually, can we hope to arrive at any valid conclusions. To this

end we have striven to make the experimental situation as simple as possible, eliminating insofar as possible factors not under direct consideration, in the hope that by such an analytical procedure facts would be
evolved which might later be synthesized into a more complete knowledge of the nature of tolerance to this drug.

Although such a reductionist approach to the study of toler-
ance/sensitization is of course a legitimate scientific strategy, it is
clear that the synthesis that Newman and Card (1937a) anticipated
has been notable by its absence, and that, with the exception of the
work of one or two laboratories (e.g., LeBlanc et al., 1973, 1975),
very few research groups have attempted to achieve the type of synthesis described. Given the expanding literature in this area, and
recognizing the tendency for reductionist science to focus on single
mechanisms, an attempt to encourage integration of research into
tolerance/sensitization in this book appears justified, if not overdue.
    Given the pattern of data shown in Fig. 2, it is clearly impossi-
ble to compile a fully comprehensive overview of tolerance/sensiti-
zation to the actions of many different types of drugs. Instead, the
editors of this text have chosen to concentrate predominately on
mechanisms in the development of tolerance, although insofar as
such mechanisms also have implications for the understanding of
sensitization, this topic is also considered in the text. This approach
has been adopted because substantially more information is available concerning tolerance to chronic drug administration. Within
this context, selected major areas of research are presented that are
directed at isolating the role of specific mechanisms in the development of tolerance and sensitization. These specific review chapters
are introduced briefly below.
    The chapters in this text are assigned into two broadly defined
subareas: those that focus primarily on behavioral mechanisms, and
those that focus primarily on molecular mechanisms. This division
is somewhat arbitrary, and indeed, all behavioral mechanisms must
eventually be attributable to molecular processes. However, much
of the research in this field can be categorized as focused primarily
on either behavioral mechanisms or on molecular mechanisms. We
have therefore assigned these contributions accordingly and incorporated, at the end of the text, two overview commentaries by indi-

viduals with expertise in behavioral analysis (Derek Blackman) and pharmacology (Harold Kalant).

The section on Behavioral Mechanisms begins with chapters describing the roles of instrumental and classical conditioning in the development of tolerance. The chapter by Wolgin summarizes evidence that, for a variety of drugs, tolerance occurs as a function of instrumental conditioning. It is now nearly 20 yr since the initial demonstration (Schuster et al., 1966) that behavioral factors can play a critical role in the development of tolerance. Wolgin joined this area of research near its inception (Carlton and Wolgin, 1971), and has subsequently continued to study the role of instrumental learning in the development of tolerance and sensitization. In his chapter, he reviews this area, citing studies with a range of drugs. His review leads him to the conclusion that, in some circumstances, the development of tolerance involves learning to suppress drug effects that interfere with efficient performance of behavioral tasks. In his chapter, Siegel explores the role of classical conditioning in the production of tolerance and sensitization to drug effects. This account of tolerance/sensitization focuses on the occurrence of conditioned responses that reduce/enhance the direct effects of drugs. Siegel has been the most active researcher in this area for a number of years. As he shows in his chapter, conditioned responses occur in a wide variety of situations for a large number of drugs, and their appearance and suppression are governed by the basic laws of classical conditioning.

As shown in the chapters by Wolgin and Siegel, behavioral processes can be critical for understanding the effects of chronically administered drugs. In the next chapter, by Barrett, Glowa, and Nader, the effect of behavioral variables on the expression of reactivity to drugs is explored. Their chapter is an unusual and important contribution, because they show that tolerance- and sensitization-like phenomena can be produced by factors other than chronic drug administration, such as the influence of behavioral history. It is therefore necessary to understand when such factors come into play if we are to distinguish a class of phenomena that represent tolerance/sensitization that is attributable specifically to chronic drug administration. Barrett and colleagues summarize their own highly original, innovative contributions to this field, as well as critically evaluating the important work of others.

The final chapter of this section, by Young and Sannerud, concerns the general problem of whether tolerance can be shown to occur to the effects of drugs as discriminative stimuli. Tolerance research is linked historically to the study of psychoactive drug effects in humans. This chapter is included because it evaluates evidence for tolerance in an animal model of human "subjective" drug effects. The number of studies using drug discrimination procedures has increased dramatically over the last decade (Stolerman and Shine, 1985), and there has been considerable controversy in the literature as to whether tolerance occurs at all to the stimulus properties of drugs. Young and Sannerud summarize and evaluate the studies relevant to this question.

There has been rapid progress in recent years in the study of molecular mechanisms mediating the acute effects of psychoactive drugs. The second section of this text concerns Molecular Mechanisms in the development of drug tolerance and sensitization. Lê and Khanna have surveyed a vast literature concerning evidence for the mediation of tolerance and sensitization through pharmacokinetic mechanisms. Such mechanisms could include altered absorption, distribution, and/or metabolism of drugs. Lê and Khanna provide a comprehensive review of the importance of these processes in the mediation of tolerance and sensitization.

The history of pharmacology is, in many respects, a history of changing concepts regarding the function of drug receptors. In the next chapter, the role of receptor modification is considered in the production of tolerance and supersensitivity. Greenshaw, Baker, and Wishart provide evidence that the functional properties of dopamine receptors are changed following chronic drug administration. More importantly, they link these changes in functional properties of receptors to the occurrence of tolerance and sensitization, specifically with respect to the effects of antipsychotic drugs. Willner covers related material from the perspective of antidepressant drugs. One of the striking aspects of antidepressants is their relatively slow onset of beneficial therapeutic effects. The question therefore arises as to whether a slow onset of action reflects the necessity for sensitization to occur to these drugs. This question is considered in detail by Willner, including a comprehensive summary of literature concerning the effects of chronic administration of antidepressant drugs on the binding characteristics of a variety of types of receptors. To conclude this section, Littleton and Little discuss drug effects that occur beyond the level of receptors. The experimental analysis of the consequences of drug-receptor interactions for the physiology of the neuron has a relatively short history. Littleton and Little provide a novel, provocative account of tolerance, dependence, and cross-dependence to a wide range of depressant drugs. This model of tolerance/dependence explores the role of adaptive drug-induced changes in Ca^{++} channels in the mediation of tolerance to drug effects and describes how such adaptive changes may be involved in the causation of withdrawal syndromes.

We are indebted to our two summary chapter authors for their synopses of critical issues raised in the preceding chapters. Our overview of this material leads us to the conclusion that tolerance and sensitization can be readily demonstrated at either the level of behavioral phenomena or at the level of molecular events. In addition, various mechanisms have been clearly described that account for the development of tolerance and sensitization. However, the inability of either type of approach to cross over and provide a complete account of tolerance and sensitization is rather striking. In other words, the integration of facts that could be "synthesized into a more complete knowledge of the nature of tolerance" presumed by Newman and Card (1937a) is still awaited. Consider, for example, the case of tolerance that is controlled by classical conditioning mechanisms. How, if at all, is it possible to explain such tolerance in terms of altered binding constants for drugs at their receptors? The question is equally problematic when posed in reverse. Indeed, is it possible to translate a reduced binding capacity of a receptor for a particular drug into the realm of stimulus control of the expression of tolerance? In the chapter by Blackman, such questions and others raised by the preceding chapters are considered from the perspective of a research worker with a distinguished career in the area of behavioral analysis of drug effects. Similarly, in the chapter by Kalant, these questions are considered from the perspective of a research worker with a distinguished career in the area of the pharmacology of drugs of abuse. It is clear from these chapters that individuals with substantial expertise in behavioral and pharmacological sciences can have very different attitudes to the fundamental question of the extent to which tolerance/sensitization phenomena can, even in principle, be explained in reductionist terms. The resolution of this critical issue for research in this area will obviously not be easy. However, it is certain that this issue will *never* be resolved unless individuals working in this area adopt a more interdisciplinary approach, such that behavioral scientists will have to consider the merits and limitations of molecular approaches, whereas scientists with more reductionist prejudices will, ultimately, have to investigate how the phenomena they are studying are related to those analyzed by behavioral scientists. Such integration across very different areas of science remains a formidable, but exciting, challenge for the future.

Acknowledgment

The authors are indebted to Professor Shepard Siegel for discussions concerning the history of drug tolerance.

References

Baker S. T., Tiffany S. T. (1985) Morphine tolerance as habituation. *Psychol. Rev.* **92,** 78–107.

Berridge V. and Edwards G. (1981) *Opium and the People* (Allen Lane, London).

Bogen E. (1936) Tolerance to alcohol; Its mechanism and significance. *California and West Med.* **44,** 262–270.

Carlton P. L. and Wolgin D. L. (1971) Contingent tolerance to the anorexigenic effect of amphetamine. *Physiol. Behav.* **7,** 221–223.

Carpenter W. B. (1855) The physiological errors of moderation. *The Scottish Temperance League*, Glasgow.

Chen C. S. (1968) A study of the alcohol tolerance effect and an introduction of a new behavioral technique. *Psychopharmacologia.* **12,** 443–448.

Courtwright D. T. (1982) *Dark Paradise: Opiate Addiction in America Before 1940.* (Harvard University Press, Cambridge, Massachusetts).

DuMez A. G. (1919) Increased tolerance and withdrawal phenomena in chronic morphinism. A review of the literature. *J. Am. Med. Assoc.* **72,** 1069–1072.

Goudie A. J. and Demellweek C. (1986) Conditioning factors in drug tolerance, in *Behavioral Analysis of Drug Dependence.* (Goldberg,

S. R. and Stolerman I. P., eds.), Academic Press, Orlando, pp 225–285.

Kalant H., Le Blanc A. E., and Gibbins R. J. (1971) Tolerance to, and dependence on, some nonopiate psychotropic drugs. *Pharmacol. Rev.* **23,** 135–191.

Khanna J. M., LeBlanc A. E., and Lê A. D. (1982) Overview: Historical overview of tolerance and physical dependence, in *Ethanol Tolerance and Dependence; Endocrinological Aspects. NIAAA Monograph*, No. **13**. (Cicero T. J., ed). U.S. Government Printing Office, Washington, D.C., pp 4–15.

Kilbey M. M. and Sannerud C. A. (1985) Models of tolerance to psychomotor stimulants: Do they predict sensitization? in *Behavioral Pharmacology: The Current Status* (Seiden, L. S. and Balster, R. L., eds.), A. R. Liss, New York. pp 295–322.

Kramer J. C. and Merlin M. D. (1983) The use of psychoactive drugs in the ancient old world, in *Discoveries in Pharmacology, Vol. 1: Psycho- and neuro-pharmacology*, (Parnham, M. J. and Bruinvels, J., eds.), Elsevier Science Publishers B.V., pp 23–47.

LeBlanc A. E., Gibbins R. J., and Kalant H. (1973) Behavioral augmentation of tolerance to ethanol in the rat. *Psychopharmacologia.* **30,** 117–122.

LeBlanc A. E., Gibbins J. J., and Kalant H. (1975) Generalization of behaviorally augmented tolerance to ethanol and its relation to physical dependence. *Psychopharmacologia.* **44,** 214–246.

Newman H. and Card J. (1937a) The nature of tolerance to ethyl alcohol. *J. Mental Disorders.* **86,** 428–440.

Newman H. and Card J. (1937b) Duration of acquired tolerance to ethyl alcohol. *J. Pharmacol.* **59,** 249–252.

Schuster C. R., Dockens W. S., and Woods, J. H. (1966) Behavioral variables affecting the development of amphetamine tolerance. *Psychopharmacologia.* **9,** 170–182.

Siegel S. (1983) Wilkie Collins: Victorian novelist as psychopharmacologist. *J. Hist. Med. Allied Sciences.* **38,** 161–175.

Siegel S. (1986) Alcohol and opiate dependence. Re-evaluation of the Victorian perspective, in *Research Advances in Alcohol and Drug Dependence*, Vol. **9**. (Cappell, H. D., Glaser, F. B., Israel, Y., Kalant, H., Schmidt, W., Sellers, E. M., and Smart, R. C., eds.), Plenum, New York, pp 279–314.

Stolerman I. P. and Shine P. J. (1985) Trends in drug discrimination research analysed with a cross-indexed bibliography. *Psychopharmacology.* **86,** 1–11.

Part 1

Behavioral Mechanisms

The Role of Instrumental Learning in Behavioral Tolerance to Drugs

David L. Wolgin

1. Introduction

Historically, the term "behavioral tolerance" has had two meanings. When used descriptively, it refers to tolerance that develops to a behavioral effect of a drug (cf. Corfield-Sumner and Stolerman, 1978; Schuster, 1978). This use of the term carries no implications for underlying mechanisms; i.e., tolerance may result from the same mechanisms that alter a physiological or pharmacological effect. However, the term has also been used mechanistically to mean that tolerance is mediated by a behavioral compensation for the initial effects of a drug (cf. Dews, 1978; Ferraro, 1978; Demellweek and Goudie, 1983b). This chapter deals with the second use of the term. More specifically, it considers the possibility that, under appropriate circumstances, tolerance is mediated by instrumental learning.

A method introduced by Chen (1968) has been particularly useful for analyzing the potential role of instrumental learning in the development of tolerance. In this method, hereafter called the "before/after" design, a group of subjects (the Before Group) is given chronic injections of a drug before being tested on a behavioral task, whereas a second group (the After Group) is given an identical number of drug injections, but after the task. A third group (the Control Group) is given injections of saline. When the Before Group develops tolerance to the initial effect of the drug, the After Group and the Control Group are given pretest injections of the drug for the first time. Because both the Before and After groups have the same drug history, any differences between the groups on this final day can be attributed to differences in behavioral experience.

An experiment by Carlton and Wolgin (1971) illustrates the use of this design. Rats in the Before Group were given injections of amphetamine (2 or 3 mg/kg) before, and injections of saline after, access to sweetened milk. Rats in the After Group received the same number of injections, but in the reverse order. As shown in Fig. 1, amphetamine initially suppressed the milk intake of the Before Group, but tolerance to this effect developed in four trials. However, when the After Group was then given amphetamine before milk for the first time, no tolerance was found. When pretest injections were then continued over the ensuing days, tolerance developed at the same rate as it had in the Before Group. Because the recovery of milk intake was contingent on behavioral experience in the drugged state, Carlton and Wolgin (1971) referred to this phenomenon as "contingent tolerance." (Contingent sensitization has also been demonstrated; *see* Kilbey and Sannerud, 1985 for a recent review.)

It is important to note that the designation "contingent tolerance" derives only from the fact that tolerance is contingent on the relation between the time of injection and testing, and does not denote a particular underlying mechanism. However, one interpretation of such tolerance is that the rats in the Before Group had learned to compensate in some way for the initial disruptive effect of the drug, prompted, presumably, by the loss of milk reward. The central role of reinforcement loss in the development of tolerance was underscored in a classic experiment by Schuster et al. (1966). In this study, rats were given injections of amphetamine and tested on a multiple schedule consisting of two components. In one, the rats were differentially reinforced for responding at a low rate (DRL), whereas in the other they were reinforced for responding after a fixed interval (FI) of time had elapsed. In two subjects, amphetamine initially caused an increased rate of responding in both components of the schedule, which resulted in a loss of reinforcement in the DRL component, but not in the FI component. When amphetamine was given chronically, tolerance developed to the rate-enhancing effect of the drug in the DRL component, but not in the FI component. In a third subject, however, amphetamine initially produced a decrease in FI responding and a loss of reinforcement. In this case, tolerance developed in the FI component.

In this experiment, the differential development of tolerance in the same subject effectively rules out an explanation of the results based on an altered distribution or metabolism of the drug and, therefore, implies that a behavioral mechanism may be involved. Generalizing from these results, Schuster et al. (1966) proposed that tolerance develops when the initial effect of a drug results in a loss of reinforcement; when the drug has no effect on reinforcement or when it increases the frequency of reinforcement, no tolerance occurs. This empirical generalization has been termed the "reinforcement density hypothesis" (Corfield-Sumner and Stolerman, 1978). Although, as Ferraro (1978) pointed out, this hypothesis does not explicitly posit a particular mechanism of tolerance, it is consistent with the view that tolerance involves instrumental learning. In brief, the rat alters its behavior to maximize reinforcement.

In the following sections, I shall critically review the literature on the role of instrumental learning in the development of tolerance. As we shall see, much of the evidence for instrumental learning is indirect and derives from studies in which tolerance is associated with an initial loss of reinforcement, as predicted from the reinforcement density hypothesis. However, not all of the evidence is consistent with this hypothesis, and so other variables must be considered as well. After reviewing the literature, I shall turn to the critical issue of what, precisely, is learned when a rat is given chronic injections of a drug. As others have noted in previous reviews (Corfield-Sumner and Stolerman, 1978; Demellweek and Goudie, 1983b), the failure to identify a specific, learned, adaptive response represents a major shortcoming in the instrumental learning theory of drug tolerance. Unfortunately, very little work has been done on this problem to date. We have recently begun to address this question in regard to tolerance to the "anorexigenic" effects of amphetamine (Salisbury and Wolgin, 1985; Wolgin et al., 1987). As we shall see, the data suggest that such tolerance involves learning to suppress responses that are incompatible with feeding. These results provide strong support for the role of instrumental learning in the development of tolerance to amphetamine.

2. Review of the Literature

The literature on behavioral tolerance is so voluminous that it is difficult to provide a comprehensive overview of all pertinent studies. Fortunately, there have been a number of excellent reviews during the last decade (e.g., Corfield-Sumner and Stolerman, 1978; Demellweek and Goudie, 1983b; Goudie and Demellweek, 1986; Krasnegor, 1978; LeBlanc and Cappell, 1977), and the reader is encouraged to consult these sources for additional information. The present review focuses on the following classes of drugs: stimulants (amphetamine, cocaine, methylphenidate), alcohol, morphine, anxiolytics (barbiturates, benzodiazepines), and marijuana. Although this list is somewhat arbitrary (it omits hallucinogens and cholinergics, for example), it covers drugs for which sufficient behavioral data exist to warrant at least tentative conclusions regarding the role of instrumental learning.

A word or two are in order concerning the criteria by which the learning theory is evaluated. Ideally, one would like to see evidence that, during chronic administration of a drug, the subject learns an identifiable response or strategy under the control of a particular reinforcer. As noted above, there is very little evidence of this sort. A less stringent criterion is to demonstrate that the development of tolerance is consistent with the reinforcement density hypothesis. In its strong form, this means that tolerance occurs if, and only if, the initial drug effect results in a loss of reinforcement; i.e., it should not occur when the initial drug effect either increases or has no effect on reinforcement. The before/after paradigm offers a convenient way to test this hypothesis, because it includes a condition in which reinforcement may be affected by the drug (the "before" condition) as well as a condition in which reinforcement is not normally affected (the "after" condition). Under both conditions, pharmacological exposure is identical. Many studies include the "before" condition, but not the "after" condition. In the absence of independent evidence for contingent tolerance, such studies provide the weakest evidence for the theory, because they do not rule out the possibility that tolerance results from chronic injections *per se*.

Although the before/after design can provide data pertinent to the reinforcement density hypothesis, it should be kept in mind that such data are still subject to alternative interpretations. As Corfield-Sumner and Stolerman (1978) pointed out, differences in the before and after groups could be the result of differential absorption, distribution, or metabolism of the drug, deprivation levels of the subjects, nonspecific effects such as stress or arousal, and retrograde effects of post-test injections, such as amnesia or conditioned aversions. Obviously, such factors must be excluded before a learning interpretation can be accepted.

Another way to address the role of reinforcement and learning is to test the subjects in more than one task. Multiple operant schedules of reinforcement offer several advantages in this regard. First, components of the schedule can be selected such that the probability of reinforcement is differentially affected by the drug. For example, in the experiment by Schuster et al. (1966) described earlier, the initial rate-increasing effect of amphetamine resulted in a decreased probability of reinforcement in the DRL component, but had no effect in the FI component; tolerance developed only to the effect on DRL responding. Second, because each subject is tested with both components, this paradigm controls for the possible effects of a host of extraneous variables. If, for example, chronic administration of amphetamine resulted in altered metabolism or distribution of the drug, tolerance would be expected to develop in both components of the schedule. The differential development of tolerance associated with an initial loss of reinforcement is therefore more consistent with a learning interpretation.

In sum, in the absence of identifying a specific learned response, the instrumental learning theory is supported by data that are consistent with the reinforcement density hypothesis. In general, such support involves the demonstration that tolerance is contingent on behavioral experience (contingent tolerance) or that it is differentially affected by the contingencies of reinforcement (differential tolerance). For this reason, the present review will focus on behavioral paradigms that contain explicit contingencies of reinforcement.

2.1. Stimulants

2.1.1. Contingent Tolerance

Perhaps the most persuasive evidence that behavioral tolerance involves instrumental learning comes from the literature on the effects of chronic administration of stimulants. Contingent tolerance has been demonstrated to a variety of such drugs, including amphetamine, cocaine, methylphenidate, and cathinone (*see* Table 1A). Although many of these studies used single doses, in several experiments (Emmett-Oglesby and Taylor, 1981; Foltin and Schuster, 1982; Woolverton et al., 1978a), the degree of tolerance was also demonstrated by a shift to the right of the dose–response function.

Several features of these studies are noteworthy. First, tolerance occurred whether the initial effect of the drug was to decrease (intake measures), or to increase (DRL), baseline responding. Second, pharmacological exposure *per se* did not appear to contribute to such tolerance. For example, posttest injections of amphetamine did not augment the rate of tolerance development when rats were later switched to pretest injections (cf. Fig. 1; *see also* Demellweek and Goudie, 1983a; Smith, 1986a). In fact, in two of the studies, chronic posttest injections of cathinone (Foltin and Schuster, 1982) and cocaine (Woolverton, et al., 1978a) resulted in *sensitization* when the subjects were subsequently given pretest injections. Third, there is no evidence that posttest injections produced a conditioned taste aversion to food (Demellweek and Goudie, 1983a). Although posttest injections of amphetamine can induce such aversions (Cappell and LeBlanc, 1973; Carey, 1978), the effect requires access to a novel food. In tolerance experiments, in contrast, the food is typically familiar. Finally, contingent tolerance is not an inevitable consequence of chronic exposure to drugs that suppress food intake. For example, tolerance has been found in rats given chronic postmeal injections of fenfluramine (Rowland et al., 1982; Rowland and Carlton, 1983), a nonstimulant anorexigen.

These findings, therefore, suggest that tolerance to stimulants is contingent on behavioral experience in the drugged state. This conclusion is supported by a number of additional experiments involving only pretest injections of stimulants in which the initial effect of the drug resulted in a loss of reinforcement (Table 1B). In these studies, tolerance developed to impairments in fixed ratio (FR) and DRL performance, as well as to the suppression of milk, water, and food intake. Conversely, other studies have shown that, when the initial effects of drug administration resulted in an increased frequency of reinforcement, tolerance did not develop (Kuribara, 1980; Leith and Barrett, 1976; Liebman and Segal, 1976; Schuster et al., 1966). In these experiments, amphetamine increased the rate of responding for food (on an FI schedule), rewarding brain stimulation, and avoidance of shock.

2.1.2. Role of Cumulative Deprivation and Body Weight Set Point

Although the results discussed above are consistent with the instrumental learning theory, alternative explanations for contingent tolerance have been proposed. For example, Panksepp and Booth (1973) noted that subjects in the Before Group may become increasingly more food deprived during the course of chronic injections of the drug. Consequently, what appears to be the development of tolerance may simply reflect an increased motivation to eat. Because subjects in the After Group do not experience such cumulative deprivation, they would not show "tolerance" when later tested under the drug. A series of experiments by Demellweek and Goudie addressed this point by systematically manipulating the level of food deprivation. In the first study (Demellweek and Goudie, 1982), two groups of rats were given chronic posttest injections of amphetamine, but one of these was pair-fed with rats in the Before Group. When these groups were subsequently tested with pretest injections of the drug, the pair-fed group ate more than the nonpair-fed group, showing that the apparent level of tolerance was influenced by cumulative deprivation. However, this level of tolerance was still far below that of the Before Group. These results were confirmed in a subsequent experiment (Demellweek and Goudie, 1983a), in which rats in the Before and After Groups were given varying degrees of food supplementation. Although the level of tolerance attained by rats in the Before Groups varied with the amount of supplementation, all of these groups acquired substantial tolerance. In contrast, none of the After Groups showed tolerance

Table 1

Summary of Studies Showing Tolerance to the Behavioral Effects of Stimulants

Authors	Drug	Dose, mg/kg	Task[a]
A. Contingent tolerance			
Campbell and Seiden, 1973	Amphetamine	1.5	DRL
Carlton and Wolgin, 1971	Amphetamine	2 and 3	Milk intake
Demellweek and Goudie, 1982	Amphetamine	2	Food intake
Demellweek and Goudie, 1983a	Amphetamine	2.5	Milk intake
Emmett-Oglesby and Taylor, 1981	Methylphenidate	15	Milk intake
Foltin and Schuster, 1982	Cathinone	4	Milk intake
Poulos et al., 1981	Amphetamine	4	Milk intake
Streather and Hinson, 1985	Amphetamine	3	Milk intake
Woolverton et al., 1978a	Cocaine	16	Milk intake
B. Studies involving only pretest injections			
Baettig et al., 1980	Amphetamine	1.5	Food intake
Branch, 1979	Amphetamine	0.1–0.56	Mult FI, FI, FI
Brocco and McMillan, 1983	Amphetamine	10	Mult FR, FI
Brown, 1965	Amphetamine	3	FR
Carey, 1978	Amphetamine	1 and 2	Food intake
Gotestam and Lewander, 1975	Amphetamine	16	Food intake
Harris et al., 1979	Amphetamine	0.3–3	FR, FI
Hoffman et al., 1987	Cocaine	1–10	Mult FR, FR, FR
Kandel et al., 1975	Amphetamine	1.5	Milk intake

Reference	Drug	Dose	Measure
Levitsky et al., 1981	Amphetamine	30	Food intake
MacPhail and Seiden, 1976	Amphetamine	4	Water intake
Pearl and Seiden, 1976	Amphetamine	2.5	Milk intake, DRL
Pearl and Seiden, 1976	Methylphenidate	20	Milk intake, DRL
Salisbury and Wolgin, 1985	Amphetamine	2	Milk intake
Sparber and Tilson, 1972	Amphetamine	1.6 and 2.5	FR
Tilson and Rech, 1973	Amphetamine	1	FR
Tilson and Sparber, 1973	Amphetamine	1	FR
Wolgin and Salisbury, 1985	Amphetamine	2, 4, and 6	Milk intake
Wolgin et al., 1987	Amphetamine	2	Milk intake
Wolgin, 1983	Amphetamine	2	Milk & food intake
Woolverton et al., 1978b	Cocaine	16 and 24	FR, DRL
C. Differential tolerance			
Emmett-Oglesby et al., 1984	Amphetamine	0.625 or 2.5	Milk intake, DRL
Schuster and Zimmerman, 1961	Amphetamine	1	DRL, activity
Schuster et al., 1966	Amphetamine	1	Mult DRL, FI
Smith and McKearney, 1977	Amphetamine	0.3–3	NT, $\overline{\text{NT}}$
Smith, 1986a	Amphetamine	1	Mult DRL, RR

[a]DRL = differential reinforcement of low rates of responding; FI = fixed interval; FR = fixed ratio; MULT = multiple schedule of reinforcement consisting of two or more components; NT = reinforcement contingent on making a fixed number of responses (N) and then pausing for a specified interval of time (T); $\overline{\text{NT}}$ = same as NT, except responding during T resets the interval; RR = random ratio.

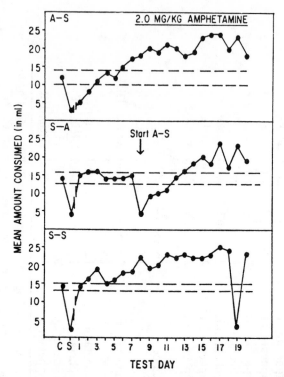

Fig. 1. Mean milk intakes of Before (A-S), After (S-A), and Saline (S-S) groups. The After Group was switched to pretest injections on test day 8. The Saline group was tested with amphetamine on test day 19. The horizontal dashed lines signify 4 standard errors of the mean. C = mean of baseline sessions; S = initial screen with 2 mg/kg *d*-amphetamine sulfate. (From ''Contingent tolerance to the anorexigenic effects of amphetamine'' by P. L. Carlton and D. L. Wolgin, *Physiol. Behav.*, 1971, **7**, 221–223. Fig. 2. Copyright 1971 by Pergamon Press. Reprinted by permission.)

when they were subsequently tested with pretest injections. Thus, contingent tolerance cannot be attributed to cumulative deprivation *per se*.

An alternative explanation of contingent tolerance derives from a novel theory of the mechanism of action of anorexigenic drugs. Stunkard (1981, 1982) has proposed that such drugs act primarily to lower a set point for the regulation of body weight and that tolerance does not develop to this effect. To explain the decrease and subsequent increase in food intake during chronic administration of the drug, he draws an analogy to the effects of lateral hypothalamic lesions on body weight regulation (cf. Powley and Keesey, 1970). According to this view, the initial suppression of feeding induced by the drug is an active mechanism to achieve a lower body weight level; the subsequent recovery of food intake serves to maintain the new weight level and does not reflect tolerance to the drug. The theory is supported by the finding that rats given chronic injections of amphetamine or fenfluramine maintain a lower weight level than saline controls, and defend that level when deprived or force-fed (Levitsky et al., 1981). Moreover, as in nondrugged rats (Peck, 1978), the level of maintained weight varies with the palatability of the diet (Wolgin, 1983).

Although there is considerable evidence that the suppressant effect of amphetamine and fenfluramine on food intake is influenced by the level of body weight (Carlton and Rowland, 1985; Wolgin, 1983; *see also* Streather and Hinson, 1986), the conclusion that these drugs lower a regulatory set point does not appear to be correct. For example, one piece of evidence cited in support of the theory is the report that prior food deprivation, which results in a loss of body weight, diminishes the suppressant effect of amphetamine on food intake (Levitsky et al., 1981). The problem with this assertion is that it is based on a comparison between deprived and nondeprived rats given the drug (cf. Fig. 2, drugged vs weight-reduced, drugged groups). However, if the intake of each drugged group in Fig. 2 is compared to that of its respective nondrugged control group, the degree of suppression in the deprived rats is at least as great as that in the nondeprived rats (cf. Fig. 2). Another cause for concern in this study is that an enormous dose of amphetamine (30 mg/kg) was used. Even though the drug was administered by gastric intubation, which might result in a slower rate of absorption than would be expected from an intraperitoneal injection, this dose is an order of magnitude higher than the highest dose used in most behavioral research. This concern is underscored by the results of a study by Carlton and Rowland (1985) on the effects of fenfluramine on rats with varying degrees of imposed weight loss. Using a more moderate dose of fenfluramine than that used by Levitsky et al., 1981 (5 mg/kg vs 20 mg/kg), they found that the degree of fenfluramine-induced suppression of intake did not vary with the level of prior weight loss.

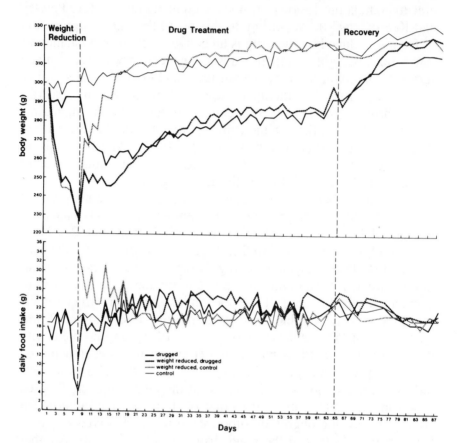

Fig. 2. Body weight (top) and food intake (bottom) of rats given daily intubations of either amphetamine (drugged) or saline (control). Prior to chronic drug treatment, half the rats in each group were food deprived in order to induce weight loss. Food deprived rats given amphetamine (weight reduced, drugged) initially ate more than nondeprived rats given amphetamine (drugged). However, when the intake of each of these groups is compared to that of its control group, the reduction in food intake is similar (compare drugged vs control and weight reduced, drugged vs weight reduced, control). Recovery = period in which amphetamine treatment was terminated. (From "Tolerance to anorectic drugs: Pharmacological or artifactual?" by D. A. Levitsky, B. J. Strupp, and J. Lupoli, *Pharmacol. Biochem. Behav.*, 1981, **14,** 661–667. Fig. 1. Copyright 1981 by Pergamon Press. Reprinted by permission.)

These shortcomings are compounded by the failure to find dose-related decreases in the level of maintained body weight in rats given chronic injections of amphetamine (Wolgin and Salisbury, 1985). According to the set point theory, rats given a higher dose of the drug would be expected to maintain a lower weight level than rats given a lower dose. As shown in Fig. 3, although rats given 4 mg/kg initially ate less and lost more weight than rats given 2 mg/kg, they eventually ingested as much, and weighed as much, as their lower dose counterparts. Extended observations with higher doses of amphetamine and saline also failed to demonstrate dose-related shifts in the level of maintained body weight. In addition, inspection of the data revealed marked variability in the intake of individual rats, which was partially masked in the group curves (Fig. 4). It seems unlikely that a physiological control system for the regulation of body weight would generate such extreme fluctuations in caloric intake.

Another problem with the set point theory is that amphetamine and fenfluramine do not show cross-tolerance. If both of these drugs acted to lower a body weight set point, then once body weight had been lowered by one of these drugs, an equipotent dose of the other should have no further effect on intake. This prediction was not confirmed, however. Amphetamine-tolerant rats were anorexic when given fenfluramine, and fenfluramine-tolerant rats were anorexic when given amphetamine (Kandel et al., 1975). Finally, the set point theory is unable to account for cases in which the development of tolerance is restricted to one task, but not to another, in the same subject. Such *differential* tolerance is discussed in the next section.

2.1.3. Differential Tolerance

Studies demonstrating differential tolerance to amphetamine are summarized in Table 1C. In general, such studies have involved two types of paradigm. In one, subjects are tested on two tasks in which the contingencies of reinforcement are differentially affected by the drug. Tolerance is then found to develop only on the task in which the drug initially produced a loss of reinforcement (e.g., Schuster et al., 1966; Schuster and Zimmerman, 1961; Smith and McKearney, 1977). In the other paradigm, subjects are given the drug in conjunction with one task and saline in conjunction with

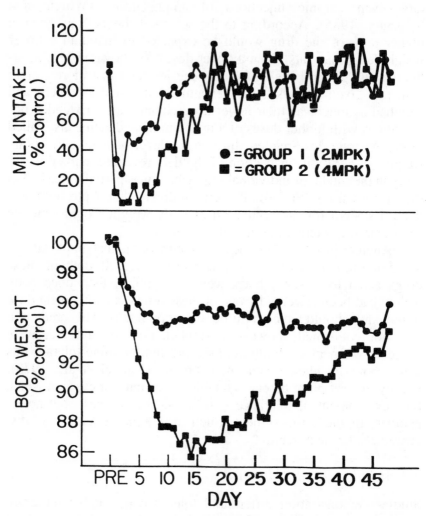

Fig. 3. Mean milk intake and body weight of rats given 2 mg/kg or 4 mg/kg *d*-amphetamine sulfate expressed as a percentage of the saline control group's data. Although initially different, the milk intake and body weight of both groups eventually converged. (From ''Amphetamine tolerance and body weight set point: A dose–response analysis'' by D. L. Wolgin and J. Salisbury, *Behav. Neurosci.*, 1985, **99,** 175–185. Fig. 2. Copyright 1985 by the American Psychological Association. Reprinted by permission of the publisher and author.)

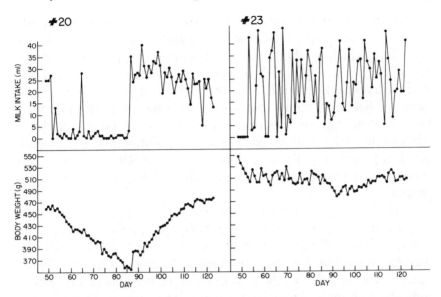

Fig. 4. Milk intake and body weight of two rats receiving chronic injections of 4 mg/kg *d*-amphetamine sulfate. Note the marked differences in the patterns of intake and the extremely variable intake of Rat 23. (From "Amphetamine tolerance and body weight set point: A dose–response analysis" by D. L. Wolgin and J. Salisbury, *Behav. Neurosci.*, 1985, **99**, 175–185. Fig. 4. Copyright 1985 by the American Psychological Association. Reprinted by permission of the publisher and author.)

another. In this case, tolerance is shown not to transfer to the other task, even though the drug initially produced a loss of reinforcement in both tasks (Emmett-Oglesby et al., 1984).

As noted earlier, it was the pioneering work of Schuster et al. (1966) on differential tolerance to amphetamine that first gave rise to the reinforcement density hypothesis. A recent paper by Smith (1986a) involving differential tolerance suggests an interesting addendum to this hypothesis. Smith gave rats chronic injections of amphetamine and tested them for 30 d on a multiple schedule of reinforcement. In one component (Random Ratio, RR), reinforcement was delivered randomly after 2.5% of the responses and, therefore, generated high rates of responding. In contrast, in the other component, the rats were differentially reinforced for responding at a low rate (DRL). In keeping with its rate-dependent properties (cf. Dews and Wenger, 1977), amphetamine initially produced a decrease in RR responding and an increase in DRL responding, which resulted in a loss of reinforcement in both components. As shown in Fig. 5, during chronic administration of the drug, tolerance developed to the rate-decreasing effect of amphetamine on RR responding, but not to the rate-increasing effect on DRL responding. However, when the rats were given additional trials on the DRL component alone (i.e., without the RR component), tolerance rapidly developed. When the RR component was subsequently reintroduced, tolerance on the DRL component disappeared.

These results have two interesting implications. First, they suggest that tolerance during the DRL component is subject to schedule-induced effects. In this case, the expression of tolerance was suppressed by the ratio component. Second, as Smith (1986a) noted, they suggest that the development of behavioral tolerance is influenced by the "global" density of reinforcement. That is, when the initial effect of the drug results in a loss of reinforcement, the relative "costliness" of that loss influences the development and/or the expression of tolerance. In the context of the multiple schedule, the proportion of reinforcers lost in the DRL component was relatively small compared to that of the ratio component. Under these conditions, tolerance was not expressed in the DRL component. However, in the absence of the ratio component, all of the reinforcers came from the DRL component; under these conditions, tolerance rapidly developed. These results provide strong support for the theory that behavioral tolerance to stimulants involves instrumental learning. Indeed, they suggest that subjects are keenly aware of the rate at which they receive reinforcement (for additional evidence on this point, *see* Carlton and Didamo, 1960; Stein et al., 1958).

An experiment by Emmett-Oglesby et al. (1984) involving differential tolerance to amphetamine provides further support for the role of learning. Groups of rats were tested on two tasks, milk intake and DRL performance, on alternate days. For one group, amphetamine was administered prior to milk drinking and saline prior to the DRL task. For the other group, the drug was given prior to the DRL task, and saline was given prior to milk drinking. During the course of chronic drug administration, tolerance developed to the initial drug-induced loss of reinforcement on each of the

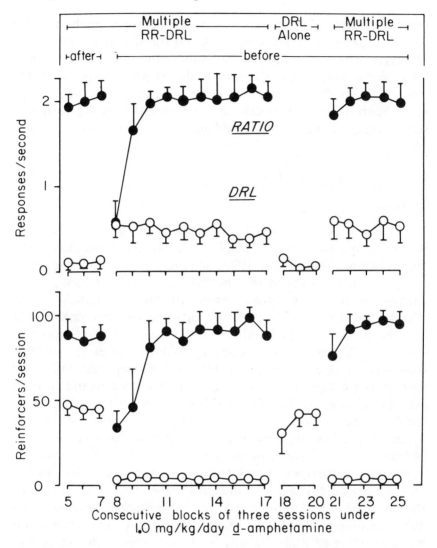

Fig. 5. Response rate (top) and reinforcement frequency (bottom) of rats given chronic injections of 1 mg/kg *d*-amphetamine and tested on random ratio (filled circles) and differential reinforcement of low rate (open circles) schedules. Tolerance on the DRL schedule occurred only when this component was tested alone. After = trials in which injections were given posttest; before = trials in which injections were given pretest. (From "Effects of chronically administered *d*-amphetamine on spaced responding maintained under multiple and single-component schedules" by J. B. Smith, *Psychopharmacology*, 1986, **88**, 296–300. Fig. 3. Copyright 1986 by Springer-Verlag. Reprinted by permission.)

tasks. However, when the rats were subsequently given the drug for the first time prior to the other task, no transfer of tolerance was observed. Thus, although amphetamine produced a loss of reinforcement in both tasks, tolerance to this effect was specific to the task in which the subjects experienced chronic injections of the drug. Such differential tolerance is consistent with the view that something quite specific is learned during the development of tolerance.

2.1.4. Dose-Specific Tolerance and State-Dependent Learning

Tolerance is often defined as a shift to the right of the dose–response function as a result of chronic administration of the drug (Carlton, 1983; Dews, 1978; Schuster, 1978). However, there are occasional reports in the literature in which tolerance is restricted to the dose given chronically (e.g., Branch, 1979; Woolverton et al., 1978b). Such reports provide additional insight into the mechanism underlying the development of tolerance to stimulants. Consider first a study by Woolverton et al. (1978b). Rats were tested on either a FR or DRL operant task, and dose–response functions were determined both before and during a period of chronic administration of cocaine (either 16 or 24 mg/kg). On the FR schedule, cocaine produced a dose–dependent increase in pausing, and during chronic administration of the drug, tolerance developed to this effect. In the group given the higher dose, the drug also produced a decrease in both the overall and the "local" rate of responding, and tolerance also developed to these effects. However, on redetermination of the dose–response function, tolerance was found to be limited to the chronic dose; no tolerance was found at lower or higher doses. Such dose-specific tolerance was also observed in the DRL condition. In this case, cocaine produced a dose-dependent increase in the rate of responding, which resulted in a decrease in reinforcement. Chronic dosing resulted in tolerance to the rate-increasing effect of the drug, but again, redetermination of the dose–response function revealed that such tolerance was specific to the dose given chronically. Moreover, the temporal pattern of responding was impaired even when the rats were retested under *saline*. A similar result was reported by Carey (1973). In this case, rats tolerant to the disruptive effect of amphetamine on DRL responding showed impaired performance (i.e., increased rates of responding) when subsequently tested without the drug.

Another example of dose-specific tolerance may be found in a study by Wolgin (1973). Groups of rats were initially screened with two doses of amphetamine (either 0.75 and 1.5 mg/kg or 1.5 and 3 mg/kg) and then given chronic injections of the higher dose (Figs. 6A and C). When tolerance had developed to the initial suppression of milk intake, half of the rats in each group were retested with the lower dose. Rats made tolerant to 1.5 mg/kg were also tolerant to the 0.75 mg/kg dose (Figs. 6C and D). However, rats made tolerant to 3 mg/kg and then switched to 1.5 mg/kg showed a loss of tolerance when retested with the lower dose; i.e., they drank *less* (Figs. 6A and B). On subsequent trials with this dose, intake returned to tolerant levels.

These examples of dose-specific tolerance are difficult to reconcile with traditional biochemical theories of tolerance, but they are compatible with an instrumental learning interpretation. It is well known that drugs like amphetamine have discriminative stimulus properties, which vary with the dose (*see* Young and Glennon, 1986 for a recent review). Consequently, when the drug is given chronically, the internal state associated with that dose may come to serve as a discriminative stimulus for an instrumentally learned adaptive response. Under these conditions, changing the dose would alter the cue properties of the drug and lead to a "generalization decrement" in responding (i.e., a loss of tolerance). From this perspective, dose-specific tolerance represents a form of state-dependent learning.

If learned tolerance can come under the discriminative control of drug-related stimuli, why are there so few reports of dose-specific tolerance in the literature? In part, the paucity of data may reflect the fact that relatively few studies of behavioral tolerance include dose–response determinations. Another potential reason derives from procedural considerations. The ability of a particular drug dose to serve as a discriminative stimulus will depend on the degree to which its cue properties become associated with the instrumentally learned response. This, in turn, will vary with the salience of the cue and with the number of trials in which it is associated with the learned response. This is to say, stimulus control will be more likely to develop if the chronic drug dose is discriminably different and if many trials are given *after the subject has become*

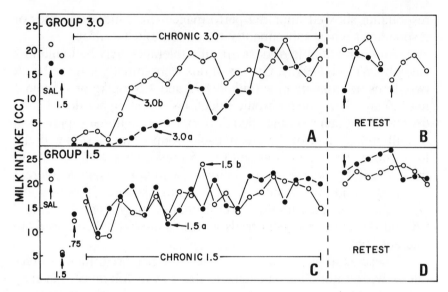

Fig. 6. Mean milk intakes of rats given various doses of *d*-amphetamine sulfate. (A) Group 3.0 was given a preliminary test with 1.5 mg/kg and then divided into 2 subgroups (3.0a and 3.0b), each of which was made tolerant to 3 mg/kg. (B) Each subgroup was then retested with 1.5 mg/kg (arrows). Note lack of tolerance with lower dose, followed by recovery. (C) Group 1.5 was given a preliminary test with 0.75 mg/kg and then divided into 2 subgroups (1.5a and 1.5b), each of which was made tolerant to 1.5 mg/kg. (D) One subgroup was then retested with 0.75 mg/kg (arrow). Note generalization of tolerance to lower dose.

tolerant. This latter requirement is particularly important because, initially, the drug cue will be associated with the acute effects of the drug, not with the instrumentally learned response. Thus, one reason that dose-specific tolerance is not routinely found may be that too few trials are given for the chronic dose to acquire stimulus control of the learned response. It is noteworthy that, in the experiment by Woolverton et al. (1978b), for example, the redetermination of the dose–response curve began after 60 d of cocaine administration.

Drug discrimination studies suggest that the discriminability of a particular drug from saline varies with the dose and that generalization of the drug cue to other doses depends upon the magnitude of the "training dose" (Young and Glennon, 1986; *see also* chapter by Young and Sannerud, this volume). Consequently, the development of dose-specific tolerance should depend upon the magnitude of the chronic dose and the discriminability of that dose from other doses. One factor that may contribute to such discriminability is whether chronic dosing is continued during the period in which the dose-response curve is redetermined. Continued exposure to the chronic dose should increase the discriminability of that dose vis-à-vis the other doses, and favor the development of dose-specific tolerance. Again, it is noteworthy that, in the study by Woolverton et al. (1978b), subjects were maintained on the daily cocaine injection regimen, while the dose–response functions were redetermined.

Differences in the discriminability of the doses appear to account for the results of the study by Wolgin (1973), in which tolerance to amphetamine "transferred" from 1.5–0.75 mg/kg, but not from 3–1.5 mg/kg (*see* Fig. 6). An important insight into the interpretation of these results was obtained by examining the initial effects of the two pairs of doses. Rats given 0.75 and 1.5 mg/kg showed relatively little difference in intake with the two doses (12.9 vs 5.6 cc) compared to the rats given 1.5 and 3 mg/kg (17.7 vs 1.1 cc). If the discriminability of these pairs of doses is proportional to their potency in suppressing intake, then the cue properties of the doses given the first group of rats (0.75 and 1.5 mg/kg) were relatively similar, whereas the cue properties of the doses given the second group (1.5 and 3 mg/kg) were relatively different. Consequently, the second group would be expected to show a generalization decrement when switched from the chronic (high) dose to the lower dose.

This interpretation is supported by a comparison of the pre- and posttolerance intakes following injection of 1.5 mg/kg by rats in the second group. Statistical analysis revealed a significant negative correlation ($r = -0.72$) between the two scores. As shown in Fig. 7, rats that had high initial intakes (and therefore were relatively insensitive to this dose) drank *less* when retested after they had become tolerant to the higher dose. For these rats, the stimulus properties of the lower dose were presumably rather different from those of the higher dose, and so tolerance did not transfer. In contrast, rats that drank little when initially tested with 1.5 mg/kg (and therefore were relatively sensitive to this dose) tended to drink more when retested following tolerance to the higher dose. In this case, the stimulus properties of the two doses were presumably

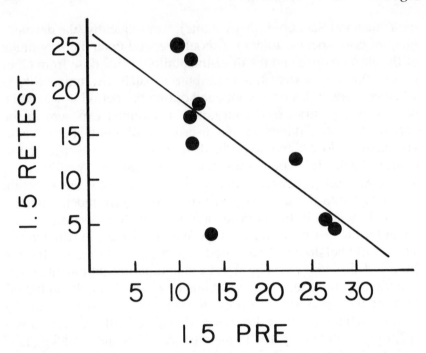

Fig. 7. Regression line relating amount of milk consumed by Group 3.0 on preliminary test with 1.5 mg/kg (1.5 PRE) and on first retest trial with that dose, after becoming tolerant to 3 mg/kg (1.5 RE-TEST). Rats that initially drank little following injection of 1.5 mg/kg showed tolerance to this dose on the retest. Rats that were not initially anorexic did not show generalization of tolerance, presumably because the stimulus properties of the two doses differed.

rather similar, and "stimulus control" was maintained during the retest with the lower dose.

The ability of drug cues to act as discriminative stimuli for an instrumentally learned response might also play a role in the phenomenon of cross-tolerance. For example, cross-tolerance has been demonstrated between various stimulant anorexigens, such as amphetamine, cocaine, methylphenidate, apomorphine, and cathinone, but not between stimulant and nonstimulant anorexigens, such as amphetamine and fenfluramine (Emmett-Oglesby and Taylor, 1981; Foltin and Schuster, 1982; Kandel et al., 1975; Pearl and Seiden, 1976; Streather and Hinson, 1985; Woolverton et al., 1978a). These results are often attributed to the fact that tolerance to the stimulants involves changes in a common catecholamine neural substrate, whereas fenfluramine tolerance involves a serotonergic mechanism (cf. Streather and Hinson, 1985; but *see* Paul et al., 1982). An alternative interpretation consistent with the instrumental learning theory is that tolerance involves a learned response that is at least partly under the control of discriminative drug cues. Cues associated with amphetamine would be shared by other stimulant anorexigens; hence, cross-tolerance would occur among these drugs. Because fenfluramine would not share these cues, however, no cross-tolerance would occur to this drug. (A complementary interpretation of cross-tolerance will be discussed later; *see* section 3. What Is Learned?)

2.1.5. Discriminative Stimuli Vs Conditioned Stimuli

The concept of discriminative stimulus control may also provide an alternative interpretation for the environmental specificity of tolerance. When rats made tolerant to amphetamine in a familiar environment are subsequently tested in a novel environment, they show a loss of tolerance. These results have been interpreted within a Pavlovian conditioning framework (Poulos et al., 1981). According to this view, amphetamine is assumed to invoke a compensatory response to the initial anorexigenic effect of the drug. During the course of chronic injections, this response becomes classically conditioned to environmental stimuli associated with the drug effect. As a result of such conditioning, the compensatory response counteracts the initial effect of the drug, resulting in tolerance. Environmental specificity results from the fact that, in a novel environment, the conditioned stimuli for the compensatory response are absent (*see* the article by Siegel in this volume for further discussion of Pavlovian conditioning and tolerance). However, an alternative interpretation of these results is that environmental cues associated with drug administration come to serve as discriminative stimuli for an instrumentally learned response. When such cues are absent, as in a novel environment, performance of the response is impaired.

Establishing the extent to which behavior is controlled by classically conditioned stimuli or discriminative stimuli has been a thorny problem for learning theorists, in part because it is so difficult to dissociate the two (Schwartz, 1984). With respect to the environmental specificity of tolerance, however, there is a simple way to determine the nature of the controlling stimulus. If tolerance is controlled by classically conditioned stimuli associated with drug administration, then injecting a tolerant rat *with saline* in the environment normally associated with the drug should unmask the conditioned compensatory response. In an experiment designed to test this prediction, however, no such response was obtained (Demellweek and Goudie, 1983a).

The failure to demonstrate directly a conditioned compensatory response is a major weakness in the Pavlovian theory of amphetamine tolerance, but it is not the only one. If tolerance involves Pavlovian conditioning, then repeated exposure to the familiar environment under the influence of the drug should be sufficient for the development of the conditioned compensatory response (i.e., tolerance). Access to milk should be irrelevant. Similarly, repeated exposure to the environment under the influence of a placebo should be sufficient to extinguish the conditioned response. Again, access to milk should be irrelevant. However, neither of these expectations was confirmed. That is, both the acquisition and extinction of tolerance were contingent on access to milk (Poulos et al., 1981).

In order to accommodate these and related findings, a modification of the classical conditioning theory of tolerance has been proposed (Poulos and Hinson, 1984; Poulos et al., 1981). According to this reformulation, the stimulus for a conditioned compensatory response is not the drug *per se*, but rather the functional effect it produces (cf. Kalant, 1977). In the case of amphetamine, the functional effect of the drug, anorexia, is said to require the presence of food. When food is available, a conditioned compensatory response is automatically recruited according to Pavlovian conditioning principles. Similarly, the extinction of tolerance requires access to food, because it is viewed as the mirror image of acquisition.

Although the reformulated version of the theory provides a rationale for the necessity of having access to food during the acquisition and extinction of tolerance, the major shortcoming of the theory remains the failure to demonstrate a conditioned compensatory response (Demellweek and Goudie, 1983a). Moreover, the importance of food availability for the development of tolerance to amphetamine is certainly compatible with the instrumental learning theory. Food provides the reinforcement for the acquisition of the learned response. On the other hand, it is not clear why, from an instrumental learning perspective, tolerance should extinguish following injections of saline and access to milk, a result that has been replicated in a number of studies (e.g., Foltin and Schuster, 1982; Wolgin and Salisbury, 1985). One possibility is that, under these conditions, the ability of the drug to act as a discriminative stimulus is lost, although the learned response itself may not be. This may explain why tolerance is reacquired so quickly once drug administration is reinstated (cf. Streather and Hinson, 1986; Wolgin and Salisbury, 1985).

It should be noted that, in the experiment by Poulos et al. (1981), which was based on the Pavlovian model, extinction involved withholding the *drug*, i.e., giving saline to tolerant animals. In contrast, in an instrumental learning paradigm, extinction would involve withholding the reinforcer (milk) following a response, but continuing the chronic injections of amphetamine. Operationally, this is difficult to accomplish with the milk drinking paradigm, because removing the milk would also eliminate the opportunity to respond. One solution might be to provide tolerant rats with an empty (or nearly empty) drinking tube or to substitute water for milk. Alternatively, an operant paradigm might be employed. In this case, reinforcement could be withheld from rats that had become tolerant to the disruptive effect of the drug on a bar-pressing task, for example. In any case, the effect of these procedures on tolerance has not yet been investigated. On the other hand, Streather and Hinson (1986) reported that giving tolerant rats injections of amphetamine without access to milk (or a drinking tube) did not result in the loss of tolerance. These results are consistent with the instrumental learning theory inasmuch as the absence of milk precluded the opportunity for drinking to extinguish.

2.1.6. Data Inconsistent with the Reinforcement Density Hypothesis

In contrast to the evidence discussed above, the results of a number of studies appear to be inconsistent with the view that tolerance to stimulants is mediated by instrumental learning. Such studies fall into three categories: (1) those that show a lack of tolerance despite an initial loss of reinforcement; (2) those that show tolerance under

circumstances in which the drug did not initially cause a loss of reinforcement; and (3) those in which tolerance developed despite an
initial increase in reinforcement.

2.1.6.1. LACK OF TOLERANCE DESPITE REINFORCEMENT
LOSS.   As noted previously by Demellweek and Goudie
(1983b), there are a number of studies involving FR schedules in
which some or all of the subjects failed to become tolerant to the
initial loss of reinforcement produced by amphetamine (e.g.,
Brown, 1965; Demellweek and Goudie, 1981; Harris and Snell,
1980). For example, in a study by Harris and Snell (1980), rats
were tested under a multiple FI, FR schedule using the before/after
design. In the Before Group, amphetamine (1 mg/kg) initially
caused a decrease in the rate of responding on both components and
tolerance did not develop to this effect over eight trials. Because the
number of reinforcements earned was not reported, it is not clear
whether the decrease in rate was accompanied by a loss of reinforcement in each component, but it seems reasonable to assume
that such was the case, at least for the FR component. Thus, tolerance failed to develop in any of the subjects despite a loss of reinforcement. It is possible, of course, that tolerance would have developed had more than eight trials been run. However, in a study by
Demellweek and Goudie (1981), some subjects failed to acquire
tolerance even after 48 trials. Considerable variability in both the
rate and degree of tolerance has also been reported with other paradigms as well (cf. Wolgin and Salisbury, 1985). Such individual
differences emphasize that group data are often unrepresentative of
the performance of individual subjects. Although there has been
some speculation regarding the variables that may contribute to individual differences in tolerance development (cf. Demellweek and
Goudie, 1983b), there has been little empirical work on this problem. At the present time, therefore, the inability to account for such
variability constitutes an important shortcoming in the reinforcement density hypothesis.

2.1.6.2. TOLERANCE WITHOUT REINFORCEMENT LOSS.
If tolerance represents a behavioral compensation for the loss of reinforcement, then no tolerance would be expected in the absence of
such a loss. Contrary to this corollary of the reinforcement density
hypothesis, two studies have reported tolerance to amphetamine

under conditions in which the drug did not produce a loss of reinforcement (Branch, 1979; Brocco and McMillan, 1983). In an experiment by Branch (1979), monkeys were tested on a three-component multiple FI schedule in which each component involved a different consequent event (reinforcer): shock, food, and escape. Dose–response curves were obtained both before and during the chronic administration of amphetamine (0.1 mg/kg). On the initial determination, amphetamine produced similar inverted U-shaped dose–response curves in each component. At the chronic dose, amphetamine produced an increased rate of responding on all three components, which had no effect on the number of reinforcements earned. Nevertheless, chronic administration of this dose resulted in the development of tolerance on all three components. Moreover, redetermination of the dose–response curves revealed a shift to the right on two of the three components (food and escape), but not on the third (shock; although responding was generally suppressed on the latter). Thus, these results seem to indicate that tolerance can develop even when the initial effect of the drug does not cause a loss of reinforcement.

One way to resolve this apparent inconsistency with the reinforcement density hypothesis is to view the effort associated with the initial increased rate of responding as a "cost" analogous to reinforcement loss, as Branch (1979) has suggested. From this perspective, tolerance resulted in a more economical response-to-reinforcement ratio. However, there is another factor involved in this experiment, which complicates the interpretation of these data. During the course of chronic injections of the drug, occasional tests in which saline was substituted for amphetamine revealed that the baseline rates of responding declined. If the data from the second dose–response determination are expressed as a percentage of this new baseline, sensitization, not tolerance, is obtained. Which interpretation of the data is correct? It is difficult to say. On the one hand, one could argue that the altered baseline represents a new control level of responding that should be used to evaluate the effects of the drug. According to this interpretation, Branch's data reveal sensitization. On the other hand, one could argue that the new baseline represents a behavioral adjustment to the initial rate-increasing effect of the drug. According to this interpretation, the data indicate tolerance. (For further discussion of Branch's data, *see* Barrett, Glowa, and Nader, this volume).

The second study showing tolerance in the apparent absence of reinforcement loss also involved a baseline shift. In an experiment by Brocco and McMillan (1983), rats were tested on a multiple FR, FI schedule for food reward and were given chronic *posttest* intubations of amphetamine (10 mg/kg). Dose–response curves were determined both before and during chronic treatment. During the initial determination, amphetamine induced a dose-related decrease in the rate of responding on both components of the schedule. At 3–5 wks later, the dose–response functions were shifted upward and to the right, indicating tolerance had developed. Because the rats were not tested under the influence of the drug during the intervening period of posttest injections, these results appear to contradict the reinforcement density hypothesis.

This conclusion is not as clear-cut as it may appear, however. During the chronic phase of the experiment, when the subjects received daily posttest intubations of the drug, both the rate of responding and the number of reinforcements earned/session continuously declined. Although brain levels of amphetamine were undoubtedly miniscule at the time of testing, this decline was related to amphetamine administration because the rates returned to normal when the drug was subsequently withdrawn. Because the effect of amphetamine is dependent on the control rate of responding (Dews and Wenger, 1977), the enhanced level of responding obtained during the dose–response redetermination may simply have reflected the lower baseline rate prevailing at that time. As in the experiment by Branch (1979), when baseline shifts are involved, it is difficult to interpret the data.

2.1.6.3. TOLERANCE DESPITE INCREASED REINFORCEMENT. Another corollary of the reinforcement density hypothesis is that tolerance should not develop if the initial effect of the drug results in an increased frequency of reinforcement. A series of studies by Leith and her colleagues involving the effects of amphetamine on rewarding brain stimulation appear to be at variance with this notion (Table 2). In these studies, rats were reinforced with electrical stimulation of the medial forebrain bundle for pressing a bar. Every 5 s, the intensity of stimulation was reduced by 5%, until 15 descending current intensities were presented, after which the current automatically reset to the highest intensity once again. Repeated trials with this paradigm typically yield response rate x current in-

Table 2
Summary of Studies Showing Tolerance or Sensitization to the Effects of Amphetamine on Rewarding Brain Stimulation

Authors	Dose, mg/kg	Schedule	Days	Electrode site[a]	Tested w/drug[b]	Result
Anderson et al., 1978	1–12	3 × daily	4	MFB	No	Tolerance
Anderson et al., 1978	1–12	3 × daily	4	Fornix, LH	No	No tolerance
Cassens et al., 1981	1–12	3 × daily	4	MFB	No	Baseline shift
Kokkinidis et al., 1980	7.5	2 × daily	10	SN	No	Baseline shift
Kokkinidis et al., 1986	5	2 × daily	10	LH	No	Baseline shift
Kokkinidis and Zacharko, 1980a	2	Daily	10	SN	Yes (0.3 mg/kg)	Sensitization
Kokkinidis and Zacharko, 1980b	7.5	2 × daily	30	SN	Yes (0.3 mg/kg)	Sensitization/BS[c]
Kokkinidis and Zacharko, 1980c	7.5	2 × daily	15	LH	No	Sensitization/BS
Leith and Barrett, 1976	1–12	3 × daily	4	MFB	No	Tolerance/BS
Leith and Barrett, 1980	5 and 10	Daily	14	MFB	No	Baseline shift
Leith and Barrett, 1981	1–12	3 × daily	4	MFB	No	Tolerance/BS
Leith and Kuczenski, 1981	1–12	3 × daily	4	MFB	No	Tolerance/BS
Leith and Kuczenski, 1981	3	Daily	6	MFB	No	Tolerance
Liebman and Segal, 1976	0.1–1.5	Daily	10–12	SN	Yes	No change
Predy and Kokkinidis, 1981	7.5	2 × daily	10	SN	No	Baseline shift
Predy and Kokkinidis, 1984	1	Daily	20	SN and NA	Yes	Sensitization
Robertson and Mogenson, 1979	1.5	Daily	9	PFC	No	Sensitization
Robertson and Mogenson, 1979	1.5	Daily	9	NA	No	No change
Robertson and Mogenson, 1979	1.5	Daily	9	SCB	No	Tolerance

[a]LH = lateral hypothalamus; MFB = medial forebrain bundle; NA = nucleus accumbens; PFC = prefrontal cortex; SCB = supracallosal bundle; SN = substantia nigra.
[b]No = subjects were not tested during chronic dosing regimen; yes = subjects were tested during chronic dosing regimen.
[c]BS = baseline shift.

tensity profiles, in which the rate of responding progressively decreases as the current intensity declines. The intensity at which responding decreases to 50% of the maximal rate defines the "reward threshold." The effect of amphetamine and/or saline on this threshold was evaluated both before and after the rats were given a series of incremental doses (1–12 mg/kg) of the drug $3 \times$/d for 4 d. The rats were not tested during this period of chronic injections.

The results of an experiment by Leith and Barrett (1976) illustrate the general findings (*see* Fig. 8). Prior to chronic treatment, amphetamine (0.5 mg/kg) had little effect on the rate of responding at the higher intensities of stimulation, but greatly facilitated responding at the lower intensities; i.e., it lowered the threshold of reward. Following chronic treatment with the drug, however, tolerance developed to this effect. That is, the rate of responding at the lower intensities declined toward the original nondrug levels. However, subsequent tests with saline revealed that these control levels had themselves decreased as a result of chronic drug treatment (Fig. 8). Similar baseline shifts were reported by other investigators as well (e.g., Cassens et al., 1981; Kokkinidis, et al., 1986).

In an experiment employing a similar dosing regimen, McCown and Barrett (1980) showed that tolerance also develops to the rewarding effects of intravenously injected amphetamine. Like humans, rats will self-administer amphetamine, presumably because it acts on the same neural circuit that mediates rewarding electrical stimulation of the brain (Stein, 1964; Wise and Bozarth, 1981). McCown and Barrett (1980) determined the rate of amphetamine self-administration before and after a series of escalating doses of the drug, as described above. Prior to chronic treatment, the rate of self-administration of amphetamine (0.25 mg/kg/injection) was quite consistent. Moreover, when the dose/injection was reduced, the rats compensated by increasing their rate of responding. Following chronic injections of the drug, the rate of responding for the standard dose (0.25 mg/kg/injection) increased by at least 45%, as if the dose had been reduced. Thus, tolerance developed to the reinforcing effect of intravenously injected amphetamine.

Do these data contradict the reinforcement density hypothesis? As previously discussed, studies that involve baseline shifts are difficult to interpret. On the one hand, Barrett (1985) has cogently

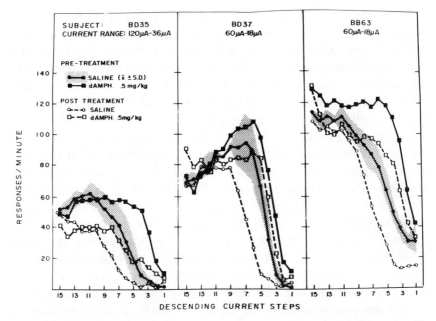

Fig. 8. Effect of *d*-amphetamine (0.5 mg/kg) and saline on self-stimulation response rates at different current intensities before (pretreatment) and after (posttreatment) chronic escalating doses (1–12 mg/kg) of amphetamine. After chronic dosing, there is a diminished effect of amphetamine and a baseline shift with saline. (From "Amphetamine and the reward system: Evidence for tolerance and post-drug depression" by N. J. Leith and R. J. Barrett, *Psychopharmacologia*, 1976, **46**, 19–25. Fig. 3. Copyright 1976 by Springer-Verlag. Reprinted by permission.)

argued in a recent review that tolerance does develop to the rewarding and reward-enhancing effects of amphetamine. He proposes that baseline shifts in reward threshold represent an adaptive, homeostatic mechanism that opposes the rewarding effect of the drug. Moreover, in conjunction with the data on tolerance to the rewarding effects of intravenous amphetamine injections, the results of these studies are consistent with human clinical evidence that tolerance occurs to the "euphoric" effects of stimulants and other abused drugs, and is accompanied by a state of dysphoria when the drug is withdrawn (Watson et al., 1972). Such dysphoria, which is analogous to a shift in an hedonic baseline, would also contribute to drug dependence. From this perspective, the argument for tolerance is certainly compelling.

On the other hand, the fact that there is a shift in the baseline means that the system as a whole has changed. If the "tolerant" level of responding is indexed to this new control level, no evidence of tolerance is found. This argument is not merely a matter of semantics. Because we are concerned with the role of learning, it is particularly important to evaluate the data with respect to the appropriate baseline. However, even if we accept the argument that tolerance has occurred, the fact that these experiments all involved baseline shifts means that they do not contradict the reinforcement density hypothesis. According to the hypothesis, animals can learn to overcome the initial disruptive effects of a drug. If the drug produces an increase in reinforcement, no learning would be expected. The foregoing experiments do not contradict the hypothesis, because the change in the rate of reinforced responding was the result of a shift in baseline, not learning. Only if the baseline remained stable would the data contradict the reinforcement density hypothesis. Therefore, at best, the data suggest that, under conditions in which learning would not be expected, other mechanisms can account for tolerance.

It should be noted, however, that not all studies involving rewarding brain stimulation have found tolerance to amphetamine. For example, a series of studies by Kokkinidis and his colleagues have shown *sensitization* to the reward-enhancing effect of the drug (Table 2). In many of these studies, a baseline shift (in the same direction as described above) was also observed following injections of saline. As summarized in Table 2, there are a number of methodological differences between the studies that report tolerance and those that report sensitization. One obvious difference is the location of the electrode. All of the studies that found tolerance had electrodes in the medial forebrain bundle; with one exception, those that found sensitization had electrodes in the substantia nigra. [The exceptional case (Kokkinidis and Zacharko, 1980c) involved electrodes in an area of the lateral hypothalamus previously found to be refractory to the development of tolerance (Anderson et al., 1978)]. Although one study involving nigral electrodes reported no change (Liebman and Segal, 1976), the combination of dose and length of treatment in this experiment was at the low end of the range. Other studies involving different electrode placements have also found tolerance, sensitization, or no change, depending on the site (Anderson, et al., 1978; Robertson and Mogenson, 1979). It is interesting that behavioral experience during the period of chronic drug administration did not appear to contribute to the development of sensitization. Some studies involved such experience (e.g., Predy and Kokkinidis, 1984), whereas others did not (e.g., Kokkinidis and Zacharko, 1980c). None of the studies that reported tolerance involved testing the rats during the chronic phase.

Inspection of Table 2 suggests that baseline shifts are quite common with the brain stimulation paradigm, whether "tolerance" or sensitization is found. Why should this be? One factor may be the dose regimen. Typically, very high doses are given several times a day, in contrast to other paradigms in which lower doses are given once daily. In several of the studies, a diminished effect was obtained when lower doses and/or fewer daily injections were used (Leith and Barrett, 1980; Leith and Kuczenski, 1981). A second factor may be the electrical stimulation itself. Such stimulation increases the turnover of neurotransmitter in the "reward circuit" (cf. Stein and Wise, 1969) and potentiates the behavioral effects of amphetamine (Eichler and Antelman, 1979). Thus, the baseline shifts observed in these studies probably reflect neurochemical alterations at the cellular level. The fact that such shifts can be blocked by pretreatments with lithium (Predy and Kokkinidis, 1981) or reversed by tricyclic antidepressants (Kokkinidis et al., 1980) supports this view. What has so far defied explanation is why these shifts are accompanied by tolerance at some electrode sites and by sensitization at others. This important issue needs to be addressed by future research.

2.1.7. Conclusions

A wide range of studies has demonstrated that tolerance to stimulants is contingent on behavioral experience in the drugged state and is differentially affected by the contingencies of reinforcement. There is also some preliminary evidence that tolerance may come under the discriminative control of both drug and environmental cues. These studies are, therefore, consistent with the view that tolerance involves instrumental learning. Alternative interpretations based upon conditioned aversions, Pavlovian conditioning, cumulative deprivation, or an altered body weight set point are unable to account for these findings. There are, however, two areas that require further research. One is the substantial individual differences that occur in both the rate and extent of tolerance. The second is the problem of how to interpret tolerance when it is accompanied by baseline shifts. On balance, however, the data are consistent with a learning interpretation.

2.2. Alcohol

2.2.1. Contingent Tolerance

Contingent tolerance to ethanol has been demonstrated in a variety of behavioral tasks in both laboratory animals and human subjects (Table 3A; *see also* Table 3B for studies involving only pretest injections). Indeed, the before/after paradigm was first introduced to demonstrate that tolerance to ethanol is contingent on behavioral experience. In this classic study by Chen (1968), subjects were trained to move down a runway to a circular track, traverse the track twice in a particular direction, and then return down the runway where food was waiting if the task was successfully completed. Thus, the task required both motoric and "cognitive" abilities. Rats in the Before Group were injected with ethanol every fourth day prior to testing, whereas rats in the After Group were injected after testing on the first three exposures to the drug, and before testing on the fourth and final exposure. On the first trial, rats in the Before Group showed an initial impairment on the maze. They completed fewer total trials and made many errors. By the fourth trial, however, they became tolerant to the drug. When rats in the After Group were then given pretest injections for the first time, they were not impaired in terms of the total number of trials completed, suggesting that tolerance had developed to the general depressant effect of ethanol. However, they were impaired in their proficiency, i.e., they completed fewer correct trials and made more errors. Chen (1968) therefore concluded that tolerance to the more cognitive impairments induced by ethanol required behavioral experience under the drug.

    The importance of reinforcement in the development of contingent tolerance to ethanol was demonstrated in a subsequent study (Chen, 1979; *see* Fig. 9). Four groups of rats were trained on a FR15 schedule for milk reward. Three of the groups were injected with ethanol (1.2 g/kg) prior to each of three 10-min sessions, but they differed in the amount of reinforcement they could earn (0, 1, or unlimited reward). The fourth group was injected with ethanol after each of these sessions. On the last trial, all of the groups were injected with a higher dose (1.5 g/kg) prior to testing. The dependent measure in this study was the latency to complete 15 responses. All three of the groups given pretest injections showed increased latencies on the first trial, but became partially tolerant on the following two trials, with the rewarded groups showing substantially greater tolerance than the unrewarded group (Fig. 9). On the final test day, when all of the groups were tested with the higher dose, only the rewarded groups showed tolerance to the drug; the nonrewarded group and the group given posttest injections were equally impaired. Thus, tolerance to ethanol was contingent, not on behavioral experience *per se*, but rather on the reinforcement derived from such experience. These results, therefore, provide strong support for the instrumental learning theory of tolerance.

Additional support for the theory comes from a series of experiments by Vogel-Sprott and her colleagues on the development of tolerance in human subjects. In one study (Mann and Vogel-Sprott, 1981), male social drinkers were given ethanol (0.84 g/kg) and then tested on a pursuit rotor task. Subjects that were given a monetary reward for exceeding their drug-free baseline levels of performance developed tolerance to the initial impairment induced by the drug, whereas subjects that were not rewarded did not become tolerant. When additional trials were run without reinforcement, the previously rewarded subjects showed a precipitous loss of tolerance, whereas the previously nonrewarded subjects were unaffected. Thus, both the acquisition and extinction of tolerance in this experiment were consistent with the instrumental learning theory.

A subsequent study by Beirness and Vogel-Sprott (1984a) explored the efficacy of different types of reinforcement on the development of tolerance to ethanol. Groups of subjects were given beer (0.84 g/kg ethanol) prior to being tested on a pursuit rotor task. Group CR (contingent reward) was given a monetary reward and verbal feedback when the group's performance met a predrug criterion. Group NCR (noncontingent reward) was also given a monetary reward and verbal feedback, but noncontingently. Group IO (information only) was given only verbal feedback, whereas Group

Table 3

Summary of Studies Showing Tolerance to the Behavioral Effects of Ethanol

Authors	Dose, g/kg	Species	Task[a]
A. *Contingent tolerance*			
Beirness and Vogel-Sprott, 1984a	0.84	Humans	Pursuit rotor
Bixler and Lewis, 1986	0.75	Rats	Wheel running
Chen, 1968	1.2	Rats	Circular maze
Chen, 1972	0.4–1.2	Rats	Circular maze
Chen, 1979	1.2 and 1.5	Rats	FR
Haubenreisser and Vogel-Sprott, 1983	0.83	Humans	Pursuit rotor
Mann and Vogel-Sprott, 1981	0.84	Humans	Pursuit rotor
Mansfield et al., 1983	2	Rats	Shock avoidance
Rawana and Vogel-Sprott, 1985	0.66	Humans	Pursuit rotor
Rumbold & White, 1987	0.85	Humans	Mult FR, DRL
Wenger et al., 1980	1.4–2.4	Rats	Shock avoidance
Wenger et al., 1981	1.6–2.2	Rats	Shock avoidance
B. *Studies involving only pretest injections*			
Bird et al., 1985	0.375–1.5	Rats	FR, VI
Le et al., 1986b	6	Rats	Shock avoidance
LeBlanc et al., 1969	3–6	Rats	Shock avoidance
Pieper and Skeen 1973	3	Rhesus monkeys	Discrimination reversal

C. Effects of mental rehearsal			
Annear and Vogel-Sprott, 1985	0.62	Humans	Pursuit rotor
Sdao-Jarvie and Vogel-Sprott, 1986	0.62	Humans	Pursuit rotor
Vogel-Sprott et al., 1984	0.66	Humans	Pursuit rotor
D. Behaviorally augmented tolerance			
DeSouza et al., 1981	2.1	Rats	FR
Kalant et al., 1978	2.2 and 6	Rats	Shock avoidance
LeBlanc et al., 1973	1.4 and 6	Rats	Circular maze, shock avoidance
LeBlanc et al., 1975a	1.4 and 2.2	Rats	Circular maze
LeBlanc et al., 1976	2.2, 2.5, and 6	Rats	Shock avoidance
Wigell and Overstreet, 1984	1.5 and 2.5	Rats	FR
E. Contingent tolerance without reinforcement loss			
Alkana et al., 1983	3.6	Mice	Hypothermia
Hjeresen et al., 1986	2.5	Rats	Hypothermia
Jorgensen and Hole, 1984	2.5	Rats	Tail flick
Jorgensen et al., 1985	2.5	Rats	Tail flick
Jorgensen et al., 1986	2.5 and 5	Rats	Tail flick
Pinel and Puttaswamaiah, 1985	1.5	Rats	Clonus
Pinel et al., 1983	1.5	Rats	Clonus
Pinel et al., 1985	1.5–5	Rats	Clonus
Traynor et al., 1976	0.8M	Aplysia	Decay of PTP
Traynor et al., 1980	0.8M	Aplysia	Decay of PTP

[a]VI = variable interval; PTP = posttetanic potentiation; for other abbreviations, *see* Table 1.

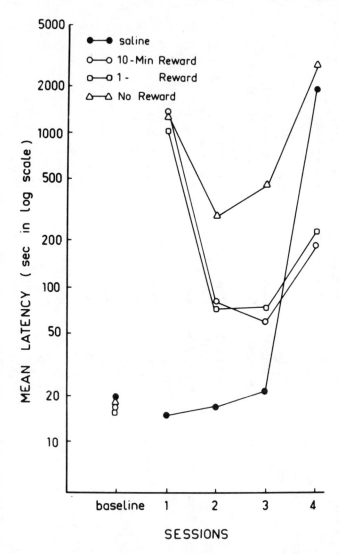

Fig. 9. Development of tolerance to ethanol as a function of amount of reinforcement. On sessions 1–3, rats in the experimental groups were given pretest injections of 1.2 g/kg; rats in the saline group were given posttest injections of the drug. On session 4, all groups received pretest injections of 1.5 g/kg. Only the rewarded groups showed tolerance on this trial. (From ''Acquisition of behavioral tolerance to ethanol as a function of reinforced practice in rats'' by C. S. Chen, *Psychopharmacology*, 1979, **63,** 285–288. Fig. 1. Copyright 1979 by Springer-Verlag. Reprinted by permission.)

NR (no reward) was given nonreinforced practice. Over the course of four trials, only this latter group failed to develop tolerance to the ethanol-induced impairment of performance. The rate of acquisition of tolerance for the other groups was: Group CR > Group IO > Group NCR. Thus, the more salient and contingent the reinforcement, the faster the rate of tolerance.

One additional feature of this experiment is noteworthy. At the conclusion of the tolerance phase, all of the groups were given an additional trial after ingesting nonalcoholic beer. Following this placebo, all of the previously tolerant groups showed better performance than they did during their drug-free baseline tests. The degree to which their performance exceeded baseline was proportional to their level of tolerance: Group CR > Group IO > Group NCR. The authors, therefore, suggested that prior exposure to ethanol induced a "drug-opposite" response (reflected in enhanced performance) that became classically conditioned to environmental cues associated with drug administration. (For reviews of the role of such conditioning in tolerance to some of the autonomic effects of ethanol, *see* Hinson and Siegel, 1980, and Siegel, this volume.) The fact that the magnitude of this putative compensatory response was correlated with the degree of tolerance induced by different levels of reinforcement seemed to imply an interaction between operant and classical conditioning mechanisms.

It is important to note that classical conditioning of a compensatory response typically requires only that the drug be given repeatedly in a consistent relation to environmental cues (cf. Siegel, this volume). In the study by Beirness and Vogel-Sprott (1984a), however, the group given repeated exposure to the drug without the opportunity for instrumental learning (Group NR) did not become tolerant and did not show enhanced performance following administration of a placebo, i.e., no evidence for classical conditioning was obtained. Therefore, the enhanced performance of the rewarded groups under placebo in this study may have simply reflected the effects of prior reinforced practice.

Finally, a study by Rumbold and White (1987) illustrates that reinforcement contributes to sensitization as well as tolerance to the effects of ethanol. Human subjects were tested on a bar-pressing task reinforced on a multiple FR, DRL schedule of monetary reward. In eight of the nine subjects, alcohol initially increased the rate of responding on the FR component, which resulted in an increase in the frequency of reinforcement. On subsequent trials, the rate of responding increased further, indicating sensitization to this effect of the drug. In one subject, however, alcohol initially caused a decrease in the rate of responding and a loss of reinforcement. This subject developed tolerance on the subsequent trials. On the DRL component, alcohol increased the rate of premature responding slightly, but this effect did not alter the frequency of reinforcement. Neither tolerance nor sensitization developed to this effect. Clearly, these examples of differential tolerance and sensitization are consistent with the reinforcement density hypothesis, and provide strong support for the role of instrumental learning in these phenomena.

2.2.2. Effect of Mental Rehearsal on Tolerance

The evidence presented so far supports the view that both animals and humans can learn to overcome the disruptive effects of ethanol, if they are reinforced for making appropriate responses. Vogel-Sprott and her colleagues have recently shown that, in human subjects, tolerance can also be acquired by mentally rehearsing the task while under the influence of the drug. In these studies (Table 3C), subjects were given more moderate doses of ethanol than in previous experiments (0.62–0.66 g/kg) and tested on the pursuit rotor task. In one study (Vogel-Sprott et al., 1984), subjects that had mentally rehearsed after drinking ethanol showed levels of tolerance comparable to subjects that had received reinforced practice under the drug. In a subsequent study (Sdao-Jarvie and Vogel-Sprott, 1986), it was shown that such tolerance was contingent on rehearsing in the drugged state. Subjects that rehearsed the task prior to drinking ethanol did not show tolerance when later tested under the drug. Finally, tolerance was greater if mental rehearsal was given in the test room than if it was given in a different room, suggesting that classical conditioning may have contributed to the development of tolerance (Annear and Vogel-Sprott, 1985). Thus, the results of these studies suggest that mental rehearsal, instrumental learning, and classical conditioning all contribute to the development of tolerance to ethanol.

It should be noted, however, that the magnitude of the initial deficit produced by the drug in these studies was quite small—a decrease of about 4 s in the dependent measure (time on the target; in the undrugged state times averaged 30–35 s). It is not clear whether mental rehearsal would be sufficient to overcome the effects of higher doses of ethanol on this task. Moreover, the effects of mental rehearsal on the impairments produced by the drug on other types of tasks have not yet been determined. Therefore, at the present time, the generality of these findings is somewhat limited. Nevertheless, the results of these experiments are certainly consistent with the general theoretical view that tolerance can be understood from a learning perspective.

2.2.3. Reinforcement and Acute Tolerance

It has long been known that the effects of alcohol associated with a particular blood alcohol concentration (BAC) are greater on the rising limb of the BAC curve than on the falling limb (Mellanby, 1919), i.e., there is acute recovery on the falling limb. This effect is not contingent on behavioral experience during the rising phase of the curve, because it has also been demonstrated using a between-subjects design (LeBlanc et al., 1975b). Nevertheless, reinforcement can facilitate such acute tolerance, as demonstrated in a recent study by Haubenreisser and Vogel-Sprott (1987). Male social drinkers were given ethanol (0.60 g/kg) and tested on a visuomotor tracking task. Half of the subjects were reinforced when their performance met or exceeded their predrug scores, whereas the other half were not reinforced. BACs were measured at regular intervals throughout the session. Compared to nonreinforced controls, subjects that were reinforced showed a delayed onset of, and a more rapid recovery from, the drug-induced behavioral impairment.

The fact that reinforcement facilitated acute recovery suggests that tolerance to the effects of both single and repeated doses of alcohol share common processes, at least to some extent. Indeed, Beirness and Vogel-Sprott (1984b) reported that the rate of recovery from the acute effects of alcohol is highly correlated with the rate at which tolerance develops to chronic administration of the drug. However, it appears that reinforcement interacts with the acute and chronic effects of alcohol in somewhat different ways. In acute recovery, reinforcement facilitates performance on both the rising and falling limbs of the BAC curve, delaying the onset and hastening the offset of the behavioral impairment (Haubenreisser and Vogel-Sprott, 1987). In contrast, reinforcement facilitates the development of tolerance to the chronic effects of the drug only when it is given during the falling limb of the BAC curve, but not when it is given during the rising limb (Haubenreisser and Vogel-Sprott, 1983).

2.2.4. Behavioral Induction or Augmentation?

Experiments utilizing the before/after design typically test the After Group with a pretest injection of the drug once only, to ascertain that prior pharmacological exposure *per se* was not sufficient to account for tolerance. As we have seen, the results of such studies are consistent with the view that tolerance to ethanol is contingent on behavioral experience (reinforced practice) under the drug. However, it is possible that, given sufficient exposure to the drug, rats in the After Group might also acquire tolerance. To test this hypothesis, LeBlanc et al. (1973) replicated Chen's (1968) study, but gave extended trials during which rats in the After Group were intermittently tested for tolerance. Rats were tested in a circular maze following a pretest injection of ethanol (1.4 g/kg) every fourth day ("test days"). On intervening days, the Before Group received pretest injections of ethanol and posttest injections of saline, the After Group received the same injections, but in the reverse order, and the Saline Group received both pretest and post-test injections of saline.

As shown in Fig. 10 (Phase I), rats in the Before Group became tolerant to the initial disruptive effect of the drug by the fourth test day. Rats in the After Group became tolerant more slowly, but reached the same level of tolerance by the sixth test day. Increased exposure to ethanol (6 g/kg by gavage) had no further effect on the level of tolerance in either of these groups (Fig. 10, Phase II). However, gavage of ethanol did promote tolerance in rats in the Saline Group (Fig. 10, Phase III), which up to that point had not shown tolerance. LeBlanc et al. (1973), therefore, concluded that behavioral experience did not induce a unique form of tolerance (i.e., "learned tolerance"), but only augmented the rate at which tolerance developed.

The generality of these findings was established in a subsequent experiment in which rats were tested on a shock-avoidance task (the moving belt test). As with the appetitive circular maze

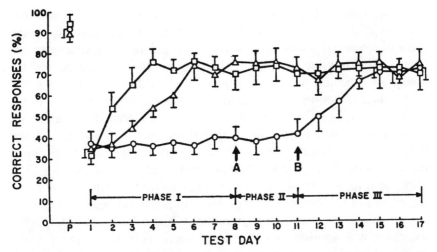

Fig. 10. Development of tolerance to ethanol on the circular maze task by rats in the Before (squares), After (triangles), and Saline (circles) groups. On test days, all three groups were given pretest injections of the drug. At arrow A, daily intubations of ethanol were given to the Before and After groups on the intervening days; at arrow B, intubations were given to the Saline group. During Phase I, tolerance developed in both the Before and After groups, but at different rates. During Phase II, no changes occurred. During Phase III, rats in the Saline Group also became tolerant. (From "Behavioral augmentation of tolerance to ethanol in the rat" by A. E. LeBlanc, R. J. Gibbins, and H. Kalant, *Psychopharmacologia*, 1973, **30,** 117–122. Fig. 1. Copyright 1973 by Springer-Verlag. Reprinted by permission.)

task, rats in the After Group became tolerant to ethanol more slowly than rats in the Before Group, but ultimately reached the same level of performance (LeBlanc et al., 1976). It was also found that daily pretest injections resulted in a faster rate of tolerance than did injections given every second or third day. Pretest injections given every fourth, sixth, or eighth day, however, did not result in tolerance. In another experiment, rats given chronic injections of ethanol and tested on the moving belt test showed tolerance not only on this task, but on the circular maze task as well (LeBlanc et al., 1975a). Because the two tasks required different sensorimotor skills, these results suggested that tolerance to ethanol did not involve task-specific learning under the drug.

One interpretation of the results discussed above is that tolerance to ethanol involves a single mechanism, presumably some form of neuronal adaptation, with which behavioral experience interacts. If this view is correct, then once tolerance is acquired by means of repeated pretest injections, it should be maintained even by posttest injections. This prediction was confirmed in a study by LeBlanc et al. (1976). Rats were given daily pretest injections of ethanol and tested on the moving belt test every fourth day. After they became tolerant, they were assigned to one of four groups. Rats subsequently given either posttest injections of ethanol (2.2 g/kg) or intubations of ethanol (6 g/kg) without testing maintained their level of tolerance. Rats not given ethanol, whether tested each day or left in the home cage, gradually lost tolerance. Thus, once tolerance was acquired with pretest injections of the drug, it could be maintained even with posttest injections. Similar results were obtained in experiments in which rats were exposed to repeated cycles of ethanol exposure and withdrawal (Kalant et al., 1978). The reacquisition of tolerance on subsequent cycles occurred at a faster rate than on the first cycle, whether the first cycle involved pretest injections and the second involved posttest injections or vice versa (*see also* DeSouza Moreira et al., 1981). A summary of these studies is presented in Table 3D.

Although the results discussed above are consistent with the view that tolerance to ethanol is augmented, but not induced, by reinforced practice under the drug, they are also subject to an alternative interpretation. As Wenger and his colleagues (Wenger et al., 1980; Wenger et al., 1981; Mansfield et al., 1983) pointed out, the gradual development of tolerance in the After Groups may have been the result of the intermittent tests for tolerance conducted every fourth day. Indeed, as shown in Table 4, when the Before and After Groups in the experiments by LeBlanc et al. (1973, 1976) are equated for pretest injections, the rate of tolerance in the two groups is almost identical.

To determine whether intermittent testing might account for tolerance, Wenger et al. (1981) modified the design used by LeBlanc et al. (1976) by adding a second After Group. Rats in the Before Group were tested on the moving belt test every day with pretest injections of ethanol. Rats in one of the two After Groups were given daily posttest injections, except when they were tested for tolerance every fourth day as in the LeBlanc et al. (1976) study.

Table 4
Level of Tolerance in Before and After Groups Equated
for Pretest Injections[a]

	Pretest injection no.		
LeBlanc et al., 1973[b]	1	5	9
Before	31	54	65
After	35	58	74
LeBlanc et al., 1976[c]			
Before	61	39	25
After	61	44	30

[a]In each experiment, rats were given pretest injections of ethanol every fourth day ("test days"; cf. Fig. 10). On intervening days, rats in the Before Group received pretest injections of ethanol, whereas rats in the After Group received posttest injections. Consequently, the fifth pretest injection occurred on test day 2 for the Before Group and on test day 5 for the After Group. Similarly, the ninth pretest injection occurred on test day 3 for the Before Group and on test day 9 for the After Group. The data for LeBlanc et al., 1973 are derived from Fig. 10. The data for LeBlanc et al., 1976 are derived from Fig. 1 of the original article.
[b]Percent correct responses.
[c]Time off belt.

Rats in the second After Group, as well as those in the Saline Group, were not tested for tolerance during this phase of the experiment. After 23 d of treatment, all of the groups were given pretest injections. As expected, rats in the Before Group became tolerant to ethanol at a faster rate than rats in the After Group that had been given intermittent pretest injections. In contrast, rats in the other After Group, which were not given intermittent pretest injections, were as impaired as rats in the Saline Group. Thus, exposure to ethanol *per se* was not sufficient for the development of tolerance. In a subsequent study, Mansfield et al. (1983) demonstrated that, although tolerance on the moving belt test was found only in the group given pretest injections, all groups exposed to ethanol were equally tolerant to the hypothermic effect of the drug. The finding that tolerance to the hypothermic effect of ethanol did not generalize to the moving belt test in rats not tested under the drug argues against a unitary, nonspecific mechanism of tolerance (*see also*

Hjeresen et al., 1986). Wenger and his associates, therefore, concluded that tolerance to ethanol is contingent on experiencing the functional effects of the drug in a context that permits learning to occur (Wenger and Woods, 1984).

A broader view of the literature suggests that this conclusion may be unwarranted, however. Consider the data in Table 5. In some cases, pretest injections of ethanol failed to induce tolerance (Results 1–3, and 7), whereas in other cases, injections outside of the behavioral task either induced (Results 4, 8, and 9) or maintained (Result 6) tolerance. Finally, in one case, pretest injections produced about the same level of tolerance as gavage of ethanol without behavioral testing (Result 5). How can we account for these inconsistent findings? An examination of the procedural details of these studies suggests that dose, frequency of injection, and number of trials may be critical variables. For example, when pretest injections failed to induce tolerance (Results 1–3), doses in the 1.4–2.5 g/kg range were administered every 4, 6, or 8 d, but saline was given on the intervening days. In contrast, when intermittent pretest injections of these doses did induce tolerance (e.g., Result 6; also, LeBlanc et al., 1976; Wenger et al. 1981), injections of ethanol were given on the intervening days. Moreover, even when no intermittent pretest injections were given, tolerance still occurred if a high dose (6 g/kg) was given daily over a 2-wk period (Result 4) or a low dose (1.5 g/kg) was given daily for a 7-wk period (Result 9).

A summary of these observations is presented in Table 6. Taken together, they suggest the following tentative conclusions:

1. Tolerance to ethanol occurs most rapidly when subjects are given daily pretest injections.
2. Tolerance produced by intermittent pretest injections requires exposure to ethanol on intervening days.
3. When subjects are not tested during chronic exposure to ethanol, the rate of tolerance varies with the dose.

Although these results appear to support the concept of behaviorally augmented tolerance, they do not rule out the possibility that learning experiences outside of the testing situation (e.g., learning to overcome ataxia) are responsible for the development of tolerance in subjects given posttest injections. Nor do they neces-

Table 5
Data Inconsistent with the Reinforcement Density Hypothesis

1. Saline Group does not become tolerant on circular maze despite intermittent pretest injections of ethanol (1.4 g/kg; LeBlanc et al., 1973).
2. Saline Group does not become tolerant on moving belt test despite intermittent pretest injections of ethanol (2.5 g/kg; LeBlanc et al., 1976).
3. Before Group does not become tolerant if injections of ethanol (2.2 g/kg) are given every fourth, sixth, or eighth day (LeBlanc et al., 1976).
4. Ethanol by gavage (6 g/kg) without behavioral testing results in partial tolerance on moving belt test (LeBlanc et al., 1975a).
5. Pretest injections of ethanol (2.2 g/kg) induce about the same level of tolerance as gavage (6 g/kg) without behavioral testing (cf 4., above; LeBlanc et al., 1975a).
6. Both gavage (6 g/kg) and posttest injections (2.2 g/kg) of ethanol prevent the loss of tolerance (LeBlanc et al., 1976).
7. After Group given intermittent pretest injections of ethanol (2 g/kg) does not become tolerant (Mansfield et al., 1983).
8. After Group shows tolerance to general depressant effect of ethanol (1.2 g/kg) on circular maze task (Chen, 1968).
9. After Group shows same level of tolerance as Before Group on a bar-pressing task following 50 posttest injections of ethanol (1.5 g/kg; Wigell and Overstreet, 1984).

Table 6
Effect of Dose, Schedule of Injections, and Duration of Treatment
on Development of Tolerance to Ethanol

Dose, g/kg	Schedule of injections	Duration, wk	Tolerance
1.4–2.5	Daily, or every 2nd or 3rd day, pretest	3	Yes
2.2	Every 4th, 6th or 8th day pretest	3	No
1.4–2.5	Every 4th day, pretest; remainder posttest	3	Yes
1.6–2.2	Daily, posttest	3	No
1.5	Daily, posttest	7	Yes
6	Daily, no tests	2	Yes

sarily mean that tolerance involves a single mechanism. Given that ethanol has both nonspecific (e.g., on cellular membranes) and specific effects on the central nervous system, multiple mechanisms of tolerance may well be involved, as others have suggested (e.g., Tabakoff et al., 1984). Indeed, tolerance to some of the effects of ethanol (e.g., sleeping time) has been found to be the result of both metabolic and pharacodynamic mechanisms (Wood and Laverty, 1979). Further support for this view will be presented in the next section.

2.2.5. Contingent Tolerance Without Loss of Reinforcement

As previously discussed, the utility of the before/after design in evaluating the role of instrumental learning in tolerance is that it allows the investigator to manipulate behavioral experience under the drug while holding drug exposure constant. When the initial drug effect involves loss of reinforcement, differences between the Before and After Groups are often attributed to the role of instrumental learning. Recently, however, contingent tolerance to ethanol has been demonstrated under conditions that do not seem to involve either reinforcement or instrumental learning. These experiments are summarized in Table 3E. As we shall see, the results of these studies require a more molecular interpretation of contingent tolerance.

Consider first a series of experiments by Pinel and his colleagues on contingent tolerance to the anticonvulsant effect of ethanol. Rats given injections of ethanol (1.5 g/kg) prior to convulsive electrical stimulation of the amygdala developed tolerance to the initial anticonvulsive effect of the drug in five trials. In contrast, rats given chronic injections of ethanol after daily convulsive stimulation, or without convulsive stimulation, did not acquire tolerance (Pinel et al., 1983). These results were confirmed in a subsequent study in which both dose level and number of trials were varied (Pinel et al., 1985). Thus, tolerance was contingent on an interaction between the drug and a behavioral effect (seizures) in a paradigm that did not involve reinforcement or instrumental behavior, at least in the traditional meaning of these terms. Nor was Pavlovian conditioning involved. Tolerance to the anticonvulsive effect of ethanol was not environmentally specific, it was not blocked by preexposure to the environment in which ethanol would later be administered, and it was not associated with a conditioned compensatory increase in seizure duration (Pinel and Puttaswamaiah, 1985).

Although the mechanism of such contingent tolerance is unclear, the results of these studies are consistent with the view that tolerance develops to the functional effect of the drug and not to the drug itself (Kalant, 1977; LeBlanc and Cappell, 1977). Further support for this conclusion may be found in the finding that tolerance to the hypothermic effect of ethanol is contingent on experiencing hypothermia. If subjects are prevented from experiencing hypothermia, either by testing them in a warm environment (Alkana et al., 1983) or by inducing hyperthermia via microwave irradiation (Hjeresen et al., 1986), tolerance either does not develop, or it develops more slowly (Le et al., 1986a).

In the experiments discussed so far, the effects of ethanol were assessed in intact organisms. However, contingent tolerance to the functional effects of ethanol has also been demonstrated in surgically simplified preparations. For example, Traynor et al. (1976) analyzed the effects of ethanol on postsynaptic potentials evoked in the isolated abdominal ganglion of *Aplysia* californica. Following repetitive electrical stimulation of the visceropleural connective, the amplitude of excitatory postsynaptic potentials recorded from cells in the abdominal ganglion in response to test stimuli was enhanced. This posttetanic potentiation gradually decayed back to normal size within 20 min. When conducted in a bath containing ethanol (0.8M), however, the rate of decay was accelerated. Repeated cycles of stimulation and recovery resulted in the development of tolerance to this enhanced rate of decay.

In a subsequent experiment (Traynor et al., 1980), such tolerance was shown to be contingent on tetanic stimulation in the presence of ethanol. Stimulation during ethanol-free periods did not result in tolerance, even though overall exposure to the drug was the same. The amount and frequency of stimulation during exposure to ethanol also played a role. At high frequencies of stimulation (e.g., trains of 100 stimuli at 60/min), tolerance developed with as few as 1000 stimuli; at low frequencies (6/min), tolerance developed after 3600 stimuli, but not after 2160 stimuli. The development of tolerance was also influenced by the concentrations of magnesium and calcium ions in the perfusing medium. These ions decrease and increase, respectively, the release of neurotransmitter from the presynaptic cell. Tolerance was blocked by increasing the concentration of magnesium ion and enhanced by increasing the concentration of calcium ion. These results, therefore, suggest that tolerance in this preparation is contingent on adequate transmitter release in the presence of ethanol.

Contingent tolerance has also been demonstrated to the effect of ethanol on the tail flick reflex in spinal rats (Jorgensen and Hole, 1984; Jorgensen et al., 1985). The dependent variable in these studies was the latency of the response elicited by radiant heat. Spinal rats given ethanol (2.5 g/kg) prior to elicitation of the reflex became tolerant in 9 d to the initial latency-increasing effect of the drug. Rats given ethanol after testing, or without testing (whether or not they were exposed to the apparatus), did not become tolerant. As with the effects of ethanol on seizure duration (cf. Pinel and Puttaswamaiah, 1985), such contingent tolerance was not environmentally specific and was not associated with a conditioned compensatory response, thereby eliminating a Pavlovian conditioning interpretation.

A subsequent series of experiments confirmed and extended these findings (Jorgensen et al., 1986). Spinal rats given posttest injections of 2.5 g/kg did not become tolerant, even when the chronic treatment was extended to 3 wk. However, rats given injections of a higher dose (5 g/kg) for 7 d without testing were tolerant when subsequently given a pretest injection of 2.5 g/kg. Thus, with a low dose of ethanol, tolerance was contingent on behavioral experience, whereas with a higher dose, repeated pharmacological exposure was sufficient. These two forms of tolerance appear to be mediated by different mechanisms. Contingent tolerance to chronic low doses of ethanol produced cross-tolerance to the latency-increasing effects of both morphine and clonidine. In contrast, pharmacological tolerance induced by injections of the higher dose did not produce cross-tolerance to these drugs.

2.2.6. Conclusions

Taken together, these studies suggest that instrumental learning cannot account for all aspects of behavioral tolerance to ethanol. Rather, both physiological and behavioral mechanisms appear to play a role. The relative contributions of these factors appears to depend on a number of variables, including dose, frequency of drug administration, and opportunities for learning and practice. The data further suggest that tolerance to ethanol can best be understood as an adaptive response to the functional effects of the drug, as previously suggested (Kalant, 1977; LeBlanc and Cappell, 1977). Adaptation to such effects can occur at multiple levels, including behavioral adjustments via instrumental (and, in some cases, Pavlovian) conditioning, practice, and mental rehearsal, as well as cellular adjustments via synaptic mechanisms.

2.3. Morphine

The literature on tolerance to morphine is currently dominated by experimental paradigms derived from the Pavlovian conditioning model (*see* chapter by Siegel, this volume). These studies are largely concerned with tolerance to the analgesic and hyperthermic effects of the drug. There are, nonetheless, a number of studies that are pertinent to the role of instrumental learning in morphine tolerance (*see* Table 7). These studies have examined the effects of chronic administration of morphine on operant responding, shock-avoidance, and rewarding-brain stimulation.

2.3.1. Operant Schedules and Shock Avoidance

The acute effect of morphine on operant responding under a multiple FR, FI schedule of reinforcement is dose-dependent (McMillan and Morse, 1967). At low doses (1 mg/kg or less), morphine increases the rate of FI responding and has no effect on FR responding. At higher doses (3 and 10 mg/kg), morphine decreases the rate of responding on both schedules. An experiment by McMillan and Morse (1967) demonstrated that tolerance develops differentially to these effects. Dose–response curves were first determined with weekly injections, and then redetermined with injections given either weekly or daily. Tolerance did not occur with the weekly schedule of injections. However, with the daily schedule, tolerance developed to the rate-decreasing effect of the higher doses, but not to the rate increasing effect of the lower doses. Thus, these data are consistent with the reinforcement density hypothesis.

A study by Smith (1979), however, suggests that behavioral experience and loss of reinforcement are not always necessary for the development of morphine tolerance. As in the experiment by

Table 7
Summary of Studies Showing Tolerance to the Effects of Morphine
on Operant Responding and Shock Avoidance

Authors	Dose, mg/kg	Species	Task[a]
Babbini et al., 1972	40	Rats	FR
Babbini et al., 1976	40	Rats	FI
Brady and Holtzman, 1980	0.3–30	Rats	VI
Dworkin and Branch, 1982	1.7 or 3	Monkeys	Sidman avoidance
France and Woods, 1985	10–100	Pigeons	FR
Heifetz and McMillan, 1971	5.6	Pigeons	Mult FI, FR
Herman et al., 1972	5–20	Rats	Shuttle box avoidance
Holtzmann, 1974	10 (2 × daily)	Rats	Sidman avoidance
McMillan and Morse, 1967	0.3–10	Pigeons	Mult FI, FR
Rhodus et al., 1974	7.5	Rats	Mult FI, FI
Sannerud and Young, 1986	0.32–32	Rats	Mult FR, TO
Smith, 1979	10	Pigeons	Mult FI, FR
Witkin et al., 1983	1.0–5.6	Monkeys	Mult FI, FI
Woods and Carney, 1978	3.2–320	Pigeons	Mult FI, FR
Woods and Carney, 1978	3.2–56	Monkeys	FR

[a]TO = Time out (response has no programmed consequence); for other abbreviations, *see* Table 1.

McMillan and Morse (1967), this study involved a multiple FR, FI schedule of reinforcement. Subjects were first given daily injections of morphine (10 mg/kg) for 4 d without testing, and were then tested with the drug on the fifth day. After a 2-wk period of drug withdrawal, they were given five more injections with daily testing. Following the initial series of injections, tolerance developed to the rate-decreasing effect of morphine on the FI component, but not on the FR component. Following the same number of pretest injections, tolerance also developed on the FR component. Smith, therefore, concluded that the rate of tolerance to morphine was affected by the opportunity to engage in the behavioral task, at least on the FR component.

The problem with this interpretation is that tolerance to the effect of morphine on FR responding might have developed in five additional trials even without behavioral experience. What is lacking, in other words, is a pharmacological control group given five injections, a 2-wk layoff, another five injections, and then a behavioral test. Without the data from such a group, we cannot conclude that morphine tolerance is augmented by behavioral experience. However, a recent experiment by Sannerud and Young (1986) provided such a control. In this experiment, rats were given morphine (10 mg/kg) either before or after daily testing on a multiple FR, TO schedule (TO = time out; during this time, a response has no programmed consequence). Cumulative dose–response curves were determined before, and at various times during, chronic administration of the drug. Rats given pretest injections of morphine developed tolerance to the rate-decreasing effect of the drug on FR responding, as evidenced by a shift in the dose–response function to the right. Rats given posttest injections did not become tolerant, despite a comparable history of drug injections.

Although these results indicate that morphine's effect on FR responding is contingent on behavioral experience under the drug, recall that in the experiment by Smith (1979), tolerance to the rate-decreasing effect of the drug on FI responding was not contingent on such experience. These data underscore an important point that is often overlooked in the literature on behavioral tolerance: The fact that tolerance is associated with an initial loss of reinforcement does not, in itself, constitute evidence for instrumental learning.

Research involving the effects of morphine on shock avoidance points to the same conclusion. In an experiment by Dworkin and Branch (1982), for example, monkeys were tested on a Sidman avoidance task consisting of three components that differed in shock density. Thus, the "cost" associated with each component was systematically varied. Dose–response curves were determined before and after subjects were given chronic injections of morphine (1.7 or 3 mg/kg) prior to testing. On the initial dose–response determination, morphine produced a dose-dependent decrease in the rate of avoidance with a concomitant increase in the number of shocks received. Chronic injections of the drug resulted in tolerance to the decreased rate of responding, which decreased the number of shocks delivered. On the dose–response redetermination, there was a small reduction in the effect of the drug on rate of avoidance responding, but this was sufficient to produce a large decrease in the number of shocks delivered. Thus, tolerance was reflected in more efficient avoidance responding.

Although these results are consistent with the reinforcement density hypothesis, they do not prove that tolerance was contingent on behavioral experience under the drug. Indeed, an experiment by Holtzman (1974) suggests that such experience is not necessary. In this study, rats were tested on a Sidman avoidance task following an injection of one of four doses (0.3–10 mg/kg) of morphine. For the next 2 d, they were given 10 mg/kg of morphine twice daily without behavioral testing. On the following day, they were again tested on the avoidance task following an injection of the same dose they had received on the first day. Thus, injections preceded testing on the first and last day, but not on the two intervening days. This procedure was then repeated every other week until all four doses were given.

On the initial dose–response trials, morphine caused an increased rate of avoidance responding at the lower doses (0.3–3 mg/kg) and a decreased rate of responding at the highest dose (10 mg/kg). Following exposure to morphine on the intervening days, tolerance developed to both of these effects. Similar results were obtained with pentazocine, an analgesic with mixed agonist and weak narcotic antagonist properties. Such tolerance was not the result of behavioral experience derived from the dose–response determinations, because control groups given injections of saline on the intervening days did not show a shift in the dose–response function. Thus, tolerance to these drugs was not contingent on behavioral experience.

The results of a study by Herman et al. (1972) on the effects of chronic morphine administration on shuttle box avoidance are consistent with this conclusion. Rats were given daily pretest injections of morphine (0, 5, 10, or 20 mg/kg) for 13 d. Both the number of avoidance responses and the latency to respond were recorded. Morphine initially caused a dose-dependent decrease in avoidance and an increase in the latency to respond at the two highest doses. However, at the lowest dose, morphine did not impair avoidance responding and actually decreased the latency to respond. In apparent agreement with the reinforcement density hypothesis, tolerance developed to the impairment in avoidance responding induced by high doses of the drug, but not to the facilitation of responding induced by the low dose. However, in a subsequent experiment, equivalent levels of tolerance were obtained when rats were given morphine in their home cages and tested only at the conclusion of the chronic phase. Thus, tolerance to the rate-decreasing effect of morphine on avoidance behavior, although consistent with the reinforcement density hypothesis, is not contingent on behavioral experience.

2.3.2. Rewarding Brain Stimulation

Morphine has a biphasic effect on the rate of operant responding for rewarding brain stimulation (Lorens and Mitchell, 1973). During the first 1–2 h, the drug decreases the rate of responding, whereas later (3–6 h), the drug increases the rate. Both the magnitude of the initial suppression and the temporal appearance of the later facilitation are dose-dependent. A large number of studies have shown that, with chronic injections of morphine in the 5–20 mg/kg dose range, tolerance develops to the initial suppression of responding, whereas sensitization develops to the later facilitation. As can be seen in Table 8, these studies involved a variety of response measures and sites of stimulation, and utilized both rate and threshold measures of reward.

There are, however, two studies that are at variance with the results discussed above. Glick and Rapaport (1974) reported that rats given 2.5 mg/kg and tested 15 min postinjection showed an initial enhancement of responding that gradually diminished to baseline levels in 4 d. When subsequently given 10 mg/kg, the rats showed suppressed responding initially, but then tolerance devel-

Table 8
Summary of Studies Showing Tolerance and Sensitization to the Effect of Morphine on Rewarding Brain Stimulation

Authors	Dose, mg/kg	Electrode Site[a]	Measure	Response
Baltzer et al., 1977	0.02 (etorphine)	LH	"On" and "off" times[b]	Shuttle box
Bush et al., 1976	15	SN	Rate	Bar-press
Esposito and Kornetsky, 1977	4–16	LH	Threshold	Wheel-turn
Esposito et al., 1979	2–40	VTA, CG, LC	Threshold	Wheel-turn
Glick and Rapaport, 1974	1.25–10	LH	Rate	Bar-press
Hand and Franklin, 1985	10	LH	Rate	Bar-press
Hand and Franklin, 1986	0.3–10	LH	Rate	Bar-press
Levitt et al., 1978	10 and 20	LH	"On" and "off" times[b]	Shuttle box
Liebman and Segal, 1976	15	SN	Rate	Bar-press
Lorens and Mitchell, 1973	5–20	LH	Rate	Bar-press
Nazzaro et al., 1981	5	VTA, SN	Threshold	Bar-press

[a]CG = central gray; LC = locus ceruleus; LH = lateral hypothalamus; SN = substantia nigra; VTA = ventral tegmental area.
[b]Amount of time spent on the side of a shuttle box in which electrical stimulation was delivered ("on" time) and on the opposite side in which it was not delivered ("off" time).

oped to this effect as well. Thus, tolerance developed to the rate-enhancing effect of a low dose and the rate-decreasing effect of a higher dose. It should be noted, however, that others have found that a comparably low dose of morphine (3 mg/kg) initially suppressed self-stimulation and that tolerance developed to this effect (Hand and Franklin, 1986). The reason for these discrepant findings is not presently clear.

The results of an experiment by Nazzaro et al. (1981) are also at variance with those of other studies. In this experiment, rats with electrodes in either the ventral tegmental area (VTA) or substantia nigra (SN) were given chronic injections of morphine (5 mg/kg), and tested in a threshold paradigm at various times postinjection. With VTA electrodes, morphine produced an initial lowering of the reward threshold, but contrary to the reinforcement density hypothesis, tolerance developed to this effect. With SN electrodes, morphine initially had no effect, but later caused a gradual increase in reward thresholds. As the authors noted, both of these results are at variance with those of other studies (e.g., Esposito et al., 1979; Liebman and Segal, 1976) and may reflect differences in dose, electrode location, or other procedural details.

In summary, the vast majority of studies on tolerance to the effects of morphine on rewarding brain stimulation are consistent with the reinforcement density hypothesis. However, this does not, in itself, provide unambiguous support for the instrumental learning theory. Two problems need to be addressed. The first is theoretical. If the initial suppression of responding induced by morphine represents a direct effect of the drug on the neural substrate for reinforcement, how does instrumental learning occur? One would expect extinction of responding, not recovery, in the absence of adequate reinforcement. The resolution of this question may be found in a study by Van der Kooy et al. (1978), in which the effect of morphine (10 mg/kg) on rewarding brain stimulation was analyzed in three tasks with different response requirements. The degree of suppression was found to be related to the performance demands of the task. Suppression was greater with a bar-pressing task than with a wall-pressing task that permitted a greater range of movements to be reinforced. No suppression was obtained in a shuttle box, where stimulation was available continuously on one side. (At higher doses, morphine produces an increase in both "on" time and "off" time, but tolerance develops to the latter effect; Levitt et al.,

1978.) Thus, the initial response suppression induced by morphine appears to reflect a performance deficit, rather than an impairment in reward.

The second problem alluded to above is the absence of controls for pharmacological exposure in all but one of these studies. Without such controls, we cannot conclude that either tolerance or sensitization is contingent on behavioral experience. Fortunately, a series of studies by Hand and Franklin (1986) included such controls and, therefore, can provide further insight into the role of instrumental learning in these phenomena. The first of these experiments investigated the effect of experiential factors in tolerance and sensitization to morphine. One group of rats (Group M-ICS) was given morphine (10 mg/kg) in the test room, and allowed to self-stimulate on a random interval schedule for 30 min at 1 and 3 h postinjection for 8 d. Two other groups were given morphine in the home cage (Group M-HC) or test cage (Group M-E) without the opportunity to self-stimulate for the first 4 d, and were then treated as Group M-ICS for the remaining 4 d. Saline control groups were also included and showed stable rates of responding throughout the experiment.

At 1 h postinjection, Group M-ICS showed an initial suppression of responding followed by recovery to baseline levels (tolerance). When Group M-E was given the opportunity to self-stimulate on the last 4 d, its rate of responding was initially only about 50% of baseline, but it gradually rose to baseline levels in three trials. In contrast, Group M-HC's rate was suppressed on all four trials (although it showed normal levels of locomotion). Thus, rats given the opportunity to self-stimulate during the first four trials (Group M-ICS) showed a more rapid rate of recovery from the initial rate-suppressant effect of morphine than rats placed in the test cages, but denied the opportunity to respond (Group M-E). These results suggest that behavioral experience contributes to the development of tolerance to morphine. However, the fact that rats injected with morphine but kept in their home cages during the first four trials (Group M-HC) differed from those placed in the test cage during the first four trials (Group M-E) suggests that factors other than instrumental learning also play a role. The nature of such factors is not presently known, but a classically conditioned compensatory response does not appear to be involved (*see* Hand and Franklin, 1986).

Behavioral experience also appeared to play a role in the development of sensitization to morphine. When the rats were tested at 3 h postinjection, Group M-ICS showed enhanced rates of responding, ultimately reaching 200% of control levels. In contrast, the other groups did not exceed baseline levels of responding. More importantly, the authors also reported that rats given morphine in the test cage with noncontingent electrical stimulation also failed to develop sensitization. Thus, neither chronic exposure to morphine, nor exposure to morphine in the test environment, nor exposure to morphine with noncontingent electrical stimulation was sufficient for the development of sensitization.

A subsequent experiment demonstrated that responding during the initial stage of morphine action was also critical for the subsequent development of sensitization. Groups of rats were given morphine (10 mg/kg) and allowed to self-stimulate either at 1 or 3 h postinjection for 5 d. They were then tested at both intervals for an additional 5 d. The group tested initially at 3 h showed neither tolerance at 1 h, nor sensitization at 3 h when subsequently tested at both intervals. In contrast, the group tested initially at 1 h showed both tolerance at 1 h and sensitization at 3 h.

Taken together, the results of these experiments seem to provide strong support for the role of instrumental learning. In keeping with the reinforcement density hypothesis, tolerance developed to the initial loss of reinforcement, and sensitization developed to the subsequent increase in response rate. Furthermore, the development of sensitization was contingent on actively engaging in self-stimulation behavior under the influence of the drug. However, despite these apparent consistencies with the theory, there is a major flaw in the interpretation of these results, at least in regard to sensitization. Recall that in this experiment rewarding brain stimulation was delivered on a random-interval schedule. In such a schedule, the probability of reinforcement is constant from moment to moment. Hence, increased rates of responding *did not* result in a higher frequency of reinforcement. Consequently, we cannot conclude that sensitization of self-stimulation is consistent with the reinforcement density hypothesis.

It is instructive to note, in this regard, that morphine also produces a biphasic effect on unconditioned locomotor activity (Babbini and Davis, 1972; Schnur et al., 1983b; Vasko and Domino, 1978). As in the case of self-stimulation behavior, tolerance

develops to the initial suppression of movement, whereas sensitization develops to the later increase (Babbini and Davis, 1972;
Mucha et al., 1981; Schnur, 1985; Schnur et al., 1983a; Vasko and
Domino, 1978). These parallel changes in reinforced (self-
stimulation) and nonreinforced (locomotion) behavior cast further
doubt on the interpretation that tolerance and sensitization are the
result of instrumental learning.

2.3.3. Conclusions

Only one study (Sannerud and Young, 1986) has provided unambiguous support for the role of behavioral experience in the development of tolerance to morphine's effect on operant performance.
In all other cases, tolerance was either shown to result from pharmacological exposure *per se*, or controls for such exposure were
not included. In the case of rewarding brain stimulation, the picture
is more complex. Although some of the evidence points to a role
for behavioral experience, other factors, poorly defined and poorly
understood, also seem to be involved.

2.4. Anxiolytics

2.4.1. Barbiturates

It is well known that chronic administration of barbiturates (e.g.,
pentobarbital and phenobarbital) induces liver microsomal enzymes
that hasten the degradation of the drug. This mechanism probably
accounts for the development of tolerance when these drugs are
given in the absence of behavioral testing, as in the report of tolerance to the suppression of avoidance behavior induced by gavage of
pentobarbital (Lê et al., 1986b). Nevertheless, a number of studies
have demonstrated that behavioral experience can contribute to the
development of tolerance to barbiturates (*see* Table 9A). For example, in an experiment by Smith and McKearney (1977), pigeons
were tested on a key-pecking task in which reinforcement was contingent on making a fixed number of responses and then waiting 30
s. In one condition, responding during the 30-s interval postponed
the delivery of reinforcement, whereas in the other condition, such
responding had no consequences. Injections of pentobarbital (1–10
mg/kg) initially increased the rate of responding during the interval,

Table 9
Summary of Studies Showing Tolerance to the Behavioral Effects of Anxiolytics

Authors	Drug	Dose, mg/kg	Species	Task[a]
A. *Barbiturates*				
Branch, 1983	Pentobarbital	1–10	Monkeys	Mult DRL, RI
Harris and Snell, 1980	Phenobarbital	50	Rats	Mult FI, FR
Kulig, 1986	Pentobarbital	35 or 50	Rats	Shock avoidance
Le et al., 1986b	Pentobarbital	12.5–50	Rats	Shock avoidance
Smith and McKearney, 1977	Pentobarbital	1–10	Pigeons	NT vs. $\overline{\text{NT}}$
Tang and Falk, 1978	Phenobarbital	20–160	Rats	FI
B. *Benzodiazepines*				
Bennett and Amrick, 1987	Diazepam	10	Rats	Mult FR, VI
Cesare and McKearney, 1980	Chlordiazepoxide	1–30	Pigeons	FI, FR
Cook and Sepinwall, 1975	Chlordiazepoxide	10	Rats	Mult FR, VI
Griffiths and Goudie, 1987	Midazolam	0.1–1.6	Rats	FR
Herberg and Montgomery, 1987	Chlordiazepoxide	7.5	Rats	VI
Le et al., 1986b	Chlordiazepoxide	15–60	Rats	Shock avoidance
Mana et al., 1986	Diazepam	2	Rats	Clonus
Margules and Stein, 1968	Oxazepam	20	Rats	Mult VI, VI
McMillan and Leander, 1978	Chlordiazepoxide	50	Rats	Mult FI, FI
Stephens and Schneider, 1985	Diazepam	0.63–40	Mice	Passive avoidance
Tizzano et al., 1986	Diazepam	1–17.5	Rats	FR

[a]RI = random interval; for other abbreviations, *see* Tables 1 and 3.

which resulted in the loss of reinforcement in the first condition, but not in the second. In keeping with the reinforcement density hypothesis, tolerance to the rate-increasing effect of the drug developed only in the former condition.

Contingent tolerance to the rate-decreasing effects of phenobarbital has also been reported (Harris and Snell, 1980). In this experiment, rats tested on a multiple FI, FR schedule were given chronic injections of the drug (50 mg/kg) either before or after daily testing. Initially, phenobarbital caused a decrease in the rate of responding on both components of the schedule. After 5 d, when the Before Group had become partially tolerant, the After Group was given pretest injections and found not to be tolerant to the drug's effect.

The experiment by Harris and Snell (1980) involved a relatively short period of drug exposure. However, a study by Tang and Falk (1978) showed that even with longer periods of exposure, sufficient to induce pharmacological tolerance, the influence of behavioral experience can still be deduced from dose–response determinations. In this study, rats were tested on a FI schedule for food reward. Dose–response curves were determined before and during a period of chronic administration of phenobarbital (80 mg/kg), which was given either before or after testing. At the chronic dose, the drug initially decreased the rate of responding, which resulted in a loss of reinforcement. With repeated injections, however, tolerance developed to these effects. When the dose–response curves were then redetermined, the After Group was as tolerant as the Before Group at the dose given chronically, suggesting that tolerance had developed as a result of pharmacological exposure. However, for higher doses, the Before Group showed a greater shift in the dose–response curve than the After Group. Thus, both behavioral and pharmacological mechanisms contributed to such tolerance.

2.4.2. Benzodiazepines

Because benzodiazepines (e.g., chlordiazepoxide, diazepam) have anxiolytic and anticonvulsant properties, they are among the most widely prescribed drugs in the world. Consequently, an understanding of the mechanism by which tolerance develops to the behavioral effects of these drugs would have considerable clinical relevance. Unfortunately, as File (1985) noted, there have been relatively few experimental studies designed to address this issue. Table 9B summarizes studies that are pertinent to the role of instrumental learning in the development of tolerance.

Tolerance has been reported to the rate-decreasing effect of chlordiazepoxide on food-reinforced FR and FI responding (Cesare and McKearney, 1980). Although these results are consistent with the reinforcement density hypothesis, they do not exclude the possibility that tolerance was the result of pharmacological exposure *per se*. Indeed, several recent studies suggest that tolerance to some of the behavioral effects of the benzodiazepines is not contingent on behavioral experience under the drug. For example, Lê et al. (1986b) reported that, following daily intubations of chlordiazepoxide (60 mg/kg), rats became tolerant to the drug-induced disruption of avoidance behavior, even though they were not tested during the chronic drug regimen.

Similar results have been obtained for the effects of other benzodiazepines on FR responding. In one experiment (Tizzano et al., 1986), rats tested on a FR schedule were given chronic injections of diazepam (either 1, 3, or 10 mg/kg) either before or after behavioral sessions. Dose–response curves were determined both before and after the chronic dosing regimen. Initially, diazepam produced a dose–dependent decrease in the rate of responding and a dose-dependent increase in the duration of each response. Following chronic dosing, tolerance developed to the rate-decreasing effect of the drug in both the Before and After Groups, as revealed by comparable nonparallel shifts in the dose–response curves. For the duration measure, however, the Before Group showed more tolerance than the After Group, and the degree of tolerance was dependent on the dose administered. Thus, behavioral experience was not necessary for the development of tolerance to the rate-decreasing effect of the drug, although it did contribute to the effect of the drug on response duration.

A study by Griffiths and Goudie (1987) on the effects of chronic administration of midazolam on FR responding also suggests that behavioral experience is not necessary for the development of tolerance. Rats in one group were given a low dose (0.1 mg/kg) of the drug prior to testing for 14 trials. On the last seven trials, they were also given posttest injections of a higher dose (0.8 mg/kg). Rats in a second group were given only pretest injections of the higher dose for seven trials. Dose–response curves were determined before and after the chronic regimens. On the initial dose–response determinations, midazolam produced an increased rate of responding at the low dose and a decreased rate of responding at higher doses. During the period of chronic dosing, the first group showed no tolerance to the rate-increasing effect of the lower dose, whereas the second group showed complete tolerance to the rate-decreasing effect of the higher dose. Note that these results appear to support the reinforcement density hypothesis. On the dose–response redetermination, however, both groups showed comparable nonparallel shifts in the dose–response curves. Similarly, in a second experiment, comparable levels of tolerance were found in groups given chronic injections of midazolam (0.5 mg/kg) either before or after testing. Thus, tolerance was not contingent on loss of reinforcement.

It is interesting to note that contingent tolerance has been demonstrated to the anticonvulsant effect of diazepam. Tolerance developed when drug injections were given prior to convulsant stimulation, but not when injections were given after such stimulation (Mana et al., 1986). This finding emphasizes once again that, even when reinforcement loss is not involved, tolerance may still be contingent on the relation between time of injection and testing.

One of the more clinically relevant effects of the benzodiazepines is their ability to increase the rate of responding that is both reinforced and punished (the "anticonflict" effect). The efficacy of a drug in this paradigm is highly correlated with its clinical anxiolytic potency (Cook and Davidson, 1973). Margules and Stein (1968; *see* Fig. 11) demonstrated that, although tolerance occurs to the sedative effects of benzodiazepines, it does not develop to the anticonflict effect. In their experiment, rats were trained on a two-component multiple schedule. In both components, bar-pressing was reinforced with milk on a variable interval (VI) schedule, but in one of the components, responding was also punished with shock. Prior to drug administration, responding occurred at a moderate rate in the unpunished component, but was totally suppressed in the punished component.

Injections of oxazepam (20 mg/kg) initially suppressed unpunished responding, and produced marked sedation and ataxia. Despite these "side effects," responding in the punished component increased under the drug. With chronic injections, tolerance

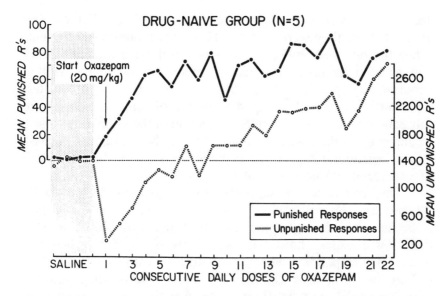

Fig. 11. Mean punished and unpunished responses of rats given daily pretest injections of oxazepam. Tolerance developed to the initial suppression of unpunished responding, but not to the increase in punished responding. (From "Increase of 'antianxiety' activity and tolerance of behavioral depression during chronic administration of oxazepam" by D. L. Margules and L. Stein, *Psychopharmacologia*, 1968, **13**, 74–80. Fig. 2. Copyright 1968 by Springer-Verlag. Reprinted by permission.)

developed to the suppressant effects of oxazepam on unpunished responding, but not to the anticonflict effect. In fact, punished responding actually increased over repeated trials (*see* Fig. 11). Similar results have been obtained in other studies (Cook and Sepinwall, 1975; McMillan and Leander, 1978). Such differential tolerance is consistent with the reinforcement density hypothesis. In the unpunished component, the drug initially caused a loss of reinforcement. In the punished component, the drug initially caused an increase in positive reinforcement (milk) and, by definition, a decrease in the negatively reinforcing effect of the shock.

Although the data described above are compatible with a learning interpretation, an experiment by Cook and Sepinwall (1975) showed that with a high dose of chlordiazepoxide, tolerance to the disruption of unpunished responding is not contingent on behavioral experience in the drugged state. Rats were given the drug (160 mg/kg) in their home cages after a daily conflict test. For the next 26 d, they were given daily tests without drug treatment. When the rats were subsequently tested following a pretest injection of a lower dose (10 mg/kg), there was no impairment in unpunished responding. In contrast, rats given 10 mg/kg posttest for two trials spaced 1 wk apart were not tolerant when they were subsequently tested with a pretest injection of the same dose. These results suggest that, whereas tolerance may be contingent on task-related behavioral experience at moderate doses of the drug, such experience is not necessary at high doses.

This conclusion is supported by the results of a study by Bennett and Amrick (1987). Compared to vehicle-injected controls, rats given a single pretreatment of diazepam (100 mg/kg) in their home cages showed higher levels of both punished and unpunished responding when tested 7 d later with a challenge dose (10 mg/kg) of the drug. Similar high levels of responding were obtained in vehicle-pretreated rats tested acutely with a nonsedating, nonmuscle relaxant partial benzodiazepine agonist (CGS 9896). These findings suggest, therefore, that the muscle relaxant and/or sedating effects of benzodiazepine administration initially interfere with responding. As tolerance develops to these effects, the levels of both punished and unpunished responding increase. Although these results demonstrate that enhanced responding can occur without task-related behavioral experience when subjects are pretreated with a high dose of the drug, it is possible that even higher levels of responding would have occurred with such experience.

Recently, several objections have been raised in regard to the conclusion that tolerance does not develop to the anticonflict effect of the benzodiazepines. One objection is that too few trials were run; i.e., tolerance might have developed with extended testing (File, 1985; Stephens and Schneider, 1985). Although many studies have used relatively short periods of treatment, in at least one study (Margules and Stein, 1968; see Fig. 11), tests were conducted over a fairly long period of time (22 d). Of course, it is possible that tolerance would have developed if even more trials had been run, but this objection can never be dispelled, because its refutation requires the confirmation of the null hypothesis.

Another problem that has been cited in regard to the interpretation of these studies is that the early period of sedation and ataxia masks the true extent of the initial anticonflict effect. Compared to a potentially higher initial baseline, the subsequent level of punished responding might actually represent an attenuation of this effect, i.e., tolerance (Stephens and Schneider, 1985). To investigate this possibility, Stephens and Schneider (1985) gave groups of mice chronic injections of either diazepam (5 mg/kg) or vehicle. After 9 d of treatment, subjects in each group were injected with a test dose of the drug (either 0.63, 2.5, 10, or 40 mg/kg) and placed in an apparatus consisting of four metal plates. For half the subjects, movement from one plate to another was punished with footshock; for the other half, such movement was unpunished. Diazepam produced a dose-dependent increase in punished activity in the group given prior injections of the vehicle. In contrast, mice previously given chronic injections of diazepam did not show this effect, presumably because they had become tolerant. Both groups showed increased levels of unpunished activity, indicating that these doses did not produce sedation. The authors therefore concluded that, when the confounding effect of sedation is eliminated, tolerance does develop to the anxiolytic effect of the benzodiazepines.

Although these findings demonstrate that tolerance can develop to the "antipunishment" effects of these drugs, it is not clear that they have direct relevance to results obtained with the "conflict" paradigm. In the latter, subjects were given repeated experience on a task that provided both reward (milk) and punishment (shock). They were therefore exposed to the contingencies of reinforcement while in the drugged state. In contrast, in the experiment by Stephens and Schneider (1985), subjects were given chronic injections without exposure to the task. Moreover, the task involved punishment, but not reward. It is quite possible that, although pharmacological exposure is sufficient for the development of tolerance to the effects of the benzodiazepines, behavioral experience can counteract this effect when the contingencies of reinforcement are appropriate.

2.4.3. Conclusions

It is clear that tolerance can develop to the behavioral effects of anxiolytic drugs as a result of pharmacological exposure *per se*. The fact that some studies have demonstrated contingent tolerance to these drugs suggests that behavioral experience may simply augment the rate at which tolerance develops. The anxiolytics therefore resemble ethanol in this regard, which is not surprising in view of the many pharmacological and behavioral properties they share with this drug.

2.5. Marijuana

Tolerance to the behavioral effects of delta-9-tetrahydrocannabinol (THC), the major active constituent of marijuana, has been demonstrated in a wide variety of behavioral tasks and in several different species (*see* Table 10). Because the initial effect of the drug produced a decrease in the frequency of reinforcement, the results of these experiments are consistent with the reinforcement density hypothesis. However, in only two cases was the before/after design utilized to control for the effects of pharmacological exposure. In one of these studies (Webster et al., 1973), rats were given chronic injections of THC (12 mg/kg) either before or after they were tested on a discriminated Sidman avoidance task. In the Before Group, the drug initially caused a disruption of avoidance responding, which led to an increase in the frequency of shock. Over subsequent trials, tolerance developed to this effect. The After Group, which was given pretest injections of the drug every fourth day to assess the development of tolerance, showed a similar course of recovery. Although these results appear to indicate that tolerance resulted from pharmacological exposure *per se*, it is possible that behavioral experience during the intermittent tests for tolerance contributed to the recovery of avoidance behavior.

In the other study (Carder and Olson, 1973), rats were given THC either before or after access to a bar-pressing task reinforced with food (Experiment 1) or water (Experiment 2). In the first experiment, which involved daily injections of a relatively low dose of THC (3 mg/kg), the level of tolerance in the Before Group was rather slight. Moreover, the interpretation of the results was complicated by the fact that the After Group showed a steady decrease in bar-pressing despite receiving posttest injections. However, in the second experiment, which involved testing on alternate days with higher doses of the drug (8–32 mg/kg), the Before Group showed complete tolerance to the initial disruption of responding. When the After Group was then tested with pretest injections, no

Table 10

Summary of Studies Showing Tolerance to the Behavioral Effects of Marijuana

Authors	Dose, mg/kg	Species	Task[a]
Abel et al., 1974	10	Pigeons	VI
Carder and Olson, 1973	3–32	Rats	CRF
Elsmore, 1976	1	Monkeys	Mult FI, DRL
Ferraro, 1972	2	Monkeys	DRL, VI
Ferraro and Grisham, 1972	1–4	Chimpanzees	DRL and visual discrimination
Frankenheim, 1974	5.62–31.6[b]	Rats	DRL
Galbicka et al., 1980	0.25 or 1	Monkeys	Mult DRL, RI
Harris et al., 1972	2–12	Monkeys	Mult FR, DRL, and Sidman avoidance
Kosersky et al., 1974	0.3	Pigeons	VI
Manning, 1976a	30	Rats	Sidman avoidance
Manning, 1976b	4, 16	Rats	DRL
McMillan et al., 1970	1.8–36	Pigeons	Mult FI, FR
Miczek, 1979	20	Rats	Food intake
Olson and Carder, 1974	6	Rats	Maze running
Smith, 1986b	0.03[c]	Pigeons	FR
Webster et al., 1973	12	Rats	Sidman avoidance

[a]CRF = continuous reinforcement; for other abbreviations, *see* Tables 1, 3, and 9.
[b]Δ[8] - THC.
[c]Nantradol.

tolerance was observed. Thus, tolerance was contingent on behavioral experience under the drug.

The demonstration of individual differences in the development of tolerance to THC provides additional evidence for the importance of behavioral experience. For example, in an experiment by Ferraro (1972), monkeys were given pretest injections of THC (2 mg/kg) and tested on a panel-pushing task reinforced on a DRL schedule. In two subjects, the drug initially caused a loss of reinforcement by either increasing or decreasing, respectively, the rate of responding. Both subjects, however, became tolerant to the effects of the drug. In contrast, no tolerance developed in a third subject, whose rate of responding gradually increased over days without affecting the rate of reinforcement. Manning (1976a) reported similar differential tolerance in rats given THC (30 mg/kg) and tested on a Sidman avoidance task. In six rats, tolerance developed to the drug-induced increase in shock frequency. In another rat, in which the drug induced a decrease in shock frequency, no tolerance was observed.

There are also, however, a number of findings that appear to be inconsistent with a learning interpretation of tolerance. For example, Elsmore (1976) reported that THC caused an increased rate of responding in both components of a multiple FI, DRL schedule of reinforcement. In apparent agreement with the reinforcement density hypothesis, tolerance developed to the rate-increasing effect of the drug on DRL performance, but not on FI performance. However, the decline in the rate of responding in the DRL component was not sufficient to restore the frequency of reinforcement to baseline levels. Moreover, the time course of the two events (decreased rate and increased reinforcement) did not correlate well. Elsmore therefore proposed that the decreased rate of responding in the DRL component may have been the result of a general weakening of response strength secondary to the loss of reinforcement, rather than a learned compensation. Such extinction would not occur in the FI component, because the increased rate of responding either had no effect on, or increased, the frequency of reinforcement in that component. However, an alternative interpretation of these results is that the increased frequency of reinforcement in the FI component reduced the ''global cost'' of the initial disruption of DRL responding (cf. Smith, 1986a). Under these conditions, the incentive to recover schedule-appropriate responding in the latter component may have been diminished.

Because tolerance develops to the disruptive effect of THC on food-motivated operant tasks, it might be expected that tolerance would also develop to the direct suppressant effect of the drug on food intake. However, the evidence on this point is equivocal. Part of the problem is that the literature on the acute effects of the drug on ingestive behavior presents a confusing picture. In humans, THC produces an increase in food intake (Foltin et al., 1986). A similar result has been reported in food-deprived rats, and in this study, intakes remained elevated over a 12-d period of chronic injections of the drug (Gluck and Ferraro, 1974). On the other hand, many other studies have found that THC decreases food intake (Dewey, 1986). Although tolerance to this "anorexic" effect of the drug has been reported (Miczek, 1979), several other studies found no recovery of intake for periods of up to 30 d with doses ranging from 0.5 to 32 mg/kg (Drewnowski and Grinker, 1978; Manning et al., 1971; Sofia and Barry, 1974). The accumulation of THC in adipose tissue (Kreuz and Axelrod, 1973) may contribute to the failure to find tolerance. However, this explanation would appear to be contradicted by the relatively rapid development of tolerance in food-motivated operant tasks (e.g., Carder and Olson, 1973).

Another finding that is difficult to reconcile with the reinforcement density hypothesis comes from an experiment by Frankenheim (1974). Rats were given chronic injections of various doses of delta-8-tetrahydrocannabinol and tested on a DRL schedule. At the lower doses (5.62 and 10 mg/kg), the drug initially induced long pauses, preceded and followed by bursts of responding. During the latter, the interresponse time distributions were shifted to the left, which resulted in a loss of reinforcement. At the higher doses (17.8 and 31.6 mg/kg), responding was completely suppressed. With chronic administration of the lowest dose, tolerance developed to both the pausing and bursting, in keeping with the reinforcement density hypothesis. However, at higher doses, tolerance to the initial suppression of responding was followed by sensitization to the increased rates of responding, even though this resulted in the continued loss of reinforcement.

It may be recalled that with both morphine (Lorens and Mitchell, 1973) and the benzodiazepines (Margules and Stein,

1968), tolerance to the depressant effect of the drug was followed by sensitization of responding. In those cases, sensitization resulted in an increased frequency of reinforcement and was, therefore, consistent with the instrumental learning theory. In the present case, however, sensitization resulted in decreased reinforcement. These contrasting outcomes underscore an important point. Changes that occur with chronic administration of a drug may occur for reasons unrelated to the contingencies of reinforcement, and yet appear to support the reinforcement density hypothesis because of the particular task or schedule being used. In the experiment by Frankenheim (1974), sensitization resulted in a loss of reinforcement, because the DRL schedule required a low rate of responding. However, if a different schedule had been used, sensitization might have resulted in an increased frequency of reinforcement. Clearly, then, the fact that tolerance or sensitization results in an increased density of reinforcement does not, in itself, constitute evidence that instrumental learning is involved.

In conclusion, tolerance to marijuana has been demonstrated in a variety of behavioral paradigms in which the initial effect of the drug was to decrease the frequency of reinforcement. In only two cases (Carder and Olson, 1973; Manning, 1976b), however, was such tolerance shown to be contingent on behavioral experience in the drugged state. Most studies have not included controls for pharmacological exposure, and in at least one study that did include such controls (Webster et al., 1973), tolerance was not contingent on behavioral experience.

3. What Is Learned?

It is clear from the preceding review of the literature that instrumental learning cannot account for all instances of tolerance (or sensitization) to the behavioral effects of drugs. Moreover, even when the data are consistent with the instrumental learning theory, the evidence is, at best, circumstantial. As others have previously noted, the major weakness of the theory has been its failure to specify what, precisely, is learned when the subject compensates for the initial disruptive effect of the drug (Corfield-Sumner and Stolerman, 1978; Demellweek and Goudie, 1983b). In this section,

I shall review some data on the development of tolerance to amphetamine-induced "anorexia" that addresses this issue.

3.1. How Does Amphetamine Suppress Feeding?

The impetus for this line of research arose from the recognition that an adequate explanation of tolerance required an understanding of the mechanism by which the drug suppressed ingestion in the first place. Historically, three different mechanisms have been proposed. At low doses (i.e., <0.5 mg/kg), amphetamine may suppress intake by means of a peripheral mechanism that involves the sympathetic innervation of the liver (Tordoff et al., 1982a; Tordoff et al., 1982b). However, at the higher doses typically used in studies of behavioral tolerance (i.e., >1 mg/kg), two other mechanisms are more likely to be involved. First, amphetamine may depress the motivation to eat by acting on feeding-related neural circuits in the brain (Leibowitz, 1975; Paul et al., 1982). Alternatively, the drug may suppress intake by inducing responses or patterns of responding (e.g., stereotyped movements) that are incompatible with feeding (Carlton, 1963; Cole, 1978; Lyon and Robbins, 1975). Of course, these mechanisms are not mutually exclusive, and both may play a role.

In order to determine the relative contributions of these factors to the suppression of feeding, amphetamine-treated rats were given sweetened milk either through an intraoral cannula or in a standard drinking tube (Salisbury and Wolgin, 1985). We reasoned that, if amphetamine acted primarily by reducing the appetite for food (i.e., producing "anorexia"), then intake would be suppressed to an equivalent degree with both methods. However, if the drug disrupted feeding in part by inducing incompatible patterns of behavior, then the suppression of intake would be greater with the drinking tube, which is the condition in which the rats must actively approach the milk. As shown in Fig. 12, amphetamine initially suppressed feeding with both methods, but the suppression was greater in bottle-fed rats. Because saline-treated rats showed almost identical intake with the two methods (Fig. 12), these results cannot be attributed to differences in the methods *per se.* Moreover, the greater intake of cannula-fed rats was not a result of their inability to avoid swallowing the milk, because rats given quinine-

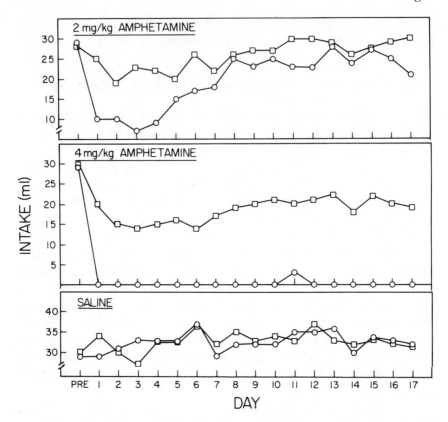

Fig. 12. Mean milk intake of cannula-fed (squares) and bottle-fed (circles) rats given chronic injections of amphetamine (2 or 4 mg/kg) or saline. Amphetamine produced greater suppression of intake in the bottle-fed rats. At the lower dose, tolerance developed. PRE = last baseline day. (Adapted from ''Role of anorexia and behavioral activation in amphetamine-induced suppression of feeding: Implications for understanding tolerance'' by J. J. Salisbury and D. L. Wolgin, Behavioral Neuroscience, 1985, **99,** 1153–1161. Copyright 1985 by the American Psychological Association. Reprinted by permission of the publisher and author.)

adulterated milk readily rejected the bitter fluid. Thus, the data suggest that, at these doses, amphetamine suppresses food intake primarily by interfering with the appetitive behavior required for feeding.

This conclusion is consistent with the results of several previous studies. For example, a reciprocal relation between feeding and activity has been found in both brain-damaged and normal rats given amphetamine (Cole, 1977, 1979; Heffner et al., 1977; Joyce and Iversen, 1984; Mason et al., 1978; Sanberg and Fibiger, 1979). Moreover, an analysis of the microstructure of feeding revealed that amphetamine-treated rats show a pattern of ingestion characterized by a long initial latency, followed by bouts of rapid eating, which were interrupted frequently by periods of activity (Blundell and Latham, 1980). Finally, the suppression of feeding induced by amphetamine is attenuated by "typical" neuroleptics, which also block stimulant-induced stereotypy, but not by "atypical" neuroleptics, which do not block such stereotypy (Burridge and Blundell, 1979).

3.2. Contingent Suppression of Stereotypy

Figure 12 also shows that, during the 17-d period of drug administration, only the rats given the lower dose (2 mg/kg) became tolerant to the suppressant effect of amphetamine. Although the bottle-fed rats drank less than the cannula-fed rats at this dose, the former group became tolerant to amphetamine more rapidly. If it can be assumed that this difference is not merely an artifact of the different initial levels of intake, this finding implies that overcoming the behavioral disruption produced by the drug played a greater role in the development of tolerance than did recovery from anorexia. In a subsequent study (Wolgin et al., 1987), we attempted to confirm this hypothesis by rating the behavioral effects of amphetamine in both bottle- and cannula-fed rats. Initially, amphetamine induced stereotyped head scanning movements along the walls and/or floors of the cages in both groups of rats. These movements occurred continuously before, during, and after the presentation of milk. With chronic injections, bottle-fed rats gradually recovered from the initial reduction of intake, and such tolerance was accompanied by a suppression of drug-induced stereotyped movements during the time that milk was available (Fig. 13). However, both prior and subsequent to milk presentation, these movements continued unabated. In contrast, cannula-fed rats, whose intakes were not seriously disrupted by stereotypy, showed no diminution of such movements throughout the course of treatment (Fig. 14).

These results suggest that, in the bottle condition, tolerance is contingent on the suppression of behaviors that are incompatible

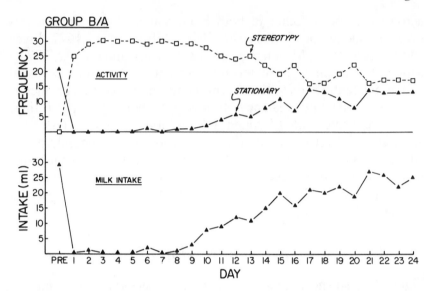

Fig. 13. Frequency of stereotyped head scanning movements and nonstereotyped stationary activity (top), and mean milk intake (bottom) of bottle-fed rats given chronic injections of amphetamine (Group B/A). Tolerance was accompanied by a suppression of stereotyped activity. PRE = last baseline day. (From "Tolerance to amphetamine: Contingent suppression of stereotypy mediates recovery of feeding" by D. L. Wolgin, G. B. Thompson, and I. A. Oslan, *Behav. Neurosci.*, 1987, **101**, 264–271. Fig. 4. Copyright 1987 by the American Psychological Association. Reprinted by permission of the publisher and author.)

with feeding. This contingency was expressed in two ways. First, only bottle-fed rats, whose feeding depended on appetitive behavior, showed suppression of stereotypy; cannula-fed rats, whose feeding was not contingent on appetitive behavior, did not. Second, bottle-fed rats showed this suppression only during the time that milk was available; during the intervals before and after milk presentation stereotypy was unabated. [Similarly, Fischman and Schuster (1974) reported that monkeys tolerant to methamphetamine suppressed stereotyped movements while performing an operant task, but reverted to stereotypy after the experimental session.] This pattern of results suggests, in turn, that bottle-fed rats may have learned to suppress stereotyped movements in order to drink the milk. The motivation for such learning may have derived from accrued deprivation during the initial period of suppressed

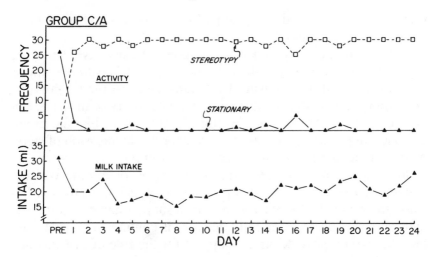

Fig. 14. Frequency of stereotyped head scanning movements and nonstereotyped stationary activity (top), and mean milk intake (bottom) of cannula-fed rats given chronic injections of amphetamine (Group C/A). In contrast to bottle-fed rats, no changes occurred in the frequency of stereotyped activity. PRE = last baseline day. (From "Tolerance to amphetamine: Contingent suppression of stereotypy mediates recovery of feeding" by D. L. Wolgin, G. B. Thompson, and I. A. Oslan, *Behav. Neurosci.*, 1987, **101**, 264–271. Fig. 3. Copyright 1987 by the American Psychological Association. Reprinted by permission of the publisher and author.)

feeding (cf. Panksepp and Booth, 1973; Demellweek and Goudie, 1982) and/or from the incentive properties of milk. The fact that cannula-fed rats showed little suppression of feeding supports the view that milk retained its reinforcing properties.

Although these data appear to provide strong support for the instrumental learning theory, it should be noted that they are not conclusive. One problem concerns the degree to which the suppression of stereotypy is under instrumental (or operant) control. As we noted previously (Wolgin et al., 1987), the behavior that eventually replaces stereotyped head scanning movements, licking, is itself a highly stereotyped response. Because drug-induced stereotypy is influenced by both environmental and experiential factors (Borberg, 1974; Ellinwood and Kilbey, 1975; Robbins, 1976; Robbins and Sahakian, 1983; Sahakian and Robbins, 1975; Schallert et al., 1980), it is quite possible that one form of stereotyped behavior was merely substituted for another. If so, the generality of these findings may be limited.

A second problem concerns the issue of causality. It may be argued that the recovery of drinking was really the result of some as yet undiscovered mechanism of tolerance and that the suppression of stereotypy was the result of its incompatibility with the act of drinking once tolerance had developed. This issue, like the previous one, hinges on the question of whether the suppression of stereotypy can be dissociated from the act of drinking from a bottle. To address this issue, we are currently trying to determine whether *cannula-fed rats* can learn to suppress stereotyped movements. Because these rats do not otherwise show a suppression of stereotypy, their ability to do so when milk delivery is contingent on such suppression would provide strong support for the role of instrumental learning.

3.3. Implications of the Theory

Assuming for the moment that the issues raised above can be resolved satisfactorily, there are several implications that follow from these data. If amphetamine-treated subjects learn to suppress stereotyped movements, then such learning should be subject to the prevailing contingencies of reinforcement and come under discriminative stimulus control. In addition, tolerance should be contingent on exposure to the task while in the drugged state. As discussed in section 2.1. above, there is considerable supporting evidence for these predictions in regard to the stimulants (e.g., Carlton and Wolgin, 1971; Schuster et al., 1966).

Other implications of the data await experimental confirmation. For example, factors that interact with stimulant-induced stereotypy should influence the rate at which tolerance develops. Consequently, tolerance should develop more readily with tasks having relatively simple response requirements that are compatible with the activated behavioral state induced by the drug (Fentress, 1976; Lyon and Robbins, 1975). On the other hand, the rate of tolerance should be retarded when the environment contains vertical surfaces and corners that "trap" the subject into a vicious cycle of stereotyped movements (Ellinwood and Kilbey, 1975; Schallert et al., 1980; Szechtman et al., 1985). In addition, there may well be individual differences in the ability of subjects to learn to suppress stereotyped movements. Such differences may account for the considerable within-group variability that is found in the rate and extent of tolerance to amphetamine (Demellweek and Goudie, 1981; Wolgin and Salisbury, 1985). Factors that might contribute to these differences include sensitivity to the drug (cf. Segal and Kuczenski, 1987), previous history with the drug and/or the task, level of motivation (e.g., food deprivation), and "innate" learning ability.

A final implication of the data concerns the phenomenon of cross-tolerance. A number of studies have shown that cross-tolerance occurs between amphetamine and other catecholaminergic drugs, such as apomorphine, cathinone, cocaine, and methylphenidate, but not between amphetamine and noncatecholaminergic drugs, such as fenfluramine or nicotine (Baettig et al., 1980; Emmett-Oglesby and Taylor, 1981; Foltin and Schuster, 1982; Kandel et al., 1975; Pearl and Seiden, 1976; Streather and Hinson, 1985; Woolverton et al., 1978a). Because drugs that show cross-tolerance interact with a common neurochemical substrate, it is often assumed that tolerance must develop to the neurochemical actions of the drug (e.g., *see* Streather and Hinson, 1985). Changes in the synthesis, release, or reuptake of the neurotransmitter, or in the sensitivity of the postsynaptic receptors, are often suggested as likely mechanisms. However, there is an alternative interpretation of these results in terms of the behavioral effects of the drugs. If tolerance to amphetamine involves learning to suppress stereotyped movements, then cross-tolerance should occur only to other drugs that produce such movements, i.e., other stimulants. Because fenfluramine, for example, does not induce stereotypy, no cross-tolerance with the stimulants would be expected. More importantly, neurochemical changes in neurons directly affected by the drug are not likely to be involved in these effects. If they were, then the suppression of stereotypy would not be restricted to periods when milk is available.

4. Summary and Conclusions

It is apparent from the foregoing review that conclusive evidence for the role of instrumental learning in drug tolerance is still lacking. At the present time, the strongest case can be made for the stimulants. In addition to numerous demonstrations of contingent and differential tolerance, there is at least preliminary evidence that rats can learn to suppress behaviors that are incompatible with reinforcement. Alternative explanations of such tolerance in terms of classically conditioned compensatory responses, cumulative deprivation, or alterations in a body weight set point cannot adequately account for these phenomena. The instrumental learning theory can also accommodate the finding that tolerance to amphetamine is environmentally specific and that cross-tolerance is limited to anorexigens that induce similar behavioral effects. Although some data appear to be inconsistent with the theory, these cases generally involve shifts in the baseline level of responding, which complicate the interpretation of the results.

The evidence with respect to ethanol is also reasonably consistent with the instrumental learning theory. Several studies have demonstrated that tolerance is influenced by various forms of feedback (including reinforcement and, in humans, even mental rehearsal) during intoxicated practice. In this case, however, the picture is more complex. Under appropriate conditions (e.g., large doses and/or frequent injections), tolerance may occur as a result of pharmacological exposure *per se*, although even then it can be augmented by behavioral experience. It still remains to be determined whether the pharmacological and behavioral "routes" represent independent mechanisms of tolerance. It is also clear that tolerance to ethanol can be contingent on behavioral experience even when reinforcement loss is not involved. This finding emphasizes that tolerance occurs to the drug's effect and not to the drug itself. Because ethanol has both specific and nonspecific effects on the central nervous system, it is likely that more than one mechanism is involved.

The evidence for instrumental learning in the development of tolerance to morphine, the anxiolytics, and marijuana is far less compelling. Depending on the behavioral paradigm, both pharmacological exposure and reinforcement loss have been implicated. Unfortunately, few of the studies included controls for pharmacological exposure *per se*, so that even where the data are consistent with the reinforcement density hypothesis, the evidence is not conclusive.

In general, the major problem in evaluating the instrumental learning theory of drug tolerance is that very few studies have tested it directly. In the vast majority of cases, support for the theory is inferred from the fact that the data are consistent with the reinforcement density hypothesis. Unfortunately, such support is inconclusive, as discussed above. Moreover, even when tolerance is shown to be contingent on behavioral experience, it is still necessary to establish what the subject has learned to do in order to overcome the initial effect of the drug. In the absence of such information, support for the theory will remain circumstantial.

Acknowledgments

I am pleased to acknowledge the able assistance of my students, Juanita Salisbury and Garrie Thompson, who put in many long hours in the laboratory. I am also grateful to Peter Carlton, who stimulated my interest in contingent tolerance many years ago, and who continues to provide guidance and constructive criticism today. Finally, I thank Maxine, Lisa, and Michael Wolgin for showing tolerance of a different sort during the preparation of this chapter.

References

Abel E. L., McMillan D. E., and Harris L. S. (1974) Δ^9-Tetrahydrocannabinol: Effects of route of administration on onset and duration of activity and tolerance development. *Psychopharmacologia* **35,** 29–38.

Alkana R. L., Finn D. A., and Malcolm R. D. (1983) The importance of experience in the development of tolerance to ethanol hypothermia. *Life Sci.* **32,** 2685–2692.

Anderson J. L., Leith N. J., and Barrett R. J. (1978) Tolerance to amphetamine's facilitation of self-stimulation responding: Anatomical specificity. *Brain Res.* **145,** 37–48.

Annear W. C. and Vogel-Sprott M. (1985) Mental rehearsal and classical conditioning contribute to ethanol tolerance in humans. *Psychopharmacology* **87,** 90–93.

Babbini M. and Davis W. M. (1972) Time-dose relations for locomotor activity effects of morphine after acute or repeated treatment. *Brit. J. Pharmacol.* **46,** 213–224.

Babbini M., Gaiardi M., and Bartoletti M. (1972) Changes in operant behavior as an index of a withdrawal state from morphine in rats. *Psychon. Sci.* **29,** 142–144.

Babbini M., Gaiardi M., and Bartoletti M. (1976) Changes in fixed-interval behavior during chronic morphine treatment and morphine abstinence in rats. *Psychopharmacologia* **45,** 255–259.

Baettig K., Martin J. R., and Classen W. (1980) Nicotine and amphetamine: Differential tolerance and no cross-tolerance for ingestive effects. *Pharmacol. Biochem. Behav.* **12,** 107–111.

Baltzer J. H., Levitt R. A., and Furby J. E. (1977) Etorphine and shuttlebox self-stimulation in the rat. *Pharmacol. Biochem. Behav.* **7,** 413–416.

Barrett R. J. (1985) Behavioral approaches to individual differences in substance abuse. Drug-taking behavior, in *Determinants of Substance Abuse. Biological, Psychological, and Environmental Factors* (Galizio M. and Maisto S. A., eds.) Plenum, New York, pp 125–175.

Beirness D. and Vogel-Sprott M. (1984a) Alcohol tolerance in social drinkers: Operant and classical conditioning effects. *Psychopharmacology* **84,** 393–397.

Beirness D. and Vogel-Sprott M. (1984b) The development of alcohol tolerance: Acute recovery as a predictor. *Psychopharmacology* **84,** 398–401.

Bennett D. A. and Amrick C. L. (1987) Home cage pretreatment with diazepam: Effects on subsequent conflict testing and rotorod assessment. *J. Pharmacol. Exp. Therap.* **242,** 595–599.

Bird, D. C., Holloway F. A., and Carney J. M. (1985) Schedule-controlled behavior as an index of the development and loss of ethanol tolerance in the rat. *Psychopharmacology* **87,** 414–420.

Bixler M. A. and Lewis M. J. (1986) Behavioral tolerance to ethanol: Investigation of a conditional compensatory response in a wheel-running task. *Soc. Neurosci. Abs.* **12,** 50.

Blundell J. E. and Latham C. J. (1980) Characterisation of adjustments to the structure of feeding behaviour following pharmacological treatment: Effects of amphetamine and fenfluramine and the antagonism produced by pimozide and methergoline. *Pharmacol. Biochem. Behav.* **12,** 717–722.

Borberg S. (1974) Conditioning of amphetamine-induced behaviour in the albino rat. *Psychopharmacologia* **34,** 191–198.

Brady L. S. and Holtzman S. G. (1980) Schedule-controlled behavior in the morphine-dependent and post-dependent rat. *Psychopharmacology* **70,** 11–18.

Branch M. N. (1979) Consequent events as determinants of drug effects on schedule-controlled behavior: Modification of effects of cocaine and *d*-amphetamine following chronic amphetamine administration. *J. Pharmacol. Exp. Ther.* **210**, 354–360.

Branch M. N. (1983) Behavioral tolerance to stimulating effects of pentobarbital: A within-subjects determination. *Pharmacol. Biochem. Behav.* **18**, 25–30.

Brocco M. J. and McMillan D. E. (1983) Tolerance to *d*-amphetamine and lack of cross-tolerance to other drugs in rats under a multiple schedule of food presentation. *J. Pharmacol. Exp. Ther.* **224**, 34–39.

Brown H. (1965) Drug-behavior interaction affecting development of tolerance to *d*-amphetamine as observed in fixed ratio behavior of rats. *Psychol. Rep.* **16**, 917–921.

Burridge S. L. and Blundell J. E. (1979) Amphetamine anorexia: Antagonism by typical but not atypical neuroleptics. *Neuropharmacol.* **18**, 453–457.

Bush H. D., Bush M. F., Miller M. A., and Reid L. D. (1976) Addictive agents and intracranial stimulation: Daily morphine and lateral hypothalamic self-stimulation. *Physiol. Psychol.* **4**, 79–85.

Campbell J. C. and Seiden L. S. (1973) Performance influence on the development of tolerance to amphetamine. *Pharmacol. Biochem. Behav.* **1**, 703–708.

Cappell H. and LeBlanc A. E. (1973) Punishment of saccharin drinking by amphetamine in rats and its reversal by chlordiazepoxide. *J. Comp. Physiol. Psychol.* **85**, 97–104.

Carder B. and Olson J. (1973) Learned behavioral tolerance to marijuana in rats. *Pharmacol. Biochem. Behav.* **1**, 73–76.

Carey R. J. (1973) Disruption of timing behavior following amphetamine withdrawal. *Physiol. Psychol.* **1**, 9–12.

Carey R. J. (1978) A comparison of the food intake suppression produced by giving amphetamine as an aversion treatment versus as an anorexic treatment. *Psychopharmacology* **56**, 45–48.

Carlton J. and Rowland N. (1985) Effects of initial body weight on anorexia and tolerance to fenfluramine in rats. *Pharmacol. Biochem. Behav.* **23**, 551–554.

Carlton P. L. (1963) Cholinergic mechanisms in the control of behavior by the brain. *Psychol. Rev.* **70**, 19–39.

Carlton P. L. (1983) *A Primer of Behavioral Pharmacology* (Freeman, New York).

Carlton P. L. and Didamo P. (1960) Some notes on the control of conditioned suppression. *J. Exp. Anal. Behav.* **3**, 255–258.

Carlton P. L. and Wolgin D. L. (1971) Contingent tolerance to the anorexigenic effects of amphetamine. *Physiol. Behav.* **7,** 221–223.

Cassens G., Actor C., Kling M., and Schildkraut J. J. (1981) Amphetamine withdrawal: Effects on threshold of intracranial reinforcement. *Psychopharmacology* **73,** 318–322.

Cesare D. A. and McKearney J. W. (1980) Tolerance to suppressive effects of chlordiazepoxide on operant behavior: Lack of cross tolerance to pentobarbital. *Pharmacol. Biochem. Behav.* **13,** 545–548.

Chen C. S. (1968) A study of the alcohol-tolerance effect and an introduction of a new behavioural technique. *Psychopharmacologia* **12,** 433–440.

Chen C. S. (1972) A further note on studies of acquired behavioural tolerance to alcohol. *Psychopharmacologia* **27,** 265–274.

Chen C. S. (1979) Acquisition of behavioral tolerance to ethanol as a function of reinforced practice in rats. *Psychopharmacology* **63,** 285–288.

Cole S. O. (1977) Interaction of arena size with different measures of amphetamine effects. *Pharmacol. Biochem. Behav.* **7,** 181–184.

Cole S. O. (1978) Brain mechanisms of amphetamine-induced anorexia, locomotion, and stereotypy: A review. *Neurosci. Biobehav. Rev.* **2,** 89–100.

Cole S. O. (1979) Interaction of food deprivation with different measures of amphetamine effects. *Pharmacol. Biochem. Behav.* **10,** 235–238.

Cook L. and Davidson A. B. (1973) Effects of behaviorally active drugs in a conflict-punishment procedure in rats, in *The Benzodiazepines* (Garattini S., Mussini E. and Randall L. O., eds.) Raven, New York, pp 327–345.

Cook L. and Sepinwall J. (1975) Behavioral analysis of the effects and mechanisms of action of benzodiazepines, in *Mechanism of Action of Benzodiazepines* (Costa E., Greengard P., eds.) Raven, New York, pp 1–28.

Corfield-Sumner P. K. and Stolerman I. P. (1978) Behavioral tolerance, in *Contemporary Research in Behavioural Pharmacology* (Blackman D. E. and Sanger D. J., eds.) Plenum, New York, pp 391–448.

Demellweek C. and Goudie A. J. (1981) A molecular analysis of the effects of chronic *d*-amphetamine treatment on fixed ratio responding in rats: Relevance to theories of tolerance, in *Quantification of Steady-State Operant Behaviour* (Bradshaw C. M., Szabadi E., and Lowe C. F., eds.) Elsevier/North Holland, Amsterdam, pp 433–437.

Demellweek C. and Goudie A. J. (1982) The role of reinforcement loss in the development of tolerance to amphetamine anorexia. *IRCS Med. Sci.* **10,** 903–904.

Demellweek C. and Goudie A. J. (1983a) An analysis of behavioural mechanisms involved in the acquisition of amphetamine anorectic tolerance. *Psychopharmacology* **79,** 58–66.

Demellweek C. and Goudie A. J. (1983b) Behavioural tolerance to amphetamine and other psychostimulants: The case for considering behavioural mechanisms. *Psychopharmacology* **80,** 287–307.

DeSouza Moreira L. F., Capriglione M. J., and Masur J. (1981) Development and reacquisition of tolerance to ethanol administered pre- and post-trial to rats. *Psychopharmacology* **73,** 165–167.

Dewey W. L. (1986) Cannabinoid pharmacology. *Pharmacol. Rev.* **38,** 151–178.

Dews P. B. (1978) Behavioral tolerance, in Behavioral tolerance: Research and treatment implications (Krasnegor N. A., ed.) *NIDA Research Monograph* **18,** 18–27.

Dews P. B. and Wenger G. R. (1977) Rate-dependency of the behavioral effects of amphetamine, in *Advances in Behavioral Pharmacology* (Vol 1) (Thompson T. and Dews P. B., eds.) Academic, New York, pp 167–227.

Drewnowski A. and Grinker J. A. (1978) Food and water intake, meal patterns and activity of obese and lean Zucker rats following chronic and acute treatment with Δ^9-tetrahydrocannabinol. *Pharmacol. Biochem. Behav.* **9,** 619–630.

Dworkin S. I. and Branch M. N. (1982) Behavioral effects of morphine and naloxone following chronic morphine administration. *Psychopharmacology* **77,** 322–326.

Eichler A. J. and Antelman S. M. (1979) Sensitization to amphetamine and stress may involve nucleus accumbens and medial frontal cortex. *Brain Res.* **176,** 412–416.

Ellinwood E. H. and Kilbey M. M. (1975) Amphetamine stereotypy: The influence of environmental factors and prepotent behavioral patterns on its topography and development. *Biol. Psychiat.* **10,** 3–16.

Elsmore T. F. (1976) The role of reinforcement loss in tolerance to chronic Δ^9-tetrahydrocannabinol effects on operant behavior of rhesus monkeys. *Pharmacol. Biochem. Behav.* **5,** 123–128.

Emmett-Oglesby M. W., Spencer D. G., Wood D. M., and Lal H. (1984) Task-specific tolerance to *d*-amphetamine. *Neuropharmacology* **23,** 563–568.

Emmett-Oglesby M. W. and Taylor K. E. (1981) Role of dose interval in the acquisition of tolerance to methylphenidate. *Neuropharmacology* **20**, 995–1002.

Esposito R. and Kornetsky C. (1977) Morphine lowering of self-stimulation thresholds: Lack of tolerance with long-term administration. *Science* **195**, 189–191.

Esposito R., McLean S., and Kornetsky C. (1979) Effects of morphine on intracranial self-stimulation to various brain stem loci. *Brain Res.* **168**, 425–429.

Fentress J. C. (1976) Dynamic boundaries of patterned behaviour: Interaction and self-organization, in *Growing Points in Ethology* (Bateson P. P. G. and Hinde R. A., eds.) Cambridge University, Cambridge, U.K., pp 135–169.

Ferraro D. P. (1972) Effects of Δ^9-trans-tetrahydrocannabinol on simple and complex learned behavior in animals, in *Current Research in Marihuana* (Lewis M. F., ed.) Academic, New York, pp 49–95.

Ferraro D. P. (1978) Behavioral tolerance to marihuana, in Behavioral tolerance: Research and treatment implications (Krasnegor N., ed.) *NIDA Res. Monog.* **18**, 103–115.

Ferraro D. P. and Grisham M. G. (1972) Tolerance to the behavioral effects of marihuana in chimpanzees. *Physiol. Behav.* **9**, 49–54.

File S. E. (1985) Tolerance to the behavioral actions of benzodiazepines. *Neurosci. Biobehav. Rev.* **9**, 113–121.

Fischman M. W. and Schuster C. R. (1974) Tolerance development to chronic methamphetamine intoxication in the rhesus monkey. *Pharmacol. Biochem. Behav.* **2**, 503–508.

Foltin R. W., Brady J. V., and Fischman M. W. (1986) Behavioral analysis of marijuana effects on food intake in humans. *Pharmacol. Biochem. Behav.* **25**, 577–582.

Foltin R. W. and Schuster C. R. (1982) Behavioral tolerance and cross-tolerance to *dl*-cathinone and *d*-amphetamine in rats. *J. Pharmacol. Exp. Ther.* **222**, 126–131.

France C. P. and Woods J. H. (1985) Effects of morphine, naltrexone, and dextrorphan in untreated and morphine-treated pigeons. *Psychopharmacology* **85**, 377–382.

Frankenheim J. M. (1974) Effects of repeated doses of 1-Δ^9-trans-tetrahydrocannabinol on schedule-controlled temporally-spaced responding of rats. *Psychopharmacologia* **38**, 125–144.

Galbicka G., Lee D. M., and Branch M. N. (1980) Schedule-dependent tolerance to behavioral effects of Δ^9-tetrahydrocannabinol when reinforcement frequencies are matched. *Pharmacol. Biochem. Behav.* **12**, 85–91.

Glick S. D. and Rapaport G. (1974) Tolerance to the facilitatory effect of morphine on self-stimulation of the medial forebrain bundle in rats. *Res. Commun. Chem. Pathol. Pharmacol.* **9**, 647–652.

Gluck J. P. and Ferraro D. P. (1974) Effects of Δ^9-THC on food and water intake of deprivation experienced rats. *Behav. Biol.* **11**, 395–401.

Gotestam K. G. and Lewander T. (1975) The duration of tolerance to the anorexigenic effect of amphetamine in the rat. *Psychopharmacologia* **42**, 41–45.

Goudie A. J. and Demellweek C. (1986) Conditioning factors in drug tolerance, in *Behavioral Analysis of Drug Dependence* (Goldberg S. R. and Stolerman I. P., eds.) Academic, New York, pp 225–285.

Griffiths J. W. and Goudie A. J. (1987) Analysis of the role of behavioural factors in the development of tolerance to the benzodiazepine midazolam. *Neuropharmacology* **26**, 201–209.

Hand T. H. and Franklin K. B. J. (1985) 6-OHDA lesions of the ventral tegmental area block morphine-induced but not amphetamine-induced facilitation of self-stimulation. *Brain Res.* **328**, 233–241.

Hand T. H. and Franklin K. B. J. (1986) Associative factors in the effects of morphine on self-stimulation. *Psychopharmacology* **88**, 472–479.

Harris R. A. and Snell D. (1980) Effects of acute and chronic administration of phenobarbital and *d*-amphetamine on schedule-controlled behavior. *Pharmacol. Biochem. Behav.* **12**, 47–52.

Harris R. A., Snell D., and Loh H. H. (1979) Effects of chronic *d*-amphetamine treatment on schedule-controlled behavior. *Psychopharmacology* **63**, 55–61.

Harris R. T., Waters W., and McLendon D. (1972) Behavioral effects in rhesus monkeys of repeated intravenous doses of Δ-9-tetrahydrocannabinol. *Psychopharmacologia* **26**, 297–306.

Haubenreisser T. and Vogel-Sprott M. (1983) Tolerance development in humans with task practice on different limbs of the blood-alcohol curve. *Psychopharmacology* **81**, 350–353.

Haubenreisser T. and Vogel-Sprott M. (1987) Reinforcement reduces behavioral impairment under an acute dose of alcohol. *Pharmacol. Biochem. Behav.* **26**, 29–33.

Heffner T. G., Zigmond M. J., and Stricker E. M. (1977) Effects of dopaminergic agonists and antagonists on feeding in intact and

6-hydroxydopamine-treated rats. *J. Pharmacol. Exp. Ther.* **201**, 386–399.

Heifetz S. A. and McMillan D. E. (1971) Development of behavioral tolerance to morphine and methadone using the schedule-controlled behavior of the pigeon. *Psychopharmacologia* **19**, 40–52.

Herberg L. J. and Montgomery A. M. J. (1987) Learnt tolerance to sedative effects of chlordiazepoxide on self-stimulation performance, but no tolerance to facilitatory effects after 80 days. *Psychopharmacology* **93**, 214–217.

Herman S. J., Freeman B. J., and Ray O. S. (1972) The effects of multiple injections of morphine sulfate on shuttle-box behavior in the rat. *Psychopharmacologia* **26**, 146–154.

Hinson R. E. and Siegel S. (1980) The contribution of Pavlovian conditioning to ethanol tolerance and dependence, in *Alcohol Tolerance and Dependence* (Rigter H. and Crabbe J. C., eds.) Elsevier/North Holland, Amsterdam, pp 181–199.

Hjeresen D. L., Reed D. R., and Woods S. C. (1986) Tolerance to hypothermia induced by ethanol depends on specific drug effects. *Psychopharmacology* **89**, 45–51.

Hoffman S. H., Branch M. N., and Sizemore G. M. (1987) Cocaine tolerance: Acute versus chronic effects as dependent upon fixed-ratio size. *J. Exp. Anal. Behav.* **47**, 363–376.

Holtzman S. G. (1974) Tolerance to the stimulant effects of morphine and pentazocine on avoidance responding in the rat. *Psychopharmacology* **39**, 23–37.

Jorgensen H. A., Berge O. G., and Hole K. (1985) Learned tolerance to ethanol in a spinal reflex separated from supraspinal control. *Pharmacol. Biochem. Behav.* **22**, 293–295.

Jorgensen H. A., Fasmer O. B., and Hole K. (1986) Learned and pharmacologically-induced tolerance to ethanol and cross-tolerance to morphine and clonidine. *Pharmacol. Biochem. Behav.* **24**, 1083–1088.

Jorgensen H. A. and Hole K. (1984) Learned tolerance to ethanol in the spinal cord. *Pharmacol. Biochem. Behav.* **20**, 789–792.

Joyce E. M. and Iversen S. D. (1984) Dissociable effects of 6-OHDA-induced lesions of neostriatum on anorexia, locomotor activity and stereotypy: The role of behavioral competition. *Psychopharmacology* **83**, 363–366.

Kalant H. (1977) Comparative aspects of tolerance to, and dependence on, alcohol, barbiturates, and opiates, in *Alcohol Intoxication and*

Withdrawal (Vol 3b) (Gross M. M., ed.) Plenum, New York, pp 169–186.

Kalant H., LeBlanc A. E., Gibbins R. J., and Wilson A. (1978) Accelerated development of tolerance during repeated cycles of ethanol exposure. *Psychopharmacology* **60,** 59–65.

Kandel D., Doyle D., and Fischman M. W. (1975) Tolerance and cross-tolerance to the effects of amphetamine, methamphetamine and fenfluramine on milk consumption in the rat. *Pharmacol. Biochem. Behav.* **3,** 705–707.

Kilbey M. M. and Sannerud C. A. (1985) Models of tolerance: Do they predict sensitization to the effects of psychomotor stimulants?, in *Behavioral Pharmacology. The Current Status* (Seiden L. S. and Balster R. L., eds.) AR Liss, New York, pp 295–321.

Kokkinidis L. and Zacharko R. M. (1980a) Enhanced self-stimulation responding from the substantia nigra after chronic amphetamine treatment: A role for conditioning factors. *Pharmacol. Biochem. Behav.* **12,** 543–547.

Kokkinidis L. and Zacharko R. M. (1980b) Response sensitization and depression following long-term amphetamine treatment in a self-stimulation paradigm. *Psychopharmacology* **68,** 73–76.

Kokkinidis L. and Zacharko R. M. (1980c) Enhanced lateral hypothalamic self-stimulation responding after chronic exposure to amphetamine. *Behav. Neural. Biol.* **29,** 493–497.

Kokkinidis L., Zacharko R. M., and Anisman H. (1986) Amphetamine withdrawal: A behavioral evaluation. *Life Sci.* **38,** 1617–1623.

Kokkinidis L., Zacharko R. M., and Predy P. A. (1980) Post-amphetamine depression of self-stimulation responding from the substantia nigra: Reversal by tricyclic antidepressants. *Pharmacol. Biochem. Behav.* **13,** 379–383.

Kosersky D. S., McMillan D. E., and Harris L. S. (1974) Δ^9-tetrahydrocannabinol and 11-hydroxy-Δ^9-tetrahydrocannabinol: Behavioral effects and tolerance development. *J. Pharmacol. Exp. Ther.* **189,** 61–65.

Krasnegor N., ed. (1978) Behavioral tolerance: Research and treatment implications. *NIDA Res. Monog.* **18,** 1–149.

Kreuz D. S. and Axelrod J. (1973) Delta-9-tetrahydrocannabinol: Localization in body fat. *Science* **179,** 391–393.

Kulig B. M. (1986) Attenuation of phenobarbital-induced deficits in coordinated locomotion during subacute exposure. *Pharmacol. Biochem. Behav.* **24,** 1805–1807.

Kuribara H. (1980) Effects of repeated administration of *d*-amphetamine on Sidman avoidance responding in rats. *Psychopharmacology* **71,** 105–107.

Lê A. D., Kalant H., and Khanna J. M. (1986a) Influence of ambient temperature on the development and maintenance of tolerance to ethanol-induced hypothermia. *Pharmacol. Biochem. Behav.* **25,** 667–672.

Lê A. D., Khanna J. M., Kalant H., and Grossi F. (1986b) Tolerance to and cross-tolerance among ethanol, pentobarbital and chlordiazepoxide. *Pharmacol. Biochem. Behav.* **24,** 93–98.

LeBlanc A. E. and Cappell H. (1977) Tolerance as adaptation: Interactions with behavior and parallels to other adaptive processes, in *Alcohol and Opiates* (Blum K., ed.) Academic, New York, pp 65–77.

LeBlanc A. E., Gibbins, R. J., and Kalant H. (1973) Behavioral augmentation of tolerance to ethanol in the rat. *Psychopharmacologia* **30,** 117–122.

LeBlanc A. E., Gibbins R. J., and Kalant H. (1975a) Generalization of behaviorally augmented tolerance to ethanol, and its relation to physical dependence. *Psychopharmacologia* **44,** 241–246.

LeBlanc A. E., Kalant H., and Gibbins R. J. (1975b) Acute tolerance to ethanol in the rat. *Psychopharmacologia* **41,** 43–46.

LeBlanc A. E., Kalant H., and Gibbins R. J. (1976) Acquisition and loss of behaviorally augmented tolerance to ethanol in the rat. *Psychopharmacology* **48,** 153–158.

LeBlanc A. E., Kalant H., Gibbins R. J., and Berman N. D. (1969) Acquisition and loss of tolerance to ethanol by the rat. *J. Pharmacol. Exp. Ther.* **168,** 244–250.

Leibowitz S. F. (1975) Catecholaminergic mechanisms of the lateral hypothalamus: Their role in the mediation of amphetamine anorexia. *Brain Res.* **98,** 529–545.

Leith N. J. and Barrett R. J. (1976) Amphetamine and the reward system: Evidence for tolerance and post-drug depression. *Psychopharmacologia* **46,** 19–25.

Leith N. J. and Barrett R. J. (1980) Effects of chronic amphetamine or reserpine on self-stimulation responding: Animal model of depression? *Psychopharmacology* **72,** 9–15.

Leith N. J. and Barrett R. J. (1981) Self-stimulation and amphetamine: Tolerance to *d* and *l* isomers and cross tolerance to cocaine and methylphenidate. *Psychopharmacology* **74,** 23–28.

Leith N. J. and Kuczenski R. (1981) Chronic amphetamine: Tolerance and reverse tolerance reflect different behavioral actions of the drug. *Pharmacol. Biochem. Behav.* **15,** 399–404.

Levitsky D. A., Strupp B. J., and Lupoli J. (1981) Tolerance to anorectic drugs: Pharmacological or artifactual? *Pharmacol. Biochem. Behav.* **14,** 661–667.

Levitt R. A., Stilwell D. J., and Evers T. M. (1978) Morphine and shuttlebox self-stimulation in the rat: Tolerance studies. *Pharmacol. Biochem. Behav.* **9,** 567–569.

Liebman J. M. and Segal D. S. (1976) Lack of tolerance or sensitization to the effects of chronic *d*-amphetamine on substantia nigra self-stimulation. *Behav. Biol.* **16,** 211–220.

Lorens S. A. and Mitchell C. L. (1973) Influence of morphine on lateral hypothalamic self-stimulation in the rat. *Psychopharmacologia* **32,** 271–277.

Lyon M. and Robbins T. (1975) The action of central nervous system stimulant drugs: A general theory concerning amphetamine effects, in *Current Developments in Psychopharmacology* (Essman W. V., and Valzelli L., eds.) Spectrum, New York, pp 78–163.

MacPhail R. C. and Seiden L. S. (1976) Effects of intermittent and repeated administration of *d*-amphetamine on restricted water intake in rats. *J. Pharmacol. Exp. Ther.* **197,** 303–310.

Mana M. J., Pinel J. P. J., and Kim C. K. (1986) Contingent tolerance to diazepam's effect on amygdaloid seizures in the rat. *Soc. Neurosci. Abs.* **12,** 1564.

Mann R. E., and Vogel-Sprott M. (1981) Control of alcohol tolerance by reinforcement in nonalcoholics. *Psychopharmacology* **75,** 315–320.

Manning F. J. (1976a) Chronic delta-9-tetrahydrocannabinol. Transient and lasting effects on avoidance behavior. *Pharmacol. Biochem. Behav.* **4,** 17–21.

Manning F. J. (1976b) Role of experience in acquisition and loss of tolerance to the effect of Δ^9-THC on spaced responding. *Pharmacol. Biochem. Behav.* **5,** 269–273.

Manning F. J., McDonough J. H., Elsmore T. F., Saller C., and Sodetz F. J. (1971) Inhibition of normal growth by chronic administration of Δ^9-tetrahydrocannabinol. *Science* **174,** 424–426.

Mansfield J. G., Benedict R. S., and Woods S. C. (1983) Response specificity of behaviorally augmented tolerance to ethanol supports a learning interpretation. *Psychopharmacology* **79,** 94–98.

Margules D. L. and Stein L. (1968) Increase of "antianxiety" activity and tolerance of behavioral depression during chronic administration of oxazepam. *Psychopharmacologia* **13**, 74–80.

Mason S. T., Sanberg P. R., and Fibiger H. C. (1978) Amphetamine-induced locomotor activity and stereotypy after kainic acid lesions of the striatum. *Life Sci.* **22**, 451–460.

McCown T. J. and Barrett R. J. (1980) Development of tolerance to the rewarding effects of self-administered S(+)-amphetamine. *Pharmacol. Biochem. Behav.* **12**, 137–141.

McMillan D. E., Harris L. S., Frankenheim J. M., and Kennedy J. S. (1970) 1-Δ^9-trans-tetrahydrocannabinol in pigeons: Tolerance to the behavioral effects. *Science* **169**, 501–503.

McMillan D. E. and Leander J. D. (1978) Chronic chlordiazepoxide and pentobarbital interactions on punished and unpunished behavior. *J. Pharmacol. Exp. Ther.* **207**, 515–520.

McMillan D. E. and Morse W. H. (1967) Some effects of morphine and morphine antagonists on schedule-controlled behavior. *J. Pharmacol. Exp. Ther.* **157**, 175–184.

Mellanby E. (1919) Alcohol: Its absorption into and disappearance from the blood under different conditions. Spec. Rep. Ser. No. 31, *Med. Res. Comm.*, London.

Miczek K. A. (1979) Chronic Δ^9-tetrahydrocannabinol in rats: Effect on social interactions, mouse killing, motor activity, consummatory behavior, and body temperature. *Psychopharmacology* **60**, 137–146.

Mucha R. F., Volkovskis C., and Kalant H. (1981) Conditioned increases in locomotor activity produced with morphine as an unconditioned stimulus, and the relation of conditioning to acute morphine effect and tolerance. *J. Comp. Physiol. Psychol.* **95**, 351–362.

Nazzaro J. M., Seeger T. F., and Gardner E. L. (1981) Morphine differentially affects ventral tegmental and substantia nigra brain reward thresholds. *Pharmacol. Biochem. Behav.* **14**, 325–331.

Olson J. and Carder B. (1974) Behavioral tolerance to marijuana as a function of amount of prior training. *Pharmacol. Biochem. Behav.* **2**, 243–247.

Panksepp J. and Booth D. A. (1973) Tolerance in the depression of intake when amphetamine is added to the rat's food. *Psychopharmacologia* **29**, 45–54.

Paul S. M., Hulihan-Giblin B., and Skolnick P. (1982) (+)−Amphetamine binding to rat hypothalamus: Relation to anorexic potency of phenylethylamines. *Science* **218**, 487–490.

Pearl R. G. and Seiden L. S. (1976) The existence of tolerance and cross-tolerance between *d*-amphetamine and methylphenidate for their effect on milk consumption and on differential-reinforcement-of-low-rate performance in the rat. *J. Pharmacol. Exp. Ther.* **198,** 635–647.

Peck J. W. (1978) Rats defend different body weights depending on palatability and accessibility of their food. *J. Comp. Physiol. Psychol.* **92,** 555–570.

Pieper W. A. and Skeen M. J. (1973) Development of functional tolerance to ethanol in rhesus monkeys (Macaca mulatta). *Pharmacol. Biochem. Behav.* **1,** 289–294.

Pinel J. P. J., Colborne B., Sigalet J. P., and Renfrey G. (1983) Learned tolerance to the anticonvulsant effects of alcohol in rats. *Pharmacol. Biochem. Behav.* **18,** 507–510.

Pinel J. P. J., Mana M. J., and Renfrey G. (1985) Contingent tolerance to the anticonvulsant effects of alcohol. *Alcohol* **2,** 495–499.

Pinel J. P. J. and Puttaswamaiah S. (1985) Tolerance to alcohol's anticonvulsant effect is not under Pavlovian control. *Pharmacol. Biochem. Behav.* **23,** 959–964.

Poulos C. X. and Hinson R. E. (1984) A homeostatic model of Pavlovian conditioning: Tolerance to scopolamine-induced adipsia. *J. Exp. Psychol.: Anim. Behav. Process.* **10,** 75–89.

Poulos C. X., Wilkinson D. A., and Cappell H. (1981) Homeostatic regulation and Pavlovian conditioning in tolerance to amphetamine-induced anorexia. *J. Comp. Physiol. Psychol.* **95,** 735–746.

Powley T. L. and Keesey R. E. (1970) Relationship of body weight to the lateral hypothalamic feeding syndrome. *J. Comp. Physiol. Psychol.* **70,** 25–36.

Predy P. A. and Kokkinidis L. (1981) Post-amphetamine depression of self-stimulation behavior in rats: Prophylactic effects of lithium. *Neurosci. Lett.* **23,** 343–347.

Predy P. A. and Kokkinidis L. (1984) Sensitization to the effects of repeated amphetamine administration on intracranial self-stimulation: Evidence for changes in reward processes. *Behav. Brain. Res.* **13,** 251–259.

Rawana E. and Vogel-Sprott M. (1985) The transfer of alcohol tolerance, and its relation to reinforcement. *Drug Alcohol Dep.* **16,** 75–83.

Rhodus D. M., Elsmore T. F., and Manning F. J. (1974) Morphine and heroin effects on multiple fixed-interval schedule performance in rats. *Psychopharmacologia* **40,** 147–155.

Robbins T. W. (1976) Relationship between reward-enhancing and stereotypical effects of psychomotor stimulant drugs. *Nature* **264,** 57–59.

Robbins T. W. and Sahakian B. J. (1983) Behavioral effects of psychomotor stimulant drugs: Clinical and neuropsychological implications, in *Stimulants: Neurochemical, Behavioral, and Clinical Perspectives* (Creese I., ed.) Raven, New York, pp 301–338.

Robertson A. and Mogenson G. J. (1979) Facilitation of self-stimulation of the prefrontal cortex in rats following chronic administration of spiroperidol or amphetamine. *Psychopharmacology* **65,** 149–154.

Rowland N., Bartness T., Carlton J., Antelman S., and Kocan D. (1982) Tolerance and sensitization of the effects of various anorectics, in *The Neural Basis of Feeding and Reward* (Hoebel B. G. and Novin D., eds.) Haer Inst. Electrophysiol Res., Brunswick, ME, pp 535–541.

Rowland N. and Carlton J. (1983) Different behavioral mechanisms underlie tolerance to the anorectic effects of fenfluramine and quipazine. *Psychopharmacology* **81,** 155–157.

Rumbold G. R. and White J. M. (1987) Effects of repeated alcohol administration on human operant behavior. *Psychopharmacology* **92,** 186–191.

Sahakian B. J. and Robbins T. W. (1975) The effect of test environment and rearing condition on amphetamine-induced stereotypy in the guinea pig. *Psychopharmacology* **45,** 115–117.

Salisbury J. J. and Wolgin D. L. (1985) Role of anorexia and behavioral activation in amphetamine-induced suppression of feeding: Implications for understanding tolerance. *Behav. Neurosci.* **99,** 1153–1161.

Sanberg P. R. and Fibiger H. C. (1979) Body weight, feeding, and drinking behaviors in rats with kainic acid-induced lesions of striatal neurons—With a note on body weight symptomatology in Huntington's disease. *Exp. Neurol.* **66,** 444–466.

Sannerud C. A. and Young A. M. (1986) Modification of morphine tolerance by behavioral variables. *J. Pharmacol. Exp. Therap.* **237,** 75–81.

Schallert T., DeRyck M., and Teitelbaum P. (1980) Atropine stereotypy as a behavioral trap: A movement subsystem and electroencephalographic analysis. *J. Comp. Physiol. Psychol.* **94,** 1–24.

Schnur P. (1985) Morphine tolerance and sensitization in the hamster. *Pharmacol. Biochem. Behav.* **22,** 157–158.

Schnur P., Bravo F., Trujillo M. (1983a) Tolerance and sensitization to the biphasic effects of low doses of morphine in the hamster. *Pharmacol. Biochem. Behav.* **19,** 435–439.

Schnur P., Bravo F., Trujillo M., and Rocha S. (1983b) Biphasic effects of morphine on locomotor activity in hamster. *Pharmacol. Biochem. Behav.* **18,** 357–361.

Schuster C. R. (1978) Theoretical basis of behavioral tolerance: Implications of the phenomenon for problems of drug abuse, in Behavioral tolerance: Research and treatment implications (Krasnegor N. A., ed.) *NIDA Res. Monog.* **18,** 4–17.

Schuster C. R., Dockens W. S., and Woods J. H. (1966) Behavioral variables affecting the development of amphetamine tolerance. *Psychopharmacologia* **9,** 170–182.

Schuster C. R. and Zimmerman J. (1961) Timing behavior during prolonged treatment with *d*-amphetamine. *J. Exp. Anal. Behav.* **4,** 327–330.

Schwartz B. (1984) *Psychology of Learning and Behavior* (2nd ed). Norton, New York.

Sdao-Jarvie K. and Vogel-Sprott M. (1986) Mental rehearsal of a task before or after ethanol: Tolerance facilitating effects. *Alcohol Drug Dep.* **18,** 23–30.

Segal D. S. and Kuczenski R. (1987) Individual differences in responsiveness to single and repeated amphetamine administration: Behavioral characteristics and neurochemical correlates. *J. Pharmacol. Exp. Therap.* **242,** 917–926.

Smith J. B. (1979) Behavioral influences on tolerance to the effects of morphine on schedule-controlled behavior. *Psychopharmacology* **66,** 105–107.

Smith J. B. (1986a) Effects of chronically administered *d*-amphetamine on spaced responding maintained under multiple and single-component schedules. *Psychopharmacology* **88,** 296–300.

Smith J. B. (1986b) Effects of fixed-ratio length on the development of tolerance to decreased responding by *l*-nantradol. *Psychopharmacology* **90,** 259–262.

Smith J. B. and McKearney J. W. (1977) Changes in the rate-increasing effects of *d*-amphetamine and pentobarbital by response consequences. *Psychopharmacology* **53,** 151–157.

Sofia R. D. and Barry H. (1974) Acute and chronic effects of Δ^9-tetrahydrocannabinol on food intake by rats. *Psychopharmacologia* **39,** 213–222.

Sparber S. B. and Tilson H. A. (1972) Tolerance and cross-tolerance to mescaline and *d*-amphetamine administered intraventricularly or peripherally. *Psychopharmacologia* **23**, 220–230.

Stein L. (1964) Self-stimulation of the brain and the central stimulant action of amphetamine. *Fed. Proc.* **23**, 836–850.

Stein L., Sidman M., and Brady J. V. (1958) Some effects of two temporal variables on conditioned suppression. *J. Exp. Anal. Behav.* **1**, 151–162.

Stein L. and Wise C. D. (1969) Release of norepinephrine from hypothalamus and amygdala by rewarding medial forebrain bundle stimulation and amphetamine. *J. Comp. Physiol. Psychol.* **67**, 89–198.

Stephens D. N. and Schneider H. H. (1985) Tolerance to the benzodiazepine diazepam in an animal model of anxiolytic activity. *Psychopharmacology* **87**, 322–327.

Streather A. and Hinson R. E. (1985) Neurochemical and behavioral factors in the development of tolerance to anorectics. *Behav. Neurosci.* **99**, 842–852.

Streather A. and Hinson R. E. (1986) Tolerance to amphetamine anorexia: Resistance to extinction and rapid reacquisition. *Soc. Neurosci. Abs.* **12**, 1451.

Stunkard A. J. (1981) Anorectic agents: A theory of action and lack of tolerance in a clinical trial, in *Anorectic agents: Mechanisms of Action and Tolerance* (Garattini S. and Samanin R., eds.) Raven, New York, pp 191–210.

Stunkard A. J. (1982) Anorectic agents lower a body weight set point. *Life Sci.* **30**, 2043–2055.

Szechtman H., Ornstein K., Teitelbaum P. and Golani I. (1985) The morphogenesis of stereotyped behavior induced by the dopamine receptor agonist apomorphine in the laboratory rat. *Neuroscience* **14**, 783–798.

Tabakoff B., Melchior C. L., and Hoffman P. (1984) Factors in ethanol tolerance. *Science* **224**, 523–524.

Tang M. and Falk J. L. (1978) Behavioral and pharmacological components of phenobarbital tolerance, in Behavioral tolerance: Research and treatment implications (Krasnegor N. A., ed.) *NIDA Res. Monog.* **18**, 142–149.

Tilson H. A. and Rech R. H. (1973) Prior drug experience and effects of amphetamine on schedule controlled behavior. *Pharmacol. Biochem. Behav.* **1**, 129–132.

Tilson H. A. and Sparber S. B. (1973) The effects of *d-* and *l*-amphetamine on fixed-interval and fixed-ratio behavior in tolerant and nontolerant rats. *J. Pharmacol. Exp. Ther.* **187**, 372–379.

Tizzano J., Bannon A., Liberto R., Anderson J., Roberts D., Muchow D. and Kallman M. (1986) Behavioral tolerance to diazepam as a consequence of the dose administered chronically. *Soc. Neurosci. Abs.* **12**, 923.

Tordoff M. G., Hopfenbeck J., Butcher L. L. and Novin D. (1982a) A peripheral locus for amphetamine anorexia. *Nature* **297**, 148–150.

Tordoff M. G., Novin D., and Russek M. (1982b) Effects of hepatic denervation on the anorexic response to epinephrine, amphetamine, and lithium chloride: A behavioral identification of glucostatic afferents. *J. Comp. Physiol. Psychol.* **96**, 361–375.

Traynor A. E., Schlapfer W. T., and Barondes S. H. (1980) Stimulation is necessary for the development of tolerance to a neuronal effect of ethanol. *J. Neurobiol.* **11**, 633–637.

Traynor M. E., Woodson P. B. J., Schlapfer W. T., and Barondes S. H. (1976) Sustained tolerance to a specific effect of ethanol on posttetanic potentiation in Aplysia. *Science* **193**, 510–511.

Van der Kooy D., Schiff B. B., and Steele D. (1978) Response-dependent effects of morphine on reinforcing lateral hypothalamic self-stimulation. *Psychopharmacology* **58**, 63–67.

Vasko M. R. and Domino E. F. (1978) Tolerance development to the biphasic effects of morphine on locomotor activity and brain acetylcholine in the rat. *J. Pharmacol. Exp. Ther.* **207**, 848–858.

Vogel-Sprott M., Rawana E. and Webster R. (1984) Mental rehearsal of a task under ethanol facilitates tolerance. *Pharmacol. Biochem. Behav.* **21**, 329–331.

Watson R., Hartmann E., and Schildkraut J. J. (1972) Amphetamine withdrawal: Affective state, sleep patterns, and MHPG excretion. *Amer. J. Psychiat.* **129**, 263–269.

Webster C. D., LeBlanc A. E., Marshman J. A. and Beaton J. M. (1973) Acquisition and loss of tolerance to 1-Δ^9-tetrahydrocannabinol in rats on an avoidance schedule. *Psychopharmacologia* **30**, 217–226.

Wenger J. R., Berlin V., and Woods S. C. (1980) Learned tolerance to the behaviorally disruptive effects of ethanol. *Behav. Neural. Biol.* **28**, 418–430.

Wenger J. R., Tiffany T. M., Bombardier C., Nicholls K. and Woods S. C. (1981) Ethanol tolerance in the rat is learned. *Science* **213**, 575–577.

Wenger J. R. and Woods S. C. (1984) Reply to Tabakoff, Melchior, and Hoffman. *Science* **224**, 524.

Wigell A. H. and Overstreet D. H. (1984) Acquisition of behaviourally augmented tolerance to ethanol and its relationship to muscarinic receptors. *Psychopharmacology* **83**, 88–92.

Wise R. A. and Bozarth M. A. (1981) Brain substrates for reinforcement and drug self-administration. *Prog. Neuro-Psychopharmacol* **5**, 467–474.

Witkin J. M., Leander J. D., and Dykstra L. A. (1983) Modification of behavioral effects of morphine, meperidine and normeperidine by naloxone and by morphine tolerance. *J. Pharmacol. Exp. Ther.* **225**, 275–283.

Wolgin D. L. (1973) An analysis of tolerance to the anorexigenic effects of amphetamine. Unpublished doctoral dissertation, Rutgers University, New Brunswick, NJ.

Wolgin D. L. (1983) Tolerance to amphetamine anorexia: Role of learning versus body weight settling point. *Behav. Neurosci.* **97**, 549–562.

Wolgin D. L. and Salisbury J. J. (1985) Amphetamine tolerance and body weight set point: A dose-response analysis. *Behav. Neurosci.* **99**, 175–185.

Wolgin D. L., Thompson G. B., and Oslan I. A. (1987) Tolerance to amphetamine: Contingent suppression of stereotypy mediates recovery of feeding. *Behav. Neurosci.* **101**, 264–271.

Wood J. M. and Laverty R. (1979) Metabolic and pharmacodynamic tolerance to ethanol in rats. *Pharmacol. Biochem. Behav.* **10**, 871–874.

Woods J. H. and Carney J. (1978) Narcotic tolerance and operant behavior, in Behavioral tolerance: Research and treatment implications (Krasnegor N. A., ed.) *NIDA Res. Monog.* **18**, 54–66.

Woolverton W. L., Kandel D., and Schuster C. R. (1978a) Tolerance and cross-tolerance to cocaine and *d*-amphetamine. *J. Pharmacol. Exp. Ther.* **205**, 525–535.

Woolverton W. L., Kandel D., and Schuster C. R. (1978b) Effects of repeated administration of cocaine on schedule-controlled behavior of rats. *Pharmacol. Biochem. Behav.* **9**, 327–337.

Young R. and Glennon R. A. (1986) Discriminative stimulus properties of amphetamine and structurally related phenalkylamines. *Med Res. Rev.* **6**, 99–130.

Pharmacological Conditioning and Drug Effects

Shepard Siegel

1. Nonpharmacological Contributions to Drug Effects

Typically, the effects of a drug, and alterations in these effects over the course of repeated administrations (i.e., tolerance and sensitization), have been attributed to wholly systemic mechanisms. For example, the effect of an exogenous opiate may be attributable to its effects at central endorphin receptors, and tolerance may be the result of the neurochemical alterations induced by repeated drug administrations. It has become apparent, however, that drug effects are importantly modulated by nonpharmacological factors. The result of the chemical stimulation depends not only on pharmacodynamic and pharmacokinetic principles, but also upon the recipient's previous experiences and expectations.

An early description of the effect of drug expectation on drug effects is contained in the 1868 novel, *The Moonstone* (Collins, 1966/1871). A character in the novel, Franklin Blake, voluntarily ingests laudanum (tincture of opium) for purposes of a "physiological experiment." Blake had previously involuntarily ingested the drug after it was surreptitiously slipped into his drink. In calculating the dose for the experiment, the phenomenon of tolerance is recognized. Moreover, this tolerance is attributed to Blake's expectation of the pharmacological stimulation on the occasion of the experiment (in contrast to his previous unwitting drugging): "On this occasion, Mr. Blake knows beforehand that he is going to take the laudanum—which is equivalent, physiologically speaking to his having (unconsciously to himself) a certain capacity in him to resist the effects" (Collins, 1966/1871; p. 465). Collins' novel was prophetic (Siegel, 1983b, 1985b). In fact, results of recent research have indicated the importance of drug expectation in drug effects: "tolerance is maximally displayed following 'expected' drug administration, but not following 'unexpected' drug administration" (Siegel, 1982a, p. 2339).

Data clearly indicating the contribution of experience and expectation to drug effects have been collected using the "balanced placebo design" (Marlatt and Rohsenow, 1982; Newlin et al., 1986). The drug typically evaluated with the balanced placebo design is alcohol. The experiments involve rather elaborate deception procedures, and use beverages, such as vodka and tonic mixtures, in which the alcohol content is difficult to detect. Independent groups of subjects are assigned to each cell of a 2×2 factorial design. One independent variable is the beverage that the subject consumes (alcoholic vs nonalcoholic), and the other independent variable is the subject's expectation concerning the beverage that is consumed (belief that the beverage is alcoholic vs belief that the beverage is nonalcoholic). Thus, the balanced placebo design consists of four groups:

(1) Subjects who consume alcohol and are correctly informed that they are consuming alcohol

(2) Subjects who consume alcohol, but are deceived into believing that they are drinking a nonalcoholic beverage

(3) Subjects who consume a nonalcoholic beverage and are correctly informed that the beverage is nonalcoholic

(4) Subjects who consume a nonalcoholic beverage, but are deceived into believing that they are drinking an alcoholic beverage.

Results of research using this procedure have indicated that a variety of effects of alcohol in humans depends on what people think they are ingesting, as well as what they really do ingest.

Studies with animals similarly indicate the importance of nonpharmacological factors in drug effects. The response to a drug is greatly affected by environmental stimuli. For example, Hinson et al. (1982) evaluated the motor-stimulatory effects of cocaine in rats with a history of pentobarbital administration. Cocaine had a considerably more pronounced effect if it was administered in the same environment in which pentobarbital was previously administered than if it was administered in an alternative environment. Speaking casually, the rat that expects the barbiturate is hyperresponsive to the stimulant.

Similar findings, indicating expectation-induced modulation of drug effects, have also been noted with other drugs: phenobarbital-induced polydipsia is exacerbated in rats administered phenobarbital in conjunction with environmental stimuli previously associated with scopolamine (Poulos and Hinson, 1984); amphetamine-induced hyperthermia is exaggerated in rats administered the stimulant in the presence of environmental cues associated with pentobarbital (Hinson and Rhijnsburger, 1984). Other examples of such apparent expectation-induced modulations of drug effects were presented by Bykov (1959), who suggested (pp. 82–83) that the phenomenon could be understood as resulting from Pavlovian conditioning.

2. The Pavlovian Conditioning Situation

Living organisms respond not only reflexively to stimuli—they also respond in anticipation of stimuli. The analysis of such anticipatory responding uses procedures and terminology developed by Ivan Pavlov, and is called Pavlovian (or classical) conditioning (Pavlov, 1927).

In the Pavlovian conditioning situation, a contingency is arranged between two stimuli; typically, one stimulus reliably predicts the occurrence of the second stimulus. Using the usual terminology, the second of these paired stimuli is termed the *unconditional stimulus* (UCS). The UCS, as the name implies, is selected because it elicits relevant activities from the outset (i.e., unconditionally), prior to any pairings. Responses elicited by the UCS are termed unconditional responses (UCRs). The stimulus signaling the presentation of the UCS is "neutral," (i.e., it elicits little relevant activity prior to its pairing with the UCS), and is termed the *conditional stimulus* (CS). The CS, as the name implies, becomes capable of eliciting new responses as a function of (i.e., conditional upon) its pairing with the unconditional stimulus.

In Pavlov's well-known conditioning research, the CS was a conveniently manipulated exteroceptive stimulus (bell, light, etc.), and the UCS was either food or orally injected dilute acid (both of which elicited a conveniently monitored salivary UCR). After a number of CS–UCS pairings, it was noted that the subject salivated not only in response to the UCS, but also in anticipation of the UCS (i.e., in response to the CS). The subject is then said to display a conditional response (CR).

2.1. Drugs as Unconditional Stimuli

A wide range of exteroceptive and interoceptive stimuli have been used in Pavlovian conditioning experiments (Razran, 1961). Drugs constitute a particularly interesting class of UCSs. After some number of drug administrations, each administration reliably signaled by a CS, pharmacological CRs can be observed in response to the CS. It was Pavlov who first demonstrated such pharmacological conditioning. He paired a tone with administration of apomorphine. The drug induced restlessness, salivation, and a "disposition to vomit." After several tone-apomorphine pairings, the tone alone "sufficed to produce all the active symptoms of the drug, only in a lesser degree" (Pavlov, 1927, p. 35).

Additional research by Krylov (reported by Pavlov, 1927, p. 35–37) indicated that, even if there was not an explicit CS (such as an auditory cue), naturally occurring predrug cues (opening the box containing the hypodermic syringe, cropping the fur, etc.) could serve as CSs. In Krylov's experiments, a dog was repeatedly injected with morphine, each injection eliciting a number of responses including copious salivation. After five or six such injections, it was observed that "the preliminaries of injection" (Pavlov, 1927, p. 35) elicited many morphine-like responses, including salivation.

2.2. The Pharmacological Conditional Response

Most pharmacological conditioning research has been greatly influenced by Pavlov's theory of CR formation. According to this theory, the CR is a replica of the UCR, and, indeed, much drug conditioning work has demonstrated CRs that mimick the drug effect (Siegel, 1985a). In contrast, in 1937 Subkov and Zilov reported that dogs with a history of epinephrine administration (each injection eliciting a tachycardiac response) displayed a conditional *brady*cardiac response. Subkov and Zilov cautioned against "the widely accepted view that the external modifications of the conditional reflex must always be identical with the response of the organism to the unconditional stimulus" (Subkov and Zilov, 1937, p. 296). Subsequent research has revealed other examples of such drug-compensatory CRs in animals. For example, in addition to its bradycardiac effect, epinephrine also decreases gastric secretion and induces hyperglycemia. Rats with a history of epinephrine administration display an *increase* in gastric secretion (Guha, et al., 1974) and *hypo*glycemia (Russek and Piña, 1962). Indeed, for many effects of many drugs, the CR is an anticipatory compensation for the drug effect; the drug-associated environmental cues elicit responses that are opposite to the drug *effect*. For example, the subject with a history of morphine administration (and its analgesic consequence) often displays a CR of hyperalgesia (Krank, 1987; Krank et al., 1981; Siegel, 1975b). Similar compensatory-CRs have been reported with respect to the thermic (Siegel, 1978), locomotor (Mucha et al., 1981; Paletta and Wagner, 1986), behaviorally sedating (Hinson and Siegel, 1983), and gastrointestinal (Raffa et al., 1982) effects of morphine. The CR seen with many nonopiate drugs is similarly opposite to the drug effect, e.g., atropine (Mulinos and Lieb, 1929), chlorpromazine (Pihl and Altman, 1971), amphetamine (Obál, 1966), methyl dopa (Korol and McLaughlin, 1976), lithium chloride (Domjan and Gillan, 1977), haloperidol (King et al., 1978), ethanol (Lê et al., 1979), and caffeine (Rozin et al., 1984).

2.3. Determination of Pharmacological CR Topography

Although both drug-mimicking and drug-compensatory CRs have been noted, the conditions that favor the expression of the two CR forms are not yet entirely clear. An important analysis of pharmacological conditioning was presented by Obál and colleagues (Obál, 1966; Obál et al., 1965, 1976). They suggested that the site of action of the drug was crucial in determining the forms of the CR. In Obál's terminology, a drug could have a central or peripheral "attack": "In the case of a stimulus with a central attack . . . the conditioned response is a reaction identical to the effect of the stimulus. On the other hand, in the case of a stimulus with a peripheral attack . . . the conditioned response is a change in the direction opposite to the effect of the stimulus" (Obál et al., 1976, pp. 249–250).

Obál's view has more recently been elaborated in a comprehensive discussion of pharmacological conditioning by Eikelboom and Stewart (1982), who suggest that there is a fundamental confusion concerning the identification of pharmacological UCSs and UCRs. This confusion arises because the observed effects of drugs, in contrast to the observed effects of most peripherally applied stimuli, may occur without the participation of the central nervous system (CNS).

In the typical (nonpharmacological) Pavlovian conditioning preparation, the UCS is an event with an afferent site of action; that is, the UCS stimulates receptors that initiate activity in the CNS. It is the effects of this CNS activity that constitute the UCR. In such a conditioning situation, the CR usually mimicks the UCR. Eikelboom and Stewart (1982) suggest that, for those drugs whose effects are mediated in a similar manner (i.e., drugs with an afferent site of action), the CR will similarly mimick the UCR.

In contrast with this situation, however, many chemical UCSs have an *efferent* site of action; the observed drug effect may be the result of direct pharmacological stimulation of the effector system. In such cases, it is accurate to consider the drug effect as the UCS (not the UCR), since it elicits (rather than results from) a CNS response. It is this central response to the drug effect that constitutes the UCR. For example, parentally administered glucose causes a rise in blood glucose concentration, and the CR seen in the animal trained with glucose is a depression in blood glucose concentration (Deutsch, 1974; Mityushov, 1954). These results have been presented as examples of a pharmacological CR (hypoglycemia) that is compensatory to the pharmacological UCR (hyperglycemia) (e.g., Siegel, 1975a; Woods and Kolkosky, 1976). In fact, the hyperglycemia noted following glucose administration is the UCS (not the UCR), with this UCS initiating CNS activities that act to compensate for the hyperglycemia; the correctly conceptualized UCR, then, is the CNS-mediated response to the glucose (i.e., homeostatic corrections for the glucose-induced hyperglycemia).

Speaking casually, it is the UCS-elicited activities of the brain that are associated with the CS. These activities may be initiated via direct afferent stimulation, or via efferent stimulation that engages feedback mechanisms to counteract the drug effect. In both cases, according to Eikelboom and Stewart (1982), the CR will mimic the UCR. In the case of pharmacological conditioning, the CR will be in the same direction as the drug effect if the drug has an afferent site of action, and the CR will be opposite in direction to the drug effect if the drug has an efferent site of action. In the case of those drugs with multiple effects, the various components of the CR would be expected either to mimic or to compensate for the drug effect, depending on the mechanism by which the effect results.

2.4. Pavlovian Conditioning and Drug Tolerance

Regardless of the mechanism whereby pharmacological CRs are frequently antagonistic to the drug effect, they would be expected to be a feature of normal drug administration procedures. In those cases in which the same drug is repeatedly administered, with discrete environmental stimuli signaling each drug administration, drug-compensatory CRs should function to increasingly attenuate the drug effect. A decreasing response to a drug, over the course of successive administrations, defines tolerance. Thus, it is likely that pharmacological conditioning contributes to tolerance.

3. Evidence for the Conditioning Analysis of Tolerance

In the last 10 years or so, there has been a great deal of research assessing the contribution of Pavlovian conditioning to tolerance. Generally, this research has evaluated the extent to which procedures known to affect Pavlovian conditioning similarly affect tolerance.

3.1. Environmental Specificity of Tolerance

On the basis of the conditioning analysis, tolerance should *not* invariably result from repeated drug administrations. Rather, tolerance should be displayed only when the drug is administered in the context of the usual drug-predictive cues, because it is these cues that elicit the drug-compensatory CRs that mediate tolerance. In fact, results of many studies have demonstrated that tolerance is more pronounced when the drug is administered in the presence of the usual predrug cues than if the drug is administered in the presence of alternative cues.

3.1.1. Environmental-Specificity of Opiate Tolerance

Early demonstrations of the contribution of predrug cues to tolerance were conducted by Mitchell and colleagues (e.g., Adams et al., 1969). In these experiments, rats responded in the expected analgesia-tolerant manner to the last of a series of morphine injections only if this final injection occurred in the same environment as the prior injections of morphine. Much subsequent research has extended these demonstrations of the importance of predrug cues in tolerance to the analgesic effect of morphine (*see* Siegel and MacRae, 1984a).

The details of the designs of experiments demonstrating the environmental specificity of morphine analgesic tolerance differed, but all incorporated two groups of rats, both receiving the drug a sufficient number of times for tolerance to develop during the initial, tolerance development, phase of the experiment. The analgesic effect of the drug was evaluated in a subsequent tolerance test phase. For one of the two groups, this test was conducted following the same cues that signaled the drug during the tolerance development phase (Same-Tested). For the second group, the tolerance test was conducted following different cues than those that signaled the drug during the tolerance development phase (Different-Tested). In addition, the design of the experiments enabled evaluation of the magnitude of the analgesic response in rats receiving the drug for the very first time (Control). Results obtained during the tolerance test in a number of experiments using this procedure are summarized in Fig. 1.

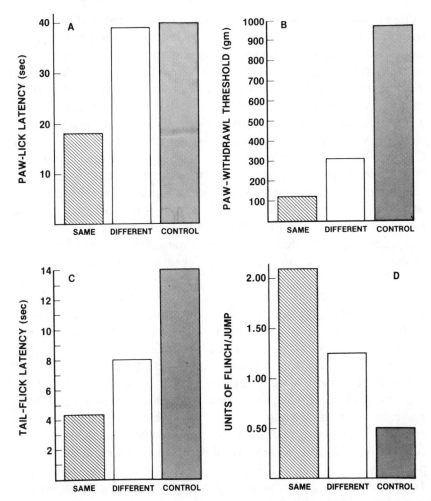

Fig. 1. Situational specificity of morphine analgesic tolerance demonstrated using the hot plate (A), paw-pressure (B), tail-flick (C), and flinch/jump (D) analgesia assessment procedures. These results are based on Siegel et al. (1978), Siegel (1976), Advokat (1980), and Tiffany and Baker (1981), respectively.

Figure 1A summarizes results of an experiment in which analgesia in rats was assessed following a tenth morphine injection (5 mg/kg), this tenth injection being paired with an audiovisual cue (Siegel et al., 1978). Pain sensitivity was measured by noting the rat's latency to lick a paw when placed on a 54°C surface (the ''hot plate'' procedure). For Same-Tested rats, the nine pretest injections were signaled by the same cue that signaled the test injection. Different-Tested rats, in contrast, received their nine prior drug injections and cue presentations in an unpaired manner. As can be seen by comparing Control rats with Same-Tested rats, tolerance to the analgesic effect of morphine was obtained: Control rats, which received the drug for the first time on the tolerance test session, were significantly less sensitive to the thermal stimulation (i.e., were more analgesic) than Same-Tested rats, which received the drug for the tenth time on the test session. However, results obtained from Different-Tested rats demonstrated that analgesic tolerance is not the inevitable consequence of repeated morphine administration. Different-Tested rats had the same pharmacological history as Same-Tested rats (i.e., they received the same dose of morphine, equally often, and at the same intervals), but Different-Tested rats were as profoundly analgesic as Control animals.

Other studies, using different drug doses and/or analgesia assessment procedures, have similarly demonstrated that Same-Tested rats are more tolerant than Different-Tested rats, although such situational specificity of tolerance is not always complete; that is, both groups of drug-experienced animals may be more tolerant than drug-naive Control animals. Figure 1B illustrates results of an experiment in which pain sensitivity was assessed with a paw-pressure analgesiometer; the rat was free to withdraw its paw from a source of gradually and constantly increasing pressure, with the amount of pressure applied before the withdrawal response occurred (i.e., the paw-withdrawal threshold) providing a measure of pain sensitivity (Siegel, 1976). As can be seen in Fig. 1B, Same-Tested rats, with a pretest history of eight morphine (5 mg/kg) injections, evidenced greater sensitivity to the paw pressure (i.e., more analgesic tolerance) than the equally drug-experienced Different-Tested rats. Thus, again, equivalent opiate exposure does not lead to equivalent levels of tolerance. As is apparent in Fig. 1B, both Same- and Different-Tested rats displayed more tolerance than Control rats, perhaps indicating a non-environmental component of tolerance as well.

A similar pattern of results may be seen in a study of analgesic tolerance, as assessed by tail-flick latency, following eight injections of morphine (7.5 mg/kg) (Advokat, 1980). As depicted in Fig. 1C, Control rats are profoundly analgesic. Although both drug-experienced groups display shorter tail-flick latencies than the

Control level, they differed significantly: Same-Tested animals were more tolerant than Different-Tested animals. This pattern of results was confirmed in a different experiment (Tiffany and Baker, 1981), using a different number of pretest sessions (five), a different dose of morphine (20 mg/kg), and a different analgesia-assessment procedure (digitalized flinch/jump magnitude to electric shocks, with smaller numbers indicating less sensitivity to shock, i.e., greater analgesia). The results of this experiment are presented in Fig. 1D. Again, Same-Tested animals were more tolerant than Different-Tested animals, although neither were as analgesic as Control animals.

The environmental specificity of tolerance to the analgesic effect of morphine is rather general, having been demonstrated in the terrestrial gastropod snail, *Cepaea nemoralis* (Kavaliers and Hirst, 1986), suggesting that such specificity "may be a general phenomenon having an early evolutionary development and broad phylogenetic continuity" (Kavaliers and Hirst, 1986, p. 1201).

The finding that opioid tolerance is more pronounced in the drug administration environment than an alternative environment has been demonstrated with respect to effects of the drug other than analgesia. It has been reported that rats tested in the context of the usual predrug cues are more tolerant to the thermic (Siegel, 1978), locomotor (Mucha et al., 1981), and behaviorally sedating (Hinson and Siegel, 1983) effects of morphine than equally drug-experienced rats tested in the context of alternative cues. Such influence of environmental cues on tolerance to the lethal effect of diacetylmorphine hydrochloride (heroin) has also been demonstrated (Siegel et al., 1982).

3.1.2. Environmental Specificity of Tolerance to Nonopiate Drugs

Environmental specificity of tolerance is rather general, and has been demonstrated with a variety of nonopiate drugs. Figure 2 summarizes the results of several studies demonstrating such specificity with respect to ethanol tolerance. As was the case in the opiate experiments, an effect of ethanol was evaluated in subjects with no prior experience with the drug (Control), or with prior experience in either the same or a different environment than that used for the final administration (Same-Tested and Different-Tested, respectively).

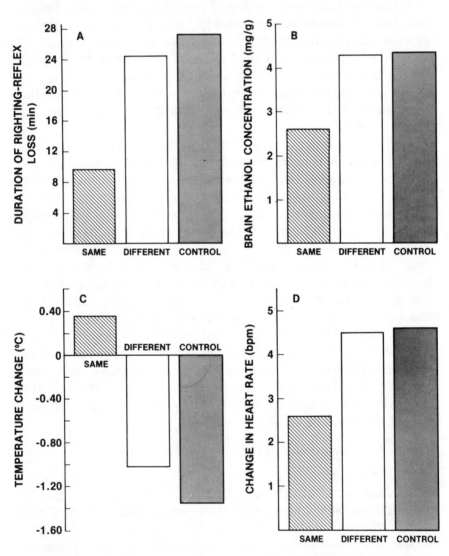

Fig. 2. Situational specificity of tolerance to a variety of effects of ethanol: righting reflex loss (A), brain ethanol concentration (B), hypothermia (C), and tachycardia (D). These results are based on Melchior and Tabakoff (1981), Melchior and Tabakoff (1985), Crowell et al. (1981), and Dafters and Anderson (1982), respectively.

Melchior and Tabakoff (1981) reported that tolerance to the sedating effect of ethanol in mice (as determined by the duration of the drug-induced righting reflex loss) was more pronounced in Same-Tested than Different-Tested subjects, with the effect of the drug in Different-Tested subjects not differing from that seen in Control animals (*see* Fig. 2A). These investigators subsequently evaluated the effects of environmental manipulations on brain ethanol concentrations 30 min after administration of 35 g/kg ethanol in mice (Melchior and Tabakoff, 1985). The results, summarized in Fig. 2B, indicated that Same-Tested mice had significantly lower brain ethanol concentrations than Different-Tested mice, with the concentration of ethanol in the brain of Different-Tested animals not differing significantly from that found in Control animals.

Figure 2C summarizes the results of a study demonstrating that Same-Tested rats are more tolerant to the hypothermic effect of ethanol than Different-Tested rats (Crowell et al., 1981). Figure 2D depicts the results of an experiment, using a within-subject design, evaluating the role of environmental cues on tolerance to the tachycardiac effect of orally ingested ethanol in humans (Dafters and Anderson, 1982). Subjects receiving the drug in the presence of cues previously present when the drug was ingested (Same-Tested) displayed more tolerance than they did when ingesting the drug in the presence of cues not previously present during alcohol consumption (Different-Tested). In the latter case, they displayed a tachycardiac response comparable to the Control level they displayed following their first session. In addition to the results summarized in Fig. 2, a considerable amount of additional data indicates substantial environmental specificity of tolerance to a variety of other effects of ethanol (*see* Siegel, 1987).

In addition to ethanol, environmental specificity of tolerance has also been demonstrated with respect to many other nonopiate drugs: pentobarbital (Cappell et al., 1981; Hinson et al., 1982), scopolamine (Poulos and Hinson, 1984), haloperidol (Poulos and Hinson, 1982), chlordiazepoxide (Cook and Sepinwall, 1975; File, 1982; Greeley and Cappell, 1985), and amphetamine (Poulos et al., 1981b). (However, for an alternative account of the associative basis of amphetamine tolerance, *see* Wolgin's chapter in this volume).

In summary, results of many studies have demonstrated environmental specificity of tolerance. That is, tolerance is more pronounced when the drug is expected than when it is not expected. It should be noted that such environmental specificity is often not absolute (i.e., some tolerance is noted in Different-Tested subjects, compared to Control subjects receiving the drug for the first time). Furthermore, there are contradictory data concerning environmental specificity of tolerance with some drugs. For example, with respect to midazolam (a short-acting benzodiazepine), King et al. (1987) have reported clear environmental specificity of tolerance to the drug's sedative effect, whereas Griffiths and Goudie (1986) reported no environmental specificity of tolerance to the drug's hypothermic effect. However, the general finding, obtained with many drugs, dosages, species, and procedural variations, is that tolerance is more pronounced when the drug is administered in the context of the usual predrug cues than when it is administered in the context of alternative cues (summarized by Siegel, 1983a, 1986, 1987). It would seem that a complete account of drug tolerance must acknowledge the importance of environmental stimuli present at the time of drug administration.

3.1.3. Sensory Preconditioning of Tolerance

On the basis of the conditioning analysis of tolerance, the drug-signaling CS elicits drug-compensatory CRs. Thus, findings that tolerance is maximally displayed only in the presence of drug-associated CSs provide support for the conditioning analysis of tolerance. It is possible, however, to present CSs and UCSs to a subject such that a particular CS (say CS_A) elicits a CR even though there were never any CS_A–UCS pairings. If the CS_A is initially paired with a second CS (CS_B), and CS_B is subsequently paired with a UCS, CRs will be elicited not only by CS_B, but also by a CS_A. The procedure is called "sensory preconditioning" (Mackintosh, 1974). On the basis of the conditioning analysis, it would be expected that tolerance should be subject to sensory preconditioning.

The multiphase design of an experiment demonstrating sensory preconditioning of tolerance is as follows:

1. Two environmental cues (corresponding to CS_A and CS_B) are initially paired with each other
2. A drug is repeatedly administered in the presence of one of these cues (CS_B) and

3. Tolerance is displayed in the presence of both CS_B (the CS paired with the drug), and CS_A (the CS paired with CS_B).

In an experiment using this design (with a tone as CS_A and a light as CS_B), Dafters et al. (1983) demonstrated sensory preconditioning of tolerance to the analgesic effect of morphine.

3.2. Retardation of Tolerance

If tolerance is a manifestation of a conditioning process, it would be expected that manipulations of the putative CS (i.e., environmental cues present at the time of drug administration) known to be effective in retarding the acquisition of CRs would similarly retard the acquisition of tolerance. Two such manipulations have been assessed: partial reinforcement and latent inhibition.

3.2.1. Partial Reinforcement of Tolerance

It has frequently been reported that, if CS-alone presentations are interspersed among paired CS–UCS presentations, the acquisition of CRs is substantially attenuated (Mackintosh, 1974). This procedure of following only a portion of the CSs with the UCS is termed "partial reinforcement."

The implication of the partial reinforcement literature for the conditioning theory of tolerance is clear: a group in which only a portion of the presentations of the drug administration cues are actually followed by the drug (i.e., a partial reinforcement group) should be slower to acquire tolerance than a group that never has exposure to environmental cues signaling the drug without actually receiving the drug (i.e., a continuous reinforcement group), even when the two groups are equated with respect to all pharmacological parameters. Such findings have been reported with respect to tolerance to several effects of morphine (Krank et al., 1984; Siegel, 1977; Siegel, 1978).

The general strategy of partial-reinforcement-of-tolerance studies may be illustrated by examination of the Krank et al. (1984) experiment. In this experiment, the effect of partial reinforcement on tolerance to the analgesic effect of morphine was evaluated. In addition, the design of the experiment permitted simultaneous evaluation of the effect of partial reinforcement on tolerance to another effect of morphine that, unlike analgesia, is not indexed by a behavioral test—reduced weight gain (*see* Mucha et al., 1978; Sanger and McCarthy, 1980).

Groups of rats received 27 injections of morphine (40 mg/kg) spaced over a 108-d period. Rats in one group were transported from their colony room (where they lived) to an alternative distinctive room prior to each morphine injection (Group Morphine-Continuous Reinforcement, MOR-CRF). This distinctive room was the CS. On days when they were not transported to the distinctive room to be injected with morphine, these MOR-CRF rats remained in their home cages in the animal colony room and were injected with saline.

Rats in the second group were injected with morphine in the distinctive room whenever rats in Group MOR-CRF received a morphine injection in the distinctive room. However, rats in this group were also transported to the distinctive room on days intervening between morphine injections. During these intervening days, rats in the group were injected with saline (Group Morphine Partial Reinforcement, MOR-PRF). Thus, the distinctive room cues were always paired with morphine for rats in Group MOR-CRF, but were paired with morphine only 25% of the time for Group MOR-PRF rats. Rats in both groups, however, had the same pharmacological history; that is, they were injected with the same dose of morphine, equally often, and at the same intervals. If tolerance is, in part, mediated by an association between predrug cues and the drug, it would be expected that the acquisition of this association, and therefore the acquisition of tolerance, should be slower in Group MOR-PRF than in Group MOR-CRF.

Two additional groups were included in the design of the Krank et al. (1984) experiment. One group received the CRF, and the other the PRF, schedule of distinctive room-injection pairings, but the injected substance was always saline (Groups SAL-CRF and SAL-PRF).

During the session corresponding to the 18th and 27th injections of morphine to rats in Groups MOR-CRF and MOR-PRF, the analgesia level of all rats was assessed (analgesia tests 1 and 2, respectively). During these analgesia tests (*see* Fig. 3), subjects received their usual injections (i.e., rats in Groups MOR-CRF and MOR-PRF were injected with morphine, and rats in Groups SAL-

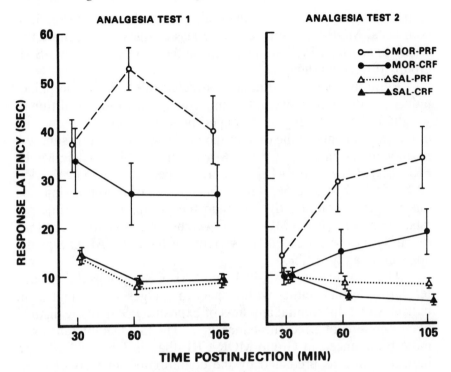

Fig. 3. Mean paw-lick latencies (± 1 SEM) during the two analgesia tests at 30, 60, and 105 min following the injection of morphine in rats
in Groups MOR-PRF and MOR-CRF, and of saline in rats in Groups
SAL-PRF and SAL-CRF [from Krank M., Hinson, R. E., and Siegel S.
(1984) Effect of partial reinforcement on tolerance to morphine-induced
analgesia and weight loss in the rat. *Behav. Neurosci.* **1**, 72–78. Copyright 1984 by the American Psychological Association. Reprinted by per-
mission].

CRF and SAL-PRF were injected with saline), and analgesia was
determined with the "hot-plate" technique. At postinjection intervals of 30, 60, and 105 min, each rat's latency to lick a paw when
placed on a 54°C surface was noted. The effect of morphine on
weight gain was assessed by taking the mean weights of the animals
in the different groups on the day of each analgesia assessment
session.

The effect of partial reinforcement on the development of tolerance to the analgesic effect of morphine can be seen in Fig. 3.
The figure depicts the mean paw-lick latencies for rats in all groups

at the indicated postinjection intervals. As can be seen in Fig. 3, in both tests, MOR-CRF subjects were less analgesic (i.e., more tolerant) than MOR-PRF subjects. In neither test did the two SAL groups differ significantly.

Reinforcement schedule also affected tolerance to morphine-induced reduction in weight gain. The groups did not differ in weight at the beginning of the experiment, but they did by the end of the experiment. At the time of the final tolerance test session, the mean weights, in grams (± 1 SEM), for groups MOR-CRF, MOR-PRF, SAL-CRF, and SAL-PRF were, respectively, 475 (± 6), 454 (± 8), 559 (± 6), and 565 (± 5). The groups of morphine-injected rats weighed significantly less than the groups of saline-injected rats, but rats in Group MOR-CRF weighed significantly more than rats in Group MOR-PRF. The weights of the two SAL groups did not significantly differ.

The fact that the SAL-PRF and SAL-CRF groups did not differ during testing, either with respect to analgesia or weight gain, indicates that differential amounts of exposure *per se* to the distinctive room did not affect these measures. The fact that tolerance was more pronounced in Group MOR-CRF than in Group MOR-PRF indicates that, as predicted by the conditioning analysis of tolerance, partial reinforcement retards the development of tolerance.

3.2.2. Latent Inhibition of Tolerance

Another procedure that, like partial reinforcement, has a deleterious effect on CR formation is preconditioning exposure to the CS. It has been reported that in many conditioning preparations, with both human and a variety of infrahuman subjects, presentations of the CS prior to the start of acquisition serve to decrease the effectiveness of that stimulus when it is subsequently paired with a UCS during conditioning. The deleterious effect of CS preexposure has been termed "latent inhibition." Although there is some controversy concerning the mechanism of latent inhibition (Lubow, 1973), the theoretical interpretation of the phenomenon is irrelevant for its exploitation as a technique to assess the conditioning theory of tolerance. According to this theory, inasmuch as tolerance results from an association between the predrug environmental CS and the pharmacological UCS, the course of tolerance acquisition should be affected by the relative novelty of environmental cues present at the time of drug administration. Thus, on the basis of the conditioning theory of tolerance, organisms with extensive experience with the administration procedure prior to its actual pairing with a drug should be relatively retarded in the acquisition of tolerance, compared with organisms with minimal prior experience with these environmental cues, despite the fact that both groups suffer the systemic effects of the same dose of drug, given the same number of times at the same interval.

Both Siegel (1977) and Tiffany and Baker (1981) have demonstrated such latent inhibition with respect to tolerance to the analgesic effect of morphine. Dyck et al. (1986) have also demonstrated latent inhibition in their study of tolerance to a drug-induced immune response. In this Dyck et al. (1986) experiment, two groups of mice received four injections of the immunomodulatory synthetic polynucleotide, poly I:C, in a distinctive environment. These four injections constituted the conditioning phase of the experiment. Mice in one group (Tolerance Group) never experienced the distinctive environment prior to the conditioning phase. Mice in another group (CS-Preexposure Group) had six prior exposures to this environment, each of which was accompanied by injection of physiological saline rather than the immunostimulatory drug, prior to the conditioning phase. Analyses of Natural Killer cell activity indicated that Tolerance Group subjects displayed tolerance to the immunostimulatory effect of poly I:C, but this tolerance was attenuated in the CS-preexposure Group. These results "may be viewed as an extension of work on morphine and ethanol tolerance to include an immunostimulatory agent" (Dyck et al., 1986, p. 29). Siegel et al. (1987) provide a further discussion of the contribution of classical conditioning to immune system functioning.

3.3. Disruption of Established Tolerance

Findings that nonpharmacological manipulations that retard CR acquisition similarly retard tolerance acquisition support the conditioning account of tolerance. Other nonpharmacological manipulations are known to decrease the strength of well-established CRs; thus, it would be expected that these procedures should similarly decrease the magnitude of tolerance. Three such manipulations have been evaluated: extinction, noncontingent reinforcement, and "external inhibition."

3.3.1. Extinction of Tolerance

Following a number of CS–UCS pairings, sufficient for CR acquisition, presentation of the CS without the UCS causes diminution of
CR strength. The phenomenon is termed "extinction." If drug tolerance is, in part, attributable to conditioning, tolerance should be
subject to extinction. In other words, it would be expected that placebo sessions would attenuate established tolerance. The results of
a number of experiments indicate that tolerance to a variety of effects of morphine is subject to extinction: analgesic (Siegel, 1975b;
1977; Siegel et al., 1980), thermic (Siegel, 1978), locomotor
(Fanselow and German, 1982), and lethal (Siegel et al., 1979).
Such extinction of morphine tolerance can be observed with a variety of routes of drug administration, including administration directly into the ventricles of the brain (MacRae and Siegel, 1987).
Extinction of tolerance has also been demonstrated with respect to
many effects of other drugs: ethanol (*see* Siegel, 1987), amphetamine (Poulos et al., 1981b), the synthetic polynucleotide, poly I:C
(Dyck et al., 1986), and midazolam (a short-acting benzodiazepine) (King et al., 1987).

3.3.2. Explicitly Unpaired Presentations of CS and UCS and Loss of Tolerance

In addition to extinction, another procedure for decreasing the
strength of established CRs is to continue to present *both* the CS
and UCS, but in an unpaired manner (*see* Mackintosh, 1974). That
is, the subject receives both conditioning stimuli, but the CS does
not signal the UCS; rather, the UCS is presented only during intervals between CS presentations. With this procedure (in contrast
with the CS–UCS pairings used to establish CRs), the CS signals a
period of UCS absence. Fanselow and German (1982) demonstrated that this procedure can be used to attenuate tolerance to the
behaviorally sedating effect of morphine in rats. Morphine was administered on a number of occasions in the presence of a distinctive
environmental cue. When tolerance was established, continued
presentation of the drug and cue in an explicitly unpaired manner
eliminated the tolerance. That is (as expected on the basis of a conditioning analysis of tolerance), despite the fact that morphine-
tolerant rats continue to receive morphine, tolerance is reversed if

the continued morphine administrations are unpaired with a cue that was initially paired with the drug.

3.3.3. External Inhibition of Tolerance

Conditional responses, once established, can be disrupted by the presentation of a novel, extraneous stimulus. The phenomenon was termed "external inhibition" by Pavlov (1927), who described its operation in the salivary conditioning situation:

The dog and the experimenter would be isolated in the experimental room, all the conditions remaining for a while constant. Suddenly some disturbing factor would arise—a sound would penetrate the room; some quick change in illumination would occur, the sun going behind a cloud; or a draught would get in underneath the door, and maybe bring some odour with it. If any of these extra stimuli happened to be introduced just at the time of application of the conditioned stimulus, it would inevitably bring about a more or less pronounced weakening or even a complete disappearance of the reflex response depending on the strength of the extra stimulus (Pavlov 1927, p. 44).

On the basis of the conditioning analysis of tolerance, it would be expected that the display of tolerance should be disrupted merely by presentation of a novel stimulus. Results of a study by Siegel and Sdao-Jarvie (1986) supported this prediction with respect to tolerance to the hypothermic effect of ethanol. The rats used in this experiment were implanted with temperature telemetry devices; thus, the time course of drug-induced temperature alterations could be monitored without the necessity of subject handling. One group of rats was injected with ethanol (1 g/kg) for 12 consecutive sessions (the intersession interval was 48 h), and a second group was injected with physiological saline for 12 sessions. The thermic effect of these substances (depicted as temperature changes from preinjection baseline) for the first and twelfth session are shown in the left and middle panel, respectively, of Fig. 4.

As is apparent in the left panel of Fig. 4, the first injection of physiological saline was followed by a transitory increase in temperature (probably as a result of the stress induced by the injection procedure). The first injection of ethanol, however, had a pronounced hypothermic effect.

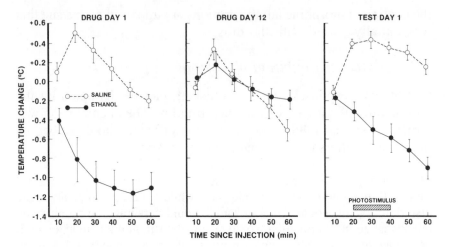

Fig. 4. Mean postinjection temperature change (± 1 SEM) following injection of saline or ethanol on Drug Day 1 (left panel), Drug Day 12
(middle panel), and the immediately subsequent Test Day (right panel).
On the test day, a novel stimulus (provided by a photostimulator) was
presented 20–40 min postinjection. [From Siegel S. and Sdao-Jarvie K.
(1986) Attenuation of ethanol tolerance by a novel stimulus.
Psychopharmacol. **88**, 258–261. Copyright 1986 by Springer-Verlag.
Reprinted by permission].

Tolerance to the hypothermic effect of ethanol was apparent
by Drug Day 12; the drug no longer affected the temperature of the
ethanol-experienced rats. As indicated in the middle panel of Fig.
4, ethanol-injected and saline-injected subjects responded with almost the same pattern of postinjection temperature alteration.

The similar postinjection temperatures seen in the two groups
on Drug Day 12, however, resulted from different processes. This
was revealed by the effects of a novel photostimulus during the test
session, which was the session following Drug Day 12. Each group
was again injected with its usual substance, ethanol or physiological saline, but, starting 20 min postinjection and continuing for 20
min, a bright strobe light flashing at 4 Hz was presented. The temperature changes seen following the injection on this test session
are depicted in the right panel of Fig. 4. As is apparent, the effect of
the photostimulus was markedly different in the two groups: it exacerbated the hyperthermic effect of the saline injection, and reinstated the hypothermic effect of the ethanol. Although the tempera-

ture of rats in the two groups did not differ significantly prior to photostimulator operation (i.e., 10 min postinjection), they diverged immediately upon photostimulator onset. Indeed, the effect of the extraneous stimulus persisted even following its termination: ethanol-injected subjects became progressively more hypothermic following the period of photostimulation application. This observation of the persistent aftereffect of an external-inhibiting stimulus confirms Pavlov's (1927) observation that such a novel stimulus "is effective not only while it lasts, but also for some time after its cessation while its after-effect lasts" (p. 45).

Although the mechanism of external inhibition is unclear, several investigators have suggested that it results because the novel stimulus elicits an orientation reflex that interferes with the expression of the CR (Mackintosh, 1974; Pavlov, 1927). It is possible, however, that the phenomenon is another demonstration of the environmental specificity of tolerance (*see* Poulos et al., 1988). There are obvious similarities between the operations used to demonstrate external inhibition of tolerance with those used to demonstrate environmental specificity of tolerance. In the former case, a subject is presented with a novel stimulus while displaying the hyporesponsivity to the drug characteristic of tolerance, and the drug effect is reinstated (i.e., the display of tolerance is disrupted). In the latter case, the novel stimulus precedes drug administration in the drug-experienced subject, and the expected level of tolerance is not displayed. Thus, in the case of the Siegel and Sdao-Jarvie (1986) experiment, presentation of the strobe may have resulted in a stimulus generalization decrement in the CS that elicited the ethanol-compensatory CR.

Finally, it should be noted that there is an alternative, nonassociative interpretation of Siegel and Sdao-Jarvie's data. Cunningham and colleagues (Cunningham and Bischof, 1987; Peris and Cunningham, 1986, 1987) have reported that some stressors, although eliciting hyperthermia in nondrugged rats, actually increase the hypothermic response to ethanol. That is, stress-induced hyperthermia often does not attenuate ethanol-induced hypothermia; rather, rats receiving ethanol for the first time display a more pronounced hypothermic response if they are stressed in conjunction with ethanol administration. Cunningham and colleagues have demonstrated that a variety of stressors (e.g., some types of handling and strobe presentation), but not all (electric shock is an exception) can augment ethanol-induced hypothermia. The mechanism of this effect is not yet clear, but an endorphinergic interpretation has been presented (Cunningham and Bischof, 1987; Peris and Cunningham, 1986). It is possible, then, that the effect of a strobe in attenuating ethanol tolerance may be the result of the hypothermia-augmenting effect of the stress induced by this novel stimulus, rather than the result of external inhibition of the compensatory CR hypothesized to mediate tolerance.

At present, there is little basis to judge between the conditioning and stress-augmentation interpretations of Siegel and Sdao-Jarvie's (1986) finding that a novel strobe attenuates established ethanol tolerance. Additional research, however, concerning attenuation of morphine tolerance by a pharmacological (rather than environmental) cue (Poulos et al., 1988) suggests that Pavlovian conditioning indeed contributes to the effect of a novel stimulus on established tolerance. This recent work is discussed in section 3.6.3.

3.4. Associative Inhibition of Tolerance

In most Pavlovian conditioning research, the CS is paired with the UCS. However, organisms can learn not only that a CS predicts the presence of the UCS, but also that a CS predicts the *absence* of the UCS. Such associations are termed "inhibitory" to distinguish them from the more commonly studied excitatory associations. (Inhibitory associations should not be confused with the just-described phenomenon of external inhibition). An example of an inhibitory training procedure is one in which the CS signals a long period free of the UCS (Rescorla, 1969). The association between the CS and UCS *absence* is not readily detectable because it does not result in overt CRs. However, the inhibitory association resulting from this "explicitly unpaired" procedure may be seen by subsequently arranging the CS to predict the presence of the UCS (i.e., the CS and UCS are paired). The prior inhibitory training will retard the acquisition of the excitatory association; that is, CRs will be slow to develop. This is a "retardation of acquisition" demonstration of conditional inhibition (Rescorla, 1969).

If morphine tolerance is, in part, attributable to a drug-compensatory CR, it should be subject to inhibitory learning. Such inhibitory learning would be an especially dramatic and counterintuitive demonstration of the contribution of learning to tolerance, as tolerance would be retarded by a procedure involving administration of the drug.

Consider the situation in which the analgesic effect of morphine is tested, in drug-experienced subjects, in the context of a distinctive environmental cue. Subjects that receive pretest cue presentations and morphine administrations in an explicitly unpaired manner (i.e., the cue always signals a long, drug-free period) should be retarded in the acquisition of tolerance when subsequently administered morphine in the presence of the cue. It has, in fact, been reported that such an explicitly unpaired technique of cue and drug presentation *does* result in an inhibitory association between the environmental and pharmacological stimuli, as evidenced by the retarded development of tolerance to the analgesic and behaviorally sedating effects of morphine (Fanslow and German, 1982; Siegel et al., 1981) and the thermic effects of pentobarbital (Hinson and Siegel, 1986). The finding that tolerance acquisition may be retarded by a treatment involving drug administration is not readily interpretable by theories of tolerance that do not acknowledge a role for learning in the development of tolerance.

3.5. Drug Tolerance and Compound Predrug Cues

Although pairing of a CS and UCS will generally promote an association between them (as evidenced by the development of CRs), it is possible to pair the two events without an association developing between them. This situation is seen in "compound conditioning," i.e., conditioning preparations in which at least two CSs simultaneously signal the UCS. It is well established that the effectiveness of any one of the CSs in becoming associated with the UCS depends on the characteristics of the stimuli with which this CS is compounded. Such compound conditioning effect may be seen in phenomena termed "blocking" and "overshadowing." On the basis of the conditioning account, drug tolerance, like other CRs, should be subject to blocking and overshadowing.

3.5.1. Blocking of Tolerance

If a CS has initially been presented such that it is a good predictor of the UCS, it will prevent a second, simultaneously presented CS from becoming associated with the UCS (*see* Kamin, 1968, 1969). For example, if a particular CS (say, CS_A) has been associated with a UCS, and CS_A is subsequently compounded with a second CS (say, CS_B), with this $CS_A + CS_B$ compound still being paired with the UCS, little is learned about CS_B. That is, prior training with one component of a compound stimulus will "block" the subsequent conditioning of a second component; in this example, CS_A blocks CS_B (CS_A is the blocking stimulus, and CS_B is the blocked stimulus). If tolerance is attributable to conditioning, it should be subject to blocking. Consider the case of a subject repeatedly administered a drug in the context of a compound environmental cue that displays tolerance in the presence of this compound CS. This subject may or may not display tolerance in the presence of each of the components of the compound CS, depending on the conditioning history of the alternative component. That is, tolerance should be displayed in the presence of the blocking CS, but not in the presence of the blocked CS. Just this finding has been reported with respect to tolerance to the analgesic effect of morphine (Dafters et al. 1983).

3.5.2. Overshadowing of Tolerance

In the case of blocking, one CS (the blocking CS) is pretrained prior to being compounded with a second CS (the blocked CS). With this procedure, subjects learn little about the added CS. Sometimes, even if there is no prior training of an element of a compound CS, subjects will still learn about only one of the elements. This occurs if one element is more "salient" than the other. Other things being equal, a group trained with a more salient CS will learn more rapidly than a group trained with a less salient CS. For example, if CS_A and CS_B are both effective CSs, but subjects learn a CS_A–UCS association faster than a CS_B–UCS association, CS_A is said to be more salient than CS_B. In this case, the effect of pairing a UCS with a $CS_A + CS_B$ compound will be to strongly associate CA_A with the UCS, with little associative strength developing between CS_B and the UCS; CS_A (the more salient CS) overshadows CS_B (the less salient CS). The overshadowing phenomenon was originally described by Pavlov (1927, pp. 142–143 and 269–270), and has been extensively investigated by Kamin (1968, 1969).

If tolerance is mediated by conditioning, it would be expected that overshadowing should be a feature of tolerance. That is, if subjects become tolerant to a drug consistently administered in the presence of a compound CS, these subjects should display more tolerance in the presence of the more salient component of the CS than they do in the presence of the less salient component. Indeed, these subjects should display less tolerance in the presence of the less salient component than they would if this less salient CS alone had signaled the drug (rather than as a part of a compound CS constructed of components differing in salience). This result, demonstrating the applicability of a compound conditioning phenomenon to tolerance to the analgesic effect of morphine, was reported by Walter and Riccio (1983).

3.5.3. Compound Predrug Cues Normally Signaling a Drug

The study of compound conditioning effects with respect to drug tolerance provides further evidence in support of the conditioning analysis of tolerance. Moreover, it is possible that drug effects are typically signaled by compound stimuli; thus, these compound conditioning effects may be especially important for understanding tolerance.

It is not difficult to specify the multiple CSs that conceivably could accompany drug administration. Although the drug may be made contingent on a single nominal CS [e.g., the tone used in Pavlov's (1927) original demonstration of pharmacological conditioning], there are typically other cues present uniquely at the time of drug administration. For example, handling an animal in conjunction with drug administration and piercing the skin with a hypodermic needle would appear to provide readily detectable signals of an impending drug effect. In fact, it is well established that such injection-ritual cues may become CSs for drugs in the absence of other, explicit CSs (*see* Siegel, 1985a). When there *is* an explicit CS, these injection-ritual cues become components of a compound CS.

3.5.3.1 INJECTION-RITUAL CUES AND THE CONDITIONING MODEL OF TOLERANCE.   Evidence summarized thus far provides substantial proof that Pavlovian conditioning contributes to tolerance. There are, however, ostensibly contrary data. For example, in contrast to previously described results indicating environmental specificity of tolerance, some investigators have reported that tolerance acquired as a result of consistent drug administration in one environment is sometimes fully displayed in a very different environment (e.g., Jørgensen and Hole, 1984; Kesner and Cook, 1983). These investigators suggested that the existence of cross-situational tolerance indicates that some tolerance is not associative. This is of course quite possible (Siegel, 1983a); however, it is also possible that a failure to demonstrate environmentally specific tolerance may result from overshadowing. That is, both drug-administration environment and injection-ritual cues are paired with the drug. Under some circumstances, the latter cues might be much more salient than the former cues; thus, the injection-ritual cues are paired with the drug. Under some circumstances, the latter cues might be much more salient than the former cues; thus, the injection-ritual cues will overshadow other, simultaneously present, environmental cues. Recently, Dafters and Bach (1985) have argued that such circumstances have prevailed in studies that have failed to demonstrate environmental specificity of tolerance. If the injection-ritual cues had overshadowed other cues, it would be expected that subjects made tolerant to morphine in one environment would display this tolerance in a different environment, because, despite the environmental alteration, the *effective* signal for the drug would have been unaltered. This effective signal is comprised of injection-ritual cues (i.e., picking up the rat and injecting it), which are similar in the test environment and in the environment in which tolerance has been acquired.

In an experiment designed to evaluate this overshadowing interpretation of instances of apparent environment-independent tolerance, Dafters and Bach (1985) reduced the salience of the injection-ritual cues. They used a latent inhibition procedure to decrease the effectiveness of this putative CS, (i.e., they repeatedly injected rats with an inert substance prior to administration of morphine). This would be expected to reduce the ability of injection-ritual cues to overshadow other environmental cues present at the time of drug administration. Dafters and Bach (1985) found that predrug exposure to the injection procedure enhanced the environmental-specificity of tolerance to the analgesic effect of morphine; indeed, in conditions in which no attempt was made to decrease the salience of the injection-ritual cues, no evidence of tolerance environmental-specificity was obtained. Although previously such transsituational tolerance has been interpreted as evidence contrary to an associative account of tolerance (e.g., Kesner and Cook, 1983), it is likely that it is attributable to overshadowing.

3.5.3.2. INTEROCEPTIVE DRUG CUES AND THE CONDITIONING MODEL OF TOLERANCE.

A second source of potential "unauthorized" CSs for a drug is provided by the interoceptive effects of the drug itself. That is, the early effect of a drug inevitably signals a later effect; thus, the maximal effects of a drug are announced by a compound CS consisting of interoceptive components, as well as exteroceptive components.

It has, in fact, been demonstrated that a drug can serve as a cue for itself, and this association may contribute to tolerance (Greeley et al., 1984). In this study, rats in one group (Paired) consistently received a low dose of ethanol (0.8 g/kg) 60 min prior to a high dose (2.5 g/kg). Another group of rats (Unpaired) received the low and high doses on an unpaired basis. When tested for the tolerance to the hypothermic effect of ethanol, Paired subjects, but not Unpaired subjects, displayed tolerance. Moreover, if the high dose of ethanol was *not* preceded by the low dose, Paired rats failed to display their usual tolerance. This tolerance, dependent on an ethanol–ethanol pairing, was apparently mediated by an ethanol-compensatory thermic CR; Paired rats, but not Unpaired rats, evidenced a hyperthermic CR (opposite to the hypothermic effect of the drug) in response to the low dose of ethanol. Moreover, the tolerance seen in Paired rats, in common with tolerance resulting from environment–drug pairings, was subject to extinction; repeated presentation of the low dose *not* followed by the high dose led to diminution of tolerance established in Paired rats.

Results of this Greeley et al. (1984) study provide convincing evidence that a small dose of a drug can serve as a signal for a larger dose of that same drug. Because a gradual increase in systemic concentration is an inevitable consequence of most drug administration procedures, such drug–drug associations may play a hitherto unappreciated role in the effects of repeated drug administrations. For example, Walter and Riccio (1983) suggested that interoceptive stimuli produced by a drug may sometimes be more salient than external drug signals, and thus may overshadow environmental cues present at the time of drug administration. Therefore, transsituational tolerance, rather than having a fundamentally different mechanism than tolerance attributable to conditioning (e.g., Goudie and Griffiths, 1984; Kesner and Cook, 1983), may be the result of relatively greater effectiveness of the interoceptive-pharmacological component of the predrug compound CS. Such differential effectiveness may arise because of procedural features of the drug administration procedure that promote overshadowing or blocking:

Tolerance controlled by internal, morphine-produced stimuli, unlike that mediated by environmental stimuli, would be expected to be relatively transsituational or 'pharmacological' in nature, even though the same underlying conditioning mechanisms would be involved. The question then becomes one of establishing the extent to which tolerance, in any given case, is controlled by one or the other of these two general classes of stimuli, rather than one of making a distinction between two different 'kinds' of tolerance'' (Walter and Riccio, 1983, p. 661).

3.5.4. Summary of Compound Conditioning Effects in Drug Tolerance

Because compound conditioning effects (blocking and overshadowing) are demonstrable in drug tolerance, and because predrug cues are typically compound cues, it is likely that many tolerance phenomena are best understood with an appreciation of compound conditioning effects. This raises the issue, of course, as to when these compound conditioning effects will be relevant. For example, evidence has previously been summarized indicating that there is substantial environmental specificity of tolerance. Evidence has also been summarized indicating that "unauthorized stimuli" (e.g., injection ritual and/or interoceptive drug cues) may overshadow simultaneously present environmental stimuli, providing a basis for cross-environmental tolerance. Why isn't such overshadowing a universal feature of drug tolerance, inasmuch as unauthorized stimuli are almost invariably part of the predrug compound signal?

The question is really part of the larger question concerning the features of the Pavlovian conditioning situation likely to promote overshadowing and blocking. The issue is of great interest to learning theorists, but a full discussion of the topic is beyond the scope of this chapter. It should be noted, however, that certain CSs are more likely than other CSs to be associated with a specific UCS (such as a drug) because of the biological predispositions of the organism forming the association (Domjan, 1983), and a role for such biologically selective associations in pharmacological learning has been discussed elsewhere (Krank, 1985). Also, there is an extensive and successful theoretical treatment of compound conditioning that addresses other conditions likely to promote overshadowing and blocking (Rescorla and Wagner, 1972). On the basis of this analysis, issues such as the amount of training, the relevant saliences of the various elements of the compound CS, and the organism's associative history must be considered in determining whether overshadowing or blocking are relevant when a drug is repeatedly administered.

3.6. Pharmacological Cues for Drugs

The previously summarized findings of Greeley et al. (1984) suggest that a drug can serve as a cue for itself, and thus, pharmacological cues, as well as environmental cues, may contribute to tolerance. There are other findings indicating that pharmacological cues, like environmental cues, can be associated with drugs and make important contributions to tolerance.

3.6.1. Drug–Drug Associations

There is abundant evidence that associations readily form between sequentially occurring drug states. For example, lithium chloride has pronounced hypothermic effects, and will induce a conditioned flavor aversion (i.e., rats will learn to avoid a flavor paired with the drug). Pentobarbital, as well, will induce hypothermia and a flavor aversion, but (over a wide dosage range) to a lesser extent than lithium. By pairing pentobarbital with lithium, pentobarbital loses its ability to induce a conditioned taste aversion (*see* Revusky, 1985), and its hypothermic effect becomes attenuated (Taukulis, 1982). It appears that pentobarbital–lithium pairings result in pentobarbital-elicited, lithium-compensatory CRs. (Lett, 1983; Taukulis, 1982).

Similarly, Taukulis (1986a) demonstrated that atropine sulfate elicits a compensatory CR of hyperthermia when paired with either of the hypothermia-inducing drugs, chlorpromazine or ethanol. Interestingly, the effective CS involves the central effects of the anticholinergic drug. A quaternary analogue of atropine sulfate, atropine methyl nitrate, was ineffective as a CS for either chlorpromazine or ethanol.

Taukulis (1986b) further demonstrated that atropine sulfate is an effective signal for an anaesthetic dose (40 mg/kg) of pentobarbital. Rats receiving atropine–pentobarbital pairings developed a hyperthermic CR (opposite to the hypothermic effect of pentobarbital) in response to atropine. Simultaneous monitoring of activity suggested that the hyperthermic CR was not secondary to any atropine-elicited conditioned hyperactivity. Moreover, the hypothermic effect of pentobarbital was attenuated when it was preceded by atropine, suggesting that such thermic tolerance resulted from an association between the anticholinergic drug and the barbiturate.

3.6.2. State-Dependent Learning of Tolerance

Results reported by Taukulis (1986b) indicate that pharmacological cues, as well as environmental cues, may contribute to tolerance. The contribution of pharmacological cues to tolerance has been further demonstrated in a recent study by Siegel (1988). This experiment was inspired by reports by Terman and colleagues (Terman et al., 1983, 1985) that rats, made tolerant to the analgesic effect of morphine, failed to display this tolerance if they were pretreated with pentobarbital. Such a finding may represent an instance of "state-dependent" learning.

There is a considerable amount of evidence that drug states in general, and the state generated by barbiturates in particular, can serve as salient stimuli (*see* Järbe, 1986; Overton, 1984). That is, learned responses acquired when the subject is *not* under the influence of a centrally acting drug, such as pentobarbital, may fail to be displayed when the subject is subsequently tested while under the influence of this drug. To the extent that tolerance to the analgesic effect of morphine is mediated by learning, it might be expected that tolerance will display such drug state dependency; pharmacological cues, such as those generated by pentobarbital, may function very much like environmental cues in affecting the display of tolerance. In other words, just as there is environmental specificity of morphine tolerance (because of associations between morphine-signaling environmental cues and the opiate), there might also be state specificity of morphine tolerance (because of associations between morphine-signaling pharmacological cues and the opiate).

Siegel (1988) confirmed the finding that pentobarbital interferes with the expression of morphine tolerance in rats that had not previously received barbiturate–opiate pairings. Additionally, the results supported the state-dependency interpretation of this interference.

This Siegel (1988) experiment consisted of six daily sessions—five tolerance training sessions followed by a tolerance test session. During each session, rats received two injections, with a 15-min interval between injections. The CS drug was administered during the first of the two injections, and the UCS drug was administered during the second. Depending on the subject's group assignment and phase of the experiment, the CS drug was either pentobarbital or physiological saline, and the UCS drug was either morphine or saline. Subjects were assigned to one of the eight cells of a $2 \times 2 \times 2$ factorial design. The groups differed with respect to the CS drug used during the training sessions (pentobarbital or saline), the UCS drug used during training sessions (morphine or saline), and the CS drug used for the test session (pentobarbital or saline). For all rats, the UCS drug used for the test session was morphine. Following the test-session morphine administration, the level of analgesia was assessed in all subjects.

Group abbreviations indicate the CS and UCS drugs used during training and testing. For example, subjects in Group Pent-MOR/Pent-MOR received pentobarbital prior to each morphine injection during training, and were also tested for the analgesic effect of morphine subsequent to pentobarbital administration. Similarly, subjects in Group Sal-SAL/Pent-MOR received physiological saline as both CS and UCS drugs during training, and received pentobarbital prior to their first morphine administration on the test session. Thirty min following the morphine administration on the test session, analgesia level was assessed in all subjects. Their tails were immersed in a 50°C water bath, and the latency to flick their tails out of the water was measured.

The results of the experiment are shown in Fig. 5. The mean response latencies (± 1 SEM) are displayed separately for groups tested with pentobarbital as the CS drug (left panel) and saline as the CS drug (right panel). Figure 5 depicts the findings obtained when the dose of pentobarbital was 20 mg/kg, although the same pattern of results was obtained with a 55 mg/kg dose of the barbiturate (*see* Siegel, 1988). As can be seen in the left panel of Fig. 5, there was no evidence of tolerance in the morphine-experienced group that received pentobarbital prior to the final morphine administration if pretest administrations were not signaled by the barbiturate (i.e., subjects in Groups Sal-MOR/Pent-MOR did not display shorter response latencies than subjects in Group Sal-SAL/Pent-MOR, despite the fact that subjects in this latter group received morphine for the first time on the test session). However, as can further be seen in the left panel of Fig. 5, signaling the test administration of morphine with pentobarbital does not inevitably interfere with morphine tolerance. Morphine-experienced subjects tested *and* trained with pentobarbital as the CS drug (Group Pent-MOR/Pent-MOR) displayed analgesic tolerance following the test administration of the opiate, i.e., they evidenced shorter response latencies than did subjects with the same exposure to the barbiturate that received morphine for the first time on the test session (Group Pent-SAL/Pent-MOR).

These results depicted in the left panel of Fig. 5 confirm and extend the findings of Terman et al. (1983, 1985). That is, rats with a history of morphine administration do not display the expected tolerance to a test injection of the opiate if it is preceded, for the first time, by pentobarbital. Results obtained from additional groups suggest that the effect of pentobarbital on morphine tolerance is a result of state-dependent learning. As would be expected from a state-dependency analysis, rats that had pretest administrations of morphine (as well as the test administration) signaled by pentobarbital, displayed substantial tolerance to the analgesic effect of morphine (i.e., subjects in Group Pent-MOR/Pent-MOR evidenced shorter response latencies than subjects in Group Pent-SAL/

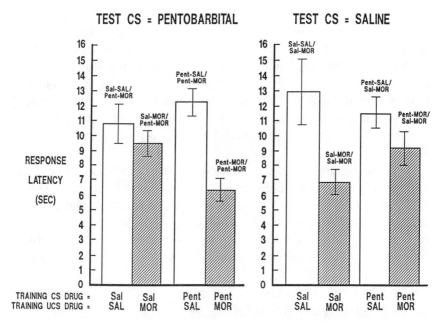

Fig. 5. Mean tail-flick latencies (± 1 SEM) following morphine administration on the tolerance test session. Subjects received 20 mg/kg pentobarbital (left panel) or physiological saline (right panel) as a CS drug prior to the test-session administration of morphine, and received the indicated combinations of CS and UCS drugs during pretest training sessions. (from Siegel S. State-dependent learning and morphine tolerance. *Behav. Neurosci.* (1988). Copyright 1988 by the American Psychological Association. Reprinted by permission.)

Pent-MOR). Although this finding is readily explicable by the state-dependency interpretation, there is an alternative explanation. It is possible that the relevant effect of pretest exposure to pentobarbital in Group Pent-MOR/Pent-MOR was to establish tolerance to pentobarbital, rather than to associate pentobarbital with morphine. Such barbiturate tolerance, rather than state-dependent learning, could conceivably account for the ineffectiveness of pentobarbital in interfering with morphine tolerance. This alternative, nonassociative interpretation in unlikely, however, considering the results obtained with subjects tested with physiological saline as the CS (*see* right panel of Fig. 5).

Examination of the right panel of Fig. 5 indicates that saline-tested subjects, like pentobarbital-tested subjects, displayed an interaction between the magnitude of tolerance and the similarity of the training CS and test CS. Greater tolerance is displayed by subjects trained with saline (Group Sal-MOR/Sal-MOR compared to Group Sal-SAL/Sal-MOR) than by subjects trained with pentobarbital (Group Pent-MOR/Sal-MOR compared to Group Pent-SAL/Sal-MOR). That is, for saline-tested subjects trained with pentobarbital as the CS drug, the *omission* of the barbiturate cue on the test session attenuated the expression of morphine tolerance.

These results further indicate that pharmacological cues, as well as environmental cues, may come to control tolerance. Furthermore, not only may a drug serve as a cue for itself and control tolerance (as demonstrated by Greeley et al., 1984), but also one type of drug (e.g., a barbiturate) may serve as a cue for another type of drug (e.g., an opiate) and similarly control tolerance.

3.6.3 External Inhibition of Tolerance by Pharmacological Stimulation

State-dependent learning of tolerance demonstrates that a particular drug (the UCS drug) can become associated with both the pharmacological state generated by prior drug (the CS drug), or the absence of such a state. Tolerance to the UCS drug is disrupted if the state signaling the drug is altered from that prevailing during prior drug administrations, just as tolerance is disrupted if environmental stimuli signaling the drug are altered. As discussed previously (Section 3.3.3), there is evidence that tolerance is also disrupted if there is a pronounced alteration in background environmental stimuli *following* drug administration (Siegel and Sdao-Jarvie, 1986). Recently, Poulos et al. (1988) demonstrated that such external inhibition of tolerance can be accomplished with pharmacological stimulation.

During the tolerance acquisition phase of the Poulos et al. (1988) experiment, two groups of rats were repeatedly injected with morphine (5 mg/kg), and became tolerant to the drug's analgesic effect. One of these groups was additionally injected with ethanol (1.2 g/kg) 15 min after each morphine injection. Following tolerance acquisition, all rats were tested for morphine analgesic tolerance with a novel state being introduced following morphine administration, i.e., they experienced either the novel introduction, or the novel omission, of the alcohol cue. Both novel states attenuated tolerance.

It would appear that, just as there is external inhibition of ethanol tolerance by a novel environmental cue (Siegel and Sdao-Jarvie, 1986), there is also external inhibition of morphine tolerance by a novel pharmacological state. As outlined previously, Cunningham and colleagues (Cunningham and Bischof, 1987; Peris and Cunningham, 1986, 1987) have provided a nonassociative interpretation of external inhibition of tolerance by a novel environmental cue. The Poulos et al. (1988) results are not readily explicable by the mechanism suggested by Cunningham and colleagues.

Cunningham et al. suggested that, inasmuch as stress exacerbates the hypothermic effect of ethanol in rats receiving ethanol for the first time, the presentation of a novel stimulus (such as the strobe used by Siegel and Sdao-Jarvie, 1986) exaggerates ethanol-hypothermia because it stresses the animal, rather than because it disrupts the expression of the ethanol-compensatory CR. Results reported by Poulos et al. (1988) do not support this alternative interpretation with respect to state-induced external inhibition of tolerance to the analgesic effect of morphine.

In the Poulos et al. (1988) experiment, there was no evidence that post-morphine ethanol augmented the analgesic effect of the opioid in rats receiving the drugs for the first time. Nevertheless, ethanol was an effective external inhibitor of morphine-analgesic tolerance. Indeed, in the case of rats receiving ethanol following each tolerance-acquisition morphine administration, the *absence* of the usual ethanol disrupted tolerance. It would seem that the phenomenon of external inhibition of tolerance is not dependent on the use of an external inhibitor that augments the effect of the drug, or that is stressful.

4. Practical Implications of the Conditioning Analysis of Tolerance

It is well established that drug-predictive signals can come to elicit pharmacological CRs. As indicated, these CRs are often opposite in direction to the drug effect, and can summate with the pharmaco-

logical UCR (yielding tolerance). There are several practical implications of the conditioning approach to tolerance.

4.1. Cross-Drug Effects

If an organism is presented with a signal for a particular drug (say,
Drug A) but actually administered an alternative drug (say, Drug
B), the observed effect of Drug B would consist of an interaction
between the Drug A CR and the Drug B UCR. When both drugs
have similar actions, and the Drug A CR is compensatory to the
drug effect, the response to Drug B will be attenuated in the presence of Drug A cues. That is, cross-tolerance will occur. Such a
role for Pavlovian conditioning in cross-tolerance has been established (Cappell et al., 1981; Lê et al., 1987; Melchior and
Tabakoff, 1985).

When Drug A and B have different effects, various results of
the CR–UCR interaction are possible. As already indicated, if the
effect of Drug B is opposite that of Drug A, the compensatory CR
to Drug A will augment the effect of Drug B (e.g., the exaggerated
convulsive effect of cocaine in rats administered the stimulant in
the presence of barbituate-associated cues [Hinson et al., 1982]). It
has been suggested that such "collisions" between CR and UCR
could be responsible for some pathological responses to drugs. For
example, Bykov (1959, pp. 82–83) reported an experiment by
Levitan in which a CS was paired with acetylcholine in a dog.
When this CS was subsequently followed by adrenalin (rather than
acetylcholine), "an astonishing result" (Bykov, 1959, p. 83) was
noted in the dog's electrocardiogram: The adrenalin induced paroxysmal tachycardia and extrasystoles. Bykov (1959) indicated that,
at the dosage used, the adrenalin did not lead to such disturbed cardiac activity if its administration was not signaled by acetylcholine-
associated cues, and he suggested that these results may be relevant
to some instances of pathological cardiac activity in humans. Appreciation of the fact that the effect of a drug is likely to be modulated by drug-anticipatory responses may provide a basis for understanding some puzzling, idiosyncratic responses to drugs.

The previously described findings of Siegel (1988), demonstrating state-dependent learning of tolerance, may be relevant to
some apparent drug interactions. Recall that pentobarbital can act

as a pharmacological signal for morphine; pentobarbital-blockage of morphine tolerance in morphine-experienced, but pentobarbital-inexperienced, subjects results from barbiturate-induced alteration in predrug cues. Some previous interpretations of pentobarbital-blockage of morphine tolerance have postulated direct pharmacodynamic interactions between the barbiturate and the opiate (*see* Pontani et al., 1985). It appears, however, that state-dependent learning, rather than pharmacodynamic interaction, best accounts for such barbiturate-opiate effects (*see* Siegel, 1988).

4.2 Control of Tolerance

In the clinical setting, pharmacological treatment regimens are often complicated by the rapid development of tolerance. As already discussed, procedures have been described, inspired by the conditioning model, that have been effective in preventing (e.g., Siegel, 1978), retarding (e.g., Siegel, 1977), reversing (e.g., Siegel, 1975b), and actively inhibiting (e.g., Fanslow and German, 1982) drug tolerance. Specifically, it has been suggested that, when a drug must be repeatedly administered, tolerance can be minimized by using a partial reinforcement procedure (Krank et al., 1984), or by using a procedure of explicitly unpaired CS and UCS presentations (Fanslow and German, 1982).

Linnoila et al. (1986) suggested that the Pavlovian conditioning analysis of tolerance may be relevant to understanding some issues related to the ethanol tolerance and drunk driving. Inasmuch as ethanol tolerance can be disrupted by the presentation of an environmental stimulus not previously associated with the drug (e.g., Siegel and Sdao-Jarvie, 1986), "it is possible that subjects who appear tolerant to ethanol's effects in a bar or party setting lose that tolerance when they move to a setting not normally associated with ethanol, such as a car" (Linnoila et al., 1986, p. 40).

4.3. Opiate Overdose

Opiates induce respiratory depression, and if a sufficiently high dose of opiate is administered, death may result. Repeated opiate administrations generally induce tolerance, such that the drug-experienced individual is capable of surviving the respiratory-depressive effects of a very large dose—a dose that would be fatal to the drug-inexperienced individual (Hug, 1972). Despite the fact that extensive tolerance develops to the lethal effect of opiates, about 1% of the American population of heroin addicts dies each year, mostly from so-called overdose (Maurer and Vogel, 1973, p. 101). In urban areas with substantial numbers of addicts, "overdose" is among the leading causes of death in young people aged 15–35 (Abelson, 1970; Helpern, 1972). Although postmortem examination of these victims reveals many symptoms, findings of pulmonary edema are routine (Jaffe and Martin, 1985). There are various interpretations of this edema, but it has usually been attributed to the effects of hypoxia resulting from drug-induced respiratory depression (*see* Duberstein and Kaufman, 1971).

Although "overdose" death is a major public health problem, its mechanisms are unclear. Although some of the fatalities undoubtedly result from pharmacologic overdose (e.g., Huber, 1974), there is evidence that many victims died after a dose that would not be expected to prove fatal in these drug-experienced and presumably drug-tolerant individuals (*see* Brecher, 1972, p. 101–114; Harvey, 1981; Reed, 1980); indeed, they sometimes die following self-administration of a heroin dose that was well-tolerated the previous day (Government of Canada, 1973, pp. 310–315). For some time, it has been recognized that these deaths are frequently not precipitated by a pharmacological "overdose," as the term is usually used. Rather, the deaths have been characterized as "an idiosyncratic reaction to an intravenous injection of unspecific material(s) and probably not a true pharmacologic overdose of narcotics" (Cherubin et al., 1972). Other investigators have similarly indicated the enigmatic feature of these deaths: "It remains unclear why a given dose of heroin will cause this reaction at one time and not at others" (Warner, 1969, p. 2277–2278); thus, "the term 'overdose' has served to indicate lack of understanding of the true mechanism of death in fatalities directly related to opiate abuse" (Greene et al., 1974, p. 175).

It is possible that the conditioning model is relevant to some instances of so-called overdose death. According to this model, environmental signals of impending pharmacological stimulation are important for tolerance because such cues enable the organism to

make timely compensatory CRs in anticipation of the unconditioned effects of the drug. The model predicts circumstances in
which a failure of tolerance may occur, and these circumstances
may prevail on occasions of "overdose." Previously described research indicates that such failures of tolerance would be expected if
the drug is administered in the context of environmental cues that
have not, in the past, been associated with the drug and, thus, that
do not elicit a drug-attenuating CR. According to this analysis, the
addict is at risk for "overdose" when the drug is administered in an
environment that, for that addict, had not previously been paired
with the drug.

Results of animal experiments indicate that conditioning contributes to tolerance to the lethal effect of opiates (Siegel et al.,
1979, 1982). Of special relevance to heroin overdose deaths are
findings that heroin-experienced rats are more likely to die following administration of a high dose of heroin if this high dose is administered in the presence of environmental cues other than those
previously associated with the drug (Siegel et al., 1982). The contribution of Pavlovian conditioning to heroin overdose has further
been supported by retrospective reports provided by human overdose survivors (Siegel, 1984). These victims frequently report that
the overdose occurred when the drug was administered in atypical
circumstances.

It is possible that the phenomenon of external inhibition is relevant to some instances of heroin overdose death. As already discussed, established tolerance (in common with other CRs) can be
disrupted by presentation of an extraneous stimulus. It is conceivable that, on occasion, a novel stimulus intrudes into the addict's
usual drug-administration ritual, thus disrupting the compensatory
CR that mediates tolerance. An example of how external inhibition
might be involved in heroin overdose is provided by an addict's description of her experience of a heroin overdose. The respondent
(E.C.) was a heavy user of heroin for three years. She usually self-
administered her first, daily dose of heroin in the bathroom of her
apartment, where she lived with her mother. Typically, E.C. would
awake earlier than her mother, turn on the water in the bathroom
(pretending to take a shower), and self-inject without arousing sus-
picion. However, on the occasion of the overdose, her mother was
already awake when E.C. started her injection ritual, and knocked

loudly on the bathroom door telling E.C. to hurry. When E.C. then injected the heroin, she immediately found that she could not breathe. She was unable to call to her mother for help (her mother eventually broke down the bathroom door and rushed E.C. to the hospital, where she was successfully treated for heroin overdose). Obviously, there are any number of reasons why E.C. may have overdosed on this occasion, but it is possible that the novel, external stimulus (mother knocking on bathroom door) disrupted the drug-compensatory CR usually elicited by drug-associated cues.

Pavlovian conditioning may not only contribute to death in illicit opiate users, but may also be relevant to some instances of death from overdose of licitly used opiates. Siegel and Ellsworth (1986) described a case of a patient receiving morphine for relief of pain from pancreatic cancer. The patient's son, N.E., regularly administered the drug in accordance with the procedures and dosage levels specified by the patient's physician. N.E. was 17 years old when he administered the fatal dose of morphine. Two years later, N.E. was a student in a class in which the Pavlovian conditioning analysis of drug tolerance was discussed. It was only then that N.E. realized the applicability of the model to his father's death, and attempted to reconstruct the circumstances of the event. Many details concerning the overdose are not accessible, and some information was forgotten over the period between the death and N.E.'s insight into the potential role of pharmacological conditioning in the death. Nevertheless, there is reason to suspect that N.E.'s interpretation of the event as another instance of exacerbation of a drug effect by environmental alteration is reasonable.

The patient was being attended at home, and received a morphine injection $4 \times /d$ (at 6-h intervals). The injections had been given for 4 wk. The patient's condition was such that he stayed in his bedroom, which was dimly lit and contained much hospital-type apparatus necessary for the his care. The morphine had always been injected in this environment. For some reason, on the day that the overdose occurred, the patient dragged himself out of the bedroom to the living room. The living room was brightly lit and different in many ways from the bedroom/sickroom. The patient, discovered in the living room by N.E., appeared to be in considerable pain. Inasmuch as it was time for his father's scheduled morphine injection, N.E. injected the drug while his father was in the living room. He had never administered the morphine in this environment before.

N.E. noticed that his father's reaction to this injection was atypical; his pupils became unusually small, and his breathing became very shallow.

Alarmed by his father's reaction to this injection in the living room, N.E. called his father's physician. The physician instructed N.E. to evaluate some indices of his father's status. Based on the information supplied by the son, the physician concluded that his father suffered an overdose of morphine. The father died some hours later.

In view of the patient's condition, there was no postmortem examination to ascertain the potential role of morphine in his death. The evidence that he died from an overdose is based on his reaction to the final morphine administration, and the physician's interpretation of the symptoms as described in his telephone conversation with N.E. at the time of the event.

Of course, there are a variety of possible explanations for this very sick patient's death. However, the symptoms immediately preceding death strongly implicate morphine overdose, and the circumstances of the death are congenial with a Pavlovian conditioning interpretation of this overdose. The patient's shallow breathing and constricted pupils are classic symptoms of opiate overdose (Jaffe and Martin, 1985). The fact that N.E. stated that he was always assiduous in preparing the morphine (including the preparation on the occasion of the apparent overdose) suggests that there was nothing unusual about the drug dosage when his father died.

It may be estimated that N.E.'s father had received over 100 prior morphine injections, each of which occurred in the bedroom. It appears that the injection in the living room, an environment not previously associated with the systemic effects of the drug, elicited an excessive, and ultimately fatal, response. The circumstances of the overdose related by N.E. are very similar to the circumstances favoring heroin overdose, as elucidated by experimental studies with rats (Siegel, et al., 1982) and interviews with drug addicts who were hospitalized for a heroin overdose (Siegel, 1984). That is, such overdoses are especially likely if the drug is administered in the context of cues not previously associated with the drug.

Finally, it should be noted that, although some retrospective reports of opiate overdose in humans support the conditioning analysis (Siegel, 1984; Siegel and Ellsworth, 1986), others do not.

Neumann and Ellis (1986) reported that their overdose-victim respondents typically reported nothing unusual about predrug cues on the occasion of the overdose.

5. Pharmacological Conditioning and Drug Dependence

Pavlovian conditioning processes have been implicated in withdrawal symptoms observed when drug use is terminated. Following a period of abstinence, drug-compensatory CRs, elicited by drug-associated environmental cues, may, in fact, *be* so-called withdrawal symptoms; that is, in many instances drug-withdrawal symptoms may be better characterized as drug-preparation symptoms (Siegel, 1983a; Hinson and Siegel, 1982). In addition, other investigators have suggested that drug-mimicking CRs may promote continued drug use by increasing the effectiveness of drug-related stimuli and the probability of drug-related thoughts (Stewart et al., 1984). The literature on the experimental, clinical, and epidemiological evidence concerning the role of pharmacological conditioning in drug dependence has been summarized elsewhere (Grabowski and O'Brien, 1981; Hinson and Siegel, 1982; Siegel, 1983a; Stewart et al., 1984; Wikler, 1980), as has the treatment implications of this research (Poulos et al., 1981a; Siegel, 1983a).

6. Unresolved Issues and Alternative Formulations

A substantial amount of data has been summarized supporting the compensatory-CR account of tolerance. There are, however, unresolved issues and other ways of interpreting the data.

6.1. Detection of the Drug-Compensatory CR

The well-documented phenomenon of environmental specificity of tolerance is expected on the basis of the conditioning analysis. It would further be expected that such environmental specificity would be correlated with drug-compensatory CRs. That is, not only should tolerance be especially pronounced in the presence of drug-associated cues, but drug-compensatory responding should be displayed when an inert substance is administered in the presence of these cues. Although such findings have been reported (e.g., Mansfield and Cunningham, 1980; Lê et al., 1979; Siegel, 1975b), there are exceptions. That is, some investigators have demonstrated the environmental specificity of tolerance, but have failed to demonstrate the compensatory CR hypothesized to mediate such tolerance (Fanselow and German, 1982; King et al., 1987; LaHoste et al., 1980; Tiffany et al., 1983). On the basis of the compensatory-CR account of tolerance, there are various reasons why the tolerance-mediating CR might not readily be dectable.

6.1.1 Sensitivity of Measure of Compensatory CR

As suggested by King et al. (1987), it is possible that drug-compensatory CRs are simply more difficult to measure than environmental specificity of tolerance. Thus, although many investigators have reported environmental specificity of tolerance to the sedative effect of morphine, a compensatory CR of hyperactivity has been reported by some (Hinson and Siegel, 1983; Mucha et al., 1981; Paletta and Wagner, 1986), but not others (Fanselow and German, 1982). Some measures of hyperactivity may simply be more sensitive than others.

Similarly, although many investigators have reported environmental specificity of tolerance to the analgesic effect of morphine, a compensatory CR of hyperalgesia has been reported by some (Siegel, 1975b; Krank et al., 1981), but not others (e.g., Palletta and Wagner, 1986; Tiffany et al., 1983). The suggestion that some measures of pain sensitivity may be more sensitive than others for revealing a hyperalgesic CR is supported by the results of a recent experiment by Krank (1987). He demonstrated that rats with a history of morphine administration (5 mg/kg) may or may not display a hyperalgesic CR, depending on the analgesia-assessment method. Rats injected with physiological saline in the context of morphine-associated cues displayed no compensatory CR with a tail-flick assessment procedure (latency to lift tail from 48°C water), but did display the hyperalgesic CR with the hot-plate procedure (latency to lick a paw or jump off a 52°C surface).

Krank's (1987) results suggest that the tail-flick technique (which has a very low baseline response latency) is less likely than the hot-plate technique (which has a higher baseline response latency) to exhibit the even further reduction in latency that would be indicitive of hyperalgesia. Indeed, when the baseline tail-flick latency of the rats in Krank's (1987) experiment was increased by injection of a small dose of morphine (1 mg/kg), a compensatory CR of hyperalgesia was seen with the tail-flick procedure as well.

6.1.2. Pharmacological Priming of the Compensatory CR

Results of several experiments have revealed an aspect of pharmacological conditioning that may be relevant to some previously reported inabilities to observe drug-anticipatory responding. It has been suggested that pharmacological CRs may not be readily detectable in some response systems because, in the absence of the drug effect they would normally attenuate, these CRs are counteracted by regulatory, homeostatic influences. A tendency to manifest a CR may result in an observable CR only if the response system is "challenged" (Hinson et al., 1982).

Examples of such challenges are found in the previously discussed findings indicating that a response to a drug can be modulated by the subject's associative history with other drugs. Thus, the exaggerated convulsive effect of cocaine seen when the stimulant is administered in the presence of pentobarbital-associated cues results from a CR (opposite to the behaviorally sedating effect of pentobarbital) summating with the unconditional stimulant effects of cocaine (Hinson et al., 1982). Similarly, the rat anticipating scopolamine (and its polydipsic consequences) is more sensitive to the adipsic effects of phenobarbital (Poulos and Hinson, 1984), and the rat anticipating pentobarbital (and its hypothermic consequences) is more sensitive to the hyperthermic effect of amphetamine (Hinson and Rhijnsburger, 1984).

6.1.3. External Inhibition of the Compensatory CR

As already discussed, the display of conditional responding is disrupted by a novel stimulus. Moreover, the novel stimulation can be provided by a pharmacological stimulus, and the omission of a usual pharmacological stimulus is a novel event that can disrupt tolerance (Poulos et al., 1988). Recall that, in one condition of the Poulos et al. (1988) experiment, rats were given a series of paired injections: morphine followed by alcohol. The usual display of tolerance to the analgesic effect of morphine was disrupted if the ethanol was not administered. That is, omission of a usual drug state following morphine acted as an external inhibitor and disrupted tolerance (presumably by interfering with the expression of the compensatory CR that mediates such tolerance). As discussed by Poulos et al. (1988), these findings may be relevant to some instances of failure to demonstrate compensatory pharmacological CRs:

> In any drug conditioning study, the initial cues produced by the drug administration are part of the stimulus constellation accompanying and predicting drug effects. Since our results show that the omission of the usual alcohol cue reduced morphine tolerance, it is not surprising that the omission of all drug cues in a placebo test occasionally results in a failure to detect compensatory responses. Placebo tests invariably involve the omission of all the usual drug cues and thus would be expected to disrupt already established compensatory responses (Poulos et al., 1988, p 415.)

6.1.4. Drug-Onset Cues and the Compensatory CR

As already discussed, it is possible that external inhibition of tolerance is really another demonstration of the environmental specificity of tolerance. On the basis of this analysis, the failure to display a drug-compensatory CR may be the result of a stimulus generalization decrement between the tolerance acquisition sessions (when a pharmacological cue is present) and CR test session (when the pharmacological cue is absent). Thus, just as pharmacological cues can control the display of tolerance (Greeley et al., 1984; Siegel, 1988; Taukulis, 1986b), the omission of these cues may change the drug-associated environment sufficiently to cause failure of conditioning responding.

Several investigators have suggested that drug onset cues may constitute an important component of the CS that elicits drug-compensatory CRs (King et al., 1987; Mackintosh, 1987; Tiffany et al., 1983; Walter and Riccio, 1983). Compensatory CR testing occurs in the absence of this component; thus, the test CS may be quite different than the training CS.

6.1.5. The Form of Conditional Antagonism of Drug Effects and the Compensatory CR

Another reason why compensatory CRs may not always be detectable is because some pharmacological CRs mediating tolerance, although effective in antagonizing the drug effect, may have no readily observable effect in the absence of the drug (*see* King et al., 1987; Mackintosh, 1987; Siegel and MacRae, 1984b). For example, in experiments concerning the contribution of environmental cues to tolerance to the behaviorally sedating effect of midazolam, King et al. (1987) reported several findings "uniquely predicted by associative models of drug tolerance" (p. 104), but found only modest evidence for a compensatory CR (conditional hyperactivity) in anticipation of the benzodiazepine. King et al. (1987) suggested a mechanism whereby the tolerance they observed was nevertheless mediated by a drug-antagonistic CR. They hypothesized that adrenocorticotropic hormone (ACTH) release was conditioned to drug predictive cues. King et al. (1987) noted that (a) exogenous ACTH administration, although having little demonstrable behavioral effect when administered alone, antagonizes the effects of benzodiazepines (e.g., Vellucci, 1984), and (b) ACTH release is conditionable (e.g., Ader, 1975). Thus, "if a release of ACTH or related peptides constitutes part of the homeostatic response to the systemic injection of midazolam, then conditional release of these peptides could contribute to the present conditional tolerance as well as to the [only] modest and transient hyperactivity. . ." (p. 113).

Similarly, Siegel and MacRae (1984b) suggested that other examples of drug-antagonistic CRs that might account for conditional tolerance, yet not clearly be evidenced in response to predrug cues alone, are conditional alterations in drug metabolism and distribution (*see* Roffman and Lal, 1974; Ritzmann et al., 1985; Melchior and Tabakoff, 1985). It is also possible, as suggested by Mackintosh (1987), that the relevant CR to morphine involves a compensatory blocking of the drug's effect, rather than an overt response antagonistic to the drug.

6.1.6. Summary of Issues Concerning Detection of the Compensatory CR

Not all experiments reporting environmental specificity of tolerance also report compensatory conditional responding. Most of the exceptions concern the hyperalgesic CR that would be expected to develop in parallel with morphine analgesic tolerance. This conditional hyperalgesia is sometimes observed and sometimes not observed.

The exceptions have been characterized as "clearly embarrassing for Siegel's account of tolerance" (Goudie and Griffiths, 1986, p. 193) and indications "that compensatory responses are not integral components of associational tolerance phenomena" (Baker and Tiffany, 1985, p. 95). In fact, failures to demonstrate overt compensatory CRs are merely assertions of the null hypothesis: "Although unequivocal positive results provide rather strong support for the theory, negative results are less crucial than authors such as Baker and Tiffany (1985) have argued" (Mackintosh, 1987, p. 92).

Several reasons for the sometimes elusive nature of this compensatory CR have been presented. These reasons, of course, are post hoc. Why would a particular analgesia assessment technique sometimes be sufficiently sensitive to reveal drug-compensatory responding, whereas sometimes an alternative, more sensitive technique must be used? Why do some compensatory CRs occur in response to predrug cues alone, and others need a "challenge" to be expressed? Although the answers to such questions are not known, they relate to larger questions that have concerned learning theorists for many years: What makes a stimulus an effective signal? Under what conditions will learning be revealed as a change in behavior? These issues are central to our understanding of Pavlovian conditioning in general, and the Pavlovian conditioning analysis of drug tolerance in particular.

Although some have cautioned that failures to demonstrate compensatory CRs do not represent major challenges to the conditioning account of tolerance (e.g., King et al., 1987; Mackintosh, 1987), others have found these exceptions especially compelling. Some investigators have attempted to develop theories that will accommodate evidence that tolerance is clearly affected by drug-associated cues, yet not require a compensatory-CR mechanism for such tolerance. These are "habituation" theories of tolerance.

6.2. Tolerance as Habituation

The term "tolerance" is used to describe the decreasing response to a drug over the course of successive administrations of the drug. The term "habituation" is used to describe the decreasing response to a peripheral stimulus over the course of successive presentations of that stimulus. Despite the obvious similarities in operations and outcomes, these two phenomena have typically been subject to quite different theoretical treatments; indeed, they have usually been studied by members of different disciplines.

An influential model of habituation has been developed by Wagner (1976, 1978). Siegel (1977, 1982b) suggested that this model may be relevant to drug tolerance, and Baker and Tiffany (1985) presented a detailed analysis of the application of Wagner's theory to tolerance. Wagner's theory has been characterized as a "comparator theory" of habituation (Mackintosh, 1987).

6.2.1. The Comparator Theory of Habituation

Detailed discussion of comparator theories in general, or Wagner's theory in particular, is beyond the scope of this chapter. Briefly, these models of habituation hypothesize that incoming stimuli are compared with a representation of these stimuli in memory. To the extent that a stimulus is already represented in memory, it is less effectively processed and less likely to elicit a response. Such representation in memory is termed "priming."

Consider the case of a rat repeatedly presented with a startle stimulus, such as a loud tone burst. The magnitude of the startle response generally habituates, i.e., it decreases over the course of repeated presentations of the auditory stimulus. According to a comparator theory, such habituation results because prior presentations of the tone serve to prime that event in memory. Such priming results because of the recent prior presentation of the stimulus ("self-generated priming"), or because cues previously associated with that stimulus retrieve a representation of the stimulus into memory ("associatively generated priming"). Habituation occurs because of both types of priming. A stimulus presented shortly after prior presentations of that stimulus is not effectively processed (and will elicit an attenuated response) because of self-generated priming; a stimulus presented long after prior presentations of that stimulus may not be effectively processed, if presented in the context of cues previously associated with that stimulus, because of associatively generated priming.

6.2.2. Comparator Theory of Drug Tolerance

As applied to drug tolerance, the theory simply treats the drug as any other stimulus. Thus, both prior drug administrations and drug-associated environmental cues prime the drug effect in memory, causing a decrease in the effectiveness of processing of the pharmacological stimulation.

The habituation and compensatory-CR models make many similar predictions (concerning, for example, environmental specificity of tolerance and extinction of tolerance). These are important theoretical distinctions, however, and Baker and Tiffany (1985), as well as Goudie and colleagues (Goudie and Demellweek, 1986; Goudie and Griffiths, 1984, 1986) have suggested that the habituation model has advantages over the compensatory-CR model. For example, as discussed above, there have been some failures to demonstrate compensatory CRs. Inasmuch as the habituation model of tolerance is silent on the mediating mechanism, a failure to detect this CR (or to detect any other measurable response to drug-predictive stimuli) is not crucial to the model.

The habituation model has shortcomings, as well as putative advantages. It is based on Wagner's (1976, 1978) comparator theory of habituation, and the evidence in support of this theory is problematic (Mackintosh, 1987). In addition, the habituation analysis, in contrast to the compensatory-CR analysis, does not provide a unitary account of tolerance and withdrawal symptoms. Indeed, the habituation analysis provides no account of withdrawal symptoms, despite the fact that they are highly correlated with tolerance (e.g., Goldstein et al., 1974). Furthermore, the habituation analysis does not address drug sensitization or the various forms of drug interactions discussed in this chapter.

In recent years, an account of habituation has been presented that is a considerable elaboration of earlier comparator models

(Wagner, 1981; Whitlow and Wagner, 1984). This "SOP" ("Sometimes Opponent Process") model, when applied to drug tolerance, incorporates aspects of both habituation and compensatory-CR analyses of the phenomenon (Paletta and Wagner, 1986; Schull, 1979).

7. Noncompensatory Pharmacological Conditional Responses

As discussed previously, not all pharmacological CRs are opposite in direction to the drug effect. Some drug CRs mimic (rather than compensate for) the drug effect. For example, the rat with a history of amphetamine injections (and the increase in locomotor activity induced by this drug) displays an amphetamine-like hyperactivity when administered an inert substance (e.g., Tilson and Rech, 1973). Indeed, sometimes with the same drug, some components of the anticipatory response mimic the drug effect, and some components are opposite to the drug effect. Siegel (1975a) found that a placebo injection in rats with a history of insulin injections elicited an insulin-like pattern of motor behavior (decreased activity, convulsions, and nonresponsiveness to peripherally applied stimulation) and, simultaneously, a compensatory *hyper*glycemic response. Lang et al. (1967) reported that, in dogs with a history of anticholinergic drug administration, a placebo elicited both a drug-like mydriatic response and a drug-compensatory hypersalivary response.

As discussed in section 2.3, the issue of the form of the CR has been addressed by an influential analysis of pharmacological conditioning (Eikelboom and Stewart, 1982). On the basis of this analysis, the CR is drug-like (rather than drug-compensatory) for those drug effects resulting from afferent (rather than efferent) stimulation. There are other theoretical interpretations of the determinants of pharmacological CR topography (*see* Mazur and Wagner, 1982).

7.1. Drug-Like CRs and Drug Sensitization

Regardless of the mechanisms by which some drug CRs look like the drug effect, such CRs (in common with compensatory CRs) would be expected to modulate the observed drug effect. In the case of drug-like CRs, the net result of the interaction between the anticipatory and central drug effects is an augmentation of the drug effect over the course of repeated administrations; that is, "reverse tolerance," or "drug sensitization," may (like tolerance) be associative.

Several recent studies have provided evidence for a Pavlovian conditioning model of sensitization. For example, rats with a history of cocaine administration display more evidence of sensitization to the behavioral effects of the drug following administration in the usual drug administration environment than they do following administration in an alternative environment (Hinson and Poulos, 1981; Post et al., 1981). Also, drug sensitization (like drug tolerance) is subject to extinction; that is, presentation of predrug cues, without the drug, to the sensitized subject decreases the magnitude of sensitization (Hinson and Poulos, 1981). Furthermore, a study of cocaine sensitization in the Brattleboro strain of rats further supports the conditioning interpretation of sensitization. Brattleboro rats have a genetic deficiency of vasopressin, a pituitary peptide that is important in learning. Rats deficient in this peptide are (compared to litter mate controls) deficient in the development of cocaine sensitization (Post et al., 1980).

7.2. Drug-Like CRs and Drug Interactions

It seems clear that drug-like CRs contribute to drug sensitization. There is also evidence that such CRs contribute to the increased responsiveness seen to new drugs (i.e., cross-sensitization). Such cross-sensitization is said to occur when a subject, sensitized to an effect of a particular drug, displays hyperresponsitivity to a new drug. That is, just as drug-compensatory CRs mediate cross-tolerance (*see* Section 4.1), drug-like CRs mediate cross-sensitization.

Results of a recent experiment by Stewart and Vezina (1987) provide evidence for the role of conditioning in cross-sensitization by demonstrating that there is environmental specificity to cross-sensitization. In this experiment, rats received repeated administrations of *d*-amphetamine, and displayed sensitization of the locomotor-enhancing effect of the drug. Rats were subsequently administered another locomotor stimulant (1.0 mg/kg morphine intraperitoneally, or 5.0 μg morphine administered directly into the ventral tegmental area of the brain). Cross-sensitization was seen when morphine was administered in the same environment as the amphetamine, but not when morphine was administered in an environment not previously associated with amphetamine. Other evidence implicating conditioning in cross-sensitization has been reported by Bennett and Krank (1985).

8. Conclusions

It is clear that the effects of a drug are importantly modulated by drug-anticipatory responses. In this chapter, this modulation has been attributed to the interaction of the CR with the drug effect.

Findings concerning the contribution of Pavlovian conditioning to the observed effect of a drug parallel Pavlov's (1910) discussion of the significance of his original "psychic secretion" observations, i.e., that digestive responses in anticipation of feeding make a significant contribution to normally observed patterns of digestive functioning. Subsequent research has demonstrated the importance of CRs in the normal and pathological functioning of a variety of physiological systems in many species including humans (e.g., Adám, 1967; Bykov, 1959). This work on the interaction of learning and physiological processes, conducted mostly by Eastern European and Soviet psychologists, is the foundation of a "synthetic physiology," ". . . a science of the course of vital processes in an integral organism *during its various natural relations with the surrounding medium*" (Bykov, 1960, p. 25; emphasis added). It would appear that inasmuch as drug administration is almost invariably predicted by a set of cues (the administration procedure, or ritual), the response of an "integral organism'" to a drug can be best understood as a combination of the direct reflexive effects of the drug as it acts on central receptor sites and the effects conditioned to the drug administration procedure.

It should be emphasized that the analysis of tolerance that stresses the importance of environment-drug associations is not an alternative to traditional interpretations. Rather, the conditioning model is complementary to views of tolerance that do not acknowledge a role for learning. Many such nonassociative analyses of tolerance emphasize the role of drug-elicited homeostatic corrections that restore pharmacologically induced physiological disturbances to normal levels. Several investigators have indicated the potential adaptive advantage if these homeostatic corrections actually antedated the pharmacological insult (e.g., see Siegel et al., 1987; Wikler, 1973). Pavlov was certainly aware of the importance of such anticipatory responding: "It is pretty evident that under natural conditions the normal animal must respond not only to stimuli that which themselves bring immediate benefit or harm, but also to other physical or chemical agencies—waves of light and the like—which in themselves only signal the approach of these stimuli" (Pavlov, 1927, p. 14). Pavlovian conditioning provides a mechanism for such anticipatory responding. On the basis of a conditioning model, the systemic alterations that mediate tolerance occur not only in response to the pharmacological stimulation but also in response to reliable environmental signals of this stimulation.

The role of Pavlovian conditioning in tolerance was recently eloquently stated by Stewart and Vezina (1987):

Environmental stimuli are always involved in the expression of the effects of centrally acting drugs in behaving animals. Conditioned environmental stimuli should not be thought of as creating abnormal, artifactual effects that get in the way of *true* pharmacological effects but rather as stimuli that, through their history of association with exposure of the organism to a drug, now participate in the *selective* expression. . .of the drug effect" (p. 152).

References

Abelson P. (1970) Death from heroin. *Science*, **168**, 1289.

Adám G. (1967) *Interoception and Behavior*. Publishing House of the Hungarian Academy of Sciences, Budapest.

Adams, W. J., Yeh S. Y., Woods L. A., and Mitchell C. L. (1969) Drug-test interaction as a factor in the development of tolerance to the analgesic effect of morphine. *J. Pharmacol. Exp. Ther.* **168**, 251–257.

Ader R. (1975) Conditional adrenocortical steroid elevations in the rat. *J. Comp. Physiol. Psychol.* **90**, 1156–1163.

Advokat C. (1980) Evidence for conditioned tolerance of the tail flick reflex. *Behav. Neural Biol.* **29**, 385–389.

Baker T. B. and Tiffany S. T. (1985) Morphine tolerance as habituation. *Psychol. Rev.* **92**, 78–108.

Bennett D. and Krank M. D. (1985) Conditioning and amphetamine-induced enhancement of tolerance to the sedative effect of morphine. *Canad. Psychol.* **26(2a)**, 28.

Brecher E. M. (1972) *Licit and Illicit Drugs.* Little, Brown, Boston.

Bykov K. M. (1959) *The Cerebral Cortex and the Internal Organs* (Translated by R. Hodes and A. Kilbey), Foreign Languages Publishing House, Moscow.

Bykov K. M. (1960) *Text-Book of Physiology* (Translated by S. Belsky and D. Myshne), Foreign Languages Publishing House, Moscow.

Cappell H., Roach C., and Poulos C. X. (1981) Pavlovian control of cross-tolerance between pentobarbital and ethanol. *Psychopharmacol.* **74**, 54–57.

Cherubin C., McCusker J., Baden M., Kavaler F., and Amsel Z. (1972) The epidemiology of death in narcotic addicts. *Am. J. Epidemiol.* **96**, 11–22.

Collins W. (1966) *The Moonstone.* Penguin Books, Harmondsworth, England (Originally published, 1871).

Cook L. and Sepinwall J. (1975) Behavioral analysis of the effects and mechanisms of action of benzodiazipines, in *Mechanisms of Action of Benzodiazipines* (Costa E. and Greengard P., eds.), pp. 1–28. Raven, New York.

Crowell C. R., Hinson R. E., and Siegel S. (1981) The role of conditional drug responses in tolerance to the hypothermic effect of ethanol. *Psychopharmacol.* **73**, 51–54.

Cunningham, C. L. and Bischof L. L. (1987) Stress and ethanol-induced hypothermia. *Physiol. Behav.* **40**, 377–382.

Dafters R. and Anderson G. (1982) Conditioned tolerance to the tachycardia effect of ethanol in humans. *Psychopharmacol.* **78**, 365–367.

Dafters R. and Bach L. (1985) Absence of environment-specificity in morphine tolerance acquired in nondistinctive environments: Habituation or stimulus overshadowing? *Psychopharmacol.* **87**, 101–106.

Dafters R., Hetherington M., and McCartney H. (1983) Blocking and sensory preconditioning effects in morphine analgesic tolerance: Support for a Pavlovian conditioning model of drug tolerance. *Q. J. Exp. Psychol.* **35B**, 1–11.

Deutsch R. (1974) Conditioned hypoglycemia: A mechanism for saccharin-induced sensitivity to insulin in the rat. *J. Comp. Physiol. Psychol.* **86**, 350–358.

Domjan M. (1983) Biological constraints on instrumental and classical conditioning: Implications for general process theory, in *The Psychology of Learning and Motivation, Volume 7* (Bower G. H., ed) pp. 215–277. Wiley, New York.

Domjan M. and Gillan D. J. (1977) After-effects of lithium-conditioned stimuli on consummatory behavior. *J. Exp. Psychol.: Anim. Behav. Processes*, **3**, 322–334.

Duberstein J. L. and Kaufman D. M. (1971) A clinical study of an epidemic of heroin intoxication and heroin-induced pulmonary edema. *Am. J. Med.* **51**, 704–714.

Dyck D. G., Greenberg A. H., and Osachuk T. A. G. (1986) Tolerance to drug-induced (poly I:C) Natural Killer (NK) cell activation: Congruence with a Pavlovian conditioning model. *J. Exp. Psychol.: Anim. Behav. Processes*, **12**, 25–31.

Eikelboom R. and Stewart J. (1982) Conditioning of drug-induced physiological responses. *Psychol. Rev.* **89**, 507–528.

Fanselow M. S. and German C. (1982) Explicitly unpaired delivery of morphine and the test situation: Extinction and retardation of tolerance to the suppressing effects of morphine on locomotor activity. *Behav. Neural Biol.* **35**, 231–241.

File S. E. (1982) Development and retention of tolerance to the sedative effects of chlordiazepoxide: Role of apparatus cue. *Eur. J. Pharmacol.*, **81**, 637–643.

Goldstein A., Arnow L., and Kalman S. M. (1974). *Principles of Drug Action: The Basis of Pharmacology.* Wiley, New York.

Goudie A. J. and Demellweek C. (1986) Conditioning factors in drug tolerance, in *Behavioral Analysis of Drug Dependence* (Goldberg S. R. and Stolerman, I. eds.), pp. 225–285. Academic, New York.

Goudie A. J. and Griffiths J. W. (1984) Environmental specificity of tolerance. *Trends Neurosci.* **7**, 310–311.

Goudie A. J. and Griffiths, J. W. (1986) Behavioral factors in drug tolerance. *Trends Pharmacol. Sci.* **7**, 192–196.

Government of Canada. (1973) *Final Report of the Commission of Inquiry into the Nonmedical Use of Drugs.* Information Canada, Ottawa.

Grabowski J. and O'Brien C. B. (1981) Conditioning factors in opiate use, in *Advances in Substance Abuse, Volume 2* (Mello N., ed.), pp. 69–121, JAI, Greenwich, CT.

Greeley J. and Cappell H. (1985) Associative control of tolerance to the sedative and hypothermic effects of chlordiazepoxide. *Psychopharmacol.*, **86**, 487–493.

Greeley J., Lê D. A., Poulos C. X., and Cappell H. (1984). Alcohol is an effective cue in the conditional control of tolerance to alcohol. *Psychopharmacol.*, **83**, 159–162.

Greene M. H., Luke J. L., and Dupont R. L. (1974) Opiate "overdose" deaths in the District of Columbia: I. Heroin-related fatalities. *Medical Annals of the District of Columbia*, **43**, 75–181.

Griffiths J. W. and Goudie, A. J. (1986) Analysis of the role of drug-predictive environmental stimuli in tolerance to the hypothermic effect of the benzodiazepine midazolam. *Psychopharmacol.* **90**, 513–521.

Guha D., Dutta S. N., and Pradhan S. N. (1974) Conditioning of gastric secretion by epinephrine in rats. *Proc. Soc. Exp. Biol. Med.* **147**, 817–819.

Harvey J. G. (1981) Drug-related mortality in an inner city area. *Drug Alc. Dep.* **7**, 239–247.

Helpern M. (1972) Fatalities from narcotic addiction in New York City: Incidence, circumstances, and pathologic findings. *Human Pathol.* **3**, 13–21.

Hinson R. E. and Poulos C. X. (1981) Sensitization to the behavioral effects of cocaine: Modification by Pavlovian conditioning. *Pharmacol. Biochem. Behav.* **15**, 559–562.

Hinson R. E. and Rhijnsburger M. (1984) Learning and cross drug effects: Thermic effects of Pentobarbital and amphetamine. *Life Sci.* **34**, 2633–2640.

Hinson, R. E. and Siegel S. (1982) Nonpharmacological bases of drug tolerance and dependence. *Journal of Psychosom. Res.* **26**, 495–503.

Hinson R. E. and Siegel S. (1983) Anticipatory hyperexcitability and tolerance to the narcotizing effect of morphine in the rat. *Behav. Neurosci.* **97**, 759–767.

Hinson R. E. and Siegel S. (1986) Pavlovian inhibitory conditioning and tolerance to pentobarbital-induced hypothermia in rats. *J. Exp. Psychol.: Anim. Behav. Processes.* **12**, 363–370.

Hinson R. E., Poulos C. X., and Cappell H. (1982) Effects of pentobarbital and cocaine in rats expecting pentobarbital. *Pharmacol. Biochem. Behav.* **16**, 661–666.

Huber D. H. (1974) Heroin deaths—Mystery or overdose? *J. Am. Med. Assoc.* **229**, 689–690.

Hug C. C. (1972) Characteristics and theories related to acute and chronic tolerance development in *Chemical and Biological Aspects of Drug Dependence* (Mulé S. J. and Brill H., eds.), pp. 307–358. CRC, Cleveland.

Jaffe J. H. and Martin W. R. (1985) Opioid analgesics and antagonists, in *The Pharmacological Basis of Therapeutics, 7th Edition* (Gilman A. G., Goodman L. S., Rall T. W., and Murad F., eds.), pp. 491–531. Macmillan, New York.

Järbe T. U. C. (1986) State-dependent learning and drug discriminative control of behavior: An overview. *Acta Neurol. Scand.* **74** (Supplement 109). 37–59.

Jørgensen H. A. and Hole K. (1984) Learned tolerance to ethanol in the spinal cord. *Pharmacol. Biochem. Behav.* **20**, 789–792.

Kamin L. J. (1968) "Attention-like" processes in classical conditioning, in *Miami Symposium on the Prediction of Behavior: Aversive Stimulation* (Jones M. R., ed.), pp. 9–31. University of Miami Press, Miami, FL.

Kamin L. J. (1969) Predictability, surprise, attention, and conditioning. in *Punishment and Aversive Behavior* (Campbell B. A. and Church R. M., eds.), pp. 279–296., Appleton-Century-Crofts, New York.

Kavaliers M. and Hirst M. (1986) Environmental specificity of tolerance to morphine-induced analgesia in a terrestrial snail; Generalization of a behavioral model of tolerance. *Pharmacol. Biochem. Behav.* **25**, 1201–1206.

Kesner R. P. and Cook D. G. (1983) Role of habituation and classical conditioning in the development of morphine tolerance. *Behav. Neurosci.* **97**, 4–12.

King D. A., Bouton, M. E., and Musty R. E. (1987) Associative control of tolerance to the sedative effects of a short-acting benzodiazepine. *Behav. Neurosci.* **101**, 104–114.

King J. J., Schiff S. R., and Bridger W. H. (1978) Haloperidol classical conditioning—Paradoxical results. *Soc. Neurosci. Abstracts*, **4**, 495.

Korol B. and McLaughlin L. J. (1976) A homeostatic adaptive response to alpha-methyl-dopa in conscious dogs. *Pavlovian J. Biol. Sci.* **11**, 67–75.

Krank M. D. (1985) Conditioning the immune system: New evidence for the modification of physiological responses by drug-associated cues. *Behav. Brain Sci.* **3**, 405–406.

Krank M. D. (1987) Conditioned hyperalgesia depends on the pain sensitivity measure. *Behav. Neurosci.* **101**, 854–857.

Krank M. D., Hinson R. E., and Siegel S. (1981) Conditional hyperalgesia is elicited by environmental signals of morphine. *Behav. Neural Biol.* **32**, 148–157.

Krank M. D., Hinson R. E., and Siegel S. (1984) The effect of partial reinforcement on tolerance to morphine-induced analgesia and weight loss in the rat. *Behav. Neurosci.*, **98**, 79–85.

La Hoste G. A., Olson R. A., Olson G. A., and Kastin A. J. (1980) Effects of Pavlovian conditioning and MIF-1 on the development of morphine tolerance in rats. *Pharmacol. Biochem. Behav.* **13**, 799–804.

Lang W. J., Ross P., and Glover A. (1967) Conditional responses induced by hypotensive drugs. *Eur. J. Pharmacol.* **2**, 169–174.

Lê A. D., Khanna J. M., and Kalant H. (1987) Role of Pavlovian conditioning in the development of tolerance and cross-tolerance to the hypothermic effect of ethanol and hydralazine. *Psychopharmacol.* **92**, 210–214.

Lê A. D., Poulos C. X., and Cappell H. (1979) Conditioned tolerance to the hypothermia effect of ethyl alcohol. *Science*, **206**, 1109–1110.

Lett B. T. (1983) Pavlovian drug-sickness pairings result in the conditioning of an antisickness response. *Behav. Neurosci.* **97**, 779–784.

Linnoila M., Stapleton J. M., Lister R., Guthrie S., and Eckhardt M. (1986) Effects of alcohol on accident risk. *Pathologist*, **40**, 36–41.

Lubow R. E. (1973) Latent inhibition. *Psychol. Bull.* **79**, 398–407.

Mackintosh N. J. (1974) *The Psychology of Animal Learning*. Academic, London.

Mackintosh N. J. (1987) Neurobiology, psychology and habituation. *Behav. Res. Ther.* **25**, 81–97.

MacRae J. R. and Siegel S. (1987) Extinction of tolerance to the analgesic effect of morphine: Intracerbroventricular administration and effects of stress. *Behav. Neurosci.* **101**, 790–796.

Mansfield J. G. and Cunningham C. L. (1980) Conditioning and extinction of tolerance to the hypothermic effect of ethanol in rats. *J. Comp. Physiol. Psychol.* **94**, 962–969.

Marlatt G. A. and Rohsenow D. J. (1982) Cognitive processes in alcohol use: Expectancy and the balanced placebo design, in *Advances in Substance Abuse: Behavioral and Biological Research* (Mello N. K., ed.) pp. 159–199, JAI, Greenwich, CT.

Maurer D. W. and Vogel V. H. (1973) *Narcotics and Narcotic Addiction*. Charles C. Thomas, Springfield, IL.

Mazur J. E. and Wagner A. R. (1982) An episodic model of associative learning, in *Quantitative Analysis of Behavior: Acquisition (Vol. 3)*

(Commons M. L., Herrnstein R. J., and Wagner A. R., eds.), pp. 3–39, Ballinger, Cambridge, MA.

Melchior C. L. and Tabakoff B. (1981) Modification of environmentally-cued tolerance to ethanol in mice. *J. Pharmacol. Exp. Ther.* **219**, 175–180.

Melchior C. L. and Tabakoff B. (1985) Features of environment-dependent tolerance to ethanol. *Psychopharmacol.* **87**, 94–100.

Mityushov M. I. (1954) Uslovnorleflektornaya inkretsiya insulina [The conditional-reflex incretion of insulin]. *Zhurnal Vysshei Nervnoi Deiatel [Journal of Higher Nervous Activity]*, **4**, 206–212.

Mucha R. F., Kalant H., and Linesman M. A. (1978) Quantitative relationships among measures of morphine tolerance and physical dependence in the rat. *Pharmacol. Biochem. Behav.*, **10**, 397,405.

Mucha R. F., Volkovsiks C., and Kalant H. (1981) Conditioned increases in locomotor activity produced with morphine as an unconditioned stimulus, and the relation of conditioning to acute morphine effect and tolerance. *J. Comp. Physiol. Psychol.* **95**, 351–362.

Mulinos M. G. and Lieb C. C. (1929) Pharmacology of learning. *Am. J. Physiol.* **90**, 456–457.

Neumann J. K. and Ellis A. R. (1986) Some contradictory data concerning a behavioral conceptualization of drug overdose. *Bulletin of the Society of Psychologists in Addictive Behaviors*, **5**, 87–90.

Newlin D. B., Thomson J. B., and Kite M. S. (1986) Opposite drug and expectation effect in the balanced placebo design: A meta-analysis. *Alc.: Clin. Exp. Res.* **10**, 98.

Obàl F. (1966) The fundamentals of the central nervous control of vegetative homeostasis. *Acta Physiol. Acad. Sci. Hung.* **30**, 15–29.

Obàl F., Vicsay M. S., and Benedek G. (1976) On the cybernetic systems of the control of vegetative functions. *Recent Developments of Neurobiology in Hungary*, **5**, 235–267.

Obàl F., Vicsay M., and Mataràsz I. (1965) Role of a central nervous mechanism in the acquired tolerance to the temperature decreasing effect of histamine. *Acta Physiol. Acad. Sci. Hung.* **28**, 65–76.

Overton D. A. (1984) State dependent learning and drug discriminations, in *Handbook of Psychopharmacol. Vol. 18* (Iversen L. L., Iversen, S. D., and Snyder S. H., eds.), pp. 59–127, Plenum, New York.

Paletta M. S. and Wagner A. R. (1986) Development of context-specific tolerance to morphine: Support for a dual-process interpretation. *Behav. Neurosci.* **100**, 611–623.

Pavlov I. P. (1910) *The Work of the Digestive Glands* (W. H. Thompson, trans.), Charles Griffin, London.

Pavlov I. P. (1927) *Conditioned Reflexes* (G. V. Anrep, trans.), Oxford University Press, London.

Peris J. and Cunningham C. L. (1986) Handling-induced enhancement of alcohol's acute physiological effects. *Life Sci.* **38**, 273–279.

Peris J. and Cunningham C. L. (1987) Stress enhances the development of tolerance to the hypothermic effect of ethanol. *Alc. Drug Res.* **7**, 187–193.

Pihl R. O. and Altman J. (1971) An experimental analysis of the placebo effect. *J. Clin. Pharmacol.* **11**, 91–95.

Pontani R. B., Vadlamani N. L., and Misra A. L. (1985) Potentiation of morphine analgesia by subanesthetic doses of pentobarbital. *Pharmacol. Biochem. Behav.* **22**, 395–398.

Post R. M., Contel N. R., and Gold P. W. (1980) Impaired behavioral sensitization to cocaine in vasopressin deficient rats. *Soc. Neurosci. Abstracts*, **6**, 111.

Post R. M., Lockfield A., Squillance K. M., and Contel N. R. (1981) Drug-environment interaction: Context dependency of cocaine-induced behavioral sensitization. *Life Sci.* **28**, 755–760.

Poulos C. X. and Hinson R. E. (1982) Pavlovian conditional tolerance to haloperidol catalepsy: Evidence of dynamic adaptations in the dopaminergic system. *Science*, **218**. 491–492.

Poulos C. X. and Hinson R. E. (1984) A homeostatic model of Pavlovian conditioning: Tolerance to scopolamine-induced adipsia. *J. Exp. Psychol.: Anim. Behav. Processes*, **10**, 75–89.

Poulos C. X., Hinson R. E., and Siegel S. (1981a) The role of Pavlovian processes in drug use: Implications for treatment. *Addit. Behav.* **6**, 205–211.

Poulos C. X., Hunt T., and Cappell H. (1988) Tolerance to morphine analgesia is reduced by the novel addition or omission of an alcohol cue. *Psychopharmacol.* **94**, 412–416.

Poulos C. X., Wilkinson D. A., and Cappell H. (1981b) Homeostatic regulation and Pavlovian conditioning in tolerance to amphetamine-induced anorexia. *J. Comp. Physiol. Psychol.* **95**, 735–746.

Raffa R. B., Porreca F., Cowan A., and Tallarida R. J. (1982) Evidence for the role of conditioning in the development of tolerance to morphine-induced inhibition of gastrointestinal motility in rats. *Fed. Proc.* **41**, 1317.

Razran G. (1961) The observable unconscious and the inferable conscious in current soviet psychophysiology: Interoceptive conditioning, semantic conditioning and the orienting reflex. *Psychol. Rev.* **68**, 81–147.

Reed T. (1980) Challenging some "common wisdom" on drug abuse. *Int. J. Addict.* **15**, 359–373.

Rescorla R. A. (1969) Pavlovian conditioned inhibition. *Psychol. Bull.* **72**, 77–94.

Rescorla R. A. and Wagner A. R. (1972) A theory of Pavlovian conditioning: variations in the effectiveness of reinforcement and non-reinforcement, in *Classical Conditioning II: Current Theory and Research* (Black A. H. and Prokasy W. F., eds.), pp. 64–99, Appleton-Century-Crofts, New York.

Revusky S. (1985) Drug interactions measured through taste aversion procedures with an emphasis on medical implications. *Ann. N.Y. Acad. Sci.* **443**, 250–271.

Ritzmann R. F., Steece K. A., Lee J. M., and DeLeon-Jones F. A. (1985) Neuropeptides differentially affect various forms of morphine tolerance. *Neuropeptides*, **6**, 255–258.

Roffman M. and Lal H. (1974) Stimulus control of hexobarbital narcosis and metabolism in rats. *J. Pharmacol. Exp. Ther.* **191**, 358–369.

Rozin P., Reff D., Mark M., and Schull J. (1984) Conditioned responses in human tolerance to caffeine. *Bull. Psychonomic Soc.* **22**, 117–120.

Russek M. and Piña S. (1962) Conditioning of adrenalin anorexia. *Nature*, **193**, 1296–1297.

Sanger D. J. and McCarthy, P. C. (1980) Differential effects of morphine in food and water intake in food deprived and freely feeding rats. *Psychopharmacol.* **72**, 103–106.

Schull J. (1979) A conditioned opponent theory of Pavlovian conditioning and habituation, in *The Psychology of Learning and Motivation. Vol 13* (Bower G. H., ed.), pp. 57–90, Academic, New York.

Siegel S. (1975a) Conditioning insulin effects. *J. Comp. Physiol. Psychol.* **89**, 189–199.

Siegel S. (1975b) Evidence from rats that morphine tolerance is a learned response. *J. Comp. Physiol. Psychol.* **89**, 498–506.

Siegel S. (1976) Morphine analgesic tolerance: Its situation specificity supports a Pavlovian conditioning model. *Science*, **193**, 323–325.

Siegel S. (1977) Morphine tolerance acquisition as an associative process. *J. Exp. Psychol.: Anim. Behav. Processes*, **3**, 1–13.

Siegel S. (1978) Tolerance to the hyperthermic effect of morphine in the rat is a learned response. *J. Comp. Physiol. Psychol.* **92**, 1137–1149.

Siegel S. (1982a) Opioid expectations modifies opioid effects. *Fed. Proc.* **41**, 2343.

Siegel S. (1982b) Pharmacological habituation and learning, in *Quantitative Analyses of Behavior: Volume III (Acquisition)* (Commons M. L., Herrnstein R., and Wagner A. R., eds.), pp. 195–217, Ballinger, Cambridge, MA.

Siegel S. (1983a) Classical conditioning, drug tolerance, and drug dependence, in *Research Advances in Alcohol and Drug Problems, Volume 7* (Israel Y., Glaser F. B., Kalant H., Popham R. E., Schmidt W., and Smart R. G., eds.), pp. 207–246. Plenum, New York.

Siegel S. (1983b) Wilkie Collins: Victorian novelist as psychopharmacologist. *J. Hist. Med.* **38**, 161—175.

Siegel S. (1984) Pavlovian conditioning and heroin overdose: Reports by overdose victims. *Bull. Psychonomic Soc.* **22**, 428–430.

Siegel S. (1985a) Drug anticipatory responses in animals, in *Placebo: Theory, Research, and Mechanisms* (White L., Tursky B., and Schwartz G., eds.), pp. 288–305. Guilford, New York.

Siegel S. (1985b) Psychopharmacology and the mystery of *The Moonstone. Am. Psychol.* **40**, 580–581.

Siegel S. (1986) Environmental modulation of tolerance: Evidence from benzodiazepine research, in *Tolerance to Beneficial and Adverse Effects of Antiepileptic Drugs* (Frey H. H., Koella W. P., Froscher W., and Meinardi H., eds.), pp. 89–100, Raven, New York.

Siegel S. (1987) Pavlovian conditioning and ethanol tolerance. *Alc. Alcoholism*, **Supplement 1**, 25–36.

Siegel S. (1988) State-dependent learning and morphine tolerance. *Behav. Neurosci.* **102**, 228–232.

Siegel S. and Ellsworth D. (1986) Pavlovian conditioning and death from apparent overdose of medically prescribed morphine: A case report. *Bull Psychonomic Soc.* **24**, 278–280.

Siegel S. and MacRae J. (1984a) Environmental specificity of tolerance. *Trends Neurosci.* **7**, 140–142.

Siegel S. and MacRae J. (1984b) Reply to Goudie and Griffiths. *Trends Neurosci.* **7**, 311.

Siegel S. and Sdao-Jarvie K. (1986) Reversal of ethanol tolerance by a novel stimulus. *Psychopharmacol.* **88**, 258–261.

Siegel S., Hinson R. E., and Krank M. D. (1978) The role of predrug signals in morphine analgesic tolerance: Support for a Pavlovian conditioning model of tolerance. *J. Exp. Psychol.: Anim. Behav. Processes*, **4**, 188–196.

Siegel S., Hinson R. E., and Krank M. D. (1979) Modulation of tolerance to the lethal effect of morphine by extinction. *Behav. Neural Biol.* **25**, 257–262.

Siegel S., Sherman J. E., and Mitchell D. (1980) Extinction of morphine analgesic tolerance. *Learn. Motiv.* **11**, 289–301.

Siegel S., Hinson R. E., and Krank M. D. (1981) Morphine-induced attenuation of morphine tolerance. *Science*, **212**, 1533–1534.

Siegel S., Hinson R. E., Krank M. D., and McCully J. (1982) Heroin "overdose" death: The contribution of drug-associated environmental cues. *Science*, **216**, 436–437.

Siegel S., Krank M. D., and Hinson R. E. (1987) Anticipation of pharmacological and nonpharmacological events: Classical conditioning and addictive behavior. *J. Drug Issues*, **17**, 83–110.

Stewart J. and Vezina P. (1987) Environment-specific enhancement of the hyperactivity induced by systemic or intra-VTA morphine injections in rats preexposed to amphetamine. *Psychobiol.* **15**, 144–153.

Stewart J., de Wit H., and Eikelboom R. (1984) Role of unconditioned and conditioned drug effects in the self-administration of opiates and stimulants. *Psychol. Rev.* **91**, 251–268.

Subkov A. A. and Zilov G. N. (1937) The role of conditioned reflex adaptation in the origin of hyperergic reactions. *Bull. Biol. Med. Exp.* **4**, 294–296.

Taukulis H. K. (1982) Attenuation of pentobarbital-elicited hypothermia in rats with a history of pentobarbital-LiCl pairings. *Pharmacol. Biochem. Behav.* **17**, 695–697.

Taukulis H. K. (1986a) Conditional hyperthermia in response to atropine associated with a hypothermic drug. *Psychopharmacol.* **90**, 327–331.

Taukulis H. K. (1986b) Conditional shifts in thermic responses to sequentially paired drugs and the "conditional hyperactivity" hypothesis. *Pharmacol. Biochem. Behav.* **25**, 83–87.

Terman G. W., Lewis J. W., and Liebskind J. C. (1983) Sodium Pentobarbital blocks morphine tolerance and potentiation in the rat. *Physiologist*, **26**, A-111.

Terman G. W., Pechnick R. N., and Liebskind J. C. (1985) Blockade of tolerance to morphine analgesia by pentobarbital. *Proc. Western Pharmacol. Soc.* **28**, 157–160.

Tiffany S. T. and Baker T. B. (1981) Morphine tolerance in the rat: Congruence with a Pavlovian paradigm. *J. Comp. Physiol. Psychol.* **95**, 747–762.

Tiffany S. T., Petrie E. C., Baker T. B., and Dahl J. (1983) Conditioned morphine tolerance in the rat: Absence of a compensatory response and cross-tolerance with stress. *Behav. Neurosci* **97**, 335–353.

Tilson H. A. and Rech R. H. (1973) Conditioned drug effects and absence of tolerance to *d*-amphetamine induced motor activity. *Pharmacol. Biochem. Behav.* **1**, 149–153.

Vellucci S. V. (1984) Chlordiazepoxide-induced potentiation of hexobarbitone sleeping time is reduced by ACTH 1-24. *Pharmacol. Biochem. Behav.* **21**, 39–41.

Wagner A. R. (1976) Priming in STM: An information processing mechanism for self-generated or retrieval-generated depression in performance, in *Habituation: Perspectives for Child Development, Animal Behavior, and Neurophysiology* (Tighe T. J. and Leaton R. N., eds.), pp. 95–128, Erlbaum, Hillsdale, NJ.

Wagner A. R. (1978) Expectancies and the priming of STM, in *Cognitive Processes in Animal Behavior* (Hulse S. H., Fowler H., and Honig W. K., eds.), pp. 179–209, Erlbaum, Hillsdale, NJ.

Wagner A. R. (1981) SOP: A model of automatic memory processing in animal behavior, in *Information Processing in Animals: Memory Mechanisms* (Spear N. E. and Miller R. R., eds.), pp. 5–47, Erlbaum, Hillsdale, NJ.

Walter T. A. and Riccio D. C. (1983) Overshadowing effects in the stimulus control of morphine analgesic tolerance. *Behav. Neurosci.* **97**, 658–662.

Werner A. (1969) Near-fatal hyperacute reaction to intravenously administered heroin. *J. Am. Med. Assoc.* **207**, 2277–2278.

Whitlow J. W., and Wagner A. R. (1984) Memory and habituation, in *Habituation, Sensitization and Behavior* (Peeke H. V. S. and Petrinovich L., eds.), pp. 103–153, Academic Press, New York.

Wikler A. (1973) Conditioning of successive adaptive response to the initial effects of drugs. *Cond. Reflex*, 8, 193–210.

Wikler A. (1980) *Opioid Dependence and Treatment*. Plenum, New York.

Woods S. C. and Kolkosky P. J. (1976) Classical conditioned changes of blood glucose level. *Psychosom. Med.* **38**, 201–219.

Behavioral and Pharmacological History as Determinants of Tolerance- and Sensitization-Like Phenomena in Drug Action

James E. Barrett, John R. Glowa, and Michael A. Nader

1. Introduction

The behavioral effects of acute drug administration quite often depend on the manner in which that behavior is and has been controlled by its environmental consequences. Experimental evidence for the importance of factors such as the schedule-controlled rate of

responding, together with other behavioral or environmental determinants of drug action, has grown steadily since Dews (1955,
1958) initially demonstrated that the effects of pentobarbital or amphetamine differed depending on the schedule of reinforcement that
maintained responding (Kelleher and Morse, 1968; McKearney and
Barrett, 1978). Since these initial studies, further work has shown
that factors other than those that currently exist, such as behavior
maintained under different stimuli (contextual determinants), as
well as circumstances that have existed previously (historical determinants), can also be of extreme significance in determining the
specific effects of a wide number of acutely administered drugs
(Barrett, 1986, 1987).

In many cases, those variables that control the way behavior is
affected by acute drug administration also appear to determine the
manner in which behavior is altered by chronic drug administration. For example, work with chronically administered amphetamine demonstrated that tolerance to the rate-increasing effects of
this compound occurred only under conditions where those drug-
induced changes in behavior were correlated with a decrease in the
rate of reinforcement (Schuster et al., 1966; see Wolgin, this text).
Schedule-controlled consequences, such as the rate of reinforcement, exert powerful, dynamic control over behavior under
nondrug conditions (Morse, 1966) and can also be extraordinarily
important under circumstances where drugs are administered
chronically. Studies demonstrating such behavioral influences on
the development of tolerance to the repeated administration of a
drug, as well as other findings relevant to the behavioral effects of
chronically administered drugs, have been reviewed by Corfield-
Sumner and Stolerman (1978) and by Goudie and Demellweek
(1986), and are treated extensively within this volume (*see* chapter
by Wolgin). These reviews, as well as the studies summarized below, provide ample evidence that the expression of tolerance to repeated drug administration can be modulated substantially by
behavioral variables such as the schedule of reinforcement and
drug-induced changes in behavioral consequences.

This chapter reviews studies of acute drug administration that
demonstrate that certain behavioral variables can alter drug action
to such an overwhelming extent that the entire dose–response function is changed. As a result of such interventions, doses of a partic-

ular drug may have either a greatly diminished effect (*tolerance-like*) or a greatly enhanced effect (*sensitization-like*) relative to what the behavioral effects of that drug would have been in the absence of such interventions. Such variables are of unquestionable importance in understanding the acute effects of drugs, and may also be important for understanding pharmacological tolerance and sensitization because they may "set" the initial effects of a drug and, as a result, alter the nature and direction of effects obtained when that drug is administered chronically.

The dominant emphasis of the material reviewed in this chapter is on historical variables, both behavioral and pharmacological, as these have been shown to influence the effects a drug has on behavior. With few exceptions, the effects of chronic drug administration, directed explicitly towards the study of behavioral mechanisms of tolerance, will not be treated except to illustrate the general significance of behavioral determinants of drug action. The next section reviews some of those determinants and describes specific experiments demonstrating that many of the same variables that affect the actions of acutely administered drugs also influence the actions of drugs that are administered chronically. Similarities in the determinants of acute and chronic drug effects provide a unifying framework within which to better understand general principles underlying the behavioral effects of drugs.

1.1. General Determinants of the Behavioral Effects of Drugs: Relationship to Tolerance and Sensitization

1.1.1. Schedule of Reinforcement

As mentioned above, the schedule of reinforcement that maintains responding has been shown repeatedly to be of critical importance in determining the acute effects of drugs. The schedule of reinforcement is also important in determining drug effects when a drug is administered chronically. For example, Smith (1986a) examined the effects of the synthetic cannabinoid *l*-nantradol on responding of pigeons maintained under either a fixed-ratio (FR) 100 or 300 schedule of food delivery. Acute doses of this compound

(0.001–0.03 mg/kg) produced similar rate-decreasing effects in all pigeons under both schedules. When the highest dose (0.03 mg/kg) was administered chronically, however, tolerance to the rate-decreasing effects developed within 20 sessions under the FR 100 schedule, but did not develop at all under the FR 300 schedule. However, when the value of the FR 300 schedule was decreased to 100 and response-independent food was given at the start of the first session's exposure to this lowered value, responding was rapidly restored to previously established nondrug levels.

Thus, whether or not tolerance was obtained to chronically administered *l*-nantradol depended on the schedule of reinforcement. Tolerance did not occur when every 300th response was reinforced, but developed rapidly at the lower value. It is also important to note that the loss of reinforcement under the FR 300 schedule, i.e., the decreased frequency of reinforcement that has often been suggested as one mechanism to account for tolerance development (*see* Wolgin, this volume), was not sufficient to influence the development of tolerance. Therefore, a drug-induced reduction in reinforcement frequency is only one of several variables that can influence tolerance development. Chronic drug effects appear to be modulated by schedule factors to a degree that parallels those found with acute drug administration.

1.1.2. Environmental Context

Much of the emphasis in studying acute and chronic drug effects on schedule-controlled behavior has been on the more immediate temporal and intensive aspects of behavior (e.g., ongoing rate of responding in the absence of the drug). As mentioned earlier, however, a number of experiments with acute drug administration have shown that the effects of various drugs can also be modified by the broader context in which behavior is maintained. Thus, the behavioral effects of drugs can depend not only on the specific and immediate rate of responding, but on other factors controlling behavior that are more remote from ongoing behavior. For example, behavior can be maintained under a single schedule of reinforcement or under conditions where one behavioral performance is maintained under one stimulus condition and alternates with a different performance maintained under a different stimulus (e.g., a multiple schedule).

Under the multiple schedule condition, drug effects on a particular behavior may differ substantially from those obtained when that behavior is studied alone. For instance, the characteristic *rate-decreasing* effects of *d*-amphetamine on responding of squirrel monkeys punished by electric shock were changed to *rate increases* when punished responding alternated with another stimulus condition in which responding postponed the delivery of electric shock (McKearney and Barrett, 1975). Similarly, the effects of ethanol on responding of pigeons maintained under a fixed-interval schedule were determined by the value of the fixed-ratio size in the alternate component of a multiple schedule (Barrett and Stanley, 1980). Changes in the effects of ethanol on fixed-interval responding were a function of the size of the fixed-ratio schedule requirement. As the fixed-ratio value was increased the effects of ethanol under the fixed-interval schedule also increased. Significantly, the rate-increasing effects of ethanol on responding under the fixed-interval schedule occurred, even though fixed-interval rates did not change with modifications in the parameter value of the fixed-ratio schedule. The effects of drugs on behavior are no less static than the behavior that is under study. Ongoing behavior is the product of multiple dynamic influences that can play a powerful role in determining the acute effects of drugs. It is of continuing interest and significance to behavioral pharmacology that drugs can magnify or reveal the influence of these factors even when such interactive influences are not manifested in behavior when the drug is not present.

Evidence for the influence of similar contextual factors on the effects of drugs administered chronically has been reported by Smith (1986b). In this study, the effects of *d*-amphetamine were examined in rats responding under a two-component multiple schedule. During one component, high rates of responding (approximately 2.0 responses/s) were maintained under a random-ratio schedule, whereas during a second component, low response rates were maintained by a schedule that reinforced low, spaced responding. When 1.0 mg/kg of *d*-amphetamine was given daily, tolerance developed both to the rate- and reinforcement-decreasing effects under the random-ratio schedule. Tolerance did not develop to the decreases in reinforcement and increases in response rate also obtained under the spaced responding schedule. However, tolerance did develop under this schedule when the random-ratio schedule was removed; furthermore, tolerance disappeared when the random-ratio schedule was reinstated. Thus, as with studies described above that demonstrated the importance of contextual factors in acute drug effects, tolerance in this study was "influenced more by global changes in response consequences during entire experimental sessions than by local changes in response consequences in single components of those sessions" (Smith 1986b, p. 299).

Similar influences, stemming from the effects produced by interactions under multiple schedule conditions, may also account for other findings where drugs were administered chronically and the results were only partially accounted for by reinforcement loss. For example, Galbicka et al. (1980) studied the effects of daily injections of Δ^9-tetrahydrocannabinol (Δ^9-THC) in squirrel monkeys that responded under a multiple schedule. In one component, food pellets were delivered under a schedule that specified a pause requirement of 28 s or longer (interresponse time > 28 s); a random-interval schedule during the second component allowed for comparable frequencies and temporal distributions of food pellets under the two components, but rates of responding in each component differed. When the drug was administered chronically, before each daily session, tolerance developed to the rate-decreasing effects of Δ^9-THC, but the rate of tolerance development did not appear to be related to the degree of reinforcement loss. Baseline frequencies of pellet delivery were recovered during the random-interval schedule, but remained below baseline levels during the interresponse-time schedule. It seems quite possible that this finding may be similar to that of Smith (1986b), described above, and that the influence of the total context in which behavior is controlled must be viewed as a significant dimension in which to evaluate the effects of drugs administered under both acute and chronic conditions.

1.1.3. Type of Event

A number of studies have shown that the type of consequent event maintaining behavior may also contribute to the particular acute effects of drugs (review by Barrett and Katz, 1981). For instance, certain drugs such as chlordiazepoxide, ethanol, and pentobarbital, as well as certain serotonin antagonists, increase responding maintained under fixed-interval schedules of food presentation, but only decrease responding similarly maintained by the presentation or termination of electric shock. Other drugs, such as *d*-amphetamine and cocaine, will increase both types of performance under these conditions. Although few studies have examined the effects of administering drugs chronically under these procedures, one study in particular illustrates that, even though differences in drug effects do not appear under conditions of acute drug administration, when administered chronically, different effects appear that are, in fact, related to the maintaining event (Branch, 1979). In this study, responding of squirrel monkeys was maintained under a three-component multiple schedule. During each component, a fixed-interval 5-min schedule was in effect; responding in the different components was maintained either by food delivery, shock presentation, or by the termination of a stimulus correlated with shock delivery (escape). The different consequent events were correlated with different visual stimuli. *d*-Amphetamine (0.03–0.3 mg/kg) increased responding under all three components when studied acutely. However, chronic administration of 0.10 mg/kg produced a shift to the right of the dose–response curves only in the food and stimulus-shock termination components; response rates under the shock-presentation schedule were actually decreased by *d*-amphetamine during the chronic period. Furthermore, when acute injections of *d*-amphetamine were studied after a period of chronic administration, performances under the shock-presentation schedule were no longer increased.

Differences in the effects of drugs on performances controlled by different events may become apparent during and after chronic administration, even though the effects of that drug on those different performances are initially the same. As Branch (1979) pointed out, these effects may be the result of drug-behavior interactions that occurred under chronic drug conditions, but did not occur when the drug was administered acutely. As will be described in more detail below, the influence of the altered behavioral consequences during acute drug exposure can, under certain conditions, be of overriding importance in changing the subsequent effects of drugs. Further studies are required to clarify these effects under conditions of both chronic and acute drug administration, and expand the range of circumstances under which they occur.

The studies described in the preceding sections point to the conclusion that many of the same variables that influence the acute effects of drugs can also be of considerable importance in determining chronic drug effects. Tolerance to repeated drug administration, as well as a restoration of prechronic drug effects, are clearly related to variables in the relatively immediate environment, such as the ongoing rate of responding and the type of maintaining event, as well as to more remote variables, such as the total environmental context in which behavior takes place. In addition to these factors, conditions that have existed previously, but that are no longer in effect, can also affect the development of tolerance. This aspect of tolerance is reviewed in the next section.

1.2. Historical Influences on Behavior: Relevance to Tolerance

As has been mentioned frequently in this chapter, one of the more fascinating aspects of the behavioral effects of drugs is that those effects can be influenced in significant ways by the environmental conditions that exist when the drug is administered. It has also become apparent that the collective influences of prior behavioral experience—even though those conditions that originally modified behavior may no longer exist—can also contribute significantly to the manner in which behavior is affected by a drug (review by Barrett and Witkin, 1986). Prior history can have enduring effects on behavior that not only transcend the time during which those events occurred but, until a drug is administered, may leave no apparent evidence that the behavior under study has been altered. Such historical factors can actually diminish the relative contribution of those conditions in the immediate environment that might normally influence drug action (e.g., event type). Prior experience can also exert a more dominant role in the expression of a drug effect than other variables, such as response rate. These experiments, discussed in more detail below, demonstrate that behavior has the unique feature of reflecting the cumulative effects of prior experience, which, together with those conditions in the current environment, can combine with the effects of drugs to produce an intriguing array of effects. When the variables contributing to these effects are properly understood, they will yield a sizeable amount of information about both drug effects and fundamental behavioral processes, such as reinforcement and punishment.

Although the majority of experiments focusing on behavioral history have concentrated on acute drug effects, it is also clear that behavioral history can also substantially alter the course and magnitude of tolerance development (Nader and Thompson, 1987; *see* Fig. 1). In this study, pigeons were trained initially to key peck under a variable-interval schedule of food delivery. Subsequently, some pigeons were studied for approximately 50 sessions under either fixed ratio or differential reinforcement of low-rate schedules, and were then returned to the original variable-interval schedule; another group of pigeons was maintained for this period under the original variable-interval schedule. Although response rates diverged considerably during exposure to the fixed ratio or differential reinforcement of low-rate schedules (average response rates stabilized at 87.79 and 8.35 responses/min in the groups of pigeons exposed to the fixed ratio and differential reinforcement of low-rate schedules, respectively), average stable rates of responding for these two groups after reexposure to the variable-interval schedules did not differ from those of the group maintained under the variable-interval schedule throughout all phases of the study (fixed-ratio history, 47.60; differential reinforcement of low-rate history, 51.95; variable-interval group, 58.75 responses/min). When methadone was chronically administered after reexposure to the original variable-interval schedule, tolerance developed more rapidly and completely in those pigeons that received training under the differential reinforcement of low-rate schedule (*see* Fig. 1). Thus, the behavioral history of an organism, apart from any specific pharmacological intervention, can alter the chronic effects of drugs.

As is true of other determinants of the behavioral effects of drugs, an understanding of those processes contributing to the expression of tolerance will broaden our understanding of the dynamic interplay between behavior and drug action considerably. It may strike many as peculiar that a particular behavioral history can alter the nature or extent, as well as the course, of tolerance, but as shown repeatedly thus far, such profound intrusions of behavioral variables on pharmacological activity clearly have ample precedent. Although a mechanism for establishing how historical influences might alter pharmacological tolerance is currently unavailable, it may be that the type of changes in behavior derived from exposure to different contingencies of reinforcement provides

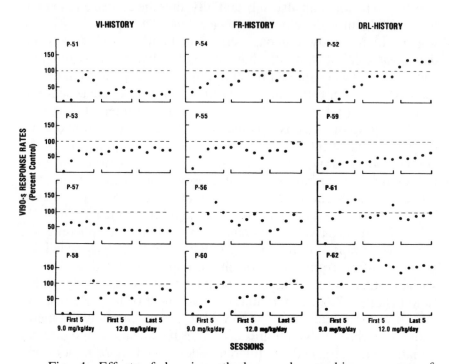

Fig. 1. Effects of chronic methadone on key-pecking responses of pigeons maintained under a variable-interval 90-s schedule. Data shown are for individual subjects. Pigeons in the VI-history column were maintained under the variable-interval schedule throughout all phases of the study. FR-history and DRL-history pigeons were trained initially under the VI 90-s schedule for at least 100 sessions and were then exposed to the respective schedules for approximately 50 sessions. Methadone was administered for at least 25 consecutive sessions to all pigeons after the history groups were returned to the VI schedule. Methadone was first administered at 9.0 mg/kg/d until rates became stable and the dose increased to 12.0 mg/kg/d for the remainder of the study (at least 12 sessions). The figure shows the first 5 sessions at 9.0 mg/kg/d, and the first and last 5 sessions under the 12.0 mg/kg/d regimen. Note that tolerance, i.e., a return to 100% of control or baseline levels prior to drug administration, occurred more rapidly and more completely in pigeons with a history of responding under the DRL schedule. In two pigeons with the DRL history (P-52 and P-62) response rates under chronic methadone exceeded those under nondrug conditions (Nader and Thompson, 1987).

a richer—or at least diverse—behavioral repertoire with which to counter the behaviorally disruptive effects of the large drug doses typically used in tolerance studies. Schedules specify relationships of a temporal and sequential nature between responding and reinforcement, generating specific performances that are a function of previous as well as current contingencies. Exposure to a differential reinforcement of low-rate schedule, which engenders long pauses and low rates of responding, may, by virtue of the modifications in behavior that develop under this schedule, provide a behavioral repertoire that then facilitates the restoration of performance (i.e., the development of tolerance) to the rate-decreasing effects of a large drug dose under a different schedule. To the extent that contingencies under certain schedules are similar to those occurring under high drug doses, previous exposure to those contingencies may facilitate the development of behavioral tolerance.

Although essential experimental evidence for these views is currently lacking, it remains that behavioral history can influence drug action under conditions where a drug is administered chronically. Further studies will undoubtedly address these issues to arrive at a better understanding of the manner in which past exposure to certain contingencies of reinforcement alters the course of tolerance to the behavioral effects of drugs. The preceding sections have reviewed briefly some of the behavioral determinants of acute and chronic drug effects, emphasizing the role of certain nonpharmacological variables commonly involved in drug action. The remainder of this chapter is devoted to reviewing studies in which behavioral or pharmacological history has modified the effects of acutely administered drugs.

2. Behavioral History and Drug Action

2.1. Early Studies

The first experiment to demonstrate the importance of past history on the effects of drugs was reported by Steinberg et al. (1961) using exploratory behavior of rats in a Y maze. In this study, "experienced" rats received twice weekly trials for 16 wk, whereas "inexperienced" rats were handled as frequently but were not exposed to the maze. At the time when drugs were administered, baseline levels of exploration for the two groups differed substantially, with the experienced rats making nearly twice as many entries into the arms of the maze than the saline control inexperienced rats (*see* Fig. 2). Amylobarbitone (15 mg/kg) and amphetamine (0.75 mg/kg) and their combination either slightly decreased or had no effect on the experienced rats. In contrast, these drugs alone produced small increases in the entries of inexperienced rats, and the mixture produced significant increases in exploration by the inexperienced rats that were nearly three times those of baseline levels (Fig. 2). As Steinberg et al. (1961) concluded, the "effects of drugs may sometimes only become manifest if the experimental subjects have an appropriate past history" (p. 555). In a subsequent extension of this study, Rushton et al. (1963) found that the attenuation of the rate-increasing effects of the drug mixture can depend on just a single prior exposure to the Y maze. Under drug-mixture conditions, the number of entries into the arms of the maze for rats that had one previous trial was approximately half that of rats that received the drug in the first exposure. In addition, activity levels on the second trial also depended on whether the first trial occurred in the presence or absence of the drug mixture. If a saline trial followed an initial trial with the drug, then activity levels were approximately twice those of rats exposed to the maze for the first time. These elevated levels stemming from initial experience with the drugs were apparent when tested without the drug up to 3 mo later, indicating the persistence of this effect (Rushton et al., 1968). However, if the initial and second trials were conducted with the drug mixture, activity levels remained elevated on both trials. These effects also were found to persist for at least 3 mo (Rushton et al., 1968). Taken together, these studies lead to the general implication that both behavioral and pharmacological experience can exert a strong influence on behavior and on the subsequent behavioral effects of drugs. As Rushton et al. (1968) concluded:

. . . behaviour under the influence of psychoactive drugs depends on striking a flexible balance between the drugs themselves, the precise circumstances under which they are administered, and the state of the recipient at the time of administration. The state of the recipient, in turn, depends on its previous experiences (p. 888).

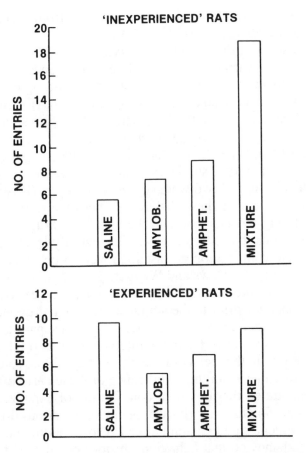

Fig. 2. Effects of drugs on exploratory behavior. The figure indicates the mean number of entries into the arms of a Y-shaped runway made by different rats. The doses of amylobarbitone sodium and amphetamine were 15 and 0.75 mg/kg, respectively. Note the differences in levels of exploration in the "inexperienced" and "experienced" rats after the mixture of amylobarbitone and amphetamine (Steinberg et al., 1961).

Subsequent experiments by Steinberg and her colleagues demonstrated the enduring nature of drug–behavior experience with other drug mixtures (chlordiazepoxide and *d*-amphetamine) using exploratory behavior of a hole-board apparatus (Dorr et al., 1971) and also extended the influence of behavioral experience to changes in the effects of morphine on activity resulting from social isolation (Katz and Steinberg, 1970, 1972).

One aspect of these studies involving exploratory and/or loco-motor behavior warrants comment. In every instance where experience was shown to alter the effects of a drug or drug combination, the experience itself also directly changed the behavior under investigation. For example, in the Steinberg et al. (1961) study, activity levels of the experienced rats after repeated exposure to the maze were almost two times higher than those of the inexperienced rats. Similarly, after only a single exposure (with saline) to the maze, activity levels on trial two were significantly lower (Rushton et al., 1963). Regardless of these differences between trial two activity levels and long-term exposure to the maze, the drugs were studied in groups of rats that differed in baseline activity levels. In view of the important contribution of baseline levels in determining the effects of drugs (e.g., Dews and Wenger, 1977), these findings must be interpreted somewhat cautiously. Thus, any experience that modifies behavior prior to the administration of a drug would demonstrate an "experiential" effect, but would not separate the effects of the experience *per se* from any intervention that produces a lasting alteration in the behavior being studied. This consideration should not detract from the significance of these early experiments that demonstrated the important contribution of experience in dramatically altering the sensitivity of behavior to various drug effects.

Subsequent studies, summarized below, addressed this issue directly and showed that behavioral history could modify drug effects even when there were no apparent differences in the behaviors being compared when a drug was administered. Nevertheless, these pioneering experiments by Steinberg and colleagues pointed quite emphatically to the conclusion that both behavioral and pharmacological experience, even when confined to a single occurrence for a brief time period, could produce lasting effects on behavior and on the behavioral effects of drugs.

A similar conclusion was reached in an experiment that investigated the effects of prior discrimination training on the effects of imipramine and chlorpromazine (Terrace, 1963). In this study (*see* Fig. 3), pigeons were trained under a discrimination procedure where key pecking produced food when the key was transilluminated with one color (S +) and, when a different color, key pecking had no consequence (extinction or S −). One group of pigeons was

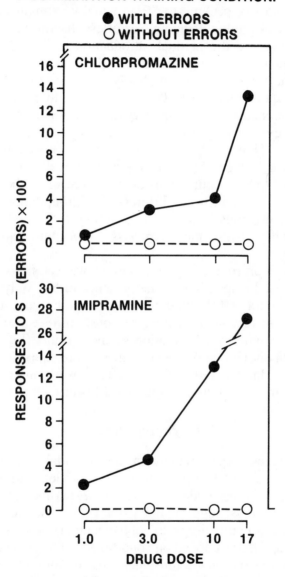

Fig. 3. Effects of chlorpromazine and imipramine on responses to S− in pigeons trained "with errors" and "without errors." Data show that pigeons trained with the errorless technique (open circles) did not respond to any drug treatment, whereas dose-dependent increases in S− responses occurred in pigeons trained with errors (closed circles). *See text* for complete description. Adapted from Terrace (1963).

trained using a conventional procedure where the key light colors alternated and key pecking to S− gradually extinguished. A second group was trained using an "errorless discrimination" procedure that involved the gradual introduction of S−. Initially S− was presented very briefly, at low levels of illumination, and the duration and intensity were gradually increased over several sessions. Using this technique, Terrace was able to establish an "intact" discrimination (i.e., responses to S+ but not to S−) without the pigeon ever having made a key pecking response in the presence of S−. Thus, at the end of the separate procedures, pigeons discriminated between S+ and S− equally well (i.e., they made no responses to S−), but the histories with regard to behavior controlled by these stimuli were different. The group trained "with errors" had a previous history of responding to S−, whereas those pigeons trained using the "errorless" procedure had never responded to S−.

When chlorpromazine or imipramine were administered, these drugs increased responding in the presence of S− only in pigeons that had been trained "with errors"; pigeons that, because of their "errorless" history never made a response to S−, also did not respond to S− after administration of these drugs (Fig. 3). Thus, even though discriminative performances were equivalent when the drugs were administered, different effects were obtained that depended on the previous discrimination history.

2.2. Recent Studies

Relatively recent experiments with squirrel monkeys, pigeons, and rats have also shown that prior history can influence the effects of a wide range of drugs under a variety of experimental conditions. These experiments support and extend those studies, described earlier, indicating that drugs do not have unitary behavioral effects. Instead, drug effects depend on conditions that have occurred in the past, as well as on the circumstances in the more immediate environment.

2.2.1. Modification of the Effects of Amphetamine

A number of experiments have now shown that *d*-amphetamine will increase punished responding of squirrel monkeys previously exposed to a shock postponement (avoidance) schedule (Fig. 4), but

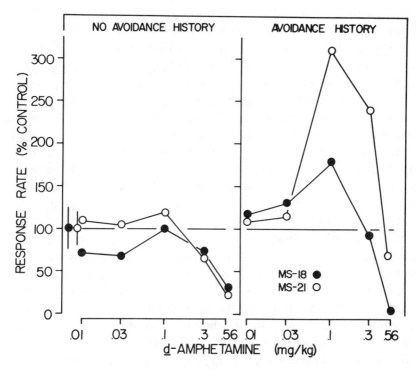

Fig. 4. Modification of the effects of *d*-amphetamine on punished behavior of squirrel monkeys by behavioral history. Responding maintained by a 5-min fixed-interval schedule of food presentation was punished by delivery of a 7 mA electric shock that followed every 30th response within each interval. Panel on the left shows the effects of *d*-amphetamine on punished responding prior to training under an avoidance or shock-postponement schedule. Panel on the right shows effects of *d*-amphetamine on punished responding after the same monkeys (MS-18 and MS-21) were exposed to an avoidance schedule for approximately 2 wk. Performances under the punishment procedure were comparable before and after avoidance training. Response rates for MS-18 before and after the avoidance history were 0.023 and 0.030 responses/s respectively; with MS-21, punished rates of responding were 0.086 and 0.082 prior to and following avoidance training, respectively. Adapted from Barrett (1977).

will only decrease responding of animals without that history (Bacotti and McKearney, 1979; Barrett, 1977). Significantly, response rates of monkeys under the punishment procedure before and after exposure to the avoidance schedule were comparable, indicating that the experience under the avoidance schedule was sufficient to alter the effects of amphetamine without directly affecting response rate. Thus, although d-amphetamine does not increase punished responding of animals trained only under punishment procedures (Hanson et al., 1967; Spealman 1979), increases in punished responding after avoidance training indicate a fundamental modification of the behavioral effects of that drug that depends on prior experience. These effects appear to persist indefinitely, unless the conditions are changed (e.g., extinction under the avoidance procedure or changes in the frequency or intensity of shock under the punishment procedure), and are the result of the specific contingencies generated by the avoidance schedule, rather than merely the delivery of shock (reviews by Barrett, 1985; Barrett and Witkin, 1986; McKearney, 1979).

Historical influences on the behavioral effects of d-amphetamine have also been reported in rats responding under a fixed-interval schedule of food delivery. Prior to the maintenance of responding under this schedule, responding was initially established in different groups under either a fixed ratio or differential reinforcement of low-rate (DRL) schedule (Urbain et al. 1978). The fixed-ratio schedule engendered high response rates (139–194 responses/min), whereas the DRL schedule generated much lower rates of responding (4–7 responses/min). When these animals were switched to the fixed-interval schedule and amphetamine was studied, this drug increased responding of rats previously trained under the DRL schedule, but did not increase responding of animals with previous exposure to the fixed-ratio schedule. Somewhat unfortunately, response rates under the fixed-interval schedule were higher in rats trained initially under the fixed-ratio schedule, and the differences in drug effect may be attributable to persistent differences in rates engendered by the fixed-ratio and DRL schedules. Nevertheless, this study demonstrated that a previous history of responding under one schedule can have enduring effects even under different experimental procedures and that these historical influences can contribute to an alteration in the behavioral effects of drugs.

In a subsequent study, also with rats, Poling et al. (1980) demonstrated that, when a history of responding under fixed-ratio or DRL schedules is given and drug effects are then studied under a variable-interval schedule, *d*-amphetamine does *not* produce effects that depend on the animal's previous history. Poling et al. (1980) attributed this finding to the greater sensitivity of fixed-interval schedules, with their varied rates and patterns of responding, to both drug effects and to historical influences on behavior. It should be noted however that, although the effects of acutely administered *d*-amphetamine did not differ under the variable-interval schedule in animals with different schedule histories, the variable-interval baseline was sensitive to historical influences when chronic methadone was administered (*see* Section 1.2., this chapter, for a description of the Nader and Thompson study). Future work addressing these issues will clarify the relative contribution of the schedule of reinforcement, type of drug, and the role of acute chronic drug administration.

2.2.2. Modification of the Effects of Morphine

Prior behavioral experience has also resulted in a change in the effects of morphine on schedule-controlled behavior. In one experiment, a chain-pulling response of squirrel monkeys was maintained under a shock-postponement (avoidance) schedule, and the effects of morphine were studied (Barrett and Stanley, 1983; *see* Fig. 5). The monkeys were then trained under a procedure where a lever-pressing response *produced* shock under a fixed-interval 3-min schedule. Shock intensity under the fixed-interval schedule was identical to that used under the avoidance schedule (10 mA) and, under the fixed-interval, maintained positively accelerated rates and patterns of responding. After approximately one month under this procedure, the chain-pulling response was again maintained under the avoidance schedule where, after stabilizing at its previous level, the effects of morphine were redetermined. As shown in Fig. 5, prior to training under the schedule of response-produced shock, morphine only decreased avoidance responding. Following interpolated training under the fixed-interval shock presentation schedule, however, morphine produced large increases in responding. As in most previous studies, these effects persisted and were not the re-

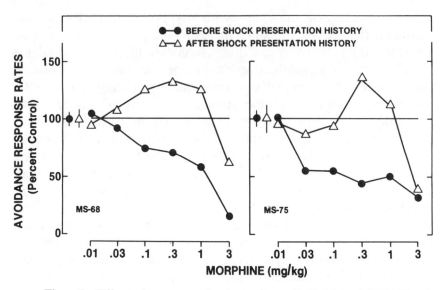

Fig. 5. Effects in two squirrel monkeys (MS-68 and MS-75) of morphine sulfate on shock-avoidance responding before (●) and after (△) exposure to a condition under which responding was maintained by response-produced shock. Morphine decreased responding prior to a history of response-produced shock, but increased responding after exposure to this condition. Unconnected points on the left of each panel denote control performances ± SE based on at least seven nondrug or saline injection sessions. Avoidance response rates before and after exposure to the shock-presentation schedule, respectively, were: MS-68: 0.157 and 0.160, MS-75: 0.540 and 0.486 responses/s. Thus, changes in the effects of morphine were not the result of shifts in avoidance baseline response rates following experience under the shock-presentation schedule. From Barrett and Stanley (1983).

sult of shifts in baseline response rates (thereby also indicating that there were no differences in sensitivity to the shock parameters).

Similar results have been obtained with chlordiazepoxide and ethanol using somewhat different procedures (Barrett and Witkin, 1986), suggesting that historical influences on the behavioral effects of drugs are not limited to a single drug class or to a few isolated experimental conditions. These behavioral manipulations, in which a specific experimental history is provided before exposure to a drug, can induce substantial modifications in the magnitude, sensitivity, and direction of the effects of a range of drugs. Appropriate behavioral histories can produce lasting changes in the qualitative effects of drugs. At least with regard to measures of behavior conventionally used to assess drug effects (e.g., rate of responding), the particular history can leave no tangible trace because rates of responding have often been identical before and after the historical intervention. Thus, the behavioral effects of a drug can change even though there is no apparent change in ongoing behavior. Possible explanations to account for these effects are given at the end of this chapter. At present, it seems clear that behaviorally relevant factors can produce marked differences in the manner in which a drug affects behavior.

These findings have certain implications for issues surrounding the etiology of drug abuse as well as issues pertaining to an individual's vulnerability or risk of abusing particular drugs. If, as seems likely, certain drugs are abused because of their effects on behavior, and those behavioral effects are related to past history, then such historical variables become exceptionally important in eventually understanding and treating, as well as preventing, drug abuse. Perhaps previous behavioral experiences generate conditions under which a drug may have quite powerful actions on behavior and on the subjective effects that drug produces; such susceptible individuals may, by virtue of their previous history, be predisposed to drug abuse. Further, if these arguments are valid, it should be possible, after achieving a better understanding of these factors, to develop behavioral strategies for "inoculating" or "immunizing" individuals against particular drug effects. Although such possibilities may seem remote at this time, it is very clear that behavioral variables can direct the effects of abused drugs in striking and significant ways. As determinants of behavior and of the behavioral effects of drugs, such factors warrant further study.

3. Pharmacological History as a Determinant of Drug Action

Experiments summarized in previous sections of this chapter have focused on studies in which behavioral history was shown to alter the "typical" effects that a drug would have on behavior. For example, *d*-amphetamine will either have no effect on or will decrease punished responding of squirrel monkeys; however, this same drug will increase punished responding following exposure to a shock-avoidance schedule. Studies in which the behavioral effects of drugs are modified by a particular behavioral history are paralleled by studies in which a pharmacological history has also been shown to modify the drug effects. These studies, described in the following section, amplify the importance of historical factors, both of a behavioral and pharmacological nature, and provide new bases for interpreting drug activity following repeated and acute drug exposure.

3.1. History with Acutely Administered Drugs

One study that systematically investigated the behavioral effects of a pharmacological history determined the ability of prior exposure to an opioid (morphine) or a central nervous system stimulant (*d*-amphetamine) to alter the behavioral effects of pentobarbital (Glowa and Barrett, 1983a; *see* Fig. 6). The effects of several drugs, including *d*-amphetamine, pentobarbital, and morphine, were studied initially on the responding of squirrel monkeys maintained under a fixed-interval stimulus-shock termination schedule (escape); responding under this schedule was suppressed by arranging that every 40th response throughout the interval also produced a shock of equal intensity and duration to that which was terminated under the escape procedure (punishment). These conditions maintained very stable low rates of punished responding that were increased by *d*-amphetamine and decreased by morphine.

At first, the effects of pentobarbital appeared variable, but with further study were found to be different depending upon the agent that had been studied immediately before. During initial exposure, or after exposure to *d*-amphetamine, pentobarbital increased responding; if morphine had been given previously, however, pentobarbital failed to increase responding. Once diminished by prior exposure to morphine, the large rate-increasing effects of pentobarbital could be restored by interposing brief exposures to *d*-amphetamine (Fig. 6). Thus, the effects of pentobarbital depended upon which drug had been given before, thereby setting up conditions under which pentobarbital would have almost opposite effects. These results illustrate that the pharmacological histories may either diminish or enhance a particular effect of a drug, much like the effects of nonpharmacological experience.

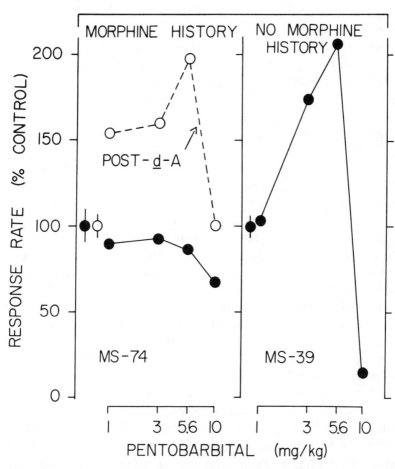

Fig. 6. Effects of pentobarbital on responding of squirrel monkeys under a procedure where responding that was maintained by the termination of a stimulus correlated with electrical shock (escape) was also punished. In monkeys with a history of morphine injections, under this procedure, pentobarbital did not increase responding (left panel, filled circles). In drug-naive monkeys, however, pentobarbital produced large increases in punished responding (right panel). In monkeys with a history of morphine, subsequent injections of *d*-amphetamine reinstated increases in responding produced by pentobarbital that were comparable to drug naive animals (*see text* for details). Adapted from Glowa and Barrett (1983a).

Prior acute drug exposure can influence a wide range of different types of behavioral responses to drugs. When drug delivery is paired with availability of novel food, consummatory behavior associated with that food is typically decreased in a dramatic fashion (Garcia and Koelling, 1966). Prior exposure to drugs can modify or even eliminate the drug-induced conditioned suppression of consummatory behavior (Vogel and Nathan, 1976). Additional studies have demonstrated that well-maintained operant performances can also be suppressed by pairing postsession drug administration with novel aspects of the maintaining event (Stolerman and D'Mello, 1978).

Studies utilizing operant performance have examined this issue in detail, as this approach allows for the intensive study of historical determinants in individual subjects. In one experiment, for example, responding by pigeons was maintained under fixed-ratio schedules of grain presentation (Glowa and Barrett, 1983b). On certain days, the grain magazine was illuminated with one color when food was delivered and saline was administered immediately following completion of the session; on other days, the grain magazine was illuminated with a different color and amphetamine was injected following the session. Key pecking was suppressed by the visual stimuli correlated with postsession administration of *d*-amphetamine. However, the dose that was necessary to suppress responding depended upon the effects of prior doses of *d*-amphetamine. When preceded by lower postsession doses that had been ineffective in suppressing responding (0.3 mg/kg), intermediate doses (1.0 mg/kg) subsequently failed to suppress responding. When preceded by higher postsession doses (3.0 mg/kg), which, however, had been effective in suppressing responding, then the intermediate dose of 1.0 mg/kg subsequently suppressed responding. Thus, the suppressive effect of stimuli paired with postsession drug administration clearly depended upon prior experience with that drug.

3.2. Drug Self-Administration

Prior exposure to drugs may also determine the subjective effects of a variety of drugs as well as the likelihood of drug abuse (Haertzen et al., 1983). Previous studies have shown that the tendency of monkeys to self-administer drugs can be influenced by the pharmacological history of the subject. For example, Schlichting et al. (1970) found that monkeys with a codeine self-administration history showed lower and more variable levels of *d*-amphetamine self-administration than did monkeys with prior histories of either pentobarbital or cocaine self-administration. Young et al. (1981) examined the ability of the antitussive agent, dextrorphan, to maintain responding that produced iv drug injections. Without any particular pharmacological history, dextrorphan did not maintain significant levels of responding. With a history of prior exposure to codeine, which did maintain high levels of responding, dextrorphan still did not maintain levels above those produced by saline. However, following exposure to ketamine, a different yet equally efficacious reinforcer, dextrorphan was effective in maintaining relatively high levels of responding. Similar effects have been found for other drugs. The dissociative anesthetic phencyclidine and the analgesic dexoxadrol have also been shown to maintain behavior when substituted for ketamine, but not codeine (Young et al., 1981). However, the abuse potential of all drugs may not be as modifiable as that of the preceding samples. For example, cyclazocine and SKF-10,047 (*N*-allyl-normetazocine) do not maintain behavior when substituted for codeine or ketamine (Hoffmeister, 1979; Young and Woods, 1980; Young et al., 1981).

Hoffmeister (1986) has shown that scheduled infusions of naloxone as low as 1.0 μg/kg can engender and maintain avoidance responding in nonphysically dependent rhesus monkeys that have previously been trained to avoid scheduled infusions of nalorphine. In these monkeys, naloxone was as potent in the nondependent monkeys as in subjects physically dependent on opiates (Downs and Woods, 1975). Although the monkeys in the Hoffmeister study had previous histories of nalorphine avoidance, responding engendered by nalorphine was also extinguished by interposing scheduled injections of cocaine before exposure to the naloxone. Thus, it is difficult to determine whether responding maintained by avoidance of naloxone infusions was the result of the nalorphine or cocaine histories. However, as Hoffmeister (1986) suggested, based on earlier studies in which responding was extinguished with saline rather than cocaine (Hoffmeister and Wüttke, 1973; Downs and Woods, 1975), it appears that the history of cocaine not that of nalorphine influenced the negative reinforcing properties of naloxone. The abuse potential and/or the behavioral effects of certain drugs can depend upon not only a subject's self-administration history, but also upon the particular drug under study, the type of drugs with which the subject has had experience, and the conditions under which present and past drugs have been available.

The preceding examples have shown that the behavioral effects of a wide range of drugs are subject to modification by prior drug experience. Not only the effects of drugs on schedule-controlled behavior in general, but also their reinforcing and, presumably, their punishing properties can be determined, in a significant manner, by such pharmacological histories. Thus, it would appear that prior pharmacological experience may act as an important determinant of the behavioral effects of some drugs. Such effects, though not yet fully understood, may provide a unique perspective on behavioral influences involved in tolerance and sensitization-like phenomena.

4. Drug–Behavior Interaction Histories

The studies reported in the previous three sections indicate that behavioral or pharmacological histories can often alter the subsequent effects of various drugs. The studies reviewed in this section will concentrate on examples where the behavioral effects produced by a drug are responsible for the subsequent modification of that drug's actions. Under these circumstances the consequences of behavior occurring when the drug is administered constitute a drug–behavior interaction history that modifies the behavioral effects of that drug under subsequent administrations.

A clear example of how the behavioral consequences occurring under the influence of a drug can alter the subsequent activity of that drug is illustrated by an experiment reported by Smith and McKearney (1977). In this study (*see* Fig. 7), key pecking by pigeons was maintained under a procedure where a key peck produced food only if at least 30 s elapsed between successive pecks. Increases in responding produced by the initial administration of *d*-amphetamine resulted in a decrease in the frequency of food presentation, because the time between responses after drug administration was typically less than the 30-s schedule requirement (Fig. 7). However, when amphetamine was administered a second time (after four additional nondrug sessions), increases in responding

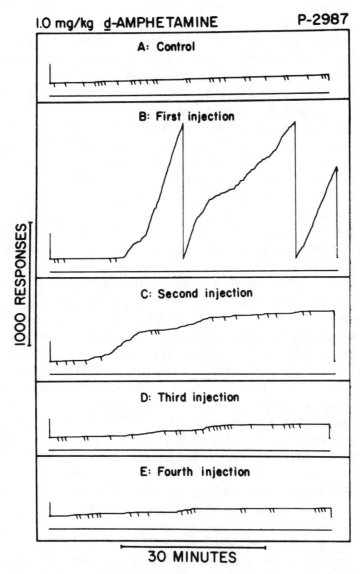

Fig. 7. Diminution of rate-increasing effects of *d*-amphetamine under a procedure where increases in responding decreased the frequency of food reinforcement. (*See text* for additional details). From Smith and McKearney (1977).

were much less than those that occurred initially. Subsequent injections of the same dose of amphetamine, separated by 4–9 nondrug sessions, resulted in still a further diminution of the rate-increasing effects such that no increases occurred upon the fourth injection.

These findings, which also occurred with pentobarbital, indicate that experience with drug-produced increases in responding that resulted in a decreased frequency of food presentation reduced the general tendency of pentobarbital and *d*-amphetamine to increase responding. When the contingency was removed during drug sessions and increases in responding did not reduce the frequency of food delivery, increases produced by both drugs persisted over subsequent administrations. Diminished drug effects, which appear to be tolerance-like phenomena, can occur when drugs produce changes in the consequences of behavior such that those altered consequences oppose the "natural" actions of the drug (cf. Smith et al., 1978).

Another experiment that showed the importance of a drug–behavior interaction found conditions under which morphine could increase punished responding of squirrel monkeys (Brady and Barrett, 1986). The effects of morphine were studied under a multiple fixed-interval schedule where responding in each of two components terminated stimuli correlated with shock. Under this schedule, shocks were delivered every 3 s after the fixed-interval 5-min period elapsed. However, a response any time after the end of the fixed interval prevented shock delivery and terminated the visual stimuli (escape). It was possible, therefore, to prevent shocks by responding immediately as the interval elapsed. As under most fixed-interval schedules, regardless of the maintaining event, responding began early in the interval and accelerated as the interval elapsed. Monkeys typically terminated the visual stimuli before any shocks were delivered under the fixed-interval schedule.

Morphine was first studied in two monkeys when responding during one component was suppressed by arranging that a shock also *followed* every 30th response during that 5-min fixed interval; this procedure reduced responding during that component (punishment), but responding remained at a high rate during the alternate component. Unexpectedly, morphine increased punished responding, an effect not previously reported for this drug, and these increases also occurred when the punishment schedule was studied alone under a single schedule condition (Fig. 8).

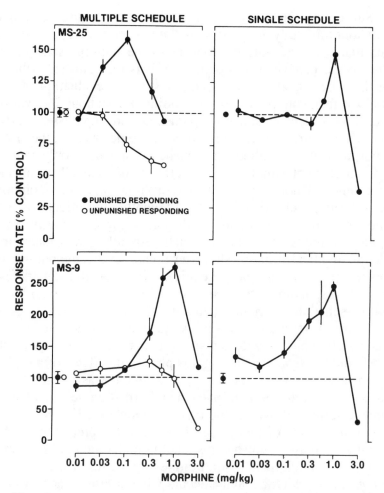

Fig. 8. Dose–response curves for two monkeys (MS-25 and MS-9) showing the effects of morphine under the multiple schedule and single-component-schedule stimulus-shock termination procedure. Unconnected points on the left represent control performances under the respective schedule conditions ± 1 SE. If the SE was less than the diameter of the point, it was not shown. The dashed horizontal line represents no change from control levels. Effects are expressed as percentage of control performances. Control response rates (responses/s) for MS-9 were 0.706 (unpunished) and 0.066 (punished) under the multiple schedule, and 0.096 under the punishment schedule alone. With MS-25, these values were 1.236 (unpunished) and 0.155 (punished) under the multiple schedule, and 0.144 under the single component punishment schedule. From Brady and Barrett (1986).

Subsequent analyses of the different procedures employed throughout this study demonstrated that the acute injections of morphine that occurred during exposure to the multiple schedule resulted in the change in morphine's effects on punished behavior (Fig. 8). These analyses involved a systematic evaluation of morphine's effects on punished responding throughout the various stages involved in establishing punished and unpunished performances maintained under the multiple schedule. At first, the effects of morphine were studied on punished responding only under a single-component stimulus-shock termination schedule. Punished responding did not increase with morphine under this condition (Fig. 9, left panel, triangles). Subsequently, the effects of morphine were studied on punished responding *after exposure* to an alternate stimulus condition in which responding was maintained under an identical stimulus-shock termination schedule but was not punished (i.e., a multiple schedule). Morphine did not increase punished responding after exposure (for 12–30 sessions) to the multiple-schedule procedure (Fig. 9, left panel, circles).

In the next phase monkeys were again exposed to the two schedule components within a single experimental session (multiple schedule). However, during this phase, the effects of morphine were determined when the multiple schedule was in effect, i.e., *during exposure* to the multiple schedule (this phase was identical to that described above for the first two monkeys, Fig. 8). In contrast to the effects of morphine obtained in the preceding phases, morphine increased punished responding under the multiple schedule (Fig. 9, middle panel, solid circles). This result indicates that the *context* in which behavior occurs when a drug is administered can interact with the effects that drug will have on behavior. Finally, when the component in which responding was not punished was removed, increases in punished responding with morphine persisted (Fig. 9, right panel). Thus, in these studies, the behavioral and pharmacological histories were important, but only when experienced together with the interactions between drug and behavioral consequences responsible for altering the drug effect.

5. Summary and Conclusions

The experiments summarized above indicate that behavioral and pharmacological histories, as well as drug-induced changes in

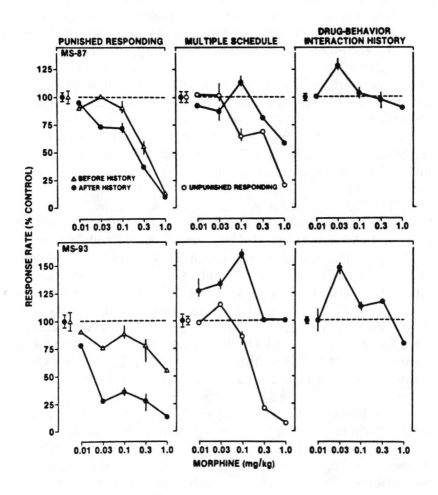

Fig. 9. Morphine dose–effect curves for two monkeys under the various experimental procedures used to examine the modification of the effects of morphine on punished responding. Unconnected points on left represent control values ± 1 SE. The dashed horizontal line represents no change from control levels. Effects are expressed as a percentage of control performances. Note that morphine did not increase punished responding before or after exposure to the multiple schedule (left panels) but did produce increases when administered under the multiple schedule. Increases with morphine continued to occur under the punishment schedule alone (right panels). From Brady and Barrett (1986).

behavioral consequences, can exert profound modifications in the fundamental dose–response relationships of several drugs. In many instances, these factors can greatly attenuate the acute effects of a drug, thereby producing an outcome resembling tolerance, whereas under other conditions, these variables can produce a qualitative modification or a heightened response of the behavioral effects of a drug that resembles sensitization. Although relatively little work has been conducted thus far on the effects of historical factors on the development and course of tolerance under chronic drug administration, it seems clear that these variables will play as critical a role in determining the chronic effects of drugs as they do in determining effects when administered acutely.

There are undoubtedly several mechanisms by which behavioral factors such as previous history can influence both the acute and chronic effects of drugs, and produce modifications of drug activity that resemble tolerance or sensitization. Modifications of the basic dose–response function, producing a change in the qualitative actions of a drug, such as when the effects of *d*-amphetamine on punished responding are changed from rate-decreasing to rate-increasing effects after a shock avoidance history, are obviously complexly determined. However, it should be possible to identify factors that are common to the several experiments that demonstrate a modification of drug activity.

First, such effects do not appear to be limited to one drug class. Modifications in the activity of drugs as diverse as amphetamine, morphine, methadone, pentobarbital, ethanol, and chlordiazepoxide have been reported. The only common factor shared by these drugs appears to be that they are all drugs of abuse. It may be that drugs of abuse have the potential for producing a variety of effects depending on some of those conditions reviewed in this chapter. Which specific behavioral effect an abused drug will produce may depend more on these conditions than is true of other drugs that have more therapeutic value. In fact, compounds such as the antipsychotic drug chlorpromazine do not show altered effects under conditions where *d*-amphetamine does (Bacotti and McKearney, 1979). As mentioned earlier in this chapter, drugs of abuse may exert their effects through multiple mechanisms in such a way that the actual effects shown will be quite susceptible to nonpharmacological variables. When compounds have malleable actions, it seems important to understand the multiple ways in which specific effects are manifested. The interplay among behavior, the environment, and pharmacological mechanisms of drug action should be a fertile area of study.

A question related to these points that requires clarification is the exact nature and extent of changes produced after various behavioral histories. For example, after a history of responding under a shock-avoidance schedule, *d*-amphetamine increases punished responding. Since only compounds that are antianxiety drugs (e.g., chlordiazepoxide, pentobarbital) typically increase punished behavior, this suggests that, after a particular history, amphetamine might be anxiolytic. It is not uncommon to discover vast individual differences in response to drugs. Potential alterations of the subjective effects of drugs by certain behavioral histories may be one method of arriving at a clearer and better understanding of some critical factors related to drug abuse, as well as to individual differences in drug action.

A second factor to consider in evaluating and understanding the influence of prior history on drug action is that of the particular history responsible for changing drug effects. In all the studies reported thus far, the effects of a particular drug on the behavior *after* the interpolated history were those that would have occurred with that drug on the interpolated behavior. Thus, since *d*-amphetamine increases avoidance responding, it seems reasonable to speculate about the relevance of this fact to the result that, after avoidance training, *d*-amphetamine increases rather than decreases punished responding. Similarly, morphine has been shown to increase responding maintained by electric shock delivery (McKearney, 1974), and exposure to this procedure changes the rate-decreasing effects of morphine under shock-avoidance schedules to rate increases. Apparently, the relevant history and altered drug effects are intimately related. Additionally, these results also lead to the possibility of reversing the experimental sequence and the direction of the drug effects. For example, after exposure to a punishment procedure, the "usual" rate-increasing effects of *d*-amphetamine on avoidance responding should be changed to rate-decreasing effects. Whether and how such effects would occur, the possible importance of parameter values and length of exposure, as well as many other variables, are all issues that must be addressed. Nevertheless, such considerations are one framework within which to evaluate and further explore these results.

A third consideration involving behavioral history and drug action is related to a point made earlier in this chapter when discussing how certain schedule histories might alter the rate and extent of tolerance to chronically administered drugs. Exposure to multiple contingencies of reinforcement produces behavioral repertoires that are, perhaps, much more malleable, sensitive, and differentially responsive to various interventions. Even performances under single schedules of reinforcement in animals without previous experimental histories are complexly and multiply determined. With extensive histories, susceptibility to other factors, such as a drug, unquestionably changes, as evidenced from the several studies reviewed in this chapter. The administration of a drug can reveal the residual effects of previous experiences in ways that have not yet been thoroughly explored with other procedures. Behavior is not a passive transducer of drug action, but appears to actively impart direction and magnitude to drug effects. Prior experience may be a more pervasive influence on the behavioral effects of drugs, under both acute and chronic conditions, than is commonly recognized. With further study, these variables may yield new information for better understanding not only tolerance and sensitization, but also general determinants of the effects of drugs.

Acknowledgments

We wish to acknowledge the assistance of Myra J. Zimmerman in the preparation of this chapter. Much of the research described in this chapter was supported by PHS Grant DA-02873.

References

Bacotti A. V. and McKearney J. W. (1979) Prior and ongoing experience as determinants of the effects of *d*-amphetamine and chlorpromazine on punished behavior. *J. Pharmacol. Exp. Ther.* **211,** 80–85.

Barrett J. E. (1977) Behavioral history as a determinant of the effects of *d*-amphetamine on punished behavior. *Science* **198,** 67–69.

Barrett J. E. (1985) Modification of the behavioral effects of drugs by environmental variables, in *Behavioral Pharmacology: The Current*

Status (Seiden L. S. and Balster R. L., eds.), pp. 7–22, Alan R. Liss, New York.

Barrett J. E. (1986) Behavioral history: residual influences on subsequent behavior and drug effects, in *Developmental Behavioral Pharmacology*, Advances in Behavioral Pharmacology Series, Vol. 5 (Krasnegor N. A., Gray D. B., and Thompson T., eds.), pp. 99–114, Lawrence Erlbaum Associates, Hillsdale, New Jersey.

Barrett J. E. (1987) Non-pharmacological factors determining the behavioral effects of drugs, in *Psychopharmacology, The Third Generation of Progress* (Meltzer H. Y., ed.), pp. 1493–1501, Raven, New York.

Barrett J. E. and Katz J. L. (1981) Drug effects on behaviors maintained by different events, in *Advances in Behavioral Pharmacology, Vol. 3* (Thompson T., Dews P. B., and McKim W. A., eds.), pp. 119–168, Academic, New York.

Barrett J. E. and Stanley J. A. (1980) Effects of ethanol on multiple fixed-interval fixed-ratio schedule performances: dynamic interactions at different fixed-ratio values. *J. Exp. Anal. Behav.* **34,** 185–198.

Barrett J. E. and Stanley J. A. (1983) Prior behavioral experience can reverse the effects of morphine. *Psychopharmacol.* **81,** 107–110.

Barrett J. E. and Witkin J. M. (1986) The role of behavioral and pharmacological history in determining the effects of abused drugs, in *Behavioral Analysis of Drug Dependence* (Goldberg S. R. and Stolerman I. P., eds.), pp. 195–223, Academic, New York.

Brady L. S. and Barrett J. E. (1986) Drug-behavior interaction history: modification of the effects of morphine on punished behavior. *J. Exper. Anal. Behav.* **45,** 221–228.

Branch M. N. (1979) Consequent events as determinants of drug effects on schedule-controlled behavior: modification of effects of cocaine and *d*-amphetamine following chronic amphetamine administration. *J. Pharmacol. Exp. Ther.* **210,** 354–360.

Corfield-Sumner P. K. and Stolerman I. P. (1978) Behavioral tolerance, in *Contemporary Research in Behavioral Pharmacology* (Blackman D. E. and Sanger D. J., eds.), pp. 391–448, Plenum, New York.

Dews P. B. (1955) Studies in behavior. I. Differential sensitivity to pentobarbital of pecking performance in pigeons depending on the schedule of reward. *J. Pharmacol. Exp. Ther.* **113,** 393–401.

Dews P. B. (1958) Studies in behavior. IV. Stimulant actions of methamphetamine. *J. Pharmacol. Exp. Ther.* **122,** 137–147.

Dews P. B. and Wenger G. R. (1977) Rate-dependency of the behavioral effects of amphetamine, in *Advances in Behavioral Pharmacology, Vol. 1* (Thompson T. and Dews P. B., eds.), pp. 167–227, Academic, New York.

Dorr M., Steinberg H., Tomkiewicz M., Joyce D., Porsolt R. D., and Summerfield A. (1971) Persistence of dose related behaviour in mice. *Nature* **231,** 121–123.

Downs D. A. and Woods J. H. (1975) Naloxone as a negative reinforcer in rhesus monkeys: effects of dose, schedule, and narcotic regimen. *Pharmacol. Rev.* **27,** 397–406.

Galbicka G., Lee D. M., and Branch M. N. (1980) Schedule-dependent tolerance to behavioral effects of Δ^9-tetrahydrocannabinol when reinforcement frequencies are matched. *Pharmacol. Biochem. Behav.* **12,** 85–91.

Garcia J. and Koelling R. A. (1966) Relation of cue to consequence in avoidance learning. *Psychonomic Science* **4,** 123–124.

Glowa J. R. and Barrett J. E. (1983a) Drug history modifies the behavioral effects of pentobarbital. *Science* **220,** 333–335.

Glowa J. R. and Barrett J. E. (1983b) Response suppression by visual stimuli paired with postsession *d*-amphetamine injections in the pigeon. *J. Exper. Anal. Behav.* **39,** 165–173.

Goudie A. J. and Demellweek C. (1986) Conditioning factors in drug tolerance, in *Behavioral Analysis of Drug Dependence* (Goldberg S. R. and Stolerman I. P., eds.), pp. 225–285, Academic, New York.

Haertzen C. A., Kocher T. R., and Miyasato K. (1983) Reinforcements from the first drug experience can predict later drug habits and/or addiction: results with coffee, cigarettes, alcohol, barbiturates, minor and major tranquilizers, stimulants, marijuana, hallocinogens, heroin, opiates and cocaine. *Drug Alcohol Depend.* **11,** 147–165.

Hanson H. M., Witoslawski J. J., and Campbell E. H. (1967) Drug effects in squirrel monkeys trained on a multiple schedule with a punishment contingency. *J. Exper. Anal. Behav.* **10,** 565–569.

Hoffmeister F. (1979) Preclinical evaluation of reinforcing and aversive properties of analgesics, in *Mechanisms of Pain and Analgesic Compounds* (Beers R. F. and Bassett E. G., eds.), pp. 447–466, Raven, New York.

Hoffmeister F. (1986) Negative reinforcing properties of naloxone in the non-dependent rhesus monkey: influence on reinforcing properties of codeine, tilidine, buprenorphine, and pentazocine. *Psychopharmacology* **90,** 441–450.

Hoffmeister F. and Wüttke, W. (1973) Negative reinforcing properties of morphine antagonists in naive rhesus monkeys. *Psychopharmacologia*, **33**, 247–258.

Katz D. M. and Steinberg H. (1970) Long-term isolation in rats reduces morphine response. *Nature* **228**, 469–471.

Katz D. M. and Steinberg H. (1972) Role of social factors in response to morphine, in *Drug Addiction, Vol. 2, Clinical and Socio-Legal Aspects* (Singh J. M., Miller L. H., and Lal H., eds.), pp. 85–97, Futura, New York.

Kelleher R. T. and Morse W. H. (1968) Determinants of the specificity of the behavioral effects of drugs. *Ergeb. Physiol. Chem. Exp. Pharmakol.* **60**, 1–56.

McKearney J. W. (1979) Interrelations among prior experience and current conditions in the determination of behavior and the effects of drugs, in *Advances in Behavioral Pharmacology, Vol. 2* (Thompson T. and Dews P. B., eds.), pp. 39–64. Academic, New York.

McKearney, J. W. (1974) Effects of *d*-amphetamine, morphine and chlorpromazine on responding under fixed-interval schedules of food presentation or electric shock presentation. *J. Pharm. Exper. Ther.* **190**, 141–153.

McKearney J. W. and Barrett J. E. (1975) Punished behavior: increases in responding after *d*-amphetamine. *Psychopharmacologia*, **41**, 23–26.

McKearney J. W. and Barrett J. E. (1978) Schedule-controlled behavior and the effects of drugs, in *Contemporary Research in Behavioral Pharmacology*, (Blackman D. E. and Sanger D. J., eds.), pp. 1–68, Plenum, New York.

Morse W. H. (1966) Intermittent reinforcement, in *Operant Behavior: Areas of Research and Application* (Honig W. K., ed.), pp. 52–108.

Nader M. A., and Thompson T. (1987) Interaction of methadone, reinforcement history and variable-interval performance. *J. Exper. Anal. Behav.* **48**, 303–315.

Poling A., Krafft K., and Chapman L. (1980) *d*-Amphetamine, operant history, and variable-interval performance. *Pharmacol. Biochem. Behav.* **12**, 559–562.

Rushton R., Steinberg H., and Tinson C. (1963) Effects of a single experience on subsequent reactions to drugs. *Br. J. Pharmacol. Chemother.* **20**, 99–105.

Rushton R., Steinberg H., and Tomkiewicz M. (1968) Equivalence and persistence of the effects of psychoactive drugs and past experience. *Nature* **220,** 885–889.

Schlichting U. U., Goldberg S. R., Wüttke W., and Hoffmeister F. (1970) *d*-Amphetamine self-administration by rhesus-monkeys with different self-administration histories. *Excerpta Medica International Congress* **220,** 62–69.

Schuster C. R., Dockens W. S., and Woods J. H. (1966) Behavioral variables affecting the development of amphetamine tolerance. *Psychopharmacologia* **9,** 170–182.

Smith J. B. (1986a) Effects of fixed-ratio length on the development of tolerance to decreased responding by *l*-nantradol. *Psychopharmacol.* **90,** 259–262.

Smith J. B. (1986b) effects of chronically administered *d*-amphetamine on spaced responding maintained under multiple and single-component schedules. *Psychopharmacol.* **88,** 296–300.

Smith J. B. and McKearney J. W. (1977) Changes in the rate-increasing effects of *d*-amphetamine and pentobarbital by response consequences. *Psychopharmacol.* **53,** 151–157.

Smith J. B., Branch M. N., and McKearney J. W. (1978) Changes in the effects of *d*-amphetamine on escape responding by its prior effects on punished responding. *J. Pharmacol. Exper. Ther.* **207,** 159–164.

Spealman R. D. (1979) Comparison of drug effects on responding punished by pressurized air or electric shock delivery in squirrel monkeys: pentobarbital, chlordiazepoxide, *d*-amphetamine and cocaine. *J. Pharmacol. Exper. Ther.* **209,** 309–315.

Steinberg H., Rushton R. and Tinson C. (1961) Modification of the effects of an amphetamine-barbiturate mixture by the past experience of rats. *Nature* **192,** 533–535.

Stolerman I. P. and D'Mello G. D. (1978) Amphetamine-induced taste aversion demonstrated with operant behavior. *Pharmacol. Biochem. Behav.* **8,** 107–111.

Terrace H. S. (1963) Errorless discrimination learning in the pigeon: effects of chlorpromazine and imipramine. *Science* **140,** 318–319.

Urbain C., Poling A., Millam J., and Thompson T. (1978) *d*-Amphetamine and fixed-interval performance: effects of operant history. *J. Exper. Anal. Behav.* **29,** 385–392.

Vogel J. R., and Nathan B. A. (1976) Reduction of learned taste aversions by pre-exposure to drugs. *Psychopharmacol.* **49,** 167–172.

Young A. M., Herling S., and Woods J. H. (1981) History of drug exposure as a determinant of drug self-administration, in *Behavioral*

Pharmacology of Human Drug Dependence (Thompson T. and Johanson C. E., eds.), pp. 75–89, NIDA Research Monograph No. 37, U. S. Government Printing Office, Washington, D. C.

Young A. M. and Woods J. H. (1980) Behavior maintained by intravenous injection of codeine, cocaine, and etorphine in the rhesus macaque and pigtail macaque. *Psychopharmacol.* **70,** 263–271.

Tolerance to Drug Discriminative Stimuli

Alice M. Young and Christine A. Sannerud

1. Introduction

Repeated encounters with a psychoactive drug create a potent opportunity for the drug to play a critical role in the development and expression of behavior. Repeated exposure can provide the behavioral conditions necessary for psychoactive drugs to function as conditional, reinforcing, or discriminative stimuli, controlling the acquisition and occurrence of complex behavioral repertoires (Thompson and Pickens, 1971). The actual role that a specific drug plays in shaping or guiding behavior depends upon the precise conditions of an individual's prior and current contact with the drug. When conditions are arranged so that one behavioral repertoire is reinforced in the presence of a drug, and a second in its absence, each repertoire may become most probable in the presence of its correlated pharmacological stimulus. That is, execution of a particular behavioral repertoire discriminates the presence or absence of the drug, and the drug is said to exert discriminative stimulus control of behavior. Once established, such discriminative stimulus control by a drug persists for extended periods without requiring progressive increases in dose. The potential for serving as a robust and persistent discriminative stimulus appears to be a common characteristic of psychoactive drugs (*see* reviews by Schuster and

221

Balster, 1977; Young, in press). Whether or not tolerance develops to the discriminative stimulus actions of psychoactive drugs is a topic of some debate. This chapter reviews the evidence.

Development and maintenance of drug discriminative stimulus control reflect a dynamic interaction of behavior, a drug, and the demands imposed by the learner's environment. Development of tolerance to drug discriminative control is also a dynamic process, shaped by interactions of the dose and chronicity of drug treatment, the behavioral conditions under which the drug is administered, and the individual's history of pharmacological stimulus control. The complexity of these multiple determinants has generated considerable controversy about whether or not drug stimulus control is subject to tolerance, and about the kinds of changes in discriminative behavior that would unambiguously document tolerance to a drug stimulus. This chapter will consider several of the issues involved in studies of tolerance to drug discriminative stimulus control. We will begin with a general description of the experimental strategies employed in studies of drug discrimination learning. Then we will describe certain pharmacological and behavioral interventions that modulate the apparent sensitivity of discriminative assays. Finally, we will review studies that have examined the nature and determinants of tolerance development.

2. Establishment and Maintenance of Drug Stimulus Control

Psychoactive drugs are potential discriminative stimuli. This potential is realized when an individual encounters a drug under conditions arranged so that the presence and absence of the drug stimulus are the only cues that uniquely co-vary with reinforcement of dissimilar behavioral repertoires. In the experimental laboratory, conditions can be arranged so that responses on one of two operanda are reinforced in the presence of a drug stimulus, and responses on the second are reinforced in its absence. Since the drug is present on the occasions when one behavioral repertoire is reinforced, and absent on the occasions when the other repertoire is reinforced, appearance of each repertoire becomes appropriate to its correlated pharmacological stimulus. That is, behavior comes under the discriminative stimulus control of the drug, and the future occurrence of each repertoire is most probable upon recurrence of the pharmacological stimulus that prevailed during previous reinforcement.

The general strategies of drug discrimination experiments are illustrated in Fig. 1. Stimulus control is achieved by an explicit differential reinforcement procedure. The details of the experimental apparatus, reinforcement presentations, and initial behavioral training are appropriate to the species selected as the experimental subject. A fixed dose of drug and an appropriate vehicle are selected as the training stimuli. We will have more to say later about the role of drug dose in discriminative control. For now, it is important to note that the training dose is generally chosen to exert behavioral actions without severely disrupting motor performance. During each discrimination training session, the subject is presented with two response options. Following presession administration of the training drug stimulus, selection or execution of one option results in reinforcer delivery. Execution of the other option may have no consequence, or may reset the requirements for reinforcement of the first option. Following presession administration of saline, execution of the second response option results in reinforcer delivery. Often intermittent schedules are used, so that later experiments can assess drug-induced changes in both stimulus control and the rates and patterns of coordinated behavioral repertoires. All other conditions (lighting, sound, type of reinforcer, and so on) are identical in every training session, so that only the presence or absence of the interoceptive drug stimulus uniquely co-varies with the contingencies of reinforcement. Training sessions are generally scheduled only once per day, to allow the effects of the drug stimulus to dissipate between successive exposures to the task.

Development of discriminative control by a psychoactive drug can be assessed conveniently by measuring the occurrence of each response repertoire in the presence and absence of the drug stimulus. The initial likelihood of each repertoire in the presence of each stimulus is determined by the individual's learning history. As the discrimination is learned, each repertoire becomes most probable in the presence of its correlated pharmacological stimulus. Figure 2 illustrates a common method of presenting evidence for development of stimulus control. Saline and the mu opioid agonist fentanyl (0.056 mg/kg) were established as discriminative stimuli for food-

Strategies employed to establish and test discriminative stimulus control of discrete responses

STIMULI   CONTINGENCIES

		Response 1	Response 2
Establishment & maintenance of control	Drug	Reinforcement	Extinction
	Saline	Extinction	Reinforcement
Tests of control Type 1	Test drug	Reinforcement	Reinforcement
Type 2	Test drug	Extinction	Extinction

Fig. 1. Diagram of the differential reinforcement contingencies used to establish and maintain discriminative stimulus control by psychoactive drugs. Formally similar differential contingencies have been employed with a wide range of drugs, species, response topographies, reinforcing events, and schedules of reinforcement.

reinforced key pecking in pigeons. During each experimental session, pigeons were presented with two response keys. Following presession administration of fentanyl, responses on only one key resulted in food delivery. Following presession administration of saline, responses on only the other key resulted in food delivery. Responses to the inappropriate key were recorded and reset the response requirement on the injection-appropriate key. Over successive sessions, the percentage of responses emitted on the key appropriate to the presession injection increased progressively, until the preset criterion of 90% accuracy was met over prolonged sequences of successive sessions (Fig. 2, closed symbols). These representative pigeons differed in the number of sessions required for establishment of stimulus control (defined in this experiment as a sequence of nine sessions of accurate discriminative performance) and in their overall rates of responding. However, both achieved and maintained highly accurate discriminative performance when exposed to appropriate learning conditions.

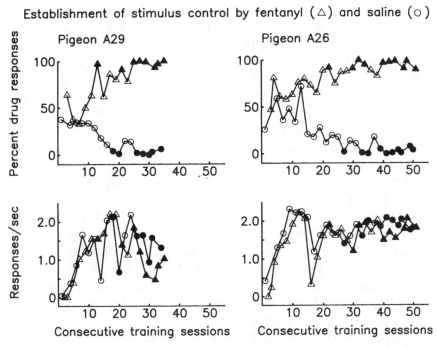

Fig. 2. Establishment of stimulus control of food-reinforced fixed-ratio performance by saline and 0.056 mg/kg fentanyl in two representative pigeons. Abscissae: consecutive training sessions. The fixed-ratio value was increased from FR 1 to FR 30 during sessions 1–21 for pigeon A29, and during sessions 1–19 for pigeon A26. Ordinate, upper panels: responses emitted on the fentanyl-appropriate key, expressed as a percentage of the total session responses. When the percentage is less than 100, the remaining responses were emitted on the saline-appropriate key. Ordinate, lower panels: overall response rate, expressed as responses/s. Circles represent performance following administration of saline; triangles, performance following administration of 0.056 mg/kg fentanyl citrate. *Filled* symbols represent performance that achieved the criteria for stimulus control: emission of fewer than 60 responses prior to the first reinforcer and completion of >90% of all session responses on the injection-appropriate key. Tests of stimulus control (not shown) began on session 39 for pigeon A29, and on session 52 for pigeon A26. (*Unpublished observations.*)

The ability to function as a discriminative stimulus appears to be a general characteristic of psychoactive drugs. Two comprehensive bibliographic surveys of drug discrimination research (Stolerman and Shine, 1985; Stolerman et al., 1982) document that drugs from a wide variety of pharmacological classes, including antidepressants, antipsychotics, cannabinoids, cholinergic agents, CNS stimulants, convulsants, dissociative anesthetics, hallucinogens, opioids, and sedative-hypnotics, have been established as discriminative stimuli under the general conditions described above. Discriminative control has also been established by certain drug withdrawal stimuli. These surveys also document the wide range of experimental conditions used in studies of pharmacological discriminative stimuli. Rats are the most common subjects, but baboons, gerbils, humans, macaque monkeys, mice, pigeons, and squirrel monkeys have also served. Response topographies, selected to match the capabilities of the subject, include key pecks, lever presses, turns in a maze, and verbal reports. Reinforcement operations include presentation of brain stimulation, food, money, or water, and postponement or termination of electric shock. The schedule of reinforcer delivery is most often a fixed ratio, but investigators have employed continuous reinforcement, direct reinforcement of low rate (DRL), fixed interval, interlocking, second-order, and variable interval schedules.

This variety demonstrates that the potential for discriminative stimulus control is not limited to a restricted range of drugs or behaviors, but rather, appears to be a highly predictable outcome of experimental conditions arranged so that the presence and absence of a drug reliably set the occasion for reinforcement of different behavioral repertoires. Furthermore, these procedures engender highly accurate discriminative behavior, as evidenced by the stringent accuracy criteria imposed by many workers. Two common criteria are, first, that prior to the first reinforcer delivery of a training session a subject must not complete the requirements for reinforcement on the incorrect operandum, and second, that a fixed percentage (often 85 or 90%) of all responses in an experimental session must occur to the injection-appropriate operandum. These criteria must be met over prolonged sequences of consecutive training sessions.

There appears to be considerable variability among drugs in the amount or duration of training required to establish discriminative stimulus control. However, direct comparisons of the speed of establishment of stimulus control by different drugs are frequently complicated by additional differences in relative dose, task parameters, and subject species. Overton and coworkers have overcome certain of these problems by systematically evaluating the number of sessions required to establish stimulus control by a wide variety of psychoactive drugs in a T-maze avoidance task in rats (*see* Overton et al., 1986 for a review). For each drug, the ''maximum usable dose'' was defined as the highest dose that did not disrupt coordinated motor behavior. Separate experiments determined the number of training sessions required to establish stimulus control by each maximum usable dose and selected lower doses. Stimulus control was defined as an injection-appropriate response on the first trial in eight out of ten (or nine out of 12) consecutive training sessions. Two conclusions from these studies are germane to the present discussion. First, an impressively wide range of drugs can be established as discriminative stimuli, but membership in a particular pharmacological class does not predict the amount of training required to establish stimulus control. Second, for individual drugs, the amount of training required is determined by the dose employed during training. For a variety of drugs, the amount of training required to establish stimulus control varies inversely with training dose, such that discriminative control by higher doses is established more rapidly than is control by lower doses (Overton, 1982a; Overton and Batta, 1979; Overton and Hayes, 1984; *see also* Colpaert et al., 1980b; Jones et al., 1976). Additionally, a greater proportion of subjects may acquire a discrimination based on high rather than lower training doses (cf. Colpaert and Janssen, 1982b; Colpaert et al., 1980b). For those subjects that do acquire stimulus control, discriminative performance with higher training doses may also be more accurate than is discriminative performance with lower training doses (De Vry and Slangen, 1986b), but such differences are not always obtained (e.g., Beardsley et al., 1987; Koek and Slangen, 1982b; White and Appel, 1982b).

The advantage conferred by a relatively high training dose may be restricted to the period of acquisition of stimulus control. After stimulus control has been established, highly accurate discriminative performance can be maintained as the training dose is decreased (Fig. 3). Overton (1979) systematically explored the ability of a wide variety of drugs to maintain discriminative control

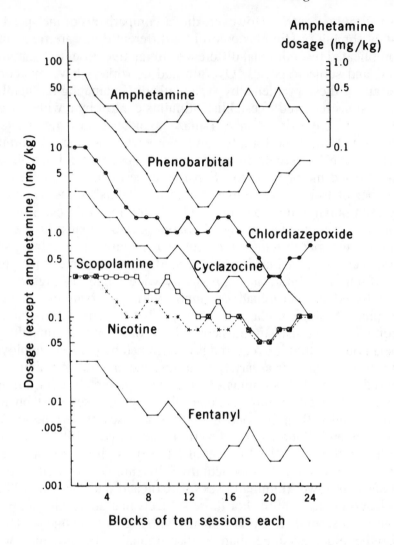

Fig. 3. Training doses at which accurate discriminative performance was achieved for ten consecutive training sessions. Saline and the initial dose of each drug were established as discriminative stimuli for water-reinforced lever presses in rats. Then, the training dose of each drug was progressively decreased according to the criteria described in the text. Each line presents training doses for one rat during successive blocks of ten sessions. Note the displaced ordinate for amphetamine. [Reprinted with permission from Overton D. A. (1979) Drug discrimination training with progressively lowered doses. *Science* **205,** 720–721. Copyright 1979 by the AAAS.]

as the training dose was progressively decreased. Saline and a "moderately high" dose of each drug were established as discriminative stimuli for water-reinforced lever presses in rats. After criterion discriminative performance was achieved, the training dose of each compound was reduced by approximately 30%, and discrimination training contingencies were maintained for ten sessions. Then, if discriminative performance remained accurate, the training dose was again decreased. If performance was not accurate, the dose was not changed, and if performance remained inaccurate for 20 sessions, the dose was raised by approximately 30%. As shown in Fig. 3, each drug maintained accurate performance as the training dose was decreased by 60–95%. At some low dose, however, discriminative control was no longer maintained. Note that control was reestablished rapidly by an increased dose.

Other experiments employing similar strategies have demonstrated that such gradual decreases in drug training dose, coupled with continued differential reinforcement of the drug- and saline-appropriate behavioral repertoires, can shift control to doses that do not initially evoke generalization with the original training dose. Such "fading" of the dose required for stimulus control has been demonstrated for hallucinogens, stimulants, sedative-hypnotics, and opioids (Beardsley et al., 1987; Colpaert and Janssen, 1982b; Colpaert et al., 1980a, Greenberg et al., 1975b, White and Appel, 1982b). Additionally, experiments with fentanyl suggest that a higher dose may be required to establish discriminative control than is required to maintain a discrimination originally established by a yet higher dose (Colpaert et al., 1980a, b). Albeit limited, these data suggest that drug discriminative control may be sustained by lower drug doses than are required for its establishment. As yet, the factors determining the lowest dose capable of sustaining discriminative performance are unknown.

The discriminative potential of psychoactive drugs is firmly established. The following section will survey briefly the pharmacological and behavioral profiles of drug discriminative stimuli.

3. Characteristics of Drug Stimulus Control

Until now, our discussion has focused on establishment and maintenance of discriminative stimulus control by a fixed dose of a single reference drug. However, drug stimulus control is not limited to the training stimulus itself. Considerable experimental attention has been paid to how changes in dose or drug change discriminative performance. Three general conclusions can be drawn from this body of research. First, abrupt decreases in the dose of the training drug can produce proportional decreases in the likelihood of drug-appropriate behavior. Second, the family of drugs that will occasion generalization with a drug training stimulus is characteristically limited to those drugs that share other pharmacological actions with the training drug. Third, patterns of discrimination and generalization appear remarkably consistent over impressively long time spans. These characteristics of drug stimulus control will be surveyed briefly.

3.1. Tests for Generalization of Stimulus Control

Tests of generalization of drug stimulus control are arranged to vary some feature of the drug stimulus and measure the resulting variation in the proportion of behavior appropriate to the presence or absence of the drug stimulus. Generalization is often explored in test sessions arranged so that discriminative behavior is not reinforced differentially (Fig. 1, "Tests of control"). This approach explicitly acknowledges the multiple stimulus elements of the experimental environment and varies only those of the drug stimulus. During test sessions, experimental contingencies are arranged to provide reinforcement either for both response options (Fig. 1, Type 1) or for neither option (extinction; Fig. 1, Type 2). An alternate strategy (Colpaert and Niemegeers, 1975; Colpaert et al., 1975a) is to identify the first option completed and arrange reinforcement for only that option throughout the remainder of the session. The possibility that reinforcer delivery or nondelivery may itself exert discriminative control, and thereby alter the outcome of generalization tests, is an unresolved issue in drug discrimination research (cf. Ator and Griffiths, 1983; Colpaert, 1977, Koek and Slangen, 1982b). In general, however, it appears that the specific testing procedure employed exerts less control over generalization patterns than do the dose and pharmacological class of the comparison stimuli. Nonetheless, the reader is cautioned that such testing variables may contribute to differences in tolerance development that we will attribute to other processes.

A generalization test has one of three outcomes:

1. The test drug can evoke a distribution of responses indistinguishable from that evoked by the training drug stimulus, an outcome interpreted as a clear generalization of drug stimulus control
2. The test drug can evoke a distribution of responses indistinguishable from that evoked by the vehicle, an outcome interpreted as a clear lack of generalization, or
3. The test drug can evoke a distribution containing both drug- and saline-appropriate behavior, an outcome with no uniformly accepted interpretation.

Several possible sources for such "partial generalization" have been identified (Colpaert, 1986; Overton, 1974; Winter, 1978). One source is the individual subject emitting responses appropriate to both stimulus conditions. However, if data from several tests are combined, apparent "partial generalization" may result from collapsing data from an individual subject that displays only drug- or only saline-appropriate behavior in each test, or from collapsing data from several subjects that individually display any of the three patterns. For these reasons, our later discussion of changes in drug stimulus control as a function of tolerance will focus primarily on changes in the doses of drug required to evoke the level of drug-appropriate behavior established as the criterion for stimulus control during training.

There are multiple ways of presenting generalization data. Some investigators (e.g., Colpaert, 1986; Shearman and Herz, 1982; Wood and Emmett-Oglesby, 1986) define a criterion for generalization, and present the proportion of animals meeting that criterion as a function of test drug and dose. Other investigators evaluate the overall distribution of drug-appropriate responses or behavioral units emitted by each subject. Among this latter group, some workers (e.g., Ator and Griffiths, 1983; Hein et al., 1981) present data for individual subjects, whereas others (e.g., Herling et al., 1980; Shannon and Holtzman, 1976; Stolerman and D'Mello, 1981) present mean values for groups of subjects. These methods yield identical results for stimuli that evoke only saline-appropriate or only drug-appropriate behavior in every subject. The methods differ in the data extracted from tests in which both discriminated operants occur and in the possible sources of "partial" generalization. However, indices of either the proportion of responses under drug stimulus control or the proportion of subjects displaying drug stimulus control can yield similar estimates of ED_{50} doses, doses required for complete generalization, and/or slopes of the generalization gradients (e.g., Koek and Slangen, 1982b; Stolerman and D'Mello, 1981; but *see* Goudie et al., 1986). These studies do suggest that the frequency of partial generalization scores derived from the proportion of drug-appropriate responses emitted by individual subjects cannot be predicted from the normal distribution, arguing cautious use of parametric statistics.

3.2. Dose Dependence

Pharmacological stimulus control is critically dependent on the dose of the test stimulus. The variation of stimulus control with drug dose can be illustrated by a study of the discriminative stimulus characteristics of pentobarbital (Kline and Young, 1986). Stimulus control of food-reinforced key pecking was established by saline and a dose of 5.6 mg/kg of pentobarbital in pigeons. After a robust discrimination was established, two generalization functions were established for each subject over a period of approximately 4.5 mo. First, a complete range of doses was tested in an unsystematic order at 2–8-d intervals. Training sessions were conducted on the intervening weekdays. Then, a period of approximately 7 wk elapsed, in which training sessions were conducted during the first 2 and last 3 wk. Finally, each dose was reexamined one or more times during a second determination of the pentobarbital generalization function. Again, training sessions were conducted on the intervening weekdays. During each test session, 20 consecutive responses on either key resulted in food delivery.

As shown in Fig. 4, stimulus control of behavior by pentobarbital varied as a function of dose. In individual pigeons, a dose of 0.32 or 1.0 mg/kg pentobarbital evoked responses exclusively to the saline-appropriate lever. As the dose of pentobarbital increased to the 5.6 mg/kg training dose, a greater percentage of responses were emitted on the drug-appropriate lever. Once a given

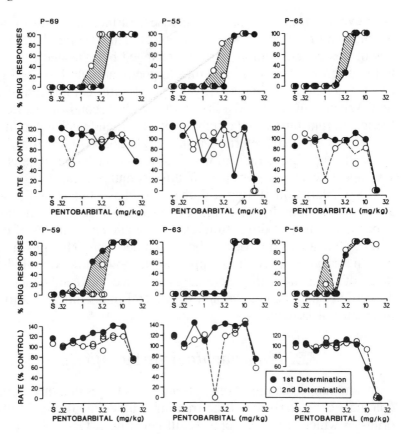

Fig. 4. Dose–response curves for generalization of stimulus control by pentobarbital in six pigeons trained to discriminate saline and 5.6 mg/kg pentobarbital in a food-reinforced task. Abscissae: mg/kg dose of pentobarbital, log scale. Ordinates, upper panels: number of responses emitted on the drug-appropriate key, expressed as a percentage of the total number of responses emitted during the session. Ordinates, lower panels: response rate, expressed as a percentage of responses during saline training sessions. Closed circles represent data obtained during the first determination of the pentobarbital generalization gradient. Open circles represent all observations obtained during the second determination; dotted lines connect the range of stimulus control values and the average of rate values. Points at "S" represent tests of saline. [Reprinted with permission from Kline F. S. and Young A. M. (1986) Differential modification of pentobarbital stimulus control by *d*-amphetamine and ethanol. *Pharmacol. Biochem. Behav.* **24,** 1305–1313. Copyright 1986 by ANKHO International, Inc.]

dose evoked drug-appropriate responses, further increases in dose maintained or improved stimulus control. The general characteristics of this generalization were not changed by extended exposure to the discrimination task, and thus to pentobarbital. A second gradient was determined approximately 7 wk after completion of the first. When compared to the first gradient, the second was shifted to the left in four birds, to the right in one bird, and unchanged in the final bird. In both determinations, high doses of pentobarbital markedly suppressed rates of responding, but such rate suppression was not a necessary correlate of drug stimulus control.

Such variation of drug stimulus control as a function of dose is a robust feature of drug discriminative stimuli in general. The dose-dependent nature of drug stimulus control allows estimates of drug potency that are as rigorous as those obtained from other pharmacological assay systems. It is important to note, however, that the range of doses that will generalize is not fixed. but can be modified by certain pharmacological and behavioral interventions (outlined in Section 4 below).

3.3. Pharmacological Specificity

A second major feature of drug stimulus control is that the pharmacological characteristics of the training drug define the set of other drugs to which stimulus control will generalize. Generalization of drug stimulus control requires considerable pharmacological specificity. Drugs defined by other criteria as being within or without an identified pharmacological class often share discriminative stimulus profiles, whereas drugs from dissimilar pharmacological classes generally do not. In general, such specificity of stimulus control appears independent of subject, species, and the details of discrimination training and testing.

Within a pharmacological class, cross-generalization patterns are frequently symmetrical. For example, among the mu opioids, stimulus control by morphine generalizes to appropriate doses of codeine and fentanyl (Gianutsos and Lal, 1976; Schaefer and Holtzman, 1977; Shannon and Holtzman, 1976), stimulus control by codeine generalizes to appropriate doses of morphine and fentanyl (Bertalmio et al., 1982; Woods et al., 1981), and stimulus control by fentanyl generalizes to appropriate doses of codeine and morphine (e.g., Colpaert et al., 1975b, c). So consistent have been the results of such cross-generalization tests for drugs within known pharmacological classes that generalization with a training drug can be used as one index of common pharmacological effect (for example, *see* Colpaert, 1986; Holtzman, 1982, 1985 for opioids; Shannon, 1983 for dissociative anesthetics; Appel et al., 1982 for hallucinogens; Young and Glennon, 1986 for CNS stimulants). However, common discriminative profiles are not unequivocal indices of other common effects. At least two lines of evidence urge caution in interpreting results of generalization studies.

First, complete generalization between drugs, or between doses of the same drug, does not necessarily indicate that the two stimuli are indistinguishable. As discussed further below, alterations in the pharmacological or behavioral context of the discriminative assay can allow behavior to discriminate among stimuli that initially occasion a common behavioral repertoire (cf. Overton, 1982b). Second, asymmetrical cross-generalization has been noted for several drug pairs (cf. Ator and Griffiths, 1983; Colpaert et al., 1976b; Jarbe and McMillan, 1983; Hirschhorn, 1977). For example, in baboons, the benzodiazepine lorazepam occasions generalization to a pentobarbital training stimulus, but pentobarbital does not occasion generalization to a lorazepam training stimulus (Ator and Griffiths, 1983). Such asymmetries urge caution in concluding that drugs share other properties or are members of a common class on the basis of one-way generalization tests.

4. Modification of the Sensitivity of the Discriminative Assay

As outlined above, discriminative control by drug stimuli co-varies with drug dose. Several lines of evidence suggest that the apparent sensitivity of the discrimination assay is a product both of the dose employed as the training stimulus and of the historical and current context in which it is encountered. The discriminative stimulus profile of any particular drug is not static, but rather the dynamic result of interactions of the drug itself and the behavioral contexts in which it has been encountered. Each of the factors highlighted below may contribute to the apparent sensitivity of the assay, and to the changes that may indicate tolerance.

4.1. Role of Training Dose

A primary determinant of the dose–effect function for generalization of stimulus control is the dose of drug employed as the training stimulus. The training dose can determine both the range of doses required for generalization and the magnitude of the change in dose required to change stimulus control, that is, the position of the generalization gradient along the dose axis and its slope. In general, proportionately lower doses exert stimulus control following training with a lower dose than do following training with a higher dose (cf. Beardsley et al., 1987; Colpaert et al., 1980a; De Vry and Slangen, 1986a, b; Jarbe and Rollenhagen, 1978; Shannon and Holtzman, 1979; White and Appel, 1982b). Furthermore, following training with a lower dose, larger changes in dose may be required to change stimulus control, resulting in a shallower slope of the gradient, than are required following training with a higher dose (cf. Colpaert, 1982b; Colpaert et al., 1980a,b, White and Appel 1982b), but this is not uniformly observed (cf. Beardsley et al., 1987; De Vry and Slangen, 1986a,b; Koek and Slangen, 1982b).

Control of the generalization gradient by training dose appears to be a general characteristic of drug discriminative stimuli, having been demonstrated for a wide variety of compounds, including CNS stimulants (Colpaert and Janssen, 1982b; Rosen et al., 1986; Stolerman and D'Mello 1981), dissociative anesthetics (Beardsley et al., 1987), hallucinogens (Greenberg et al., 1975b; White and Appel, 1982a,b), opioids (Colpaert, 1980a,b; Koek and Slangen, 1982b; Shannon and Holtzman, 1979), and sedative hypnotics (De Vry and Slangen, 1986a,b). Moreover, such control appears largely independent of the particular procedures used in training and testing, having been observed under a wide range of reinforcement schedules, testing paradigms, and measures of stimulus control (e.g., Colpaert et al., 1980a,b; Koek and Slangen, 1982b; Stolerman and D'Mello, 1981; Shannon and Holtzman, 1979; White and Appel, 1982b). Importantly, control of the generalization gradient by the current training dose is observed both when different groups of subjects are trained with different doses, and when the same subjects are tested first after training with a high dose and again after training with a lower dose (cf. Colpaert et al., 1980a,b).

The general relation between training dose and the resulting generalization gradient under the latter condition is illustrated in Fig. 5 (derived from Rosen et al., 1986). Amphetamine and saline were established as discriminative stimuli for food-reinforced lever presses maintained under fixed-ratio schedules in rats. Stimulus control was defined as the individual subject completing a ratio on the injection-appropriate lever before completing a ratio on the inappropriate lever, and distributing at least 90% of all responses to the injection-appropriate lever. Initially, the training dose of amphetamine was 1.0 mg/kg. After development of stimulus control, various doses of amphetamine and other stimulants were tested for generalization. As shown by the closed circles, under the high training dose condition, an amphetamine dose of 1.0 mg/kg was required for stimulus generalization by every subject. Lower doses occasioned progressively less drug-appropriate behavior. A dose higher than 1.0 mg/kg abolished responding in every subject.

After completion of all generalization tests, the training dose of amphetamine was progressively decreased over successive training sessions. The dose of amphetamine was reduced in decrements of approximately 30%, and subjects were exposed to at least eight training sessions at each dose. Each subject was then maintained at the lowest dose that sustained discriminative performance (0.17–0.32 mg/kg in individual rats), and generalization tests were again conducted. Lowering the training dose of amphetamine lowered the minimal dose required for stimulus control in every subject to 0.32 mg/kg (Fig. 5, open circles). Note, however, that doses higher than the new training dose (i.e., 0.56 and 1.0 mg/kg) continued to evoke drug stimulus control. Note also that the decrease in the dose required for stimulus control was not accompanied by a decrease in the dose required to suppress response rates (Fig. 5, lower panel). This dissociation of the discriminative and rate-altering potency of amphetamine suggests that behavioral disruption was not an essential element of the amphetamine stimulus.

A more complete demonstration of a dissociation of the discriminative and rate-altering potencies of a drug stimulus has been

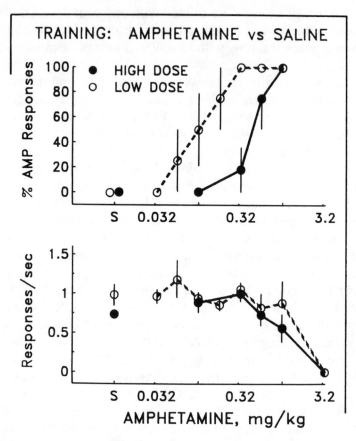

Fig. 5. Dose–response functions for generalization of stimulus control by amphetamine in rats before and after training with a progressively lowered dose. Closed circles represent the initial function obtained in rats trained to discriminate saline and 1.0 mg/kg amphetamine. Open circles represent the function obtained in the same rats after the training dose was progressively decreased to the lowest dose that sustained discriminative performance (0.17–0.32 mg/kg). Abscissae: mg/kg dose of amphetamine, log scale. Ordinates, upper panels: number of responses emitted on the drug-appropriate lever, expressed as a percentage of the total number of responses emitted during the session. Ordinates, lower panels: response rate, expressed as a percentage of responses during saline training sessions. Data points represent the mean of a single determination in each of four rats. [Derived from Rosen J. B., Young A. M., Beuthin F. C., and Louis-Ferdinand R. T. (1986) Discriminative stimulus properties of amphetamine and other stimulants in lead-exposed and normal rats. *Pharmacol. Biochem. Behav.* **24,** 211–215.]

provided by Beardsley et al. (1987), who examined dose–response functions for the discriminative and rate-altering effects of the dissociative anesthetic phencyclidine, as the training dose was systematically decreased from 3.0 to 0.375 mg/kg in rats. Decreases in the training dose were accompanied by proportional decreases in both the minimal dose required to evoke stimulus control and the ED_{50} dose. Thus, the ratio of the ED_{50} dose to the training dose remained constant across a roughly 10-fold range of training doses. In contrast, decreases in the training dose did not alter the rate-decreasing potency of phencyclidine, with the result that the ratio of the ED_{50} dose for rate suppression to the training dose increased by approximately 15-fold. These data convincingly demonstrate that the discriminative and rate-altering potencies of psychoactive drug stimuli are separable and suggest that changes in sensitivity to one facet of a drug's action should not be used *a priori* to infer a parallel change in sensitivity to its stimulus actions (cf. Section 5.1.2. below).

In the studies by Rosen et al. (1986) and Beardsley et al. (1987) described above, the training dose was progressively decreased. To date, one published study has reported the effects of increasing the training dose (White and Appel, 1982b). Saline and a dose of 0.08 mg/kg LSD were established as discriminative stimuli for water-reinforced lever pressing in rats. After all subjects acquired the discrimination, the training dose was progressively decreased to 0.02 mg/kg in one subgroup, increased to 0.32 mg/kg in a second subgroup, or unchanged in a third subgroup. Both changes in the training dose produced a proportional change in the doses required for stimulus control. Of particular interest here, the original training dose of 0.08 mg/kg LSD did not evoke complete generalization after retraining with the higher 0.32 mg/kg dose. Thus, it appears that the range of doses capable of exerting stimulus control in the drug discrimination assay is determined by current rather than historical training conditions. As will be discussed later, this pattern has led to controversy about the interpretation of experiments designed to evaluate tolerance to drug stimulus control.

4.2. Role of Behavioral Contingencies

The dose of a drug stimulus employed for discrimination training is not the only determinant of the obtained generalization gradient. The range of doses capable of exerting drug stimulus control can also be altered by the behavioral contingencies imposed during discrimination training. In the usual discrimination experiment, symmetrical differential reinforcement contingencies are in effect in the presence of drug and saline (cf. Fig. 1). Studies with fentanyl and phencyclidine (Colpaert and Janssen, 1981; De Vry et al., 1984; Koek and Slangen, 1982a, b; McMillan and Wenger, 1984) suggest that changes in the symmetry of these contingencies can change the apparent sensitivity of the discrimination assay. First, differentially punishing one discriminated response repertoire increases the likelihood that novel test doses will evoke the alternate repertoire (Colpaert and Janssen, 1981). Specifically, punishing saline-appropriate responses in the presence of fentanyl resulted in a shallower generalization gradient and lower ED_{50} for stimulus control than were obtained after exposure to symmetrical training contingencies. Conversely, punishing drug-appropriate responses in the presence of saline resulted in a steeper gradient and higher ED_{50} than were obtained after exposure to symmetrical contingencies. Similar changes in dose–response gradients were obtained in cross-generalization tests with morphine and sufentanil.

A second alteration in symmetry known to modify generalization gradients is a change in the relative frequency or density of reinforcement for the drug- or saline-appropriate repertoire. Increasing the relative frequency of reinforcement for one discriminated repertoire increases the likelihood that novel stimuli will also evoke that repertoire (De Vry et al., 1984; Koek and Slangen, 1982a, b; McMillan and Wenger, 1984). For example, Slangen and colleagues (De Vry et al., 1984; Koek and Slangen, 1982a) reported that increasing the frequency of reinforcement during fentanyl training sessions, relative to that available during saline training sessions decreases the ED_{50} for fentanyl stimulus control. Conversely, increasing the frequency of reinforcement during saline training sessions increases the ED_{50}. Such control of a generalization gradient by differences in the relative frequency of reinforcement during drug and saline training sessions may also result from the direct behavioral effects of the training drug itself. Several investigators have noted that, for many drugs, doses commonly employed as training stimuli can have pronounced rate-decreasing effects, which may act to decrease the relative frequency of reinforcement during drug training sessions and thereby inflate the apparent ED_{50} for generalization of stimulus control (e.g., Colpaert, 1982b; Koek and Slangen, 1982a,b; cf. De Vry et al., 1984).

A final manipulation of the discrimination context that alters the apparent sensitivity of the assay is establishment of a discrimination between, not one dose of a drug and saline, but two doses of the same drug. Such training can increase the apparent ED_{50} for generalization of stimulus control by the higher drug dose (Colpaert, 1982a; Colpaert and Janssen, 1982a; Koek and Slangen, 1982b). Parametric experiments suggest that the ratio between the two training doses controls the slope of the gradient, with a relatively small ratio generating a steeper gradient than does a saline–drug discrimination (Colpaert and Janssen, 1982a).

Overall, such patterns highlight the interactive control that pharmacological and behavioral variables exert over the apparent sensitivity of the discrimination assay. These patterns also demonstrate that changes in the dose required for drug stimulus control can occur in the absence of changes in the dose or frequency of repeated drug administration. This dissociation has led to controversy in interpreting the results of experiments designed to assess tolerance processes.

4.3. Modulation of Generalization Profiles

The pharmacological and behavioral factors outlined above may also control membership in the set of other drugs that will occasion generalization with a drug discriminative stimulus. Some workers (e.g., Colpaert and Janssen, 1981) have suggested that discrimination training conditions that generate dose–response gradients with low ED_{50} values and shallow slopes will engender generalization sets that include not only a greater range of doses of the training drug itself, but also a larger variety of other psychoactive drugs. For example, the training dose of an opioid stimulus appears to determine the set of other drugs that will occasion drug-appropriate responses (e.g., Colpaert et al., 1980b; Holtzman, 1982; Shannon and Holtzman, 1979; Teal and Holtzman, 1980a, b). In particular, certain agonist-antagonist opioids evoke less morphine-appropriate behavior when the training dose of morphine is 5.6 or 10 mg/kg than when it is 1.0 mg/kg (Holtzman, 1982; Shannon and

Holtzman, 1979; cf. Colpaert et al., 1980b). The nonopioid
d-amphetamine may also evoke less generalization to a higher than
a lower morphine dose (Colpaert et al., 1980b; but *see* Koek and
Slangen, 1982b).

Although the set of drugs that will evoke generalization with a
low morphine dose is larger than the set that evokes generalization
with a higher dose, it is clearly not infinite, as behaviorally active
doses of naltrexone, scopolamine, and phencyclidine occasion little
generalization to any training dose of morphine (Holtzman, 1982).
Taken together, these patterns suggest that changes in generalization as a function of training dose may reflect changes in the promi-
nence of separable stimulus elements of a drug. They also suggest
that changes in the intensity of a drug stimulus may be reflected in
changes in the set of drugs that evoke generalization.

5. Development of Tolerance to Drug Stimulus Control

The literature reviewed above has established that development and
maintenance of drug discriminative stimulus control reflect a dynamic interaction of behavior, drug, and the demands and possibilities imposed by the learner's environment. Questions about the occurrence and nature of tolerance to drug stimulus control focus
attention on the role of drug exposure *per se* in maintaining or
changing discriminative performance. However, defining the contribution of frequent drug exposure to the potency of a drug as a
discriminative stimulus requires attention not only to the frequency,
duration, and dose of drug administered, but also to the requirements of the discrimination task and the outcomes of discriminative
behavior during such drug exposure. As discussed in this section,
the evidence for development of tolerance to drug stimulus control
has frequently given rise to conflicting interpretations.

Tolerance can be defined as the decrease in sensitivity to a
drug that results from prior exposure to that or a related drug
(Goldstein et al., 1974). Thus, tolerance is reflected in a decreased
responsiveness to a fixed dose of a drug, with a concomitant increase in the dose necessary to achieve its original effects. In gen-

eral, tolerance is viewed as a decrease in the potency of a drug, not in its efficacy. However, under certain conditions, the original effects cannot be evoked by any dose in the tolerant organism.

Different investigators have called attention to different possible outcomes of the development of tolerance to a drug discriminative stimulus. One outcome of tolerance development could be a progressive deterioration in discriminative control by an initially effective training dose, reflected in an increased probability of saline-appropriate responses after drug administration (cf. Colpaert et al., 1980a). Such deterioration might be produced by repeated exposure to the training dose itself, or by exposure to supplemental doses administered outside the confines of the discrimination task. A second outcome of the development of tolerance could be a change in the range of doses capable of evoking drug-appropriate behavior, with or without a change in control by the training dose itself. Thus, an increase in the dose required for stimulus control or in the ED_{50} for stimulus control might reflect tolerance. A third outcome of tolerance could be a concurrent development of cross-tolerance to drugs within the same pharmacological class as the training drug. A final outcome of tolerance development could be a change in the set of drugs that evoke generalization. This section will review the conditions of chronic drug exposure that do, and do not, yield tolerance to drug stimulus control.

5.1. Does Extended Exposure to a Drug Discrimination Task, and Thus to a Drug, Confer Tolerance to Drug Stimulus Control?

It should be evident that repeated administration of a drug, *per se*, does not prevent establishment or maintenance of drug stimulus control, but rather is required for revealing a drug's discriminative potential. However, establishment and maintenance of discriminative control may coincide with apparent development of tolerance to other effects of the drug stimulus. Such lack of congruence in the development of tolerance to multiple effects of the same drug has engendered controversy, but is not unique to drug stimulus control, as discussed in the chapter by Wolgin in the present volume.

5.1.1. Maintenance of Stimulus Control during Extended Training

It is well established that an initially effective training dose will continue to exert stimulus control even after prolonged exposure to the discrimination task and its drug stimulus. Thus, maintenance of drug stimulus control does not require regular increases in dose. Moreover, several reports suggest that the doses that generalize to a particular training dose do not become progressively lower or higher as the duration of exposure to the discrimination task, and thus to drug, increases.

Several investigators have evaluated the stability of drug stimulus control by constructing repeated dose-response functions during sustained exposure to a discrimination task (cf. Fig. 4; Colpaert et al., 1976a; Colpaert et al., 1978a, b, d; Hein et al., 1981; McMillan, 1987, Sannerud and Young, 1987, Shannon and Holtzman, 1976). In each of these experiments, the presence and absence of the drug training dose continued to exert excellent stimulus control throughout prolonged discrimination training. Repeated tests of other doses allowed estimates of any changes in the range of doses capable of exerting drug stimulus control across the same period. In a representative experiment, Colpaert et al. (1978a) generated dose–response functions for a 0.04 mg/kg fentanyl stimulus by testing selected doses of fentanyl in successive daily sessions. Two weeks of discrimination training sessions separated each of five test sequences. The minimal dose of fentanyl required for stimulus control in individual animals oscillated by a factor of 2–4 over successive tests, but did not progressively increase or decrease over the course of the 17-wk experiment. The minimal dose of morphine required for stimulus control in cross-generalization tests also oscillated by a factor of 2–4.

Similarly, in tests of a discrimination established by 10 mg/kg cocaine under similar experimental conditions, the minimal dose of cocaine required for stimulus control oscillated around an apparently stable baseline value (Colpaert et al., 1978b). In a second test of the stability of opioid stimulus control, Sannerud and Young (1987) generated repeated generalization functions for a 3.2 mg/kg morphine stimulus at 1- to 2-mo intervals for periods of 9–6 mo. In individual animals, the minimal doses required for stimulus control varied by a factor of 0.25–0.75 common log unit over all tests, and by a factor of 0.25 or 0.5 log unit over successive tests (Fig. 6). Again, the minimal doses required for stimulus control did not progressively increase or decrease, but rather oscillated around an apparently stable baseline value.

As an extreme example, McMillan (1987) evaluated dose–response gradients for phencyclidine in pigeons over a period of 5 yr, during which time subjects were tested with a wide variety of drugs and behavioral manipulations. In three of five birds, the dose required for stimulus control showed very little variation, although in the remaining two birds, the dose varied but did not systematically rise or fall. Taken together, these experiments reinforce the general observation that an initially effective training dose maintains stimulus control even after prolonged exposure to a discrimination task. Additionally, these experiments demonstrate that, although the dose of a drug required for stimulus control does not appear to change progressively over the course of sustained training, generalization gradients for drug stimulus control in individual subjects are not static. Therefore, experiments to evaluate changes in individual sensitivity to drug stimulus control should employ multiple observations to estimate the "normal" oscillation in the doses required for stimulus control under training conditions.

The robust stimulus control exerted by the initial drug training dose over prolonged discrimination training has lead some workers to argue that drug stimulus control is not subject to tolerance development. In particular, Colpaert and colleagues (Colpaert 1978a, b; Colpaert et al., 1978d) have argued that sensitivity to drug stimulus control, as defined operationally by the dose required for drug-appropriate behavior, is not modulated by pharmacodynamic tolerance processes. This argument proposes that the dose required for stimulus control is set by the training dose, modulated by the subject's physiological sensitivity and by a phasic sensory gating process, and responsive to the continued differential reinforcement imposed during drug and saline training sessions. In contrast to its responsiveness to behavioral processes, the system is proposed to be resistant to modulation by the pharmacodynamic consequences of either frequent exposure to the training drug or supplemental drug maintenance regimens. The effects of the latter interventions are discussed below. Unraveling any altered sensitivity to a drug

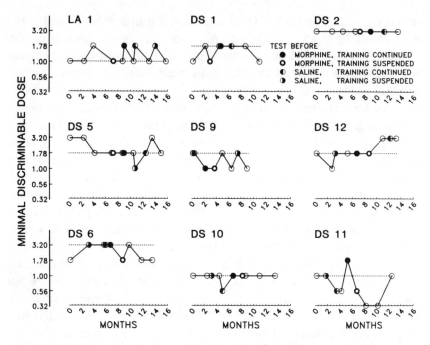

Fig. 6. Lack of change in the minimal dose to occasion drug-appropriate responding in tests of stimulus control conducted periodically during the 16 mo after establishment of discriminative control by 3.2 mg/kg morphine in rats. Each symbol represents the MDD (Minimal Discriminable Dose; lowest dose in a generalization gradient to engender greater than 90% drug-appropriate responding) from a single cumulative dose generalization test. Filled symbols represent cumulative dose tests conducted immediately before each of four experimental phases, as noted in the legend. The dotted line represents the median MDD value calculated over all tests. Note that the MDD for eight of the nine subjects varied by a maximum of 0.25–0.50 log unit throughout the study. [Reprinted with permission from Sannerud C. A. and Young A. M. (1987) Environmental modification of tolerance to morphine discriminative stimulus properties in rats. *Psychopharmacology* **93**, 59–68. Copyright 1987 by Springer-Verlag New York, Inc.]

stimulus produced by the repeated training that characterizes most drug discrimination procedures will require experiments that halt discrimination training for a period and then reassess control by various doses of the training drug. Any changes in the dose required for stimulus control should be compared to changes observed in matched subjects for whom discrimination training is not halted. Further, in order to examine the contribution of frequent drug treatment itself, repeated exposure to the training drug stimulus should be discontinued when training is halted in one subgroup and continued in the absence of training in another subgroup.

To our knowledge, no experiment has examined all parts of this question. In particular, experiments have not evaluated whether maintenance of stimulus control by a frequently encountered drug dose requires continued training, with its explicit differential reinforcement of continued discrimination. However, two recent experiments have explored the role of frequent drug administration by halting both drug exposure and discrimination training for a time, and then reevaluating the range of doses required for stimulus control (McMillan, 1987; Sannerud and Young, 1987). Both suggest that halting drug exposure yields few changes in the range of doses that will exert drug stimulus control. In rats trained to discriminate saline and a dose of 3.2 mg/kg morphine, dose–response functions determined after discrimination training and morphine administration were suspended for 2 wk were generally similar to those determined during sustained exposure to the task (Sannerud and Young, 1987; compare the open and closed triangles in Fig. 7). Similarly, in pigeons trained to discriminate saline and phencyclidine (1.2 or 1.5 mg/kg), dose–response functions determined after discrimination training and its correlated drug exposure were suspended for 2 mo were no different from those determined during sustained exposure to the discrimination task (McMillan, 1987). Albeit limited, these data support the suggestion by Colpaert (1978a,b) that the frequent drug exposure required by most discrimination protocols does not maintain a "background" level of tolerance to the training stimulus.

5.1.2. Development of Tolerance to Other Effects of a Drug Stimulus

Continued discriminative control by a drug stimulus may coincide with development of tolerance to other effects of the same drug and dose. Several workers have noted that development of tolerance to drug-induced disruption of motor performance may occur without loss of discriminative accuracy (i.e., Bueno and Carlini, 1972,; Greenberg et al., 1975a; Henriksson et al., 1975; Kuhn et al., 1974). For example, Colpaert and coworkers demonstrated differ-

CHRONIC SALINE TREATMENT

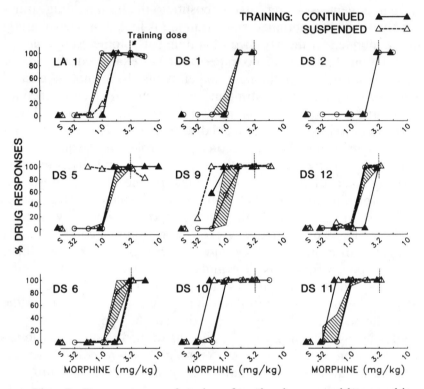

Fig. 7. Dose–response functions for stimulus control by morphine in individual rats before and during twice daily administration of saline. Abscissae: mg/kg dose of morphine (log scale). Ordinates: responses to the morphine-appropriate lever, expressed as a percentage of total responses. Morphine (3.2 mg/kg) and saline were established as discriminative stimuli for lever presses maintained under fixed-ratio schedules of food presentation in rats. *Open circles and shaded areas* represent the median and ± absolute average deviation for control tests conducted throughout the experimental period. *Closed triangles* represent tests conducted after discrimination training was continued during the 2-w saline treatment. *Open triangles* represent tests conducted after training was suspended during the 2-w saline treatment. The vertical dotted line indicates the training dose of morphine. *See* text for further details. [Reprinted with permission from Sannerud C. A. and Young A. M. (1987) Environmental modification of tolerance to morphine discriminative stimulus properties in rats. *Psychopharmacology* **93,** 59–68. Copyright 1987 by Springer-Verlag New York, Inc.]

ential development of tolerance to the stimulus, analgesic, and rate-altering effects of fentanyl (Colpaert et al., 1976a; cf. Colpaert et al. 1978c, d). These investigators established 0.04 mg/kg fentanyl and saline as discriminative stimuli for food-maintained responses in rats. Dose–response functions were generated by testing selected doses of fentanyl over successive daily sessions. Two weeks of discrimination training sessions separated each of five dose–response tests of fentanyl and three tests of morphine. Analgesic tolerance was assessed by a warm water tail withdrawal assay conducted after completion of all generalization tests. Establishment of stimulus control required a median of 40 sessions, during which time the response rates during both saline and fentanyl sessions increased markedly, probably as a function of improved ratio schedule control. Following establishment of stimulus control, there were no further systematic changes in either the discriminative or rate-altering effects of 0.04 mg/kg fentanyl or saline over 15 wk. During dose–response tests, neither the ED_{50} values for stimulus generalization nor the rate-altering effects of fentanyl and morphine changed as a function of time. In contrast, at the end of the 17-wk test period, a fixed iv dose of 0.02 mg/kg fentanyl produced a significantly shorter duration of analgesia in discrimination subjects than in control subjects who had received no prior opioid exposure. Miksic and Lal (1977) extended this observation by evaluating the dose–response functions for the analgesic effects of morphine in naive rats and in rats administered 10 mg/kg morphine approximately every other day during discrimination training. The analgesic potency of morphine was at least fourfold lower in the discrimination subjects than in the control subjects. Thus, sustained discriminative stimulus control by a fixed dose of a drug can be accompanied by development of tolerance to other effects of that dose.

In summary, the repeated exposure to a drug required to demonstrate discriminative control does not appear to confer tolerance to drug stimulus control. Maintenance of drug stimulus control does not require progressive increases in dose, nor does the range of doses required for generalization change during extended drug exposure or after periods of drug abstinence. However, sustained discriminative control can be accompanied by development of tolerance to other effects of the training dose. It remains to be determined whether maintenance of stimulus control by the original training dose is fostered by the explicit reinforcement contingencies imposed during discrimination training.

5.2. Does Supplemental Drug Administration Confer Tolerance to Drug Stimulus Control?

A fixed dose of drug maintains robust discriminative control of behavior over prolonged exposure. A number of experiments have evaluated whether establishment or maintenance of this discriminative control can be modulated by administration of supplemental drug doses that are not paired with the demands of the discrimination task. Three experimental strategies have been used to examine the ability of supplemental drug administration to confer tolerance. One strategy has been to preexpose subjects to a drug before beginning discrimination training and evaluate any changes in the speed or accuracy of acquisition. The second strategy has been to establish stimulus control by a fixed training dose of a drug and then concurrently expose subjects both to the training dose administered before drug training sessions and to supplemental doses administered without correlated exposure to the discrimination task, followed by tests for any changes in the doses required for generalization. The final strategy has been to establish stimulus control by a fixed training dose and then suspend discrimination training, expose subjects to supplemental doses, and evaluate any changes in the doses required for generalization.

5.2.1. Preexposure Effects

Preexposure effects have been examined for cannabinoids, antimuscarinics, sedative-hypnotics, and opioids. It appears that prior exposure to a drug stimulus, even to an intensity slightly higher than that to be used in later discrimination training, may not alter the ability of differential reinforcement contingencies to establish discriminative control by that stimulus. For example, Bueno and Carlini (1972) demonstrated that rats that had developed tolerance to the motoric effects of daily injections of cannabis extract were nonetheless able to acquire and maintain a discrimination between an identical dose of the extract and its vehicle. Using a conceptually similar strategy, Colpaert and coworkers (1978d) compared the establishment of stimulus control by a 0.04 mg/kg dose of fentanyl in rats administered daily supplemental doses of either saline or 0.06 mg/kg fentanyl for 30 d before initiation of discrimination training. Preexposure to fentanyl yielded tolerance to the analgesic effects of 0.04 mg/kg fentanyl, but did not alter either the number of sessions required to establish stimulus control by 0.04 mg/kg fentanyl or the ED_{50} of later fentanyl generalization gradients. Indeed, within the first 10 discrimination training sessions, rats were more likely to emit the fentanyl-appropriate response when given saline than the saline-appropriate response when given fentanyl.

Taken together, these experiments suggest that preexposure to a drug stimulus, even to an intensity slightly higher than that to be used in later discrimination training, does not impair the ability of differential reinforcement contingencies to establish discriminative control by the drug. Discrepant results have been reported, however. Preexposure to a 5 mg/kg training dose of THC has been reported to impair acquisition slightly (Jarbe and Henriksson, 1973), whereas preexposure to a 3 mg/kg training dose of scopolamine has been reported to improve acquisition slightly (McKim, 1976). Although the factors underlying these different outcomes are unclear, the small differences reported between preexposed and control groups suggest that any effects of preexposure to the training dose are minimal.

Several experiments have extended the nonpaired supplemental drug administration into the period of initial discrimination training. In a study with pentobarbital, Jarbe and Holmgren (1977) pretreated separate groups of gerbils with saline or doses of 10 or 20 mg/kg pentobarbital for 20 d. Preexposure to pentobarbital resulted in alterations in its thermic and motoric effects. Next, all subjects were trained to discriminate saline and either 10 or 20 mg/kg pentobarbital in a shock-escape task. Subjects in the drug preexposure groups continued to receive daily injections of pentobarbital, before drug training sessions and after saline training sessions. Nonetheless, establishment of stimulus control by pentobarbital required no longer in these subjects than in subjects exposed to pentobarbital before drug training sessions only. Additionally, the groups did not differ in the asymptotic accuracy of the maintained discrimination or in susceptibility to bemegride antagonism.

In two conceptually similar studies with opioids (Gaiardi et al., 1986; Overton and Batta, 1979), rats initially were administered doses of morphine in the home cage for one or more weeks. Then, discrimination training began, and discrete operant responses were brought under stimulus control of saline and a dose of 10 or 15 mg/kg morphine. In subjects administered a supplemental dose of morphine equal to or double the training dose, establishment of stimulus control required no more sessions than were required in subjects administered morphine only before drug training sessions. Additionally, the accuracy of the final discrimination was not altered by daily administration of these supplemental doses of morphine. Thus, for the drugs studied to date, repeated supplemental administration of doses equal to or roughly twofold higher than the training dose does not alter either establishment or maintenance of drug stimulus control.

Higher supplemental doses can alter establishment of stimulus control. For morphine, administration of supplemental doses 10- to 40-fold higher than the training dose retards acquisition of drug stimulus control and decreases the accuracy of the established discrimination (Overton and Batta, 1979). Similar disruptions have been reported for the antimuscarinic scopolamine (Overton, 1977). The frequent administration of high supplemental drug doses may be functionally equivalent to decreasing the training dose of a drug stimulus. If this is indeed the case, drugs established as discriminative stimuli during repeated supplemental treatment with high doses might be expected to engender dose–response gradients or generalization sets different from those controlled by equal training doses established in the absence of supplemental dosing. To date, few published experiments have evaluated this possibility. An early paper by McKim (1976) reporting that higher doses of scopolamine are required for stimulus control in rats preexposed to the training dose (3.0 mg/kg) than in rats preexposed to saline suggests that such experiments may provide interesting information. Further work is required to determine whether repeated supplemental administration of high doses before or during initial discrimination training alters the establishment of discriminative control by drugs from other important pharmacological classes, whether such treatment changes the apparent sensitivity or specificity of the established discrimination, and whether the effects of early exposure to supplemental drug doses differ from the effects of supplemental doses administered after strong discriminative control has been established (*see below*).

5.2.2. Alteration of an Established Drug Discriminative Stimulus

Several experiments have examined whether supplemental drug administration will confer tolerance to an established drug discriminative stimulus. These studies have established discriminative control by a fixed dose of drug and then determined the ability of supplemental drugs, administered outside the confines of the discrimination task, to alter stimulus control both by the training dose itself and by lower and higher doses. These experiments have shown both that such supplemental treatment can confer tolerance to an established drug stimulus, and that the behavioral processes of discrimination learning can exert considerable control over tolerance development.

5.2.2.1. DISCRIMINATION TRAINING CONTINUED DURING SUPPLEMENTAL DRUG EXPOSURE. Two initial studies of the ability of supplemental drug exposure to alter stimulus control by an established drug stimulus continued discrimination training during the period of supplemental drug administration. In the first, Hirschhorn and Rosecrans (1974) established saline, and either 10 mg/kg morphine or 4 mg/kg THC as discriminative stimuli for milk-reinforced responses in separate groups of rats. At the end of the initial 40 training sessions, rats were emitting over 75% of their responses to the drug-appropriate lever when administered their training drug, and less than 30% of their responses to that lever when administered saline. Each rat was then administered supplemental daily doses of morphine or THC in one or two injections within 6 h of each daily training session. The supplemental dose was increased from twice the training dose to 8 or 16 times the training dose at 8- or 16-d intervals. Discriminative control by 10 mg/kg morphine and saline was not altered by supplemental administration of 20 mg/kg/d morphine, deteriorated during administration of 40 and 80 mg/kg/d, but recovered to initial patterns during the final phase of 160 mg/kg/d. Similarly, discriminative control by

THC was not altered by supplemental administration of 8 or 16 mg/kg THC/d, but deteriorated during administration of 32 mg/kg/d.

In a conceptually similar experiment, Colpaert and coworkers (1978d) examined the ability of supplemental drug administration to alter stimulus control by fentanyl. Fentanyl (0.04 mg/kg) and saline were established as discriminative stimuli, and dose–response functions were assessed in successive 3-wk blocks. During evaluation of tolerance, separate groups of rats received either saline or supplemental doses of fentanyl approximately 6 h after each daily session. The supplemental dose of fentanyl was initially 0.06 mg/kg (1.5 times the training dose), and was increased to 0.08 mg/kg, and again to 0.16 mg/kg (four times the training dose) for successive 3-wk test blocks. Neither stimulus control by 0.04 mg/kg fentanyl and saline nor the ED_{50} doses for generalization were altered by these supplemental doses of fentanyl.

Taken together, these early experiments suggested that well-established drug stimulus control might be insensitive to tolerance processes. Indeed, the maintenance of control by an established drug discriminative stimulus during supplemental treatment with high drug doses has been interpreted as evidence that the stimulus actions of a drug are not subject to tolerance development (Colpaert, 1978a, b). However, as reviewed above, continued stimulus control by a particular dose of a training drug is a product not only of the potency of the drug, but also of the behavioral outcomes of discriminative learning. As noted by Hirschhorn and Rosecrans (1974, p 251), the discrimination training conducted during supplemental drug administration may have obscured tolerance development by establishing stimulus control by progressively lower doses of the training drug. Continued differential reinforcement in the presence and absence of the original training dose may have overcome any changes in the potency of that dose, with the observed result of a maintenance of drug stimulus control.

Evaluation of the contributions of behavioral and pharmacodynamic processes to any development of tolerance to a well-established drug stimulus has required designs that separate the influence of behavioral and drug exposure factors. One such approach is provided by studies in which drug discrimination training is suspended during the period of supplemental drug administration, as discussed below.

5.2.2.2. DISCRIMINATION TRAINING SUSPENDED DURING SUPPLEMENTAL DRUG EXPOSURE. A second approach to evaluation of tolerance to an established drug discriminative stimulus is to suspend discrimination training during the period of supplemental drug administration. In contrast to the above results, if the discriminative relationship between a drug stimulus and behavior is suspended during the period of supplemental drug treatment, marked tolerance can develop to drug stimulus control. In the first experiment of this type, Shannon and Holtzman (1976) established stimulus control by 3.0 mg/kg morphine and saline in a discrete-trial shock avoidance task in rats. At biweekly intervals, a test dose was evaluated for stimulus control. Then, training was suspended for 3 d, and a dose of 10 mg/kg morphine (3.3 times the training dose) was administered at 12-h intervals, followed by a reevaluation of stimulus control by the test dose. This procedure was repeated four times to generate a dose–response curve. This brief period of supplemental morphine treatment increased the dose of morphine required for stimulus control by threefold, to 10 mg/kg. This increase in the dose of morphine required for stimulus control was not the result of suspending training, as three days of saline administration did not change control by the 3.0 mg/kg training dose.

In order to determine the pharmacological generality of such tolerance, other experiments assessed cross-tolerance to methadone and pentobarbital. Administration of the same supplemental dose of morphine conferred cross-tolerance to methadone, increasing the dose of methadone required for generalization of morphine stimulus control. In contrast, administration of 17.5 mg/kg pentobarbital twice daily for 3 d, a regimen sufficient to produce tolerance to the hypnotic effects of each pentobarbital challenge, did not confer tolerance to morphine stimulus control. The investigators concluded that morphine stimulus control is subject to tolerance development when training is suspended during the period of supplemental drug administration.

In a conceptually similar experiment, Miksic and Lal (1977) established 10 mg/kg morphine and saline as discriminative stimuli for food-reinforced behaviors in rats. After an initial dose–response function was determined, discrimination training was suspended for 11 d, during which time supplemental doses of morphine were administered, beginning at 10 mg/kg and increasing 10 mg/kg/d to a final dose of 110 mg/kg. On the 12th d, stimulus control was evaluated by testing increasing doses of morphine across consecutive experimental sessions. In a second experiment, other rats were exposed to the same supplemental morphine regimen, but tested with only a single dose of morphine. In both designs, the morphine dose required for stimulus control was increased to 20 or 40 mg/kg in individual rats, demonstrating tolerance.

Taken together, these experiments suggest that tolerance, as reflected by the requirement for an increased dose for stimulus control, can develop to a drug stimulus if discrimination training is suspended during the period of supplemental drug administration. As will be discussed in Section 5.2.2.4., below, similar studies with several other drugs have now established that, under conditions of suspended training, supplemental drug administration can markedly increase the dose required for stimulus control (Barrett and Leith, 1981; Emmett-Oglesby et al., 1988; Holtzman, 1987; McKenna and Ho 1977; Schechter, 1986; Schechter and Rosecrans, 1972; Wood and Emmett-Oglesby, 1986; Wood et al., 1984). However, interpretation of such patterns is not without controversy. As discussed in the next section, a change in the dose required for stimulus control may not, by itself, be unequivocal evidence of tolerance development.

5.2.2.3. COMPARISON OF TOLERANCE UNDER CONDITIONS OF CONTINUED OR SUSPENDED DISCRIMINATION TRAINING.   A change in the dose of a drug required for stimulus control may reflect a pharmacodynamic change in sensitivity conferred by a period of repeated drug administration but, as noted by Colpaert (1978a, b), may also result from the behavioral demands of the discrimination task itself. A requirement for an increased dose may result either from development of pharmacodynamic tolerance, or from a change in the reference standard for the discrimination (i.e. an increase in the training dose; *see* Section 4.1., above). Inasmuch as a similar outcome can be produced by disparate interventions, the observation that an increased dose is required for stimulus control does not provide unequivocal evidence that a particular treatment regimen produces tolerance. Additionally, the studies reviewed above employed different drugs, doses, and frequencies of supplemental drug administration, as well as different behavioral conditions during such treatment, so that multiple processes may underlie the different patterns of final discriminative performance.

Three recent studies have attempted to isolate the roles of selected behavioral and pharmacological variables in modulating the development of tolerance to drug stimulus control. These studies highlight the multiple determinants of tolerance processes. In one study, Emmett-Oglesby et al. (1988) evaluated whether suspending discrimination training would allow supplemental doses of fentanyl to confer tolerance to fentanyl or cross-tolerance to morphine. Saline and 0.04 mg/kg fentanyl were established as discriminative stimuli for food-maintained behaviors in rats. The range of doses of fentanyl or morphine capable of exerting stimulus control was evaluated by administering increasing doses before consecutive test sessions conducted at 40-m intervals. Then, training was suspended for 3–7 d, and a supplemental dose of fentanyl or morphine administered at selected intervals, followed by a redetermination of the doses required for stimulus control. The treatment doses of both fentanyl and morphine were chosen to be twice the test dose required for complete generalization of fentanyl stimulus control. In agreement with the earlier report by Colpaert et al. (1978d), supplemental treatment with 0.16 mg/kg/d fentanyl, administered either as two daily injections of a dose of 0.08 mg/kg (two times the training dose) or, in a subsequent week, as one daily injection of a dose of 0.16 mg/kg, did not confer tolerance to stimulus control by either fentanyl or morphine. This outcome contrasts with the ability of lower maintenance doses of fentanyl, administered with similar frequency, to confer tolerance to the analgesic actions of fentanyl (cf. Colpaert et al., 1980c).

These patterns suggest that the failure of supplemental treatment with 0.16 mg/kg fentanyl to confer tolerance is specific to stimulus control and independent of the behavioral contingencies imposed during that treatment. The next experiment by Emmett-Oglesby et al. (1988), however, demonstrated that the lack of tolerance to fentanyl stimulus control was probably the result of an insufficient regimen of drug treatment. Supplemental administration of morphine did confer tolerance to fentanyl stimulus control. Specifically, twice daily administration of 8.0 mg/kg morphine, a dose twice that required for generalization of fentanyl stimulus control, increased by approximately twofold the doses of both fentanyl and morphine required for stimulus control. The investigators suggest that the lack of tolerance conferred by fentanyl may be related to its short duration of action. In agreement with this possibility, frequent administration of a maximally tolerable dose of fentanyl (0.12 mg/kg injected every 6 h) did increase by approximately twofold the doses of both fentanyl and morphine required for stimulus control. These results demonstrate the critical importance of the pharmacological details of the supplemental drug regimen used to examine development of tolerance to drug stimulus control. They also demonstrate that tolerance to a drug's discriminative effects may not appear under treatment regimens sufficient to produce tolerance to other behavioral or physiological effects of the drug.

A second study evaluated the prediction that an increase in the dose required for stimulus control following supplemental drug treatment reflects the supplemental treatment acting to retrain the subject with a higher dose (cf. Colpaert, 1978a). One prediction of this explanation would be that the requirement for a higher dose should outlast the period of supplemental drug treatment, a prediction examined by Wood and Emmett-Oglesby (1986). As part of a larger examination of the development of tolerance to cocaine stimulus control in rats, these investigators suspended training and administered 60 mg/kg/d of cocaine for 12 d. During supplemental treatment, the dose of cocaine required for stimulus control was increased approximately twofold, such that the percentage of subjects displaying stimulus control by the 10 mg/kg training dose decreased from 91 to 37%. After supplemental treatment was terminated, stimulus control by the training dose was assessed in abbreviated sessions every 3 d. Stimulus control by the training dose "spontaneously recovered" within 18 d, suggesting that the reference point for the discrimination had not been reset. Albeit a single report, these results suggest that an increase in the dose required for stimulus control during supplemental drug treatment is not accompanied by an altered sensitivity under normal conditions, consistent with an argument that the observed change reflects pharmacological tolerance rather than a discrimination reset with a higher dose.

A third study evaluated the roles of behavioral processes in modulating the tolerance developed during repeated administration of high, relatively frequent morphine doses. In order to evaluate explicitly the contributions of discriminative learning processes to the development of tolerance, Sannerud and Young (1987; *see* Fig.

8) assessed changes in morphine stimulus control as a function of different training conditions during identical supplemental morphine treatment regimens. Saline and 3.2 mg/kg morphine were established as discriminative stimuli in rats. Each training session consisted of several 5-min experimental components separated by 15-min time-out periods. At the start of each time-out period, a subject received an injection of either saline or 3.2 mg/kg morphine. After the 15 min elapsed, reponses on only the lever appropriate to the presession injection produced food under an FR 30 schedule. Morphine generalization gradients were generated by administering increasing doses of morphine before successive time-out periods. To evaluate the role of behavioral processes in tolerance development, identical supplemental doses of morphine (17.8 mg/kg) or saline were administered twice daily, and discrimination training was either continued or suspended.

When discrimination training was continued throughout the period of supplemental morphine administration, the training dose of morphine continued to evoke excellent stimulus control. Additionally, during cumulative dose tests, the training dose of 3.2 mg/kg was sufficient for stimulus control in one-third of the subjects, indicating no tolerance (Fig. 8, closed triangles). The remaining 6 subjects required 5.6 mg/kg for stimulus control, suggesting some development of tolerance. This increase in the dose required for generalization when training continued was, however, less than that observed when discrimination training was explicitly suspended during supplemental drug administration (*see below*).

In contrast to the small degree of tolerance observed when discrimination training was continued, suspending training during supplemental drug treatment increased the morphine dose required for stimulus control to 10 mg/kg in seven out of nine subjects (Fig. 8, open triangles). These differences between the treatment conditions are preserved when the data for individual subjects are collapsed, and presented as either the mean percentage of drug-appropriate responses emitted by each subject (Fig. 9, upper panel) or the percentage of individual subjects displaying drug-appropriate behavior (Fig. 9, lower panel). Moreover, the tolerance developed under conditions of suspended training did not reflect a permanent change in stimulus control, as original morphine generalization gradients were recaptured after termination of supplemental treatment. Neither did the change in stimulus control appear to be the result of

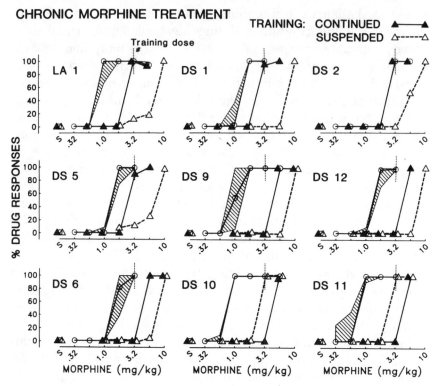

Fig. 8. Dose–response functions for stimulus control by morphine in individual rats before and during twice daily administration of 17.8 mg/kg morphine. *Open circles and shaded areas* represent the median and ± absolute average deviation for control tests conducted throughout the experimental period. *Closed triangles* represent tests conducted after discrimination training was continued during the 2-w morphine treatment. *Open triangles* represent tests conducted after training was suspended during the 2-w morphine treatment. Other details as in Fig. 7. [Reprinted with permission from Sannerud C. A. and Young A. M. (1987) Environmental modification of tolerance to morphine discriminative stimulus properties in rats. *Psychopharmacology* **93**, 59–68. Copyright 1987 by Springer-Verlag New York, Inc.]

recent high-dose morphine exposure, because acute pretreatments of 17.8 mg/kg morphine, given 12 h before a cumulative dose test, did not change the minimal dose required for stimulus control. The change in stimulus control did, however, require supplemental morphine administration. Supplemental saline administration did not alter the discriminative control by morphine (*see* Fig. 7), indi-

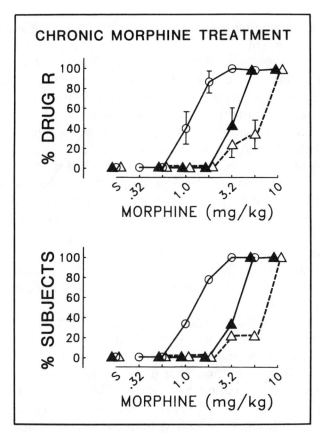

Fig. 9. Group dose–response functions for stimulus control by morphine in the presence or absence of twice daily administration of supplemental doses of 17.8 mg/kg morphine. Abscissae: mg/kg dose of morphine (log scale). Ordinate, upper panel: responses to the morphine-appropriate lever, expressed as a percentage of total responses and averaged for nine subjects. Vertical bars represents ± 1 SEM. Ordinate, lower panel: percentage of subjects ($N = 9$) emitting 90% or more of their total responses to the morphine-appropriate lever. *Circles* represent discriminative performance during control tests conducted before and at least 2 wk after supplemental morphine treatment. *Closed triangles* represent performance when discrimination training was *continued* during the period of supplemental morphine treatment. *Open triangles* represent performance when discrimination training was *suspended* during the period of supplemental morphine administration. [Derived from Sannerud C. A. and Young A. M. (1987) Environmental modification of tolerance to morphine discriminative stimulus properties in rats. *Psychopharmacology* **93,** 59–68.]

cating that suspending training, in itself, was not sufficient for tolerance development.

Thus, at least under the particular supplemental morphine treatment regimen we employed, suspending training resulted in a greater degree of tolerance to a morphine discriminative stimulus than did continued training. The control of tolerance development by the behavioral contingencies imposed during supplemental drug treatment clearly was not absolute. Rather, these data suggest that development of tolerance to a drug discriminative stimulus reflects an interaction of supplemental drug treatment and the training conditions employed during that treatment. At appropriate doses and frequencies, repeated drug treatment while discrimination training is suspended can confer marked tolerance to drug stimulus control. In contrast, tolerance to a drug discriminative stimulus may be retarded or prevented if the same regimen of supplemental drug treatment is accompanied by continued differential reinforcement of the discriminated operant.

As was originally suggested by Hirschhorn and Rosecrans (1974), less tolerance when supplemental drug treatment is accompanied by continued discrimination training may reflect an interplay of behavioral and pharmacodynamic processes that approximates a dose fading procedure (Colpaert et al., 1980a; Greenberg et al., 1975b; Hirschhorn and Rosecrans, 1974; Overton, 1979), under which continued discrimination training with progressively lower doses transfers stimulus control to doses lower than those originally required for generalization (*see* Section 4.1., above). Such shaping of discriminative control by functionally lower doses during supplemental drug treatment would obscure any development of tolerance to the originally discriminable doses. Although this interpretation of the apparent lack of tolerance during continued discrimination training is attractive, no studies have yet demonstrated that rapidly decreasing training doses can maintain stimulus control over a period as short as two weeks. Further, such a shaping mechanism would predict that the transfer of control to a lower dose should be retained after cessation of supplemental treatment. Thus, if these experiments were to be repeated, one might expect that the doses required for stimulus control would be decreased after termination of supplemental drug administration. No experiments have yet evaluated this prediction by halting supplemental drug treatment, waiting for a period of time sufficient to allow drug washout and resolution of withdrawal processes, and then retesting for sensitivity of drug stimulus control.

The three studies reviewed in this section highlight the interactive control exerted by behavioral and pharmacodynamic processes in the development of tolerance to drug stimulus control. Development of tolerance requires the interplay of both maintenance regimens appropriate to the agent under study and behavioral conditions that may limit the organism's ability to learn a new discrimination. When both requirements are met, tolerance does develop to drug stimulus control.

5.2.2.4. CHARACTERISTICS OF TOLERANCE TO DRUG STIMULUS CONTROL CONFERRED BY SUPPLEMENTAL DRUG ADMINISTRATION.

The experimental work reviewed above has established that supplemental drug administered outside the confines of the discrimination task can confer marked tolerance to a drug discriminative stimulus. This pattern has been obtained for amphetamine, caffeine, cocaine, fentanyl, morphine, and nicotine, arguing considerable pharmacological generality (Barrett and Leith, 1981; Emmett-Oglesby et al., 1988, Holtzman, 1987; McKenna and Ho, 1977; Miksic and Lal, 1977; Schechter and Rosecrans, 1972; Sannerud and Young, 1987, Shannon and Holtzman, 1976; Wood and Emmett-Oglesby, 1986; Wood et al., 1984). Such tolerance to drug discriminative control has three main characteristics: the dose of drug required to evoke stimulus control is increased, the requirement for an increased dose is limited to the period of supplemental drug treatment, and the change in dose may require suspension of discrimination training.

Importantly, the emerging body of research suggests that the patterns of tolerance and cross-tolerance for discriminative control may parallel the patterns observed for other effects of a drug. However, it appears that the degree of tolerance developed to a drug's discriminative stimulus actions may differ from that observed for other physiological and behavioral endpoints. Additionally, many of the parametric questions about tolerance to drug discriminative stimulus control are as yet unanswered. In particular, there is only limited information about the range of supplemental doses or the frequency and duration of administration required to confer tolerance to drug stimulus control. Moreover, it is unclear whether administration of a fixed supplemental dose will confer an equal degree of tolerance to a low and a high training dose of a drug, or whether the degree of tolerance conferred is determined by the proportional relationship between the training dose and the daily maintenance dose. Studies of such issues will clarify any unique behavioral or pharmacological characteristics of tolerance to drug stimulus control.

The largest body of work evaluating tolerance to drug stimulus control has focused on the patterns of tolerance and cross-tolerance within only two groups of drugs, the opioids and the behavioral stimulants. For both groups, the magnitude of tolerance appears to be an increasing function of the total daily dose administered. A parametric assessment of the change in the dose required for stimulus control as a function of the maintenance dose is available for the psychomotor stimulant cocaine. Wood and colleagues (Wood and Emmett-Oglesby, 1986, 1987; Wood et al. 1984, 1987) established saline and 5 or 10 mg/kg cocaine as discriminative stimuli for food-maintained behaviors in rats. After initial generalization gradients were determined, training was suspended for 7–16 d, and supplemental doses of 5, 10, or 20 mg/kg of cocaine were administered three times a day to separate groups of rats, resulting in daily maintenance doses of 15, 20 or 60 mg/kg/d. The degree of tolerance was determined by the daily maintenance dose. For example, in rats trained with 10 mg/kg cocaine, cocaine maintenance doses of 30 or 60 mg/kg/d (filled and open triangles, respectively, in Figure 10) increased the cocaine dose required for generalization by twofold, whereas a maintenance dose of 15/mg/kg/d (filled circles in Fig. 10) did not confer tolerance. Importantly, stimulus control by the original training dose of cocaine reappeared without explicit retraining after termination of chronic administration, demonstrating that the change during chronic drug administration was not the result of acquisition of control by a lower or higher dose.

Tolerance to stimulus control by caffeine and morphine has been studied under a more limited range of maintenance doses. For caffeine, a maintenance dose of 60 mg/kg/d (given in two divided doses) increased the dose of caffeine required for stimulus control fourfold in rats trained to discriminate 10 mg/kg caffeine, and threefold in rats trained to discriminate 30 mg/kg caffeine (Holtzman, 1987). For morphine, maintenance doses of 20 to 36 mg/kg (representing 7–11-fold increases in a 3.0 or 3.2 mg/kg

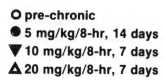

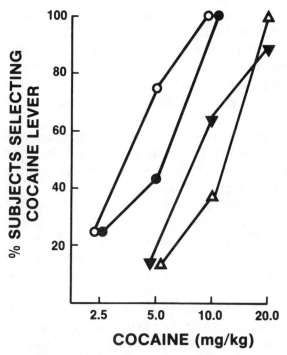

Fig. 10. Dose–response functions for stimulus control by cocaine before and after repeated treatment with supplemental doses of cocaine. Abscissae: mg/kg dose of cocaine (log scale). Ordinate: percentage of subjects ($N = 8$ at all points) completing ten responses on the cocaine-appropriate lever before competing ten responses on the saline-appropriate lever. *Open circles* represent discriminative performance during control tests conducted before supplemental cocaine treatment. *Closed circles, closed triangles, and open triangles* represent performance after treatment with the indicated dose of cocaine, injected every 8. h for the specified duration. [Reprinted with permission from Wood D. M. and Emmett-Oglesby M. (1986) Characteristics of tolerance, recovery from tolerance and cross-tolerance for cocaine used as a discriminative stimulus. *J. Pharmacol. Exp. Ther.* **237**, 120–125, copyright by the American Society for Pharmacology and Experimental Therapeutics, 1986.]

training dose) increased the dose required for stimulus control by three–sixfold (Sannerud and Young, 1987; Shannon and Holtzman, 1976). Similarly, a maintenance dose of 110 mg/kg (representing an 11-fold increase in the 10 mg/kg training dose) increased the morphine dose required for stimulus control by fourfold (Miksic and Lal, 1977). Albeit limited, these data suggest that the proportional relation between the training dose of a drug stimulus and the supplemental maintenance dose used to induce tolerance may control the degree of tolerance developed. One discrepant report has appeared. In pigeons, once-a-day administration of 100 mg/kg morphine did not confer tolerance to a 5.6 mg/kg morphine stimulus (France et al., 1984).

To date, few studies have compared the degree of tolerance developed to drug stimulus control with that developed to other behavioral effects of the drug under similar drug maintenance regimens. For morphine, a maintenance regimen of 36 mg/kg/d conferred a threefold tolerance to morphine stimulus control, but little tolerance to the rate-decreasing effects of morphine (Sannerud and Young, 1987). A similar dissociation was observed for cocaine (Wood et al., 1984). The contributions of behavioral and pharmacological factors to such differential tolerance remain to be clarified.

Studies with both the opioids and the behavioral stimulants have demonstrated that tolerance to drug discriminative control displays considerable pharmacological specificity. Tolerance to a morphine discriminative stimulus is conferred by supplemental administration of fentanyl or morphine, but not of pentobarbital (Emmett-Oglesby et al., 1988; Shannon and Holtzman, 1976). For the behavioral stimulant cocaine, tolerance develops during supplemental administration of cocaine, amphetamine, or apomorphine, but not during supplemental administration of morphine (Wood and Emmett-Oglesby, 1986, 1987). Additionally, supplemental administration of cocaine confers cross-tolerance to methylphenidate, phenmetrazine, and phentermine, increasing the doses of these stimulants required for generalization to a cocaine training stimulus (Wood and Emmett-Oglesby, 1988). Interestingly, administration of a standard maintenance dose of cocaine produces a similar increase in the dose of each drug required for stimulus control. Such cross-tolerance may not always be symmetrical, however. Supplemental administration of apomorphine confers cross-tolerance to a cocaine training stimulus, but supplemental administration of cocaine produces an insurmountable suppression of apomorphine generalization to cocaine (Wood and Emmett-Oglesby, 1987). It is as yet unknown whether development of tolerance to other drug discriminative stimuli will alter the set of drugs that can occasion generalization.

Such assessments of cross-tolerance may provide a particularly powerful way to assess the pharmacological determinants of tolerance to drug discriminative control. For example, Holtzman (1987) has demonstrated that the behavioral stimulants caffeine and methylphenidate share common discriminative stimulus properties, including the ability to confer symmetrical cross-tolerance. Supplemental administration of methylphenidate confers tolerance to a caffeine training stimulus, and supplemental administration of caffeine confers tolerance to methylphenidate generalization to caffeine. Interestingly, similar cross-tolerance is not obtained for the locomotor effects of caffeine and methylphenidate (Finn and Holtzman, 1986). Such disparate patterns may provide useful clues to the commonalities and differences in the pharmacological substrates underlying the discriminative and other effects of important psychoactive drugs.

In summary, drug stimulus control does appear to be subject to tolerance development. However, the complex interactions of behavior and drug, which are required for development and maintenance of stimulus control, also profoundly modify the development and expression of tolerance. Further explorations of a wider range of training drugs, training doses, and maintenance regimens will be needed to provide a full picture of this dynamic process.

6. Conclusions

The potential of psychoactive drugs to function as discriminative stimuli is firmly established. Such potential is not limited to a restricted range of drugs or behavioral conditions, but rather appears to be a highly predictable outcome of reinforcement conditions arranged so that the presence and absence of a psychoactive drug reliably set the occasion for reinforcement of discrete behavioral repertoires. Once established, drug discriminative control is easily maintained and lawfully generalized within a multiply determined range of stimulus intensities and qualities (i.e., doses and pharmacological classes). The experiments surveyed above suggest that both pharmacological and behavioral factors will modulate the acquisition, maintenance, and generalization of drug stimulus control. This work also suggests that both behavioral and pharmacological factors will modulate the development of tolerance to drug stimulus control. The dynamic discriminative relation between a drug and behavior continues during chronic administration of supplemental drugs, shaping changes in the intensity, and perhaps quality, of the drug required to sustain continued discriminative control.

Tolerance to a drug discriminative stimulus represents a dynamic interplay of behavior, drug, and the demands and possibilities of the individual's learning environment. The discriminative potential of psychoactive drugs is realized when the learner encounters conditions of differential reinforcement with respect to the presence and absence of a drug. In the experimental laboratory, such conditions are most often accompanied by relatively frequent drug exposure. Discriminative control thus represents one condition under which continued sensitivity to the effects of a frequently encountered drug will improve behavioral outcomes. The range of chronic drug regimens under which discriminative control can be established and maintained are, however, as yet unknown. We also know little about how "drug holidays" alter the accuracy and dose–response characteristics of an established drug discrimination.

Well-established drug discriminative performance remains highly sensitive to changes in both drug exposure and the demands of the behavioral situation in which a drug is encountered. Development of tolerance is critically dependent upon the details of the supplemental drug administration regimen, including the particular agent employed and its dose and frequency of administration. Most of the details of such control are as yet unknown. The range of chronic administration regimens capable of conferring tolerance, the speed of their effect, and the upper limits of tolerance development are unexplored. Behavioral processes also modulate tolerance development, with the dynamic relationship between a drug and the requirements of a discrimination task acting to control the degree of tolerance conferred by a particular supplemental treatment regimen. A treatment regimen sufficient to produce tolerance while discrimination training is suspended may be markedly less effective if accompanied by continued differential reinforcement of the discriminated operant. The limits of such behavioral influences are unexplored. Tolerance to drug stimulus control is multiply determined, and characterization of its domain will require multiple experimental approaches.

Studies of the factors that control development or expression of tolerance to the discriminable effects of drugs may have considerable practical importance. Inasmuch as the subjective effects of drugs may play an important role in their abuse potential (i.e. Fraser, 1968; Jasinksi, 1977), discrimination procedures may provide a laboratory model of how chronic drug exposure modifies the subjective effects underlying abusive use. Drug discrimination techniques may provide an exquisitely sensitive model of the factors that modulate psychoactive drug effects in humans with a range of drug experience. Inasmuch as human drug use, especially abusive use, occurs over extended time periods, it is likely that the different social environments attending intoxication and sobriety provide the reinforcement conditions necessary for establishment of drug stimulus control. An individual may learn to execute certain behaviors or expect certain reinforcement contingencies in the presence of drug, and a different set in the absence of drug. Such discriminative control may in turn modulate the range of situations in which an individual engages in particular behavioral repertoires. Laboratory studies suggest that frequent drug exposure does not inevitably lead to tolerance to a drug's discriminative effects. If this is indeed the case, these studies suggest that a need to overcome tolerance to a desired subjective effect is not a universal explanation for the progressive increase in the self-administered dose that characterizes much abusive drug use.

The potential of the drug discrimination assay has only begun to be exploited. The robust behavioral and pharmacological control afforded by the assay promise continuing fruitful explorations of the drug–behavior interactions that mold tolerance development.

Acknowledgments

Preparation of this chapter was supported in part by USPHS grants DA 03796 and RR 08167 and a Wayne State University Career Development Award. The authors thank Carla Coleman and Ann Perkins for excellent secretarial support.

References

Appel J. B., White F. J. and Holohean A. M. (1982) Analyzing mechanisms of hallucinogenic drug action with drug discrimination procedures. *Neurosci. Biobehav. Rev.* **6,** 529–536.

Ator N. A. and Griffiths R. R (1983) Lorazepam and pentobarbital drug discrimination in baboons: Cross-drug generalization and interaction with Ro 15-1788. *J. Pharmacol. Exp. Ther.* **226,** 776–782.

Barrett R. J. and Leith N. J. (1980) Tolerance to the discriminative stimulus properties of d-amphetamine. *Neuropharmacology* **20,** 251–255.

Beardsley P. M., Balster R. L., and Salay J. M. (1987) Separation of the response rate and discriminative stimulus effects of phencyclidine: Training dose as a factor in phencyclidine-saline discrimination. *J. Pharmacol. Exp. Ther.* **241,** 159–165.

Bertalmio A. J., Herling S., Hampton R. Y., Winger G., and Woods J. H. (1982) A procedure for rapid evaluation of the discriminative stimulus effects of drugs. *J. Pharmacol. Methods* **7,** 289–299.

Bueno O. F. A. and Carlini E. A. (1972) Dissociation of learning in marihuana tolerant rats. *Psychopharmacologia (Berl.)* **25,** 49–56.

Colpaert F. C. (1977). Drug-produced cues and states: Some theoretical and methodological inferences, in *Discriminative Stimulus Properties of Drugs* (Lal H., ed.) Amsterdam, New York; pp. 5–21.

Colpaert F. C. (1978a) Discriminative stimulus properties of narcotic analgesic drugs. *Pharmacol. Biochem. Behav.* **9,** 863–887.

Colpaert F. C. (1978b) Narcotic cue, narcotic analgesia, and the tolerance problem: The regulation of sensitivity to drug cues and to pain by an internal cue processing model, in *Stimulus Properties of Drugs: Ten Years of Progress.* (Colpaert F. C. and Rosecrans J. A., eds.) Elsevier/North-Holland Biomedical Press, Amsterdam, pp. 301–321.

Colpaert F. C. (1982a) Increased naloxone reversibility in fentanyl dose-dose discrimination. *Eur. J. Pharmacol.* **84,** 229–231.

Colpaert F. C. (1982b) The pharmacological specificity of opiate drug discrimination, in *Drug Discrimination: Applications in CNS Pharmacology* (Colpaert F. C. and Slangen J. L., eds.) Elsevier Biomedical Press, Amsterdam, pp. 3–16.

Colpaert F. C. (1986) Drug discrimination: Behavioral, pharmacological, and molecular mechanisms of discriminative drug effects, in *Behavioral Analysis of Drug Dependence,* (Goldberg S. R. and Stolerman I. P., eds.) Academic Press, Inc., Orlando, pp. 161–193.

Colpaert F. C. and Janssen P. A. J. (1981) Factors regulating drug cue sensitivity: The effect of frustrative non-reward in fentanyl-saline discrimination. *Arch. Int. Pharmacodyn.* **254,** 241–251.

Colpaert F. C. and Janssen P. A. J. (1982a) Factors regulating drug cue sensitivity: The effects of dose ratio and absolute dose level in the case of fentanyl dose-dose discrimination. *Arch. Int. Pharmacodyn.* **258,** 283–299.

Colpaert F. C. and Janssen, P. A. J. (1982b) Factors regulating drug cue sensitivity: Limits of discriminability and the role of a progressively decreasing training dose in cocaine-saline discrimination. *Neuropharmacology* **21,** 1187–1194.

Colpaert F. C. and Niemegeers C. J. E. (1975) On the narcotic cuing action of fentanyl and other narcotic analgesic drugs. *Arch. Int. Pharmacodyn.* **217,** 170–172.

Colpaert F. C., Lal H., Niemegeers C. J. E., and Janssen P. A. J. (1975a) Investigations on drug produced and subjectively experienced discriminative stimuli 1. The fentanyl cue, a tool to investigate subjectively experienced narcotic drug actions. *Life Sci.* **16,** 705–716.

Colpaert F. C., Niemegeers C. J. E., and Janssen P. A. J. (1975b) The narcotic discriminative stimulus complex: Relation to analgesic activity. *J. Pharm. Pharmac.* **28,** 183–187.

Colpaert F. C.. Niemegeers C. J. E., and Janssen, P. A. J. (1975c) The narcotic cue: Evidence for the specificity of the stimulus properties of narcotic drugs. *Arch. Int. Pharmacodyn.* **218,** 268–276.

Colpaert F. C., Kuyps J. J. M. D., Niemegeers C. J. E., and Janssen P. A. J. (1976a) Discriminative stimulus properties of fentanyl and morphine: Tolerance and dependence. *Pharmacol. Biochem. Behav.* **5,** 401–408.

Colpaert F. C., Niemegeers C. J. E., and Janssen P. A. J. (1976b)

Fentanyl and apomorphine: Asymmetrical generalization of discriminative stimulus properties. *Neuropharmacology* **15**, 541–545.

Colpaert F. C., Niemegeers C. J. E., and Janssen P. A. J. (1978a) Changes of sensitivity to the cuing properties of narcotic drugs as evidenced by generalization and cross-generalization experiments. *Psychopharmocology* **58**, 257–262.

Colpaert F. C., Niemegeers C. J. E., and Janssen P. A. J. (1978b) Factors regulating drug cue sensitivity. A long-term study of the cocaine cue, in *Stimulus Properties of Drugs: Ten Years of Progress,* (Colpaert F. C. and Rosecrans J. A., eds.) Elsevier/North-Holland Biomedical Press, Amsterdam, pp. 281–299.

Colpaert F. C., Niemegeers C. J. E., and Janssen P. A. J. (1978c) Narcotic cuing and analgesic activity of narcotic analgesics: Associative and dissociative characteristics. *Psychopharmacology* **57**, 21–26.

Colpaert F. C., Niemegeers C. J. E., and Janssen P. A. J. (1978d) Studies on the regulation of sensitivity to the narcotic cue. *Neuropharmacology* **17**, 705–713.

Colpaert F. C., Niemegeers C. J. E. and Janssen P. A. J. (1980a) Factors regulating drug cue sensitivity: Limits of discriminability and the role of a progressively decreasing training dose in fentanyl-saline discrimination. *J. Pharmacol. Exp. Ther.* **212**, 474–480.

Colpaert F. C., Niemegeers C. J. E. and Janssen P. A. J. (1980b) Factors regulating drug cue sensitivity; The effect of training dose in fentanyl-saline discrimination. *Neuropharmacology* **19**, 705–713.

Colpaert F. C., Niemegeers, C. J. E., Janssen P. A. J. and Maroli A. N. (1980c) The effects of prior fentanyl administration and of pain on fentanyl analgesia: Tolerance to an enhancement of narcotic analgesia. *J. Pharmacol. Exp. Ther.* **213**, 418–424.

De Vry J. and Slangen J. L. (1986a) Effects of chlordiazepoxide training dose on the mixed agonist-antagonist properties of benzodiazepine receptor antagonist Ro 15–1788, in a drug discrimination procedure. *Psychopharmacology* **88**, 177–183.

De Vry J. and Slangen J. L. (1986b) Effects of training dose on discrimination and cross-generalization of chlodiazepoxide, pentobarbital and ethanol in the rat. *Psychopharmacology.* **88**, 341–345

De Vry J., Koek W., and Slangen J. L. (1984) Effects of drug-induced differences in reinforcement frequency on discriminative stimulus properties of fentanyl. *Psychopharmacology* **83**, 257–261.

Emmett-Oglesby M. W., Shippenberg T. S., and Herz A. (1988) Tolerance and cross-tolerance to the discriminative stimulus properties of fentanyl and morphine. *J. Pharmacol. Exp. Ther.* **245**, 17–23.

Finn I. B., and Holtzman S. G. (1986) Tolerance to caffeine-induced stimulation of locomotor activity in rats. *J. Pharmacol. Exp. Ther.* **238,** 542–546.

France C. P., Jacobson A. E., and Woods J. H. (1984) Discriminative stimulus effects of reversible and irreversible opiate agonists: Morphine, oxymorphazone and buprenorphine. *J. Pharmacol. Exp. Ther.* **230,** 652–657.

Fraser H. F. (1968) Methods for assessing the addiction liability of opioids and opioid antagonists in man, in *The Addictive States,* (Wikler, A. ed.) William and Wilkins, Baltimore, pp. 176–187.

Gaiardi M., Bartoletti M., Gubellini C., Bacchi A., and Babbini M. (1986) Sensitivity to the narcotic cue in non-dependent, morphine-dependent and post-dependent rats. *Neuropharmacology* **2,** 119–123.

Gianutsos G. and Lal H. (1976) Selective interaction of drugs with a discriminable stimulus associated with narcotic action. *Life Sci.* **19,** 91–98.

Goldstein A., Aronow L., and Kalman S. M. (1974) *Principles of Drug Action: The Basis of Pharmacology (2nd Edition)* John Wiley and Sons, New York.

Goudie A. J., Atkinson J., and West C. R. (1986) Discriminative properties of the psychostimulant *dl*-cathinone in a two lever operant task. *Neuropsychopharmacology* **25,** 85–94.

Greenberg I., Kuhn D., and Appel J. B. (1975a) Comparison of the discriminative stimulus properties of delta-9-THC and psilocybin in rats. *Pharmacol. Biochem. Behav.* **3,** 931–934.

Greenberg I., Kuhn D. M., and Appel J. B. (1975b) Behaviorally induced sensitivity to the discriminable properties of LSD. *Psychopharmacologia* (Berl.) **43,** 229–232.

Hein D. W., Young A. M., Herling S., and Woods J. H. (1981) Pharmacological analysis of the discriminative stimulus characteristics of ethylketazocine in the rhesus monkey. *J. Pharmacol. Exp. Ther.* **218,** 7–15.

Henriksson B. G., Johansson J. O., and Jarbe T. U. C. (1975) delta-9-Tetrahydrocannabinol produced discrimination in pigeons. *Pharmacol. Biochem. Behav.* **3,** 771–774.

Herling S., Coale E. H., Jr., Valentino R. J., Hein D. W., and Woods J. H. (1980) Narcotic discrimination in pigeons. *J. Pharmacol. Exp. Ther.* **214,** 139–146.

Hirschhorn I. D. (1977) Pentazocine, cyclazocine, and nalorphine as discriminative stimuli. *Psychopharmacology* **54,** 289–294.

Hirschhorn I. D. and Rosecrans J. A. (1974) Morphine and delta-9-tetrahydrocannabinol: Tolerance to the stimulus effects. *Psychopharmacologia* (Berl.) **36**, 243–253.

Holtzman S. G. (1982) Discriminative stimulus properties of opioids in the rat and squirrel monkey, in *Drug Discrimination: Applications in CNS Pharmacology* (Colpaert F. C. and Slangen J. L., eds.) Biomedical Press, Amsterdam, pp. 17–36.

Holtzman S. G. (1985) Discriminative stimulus properties of opioids that interact with mu, kappa and PCP/sigma receptors, in *Behavorial Pharmacology: The Current Status* (Seiden L. S. and Balster R., eds.) Neurology and Neurobiology, **Vol. 13**, Alan R. Liss, Inc., New York, pp. 131–147.

Holtzman S. G. (1987) Discriminative stimulus effects of caffeine: Tolerance and cross-tolerance with methylphenidate. *Life Sci.* **40**, 382–389.

Jarbe T. U. C. and Henriksson B. G. (1973) Open-field behavior and acquisition of discriminative response control in delta-9-THC tolerant rats. *Experimentia* **29**, 1251–1253.

Jarbe T. U. C. and Holmgren B. (1977) Discriminative properties of pentobarbital after repeated noncontingent exposure in gerbils. *Psychopharmacology* **53**, 39–44.

Jarbe T. U. C. and McMillan D. E. (1983) Interaction of the discriminative stimulus properties of diazepam and ethanol in pigeons. *Pharmacol. Biochem. Behav.* **18**, 73–80.

Jarbe T. U. C. and Rollenhagen C. (1978) Morphine as a discriminative cue in gerbils: Drug generalization and antagonism. *Psychopharmacology* **58**, 271–275.

Jasinski D. R. (1977) Assessment of the abuse potentiality of morphinelike drugs (Methods used in man), in *Drug Addiction I. (Handbook of Experimental Pharmacology, Vol. 45)* (Martin W. R., ed.) Springer-Verlag, Berlin, pp. 197–258.

Jones C. N., Grant L. D., and Vospalek D. M. (1976) Temporal parameters of *d*-amphetamine as a discriminative stimulus in the rat. *Psychopharmacologia (Berl.)* **46**, 59–64.

Kline F. S. and Young A. M. (1986) Differential modification of pentobarbital stimulus control by *d*-amphetamine and ethanol. *Pharmacol. Biochem. Behav.* **24**, 1305–1313.

Koek W. and Slangen J. L. (1982a) Effects of reinforcement differences between drug and saline sessions on discriminative stimulus properties of fentanyl, in *Drug Discrimination: Applications in CNS Phar-*

macology (Colpaert F. C. and Slangen J. L., eds.) Elsevier Biomedical Press, Amsterdam, pp. 343–354.

Koek W. and Slangen J. L. (1982b) The role of fentanyl training dose and the alternative stimulus condition in drug generalization. *Psychopharmacology* **76**, 149–156.

Kuhn D. M., Greenberg I., and Appel J. B. (1974) Differential effects on lever choice and response rates produced by *d*-amphetamine. *Bull. Psychon. Sci.* **3**, 119–120.

McKenna M. and Ho B. T. (1977) Induced tolerance to the discriminative stimulus properties of cocaine. *Pharmacol. Biochem. Behav.* **7**, 273–276.

McKim W. A. (1976) The effects of pre-exposure to scopolamine on subsequent drug-state discrimination. *Psychopharmacology* **47**, 153–155.

McMillan D. E. (1987) On the stability of phencyclidine discrimination in the pigeon. *Alcohol Drug Res.* **7**, 147–151.

McMillan D. E. and Wenger G. R. (1984) Bias of phencyclidine discrimination by the schedule of reinforcement *J. Exp. Anal. Behav.* **42**, 51–66.

Miksic S. and Lal H. (1977) Tolerance to morphine-produced discriminative stimuli and analgesia. *Psychopharmacology* **54**, 217–221.

Overton D. A. (1974) Experimental methods for the study of state-dependent learning. *Fed. Proc.* **33**, 1800–1813.

Overton D. A. (1977) Discriminable effects of antimuscarinics: Dose response and substitution test studies. *Pharmacol. Biochem. Behav.* **6**, 659–666.

Overton D. A. (1979) Drug discrimination training with progressively lowered doses. *Science* **205**, 720–721.

Overton D. A. (1982a) Comparison of the degree of discriminability of various drugs using the T-maze drug discrimination paradigm. *Psychopharmacology* **76**, 385–395.

Overton D. A. (1982b) Multiple drug training as a method for increasing the specificity of the drug discrimination procedure. *J. Pharmacol. Exp. Ther.* **221**, 166–172.

Overton D. A. and Batta S. K. (1979) Investigation of narcotics and antitussives using drug discrimination techniques. *J. Pharmacol. Exp. Ther.* **211**, 401–408.

Overton D. A. and Hayes M. W. (1984) Optimal training parameters in the two-bar fixed-ratio drug discrimination task. *Pharmacol. Biochem. Behav.* **21**, 19–28.

Overton D. A., Leonard W. R. and Merkle D. A. (1986) Methods for measuring the strength of discriminable drug effects. *Neurosci. Biobehav. Rev.* **10,** 251–263.

Rosen J. B., Young A. M., Beuthin F. C., and Louis-Ferdinand R. T. (1986) Discriminative stimulus properties of amphetamine and other stimulants in lead-exposed and normal rats. *Pharmacol. Biochem. Behav.* **24,** 211–215.

Sannerud C. A. and Young A. M. (1987) Environmental modification of tolerance to morphine discriminative stimulus properties in rats. *Psychopharmacology* **93,** 59–68.

Schaefer G. J. and Holtzman S. G. (1977) Discriminative effects of morphine in the squirrel monkey. *J. Pharmacol. Exp. Ther.* **201,** 67–75.

Schechter M. D. (1986) Induction of and recovery from tolerance to the discriminative stimulus properties of *l*-cathinone. *Pharmacol. Biochem. and Behav.* **25,** 13–16.

Schechter M. D. and Rosecrans J. A. (1972) Behavioral tolerance to an effect of nicotine in the rat. *Arch. Int. Pharmacodyn.* **195,** 52–56.

Schuster C. R. and Balster R. L. (1977) The discriminative stimulus properties of drugs, in *Advances in Behavioral Pharmacology, Vol. 1.* (Thompson T. and Dews P. B., eds.) Academic Press, New York, pp. 85–138.

Shannon H. E. (1983) Discriminative stimulus effects of phencyclidine: Structure-activity relationships, in *Phencyclidine and Related Arylcyclohexylamines: Present and Future Applications* (Kamenka J. M., Domino E. F., and Geneste P., eds.) NPP Books, Ann Arbor, MI, pp. 311–335.

Shannon H. E. and Holtzman S. G. (1976) Evaluation of the discriminative effects of morphine in the rat. *J. Pharmacol. Exp. Ther.* **198,** 54–65.

Shannon H. E. and Holtzman S. G. (1979) Morphine training dose: A determinant of stimulus generalization to narcotic antagonists in the rat. *Psychopharmacology* **61,** 239–244.

Shearman G. T. and Herz A. (1982) Evidence that the discriminative stimulus properties of fentanyl and ethylketocyclazocine are mediated by an interaction with different opiate receptors. *J. Pharmacol. Exp. Ther.* **221,** 735–739.

Stolerman I. P. and D'Mello G. D. (1981) Role of training conditions in discrimination of central nervous system stimulants by rats. *Psychopharmacology* **73,** 295–303.

Stolerman I. P. and Shine P. J. (1985) Trends in drug discrimination research analysed with a cross-indexed bibliography, 1982–1983. *Psychopharmacology* **86**, 1–11.

Stolerman I. P., Baldy R. E., and Shine P. J. (1982) Drug discrimination procedure: A bibliography, in *Drug Discrimination: Applications in CNS Pharmacology*, (Colpaert F. C. and Slangen J. F. eds.) Elsevier Biomedical Press, Amsterdam, pp. 401–424.

Teal J. J. and Holtzman S. G. (1980a) Discriminative stimulus effects of prototype opiate receptor agonists in monkeys. *Eur. J. Pharmacol.* **68**, 1–10.

Teal J. J. and Holtzman S. G. (1980b) Discriminative stimulus effects of cyclazocine in the rat. *J. Pharmacol. Exp. Ther.* **212**, 368–376.

Thompson T. and Pickens R. (eds.) (1971) *Stimulus Properties of Drugs*. Appleton-Century Crofts, New York.

White F. J. and Appel J. B. (1982a) Lysergic acid diethylamide [LSD] and lisuride: Differentiation of their neuropharmacological actions. *Science* **216**, 535–537.

White F. J. and Appel J. B. (1982b) Training dose as a factor in LSD-saline discrimination. *Psychopharmacology* **76**, 20–25.

Winter J. C. (1978) Drug-induced stimulus control, in *Contemporary Research in Behavioral Pharmacology* (Blackman D. E. and Sanger D. J., eds.) Plenum Publishing Corporation, New York, pp. 209–237.

Wood D. M. and Emmett-Oglesby M. (1986) Characteristics of tolerance, recovery from tolerance and cross-tolerance for cocaine used as a discriminative stimulus. *J. Pharmacol. Exp. Ther.* **237**, 120–125.

Wood D. M. and Emmett-Oglesby M. W. (1987) Evidence for dopaminergic involvement in tolerance to the discriminative stimulus properties of cocaine. *Eur. J. Pharmacol.* **138**, 155–157.

Wood D. M. and Emmett-Oglesby M. W. (1988) Substitution and cross-tolerance profiles of anorectic drugs in rats trained to discriminate cocaine. *Psychopharmacology* **95**, 364–368.

Wood D. M., Lal H., and Emmett-Oglesby M. (1984) Acquisition and recovery of tolerance to the discriminative stimulus properties of cocaine. *Neuropharmacology* **23**, 1419–1423.

Wood D. M., Retz K. C., and Emmett-Oglesby M. W. (1987) Evidence of a central mechanism mediating tolerance to the discriminative stimulus properties of cocaine. *Pharmacol. Biochem. Behav.* **28**, 401–406.

Woods J. H., Herling S., and Young A. M. (1980) Comparison of discriminative and reinforcing stimulus characteristics of morphine-like opioids and two met-enkephalin analogues. *Neuropeptides* **1,** 409–419.

Young A. M. Discriminative stimulus profiles of psychoactive drugs, in *Advances in Substance Abuse, Vol. 4.* (Mello, N. K., ed.) JAI Press, Greenwich, CN, in press.

Young R. and Glennon R. A. (1986) Discriminative stimulus properties of amphetamine and structurally related phenylalkylamines. *Med. Res. Rev.* **6,** 99–130.

Part 2
Molecular Mechanisms

Dispositional Mechanisms in Drug Tolerance and Sensitization

A D. Lê
and Jatinder M. Khanna

1. Introduction

It is well documented that tolerance and sensitization are two phenomena associated with chronic administration of psychoactive drugs. Depending on the drug examined, tolerance to some effects and sensitization to other effects can be demonstrated within a single organism. In general, it can be said that tolerance occurs to most psychoactive drug, whereas sensitization is predominant only with some psychostimulant drugs. Of the two phenomena, tolerance has been investigated much more extensively and, therefore, has been much better characterized than has sensitization.

Generally, chronic tolerance is defined as a reduction, and sensitization or reverse tolerance as an increase, in the effect induced by a given dose of the drug upon repeated administration. Within this broad definition, several subdefinitions may be used.

Initial or innate tolerance reflects individual variation in the acute response to the drug, and acute tolerance or within session tolerance refers to the development of tolerance within the duration of a single exposure to the drug. Although there are individual variations in the response to drugs, or the effects of drugs, that become sensitized following chronic treatment (Meier et al., 1963; Moisset and Welch, 1973), there has been no report concerning acute sensitization as paralleled to that of acute tolerance.

Mechanistically, chronic tolerance and sensitization can be divided into two classes. The first is dispositional tolerance or dispositional sensitization. This includes changes in the pharmacokinetic parameters of the drug that might lead to a decrease (tolerance) or increase (sensitization) in the concentration and/or the duration of the drug in the tissue on which the drug exerts its pharmacological actions. The second is functional tolerance and sensitization, or what is simply referred to as central nervous system (CNS) or tissue tolerance and sensitization. This includes changes in the properties and functions of the target tissue that render it less (tolerance) or more (sensitization) reactive to the same degree of exposure to the drug. Similarly, functional and dispositional components can be involved in both cross-tolerance and cross-sensitization.

2. Scope of This Review

In this chapter, we will examine the changes in pharmacokinetic parameters of various drugs following chronic treatment, and see how these changes might account for or contribute to the development of tolerance and sensitization and cross-tolerance and cross-sensitization. Several drug classes, including alcohol, barbiturates, opiates and psychostimulants will be examined, since tolerance is a predominant phenomenon in some drug classes, whereas sensitization is predominant in others. We will first provide a general background in the pharmacokinetic principles. This will then be followed by a discussion of the important changes in pharmacokinetic properties of prototype drug(s) for each drug class following chronic administration. When necessary, we will discuss other drugs to illustrate some important principles. Various factors ranging from aging, diet, and so on, to behavioral manipulations that might affect drug disposition will then be examined. Finally, we will discuss the relevance of dispositional factors in tolerance and sensitization, particularly in relation to the validity of some of the common assumptions that have been made to rule out the occurrence of dispositional factors in tolerance and sensitization.

3. Mechanisms of Dispositional Tolerance

In general, the intensity of a drug's effect is related to the concentration, whereas the duration of the effect is dependent on the rate of removal of the drug from its site of action. Upon administration, the concentration that can be achieved and the length of time that the drug remains at its site of action are generally determined by four main processes: absorption, distribution, metabolism or biotransformation, and excretion. Following chronic treatment, these processes may be subject to many changes that will alter the amount and time the drug is available at its site of action. For psychoactive drugs, the effects of interest are those mediated centrally and, therefore, the concentration and the persistence of the drug in the brain are of primary concern.

There are many inherent factors in a drug that govern its absorption, distribution, metabolism, and excretion. Chemical properties of the drug, such as its molecular size, lipid solubility, the degree of ionization, and protein binding, are some important examples. Such information is available in many pharmacological textbooks (Gilman et al., 1985; Kalant et al., 1985) and will not be discussed here. In this section, a brief overview of the potential changes in those pharmacokinetic parameters that have important implications in dispositional tolerance or sensitization will be outlined.

3.1. Change in Absorption

In order to reach its primary site of action, the CNS, the psychoactive drug administered has first to be absorbed into the general circulation from the site of injection. For any given dose of a drug, the rate of drug absorption is dependent upon the route of administration employed. Common routes of administration employed for chronic treatment are oral, subcutaneous (sc), intravenous (iv), intraperitoneal (ip), and inhalation. For determination of tolerance and sensitization, ip, sc, or iv routes of administration are commonly used, because the amount of drug being delivered or the concentration achieved in the circulation is easier to control.

For iv administration, the drug is delivered directly to the general circulation and is, therefore, not influenced by many variables that can affect absorption. Similarly, the peritoneal cavity, being highly vascularized and having a large absorbing surface, allows rapid absorption after ip route of administration. In contrast, following oral administration, the rate of gastric emptying plays a crucial role in drug absorption. For both oral and ip routes, the drug passes across the membranes of the gastrointestinal tract into the portal vein, through the liver, and finally to the general circulation. During this passage, the drug may be metabolized in the cell membranes of the gut wall or in the liver or both. This phenomenon is commonly referred to as the first pass effect. The extent and the significance of the first pass effect are dependent on the drug employed and the dose administered, because the first pass metabolism is saturable. Finally, absorption from the sc injection site occurs by simple diffusion through the capillary membranes into the blood and is related to the lipid solubility of the drug. Because of low blood flow in subcutaneous tissue, absorption tends to be relatively slow from this site.

The concentration of the drug at the site of action has been often regarded as the critical determinant for the intensity of the effect. However, for psychoactive drugs, the rate at which the drug reaches its target tissue is also an important variable. In other words, for a given concentration of the drug at its site of action, the intensity of the drug effect is dependent on the rate at which such concentration is achieved. For psychoactive drugs, it is believed that, when the drug reaches the brain at a low rate, compensation in the nervous system or acute tolerance develops, and therefore, a much higher dose is required to achieve the desired effect. The importance of the rate of absorption and drug effects can best be illustrated from the study of Mirsky et al. (1941), in which the extent of ethanol intoxication in hepatectomized rabbits was dependent on their rate of iv infusion of ethanol. When the rate of infusion was low, a much higher blood ethanol level was required to produce the same degree of intoxication as that induced by a lower level of ethanol that was reached more rapidly.

It is also common knowledge that the dose required to produce a given degree of drug effect after iv or ip routes is much smaller than that required for oral or other routes with slow absorption rates. As pointed out earlier, these other routes of administration might produce a lower concentration of the drug than iv or ip. This fact alone, however, may not be able to account for the large difference in the dose requirement for certain drugs. It is possible that the differences in the dosage requirement with these different routes may be related to differences in rates of absorption and acute tolerance development.

Variation in the rate of absorption, therefore, can play a critical role in tolerance and sensitization. When one considers the role of absorption in tolerance and sensitization, the amount as well as the rate of absorption has to be examined. If a drug acutely affects the rate of gastric emptying, intestinal motility, or vascular blood flow, these effects might develop tolerance following chronic administration. Such tolerance, depending on the route employed for giving the test dose to assess tolerance or sensitization, would influence the rate and amount of the drug absorbed.

3.2. Change in Volume of Distribution

Through the vascular system, drugs can be distributed to various fluid and tissue compartments before gaining access to their sites of action. Ethanol, for example, distributes throughout the total body water, which consists of extra- and intracellular compartments. The extracellular compartment is comprised of plasma volume and interstitial fluid.

Some drugs are simply dissolved in serum, but many drugs are bound to blood constituents such as albumin, globulin, lipoprotein and erythrocytes. Albumin, in general, is the most important macromolecule and accounts for most of the drug binding. Plasma protein binding influences the activity and the fate of the drug in the body. Only the unbound or free drug diffuses through the capillary walls, reaches the site of action, and is subject to biotransformation and elimination from the body. The binding between drug and albumin is reversible, and the equilibrium between bound and free forms is constantly maintained.

Certain drugs are taken up specifically by tissue constituents. Highly lipid-soluble drugs, such as thiopental, tetrahydrocannabinol, and DDT, are typical drugs that concentrate in adipose tissue. Since the blood flow to fat is relatively low, a long period of time will be required for equilibrium to be achieved between the unbound drug in plasma and adipose tissue. The redistribution of the drug from active sites to adipose tissue serves to diminish the pharmacological effects of the drugs. Drugs stored in body fat may also require a long time to be removed from the body, and thus, body fat serves as a site of storage after prolonged exposure to the drug. Besides adipose tissue, some drugs bind or sequester into other tissues. For example, amphetamine is sequestered and attains highest concentrations in the lungs and liver (Lemberger and Rubin, 1976).

In most areas of the brain, the capillary cells are connected together by occluding zonulae without space in between and constitute the blood–brain barrier (BBB). This barrier acts to reduce or restrict the entry of the drug into the brain. The epithelial cells of the choroid plexus are also connected by occluding zonulae. Drugs can pass from the CSF to the brain without any restriction, since the cells that constitute the CSF–brain barrier do not connect together by occluding zonulae. Lipid-soluble substances diffuse across the BBB at rates determined by their lipid/water partition coefficients. Since the brain receives about one-sixth of the total amount of blood leaving the heart, most drugs of abuse are distributed to brain tissue quite rapidly because they are generally quite lipid-soluble. Water-soluble material, such as the amino acids, can also gain rapid access to the brain through active transport mechanisms.

3.3. Biotransformation or Change in Rate of Metabolism

Biotransformation is the modification or transformation of a drug to another chemical structure by the body. This process may have several consequences with respect to the pharmacological activity of the drug. Mostly, an active drug is converted to an inactive compound, or in some instances, an active drug can be converted into another compound that may be more or less active than the parent compound. In addition, an inactive precursor can be converted into a pharmacologically active drug; an example is the conversion of L-dopa to dopamine. In general, the body metabolizes drugs that are lipid soluble, nonpolar compounds to more water-soluble and polar compounds that can be excreted from the body.

Although drug biotransformation may occur in many tissues in the body and at almost any subcellular site within those tissues, the liver, by far, is the principal organ for the metabolic conversion of most drugs in mammals. Within the liver, the smooth endoplasmic reticulum is the most important site for biotransformation. Extrahepatic tissues such as lung, kidney, brain, and blood may also play an important role in the biotransformation of drugs. For example, heroin is deacetylated rapidly to morphine in the blood.

In general, drug biotransformation reactions can be grouped into two phases. Phase 1 includes oxidation, reduction, or the hydrolysis reaction, which may activate or inactivate the drug, or leave its activity unchanged. The basic process in phase 1 involves the microsomal enzymes system. Phase 2 reactions include synthetic reactions or conjugation, which involve the chemical combination of a compound with a molecule provided by the body. Phase 2 reactions almost always result in inactivation of the drug. Although microsomal enzymes are important in the metabolism of many drugs, they are by no means the only enzymes involved in their metabolism. Ethanol, for example, is metabolized primarily by a nonmicrosomal enzyme system involving alcohol dehydrogenase and NAD.

The activity of microsomal enzymes or nonmicrosomal enzymes can be stimulated or inhibited by chemical agents or pathological conditions. A barbiturate such as pentobarbital or phenobarbital can readily stimulate the microsomal enzymes. This will lead to an increase not only in its own metabolism, but also that of other drugs. Numerous environmental contaminants such as polycyclic aromatic hydrocarbons are also good enzyme inducers. There are also many other factors that can affect or influence the activity of these enzymes including diet, age and sex; these will be discussed in a subsequent section.

3.4. Change in Excretion

Although drugs can be excreted in any media eliminated from the body, the principal route of drug excretion is by way of the kidney. As a major excretory organ, the kidney is responsible for eliminating the majority of nonvolatile, water-soluble substances produced or required by the body. The rate at which a drug will be eliminated through the kidney is the net result of three processes: glomerular filtration, tubular secretion, and tubular reabsorption. For glomerular filtration, the free drug is excreted by simple diffusion through the glomerular membrane. Since the free and bound drug exist in equilibrium, when the free drug is excreted, the still protein-bound drug dissociates rapidly and more drug is thus available to be excreted. Drug reabsorption can occur primarily in the proximal tubule, where there are specialized cells that have the ability to reabsorb a drug by either diffusion or active transport mechanisms. For example, amphetamine exists in the alkaline urine primarily in an unionized form, and it can therefore diffuse back from the tubular lumen into the tubular epithelium and from there it can reenter the circulation. Some drugs can be actively secreted from the blood to the tubular lumen. Penicillin, salicylic acid, and quaternary compounds such as *n*-methylnicotine can be actively secreted into the renal tubule.

The liver also secretes drugs and their metabolites into the bile. Not all drugs reaching the small intestine by this route are subsequently excreted in the feces. They can be reabsorbed from the intestinal tract into the circulation. These agents then remain in the enterohepatic circulation until they are excreted in the urine.

In addition to renal and biliary excretion, drugs can be also excreted through the lungs or skin. Volatile solvents or general anesthetics can be eliminated in part from the body through expired air. Drugs are excreted in sweat, usually by means of simple diffusion through the sweat gland.

3.5. Summary

In summary, dispositional tolerance could, theoretically, result from any of the factors listed below:

1. Decreased absorption of the drug from the gastrointestinal tract

2. Increased destruction in the gastrointestinal tract before absorption
3. Storage of the drug in some kind of a reservoir somewhere in the body (e.g., protein-bound or highly lipid-soluble drugs), thus limiting its availability to the general circulation
4. Increase in body water content resulting in a decrease in the effective concentration of the water-soluble drugs
5. Increased elimination of the drug because of induction of drug-metabolizing enzymes or changes in blood flow
6. More rapid excretion of the drug in urine, expired air, or sweat, and so on
7. Decreased penetration of the drug to the brain because of changes in BBB as a result of chronic treatment.

4. Dispositional Factors in Tolerance and Sensitization to Various Drug Classes

4.1. Dispositional Tolerance to Ethanol

4.1.1. General Aspects of Alcohol Tolerance

Chronic treatment with ethanol has been shown to produce tolerance to a variety of effects. Tolerance usually develops to the depressant and aversive actions of ethanol, and no tolerance seems to develop to its stimulant or euphoriant action. The assortment of methods that have been used to quantify tolerance has included tests of motor performance (moving belt test, tilting plane, rotarod, and so on), different types of learning tasks (maze running, and so on), and various types of physiological measures such as body temperature, loss of righting reflex (LRR), EEG and LD_{50}, and so on (*see* Kalant et al., 1971, and the chapter by Wolgin, this volume). The choice of method is usually dictated by the predominant action of the drug, as well as the ease with which it could be quantified. For example, analgesia has been one of the key tests for quantifying tolerance to opiates, whereas hypothermia has been used extensively for testing tolerance to ethanol and other hypnosedatives.

In general, tolerance to ethanol appears to develop fairly rapidly, within days or even within a single exposure. There are major differences between the receptor-mediated (e.g., opiates and benzodiazepines) and nonreceptor-mediated drugs (such as alcohol and barbiturates) with respect to certain features and characteristics of tolerance development. In the case of ethanol and barbiturates, the extent of tolerance is usually not higher than 100%, whereas a considerably greater degree of tolerance (5–20-fold) develops to benzodiazepines and opiates. No tolerance to lethality appears to develop to alcohol and barbiturates, but is seen readily in opiate-tolerant subjects. Tolerance to alcohol and barbiturates does not result in any changes in the maximum response. However, there is a progressive decrease in the maximum response that could be attained with opiates and benzodiazepines (Mucha et al., 1978; Lê et al., 1986).

4.1.2 Absorption, Distribution, and Excretion

The absorption of ethanol from the gastrointestinal tract, its distribution in the body, and excretion through the lungs, kidneys, and sweat glands have been reviewed extensively (*see* Wallgren, 1970; Wallgren and Barry, 1970; Kalant, 1971). In general, after oral ingestion, the absorption of ethanol into the circulation proceeds rapidly by diffusion across the gastric and intestinal mucosa.

After absorption, alcohol is rapidly distributed in the water phase of the body, and the concentration of alcohol in blood and other tissues is proportional to the water content of the tissue. In fact, alcohol measurements have been used to determine total body water. Only a small proportion of ethanol (2–10%) is eliminated unchanged via the lungs, kidneys, and sweat. The contribution of these routes in the overall elimination of ethanol from the body is, therefore, not very significant. Ethanol also diffuses rapidly across the BBB and no carrier-mediated systems are required to transport it from the blood to the brain. The concentration in various parts of the brain, like the rest of the body, is dependent upon blood circulation, water content of the area, degree of vascularization, and blood flow in the area (Chin, 1979).

Systematic studies directed towards direct examination of changes in absorption, distribution, or excretion accompanying the development of tolerance to ethanol have not been undertaken in recent years. Therefore, most of the information available on this topic is either based on work done prior to 1960 or some indirect or fragmentary data from recent studies.

A decrease in ethanol absorption as an explanation for tolerance to ethanol has been invoked in only one single study. Troshina (1957) reported that after 5 mo of chronic ethanol treatment, only 42% of the administered alcohol was absorbed 3 h after administration compared to 84% on the first day. However, no support for decreased absorption in tolerant subjects is available from other studies (Levy, 1935; Newman and Lehman, 1938). Similarly, changes in ethanol excretion do not contribute significantly to ethanol tolerance (Levy, 1935; Newman and Lehman, 1938). The possibility that, under certain specific conditions (conditioning, and so on), the absorption of alcohol may be retarded is discussed in Section 6.1.

The possibility that tolerance may be the result of reduced entry of alcohol into the brains of habituated animals is also not supported by experimental data, since either no differences or differences in the opposite direction (i.e., a faster entry of alcohol into the CNS) have been reported between blood and brain in alcohol-treated subjects compared to controls. Fleming and Stotz (1935,1936) reported a higher concentration of ethanol in CSF in alcoholic patients than in nonalcoholics after a test dose of ethanol. Similarly, Newman and Lehman (1938) found higher brain levels in ethanol-treated rats (10% ethanol as the sole source of fluid for 163 d) than in the water-receiving controls.

4.1.3 Metabolism

Considerable attention has been devoted to investigations on alcohol metabolism, and many excellent reviews on this topic are available (Hawkins and Kalant, 1972; Lieber, 1977; Khanna and Israel, 1980; Crow, 1985). Although small amounts of ethanol could be oxidized by many tissues, such as the gastrointestinal tract, kidney, lungs, and even the brain, the liver is by far the key organ responsible for the major portion of ethanol metabolism. Three main enzyme systems that are capable of oxidizing ethanol to acetaldehyde have been described. These are

1. Alcohol dehydrogenase (ADH)
2. Catalase and
3. The microsomal ethanol oxidizing system (MEOS).

The role of catalase and MEOS in the oxidation of ethanol remains controversial, whereas there is little doubt that the ADH pathway is the major route of ethanol metabolism.

Since ADH is the principal enzyme responsible for ethanol metabolism and this enzyme system is saturated at a very low ethanol concentration (0.5–2.0 mM), it is not surprising that ethanol metabolism follows a zero order kinetics, and is independent of the dose or concentration of ethanol. Evidence for the zero elimination kinetics was provided by the very early researchers in this field (Widmark, 1932; Carpenter, 1940; Jacobsen, 1952; Newman et al., 1952), as well as by some modern researchers (Guynn and Pieklik, 1975; Kalant et al., 1975; Khanna et al., 1977; *see also* reviews by Hawkins and Kalant, 1972; Khanna and Israel, 1980). However, some recent researchers have challenged the notion of zero-order kinetics, probably because of the involvement of MEOS, and suggested that a multicompartment model provides a better fit for ethanol elimination (for references, *see* Wilkinson, 1980). Similarly, others have challenged the concept of rate constancy of ethanol metabolism (Feinman et al., 1978; Pikkarainen and Lieber, 1980; Shigeta et al., 1983, 1984).

Although metabolic tolerance to ethanol was reported as early as 1908 by Pringsheim and this finding was confirmed by Schweisheimer (1913) and Gettler and Freireich (1935), other researchers (Levy, 1935; Newman and Lehman, 1938) failed to confirm this observation. In fact, this issue was in dispute until the mid-60s (*see* Mendelson and Mello, 1964; Mendelson et al., 1965; Hawkins et al. 1966). Considerable evidence, however, has been put forward during the last two decades, and metabolic tolerance to ethanol has been demonstrated convincingly both in humans and in many different animal species (for references, *see* Hawkins and Kalant, 1972; Khanna and Israel, 1980). Increases in alcohol metabolism have been demonstrated both when alcohol was administered by intubation and when given in liquid diets (Hawkins et al., 1966; Khanna et al., 1972). Factors such as insufficient amount of ethanol intake, inadequate duration of alcohol treatment, impaired general well-being of the animal, and age of the test animals may account for the failure to observe metabolic tolerance in some studies (for a discussion of this issue, *see* Hawkins et al., 1966).

Several investigators have put forward evidence in favor of an adaptation in the ADH system (either increase in enzyme activity or increase in NADH reoxidation), whereas others have implicated MEOS in the increase in rate of ethanol metabolism as a result of chronic ethanol treatment. There are several reviews of the evidence both for and against ADH and MEOS in metabolic tolerance to ethanol, and the interested reader should consult these reviews (Hawkins and Kalant, 1972; Lieber, 1977; Khanna and Israel, 1980; Thurman et al., 1975).

The time-course kinetics of acquisition and loss of metabolic tolerance to ethanol have not been studied extensively. In most recent studies, metabolic tolerance was examined after 2–4 wk of chronic administration of alcohol, and a 25–50% increase in ethanol metabolic rate is usually reported. It appears, however, that tolerance acquisition is fairly rapid. Tobon and Mezey (1971) reported an increase in ethanol metabolism of 44 and 71%, respectively after 3 and 7 d of ethanol feeding in a liquid diet. The increase in ethanol metabolism reported in this study after 3 d of ethanol feeding was not markedly different than that seen after 14 d (41%) of ethanol administration by these authors (Mezey, 1972). In another study (Wood and Laverty, 1979), the elimination rate of ethanol was not significantly higher from controls until 7 d of chronic ethanol treatment, and it reached a maximum after 16 d.

Very few studies have examined the time course of decay of metabolic tolerance. However, it appears that the loss of metabolic tolerance is fairly rapid. Mezey (1972) reported that the increased ethanol metabolic rates returned to control values in 2 d. In the study of Wood and Laverty (1979), increased ethanol metabolism decayed by approximately 50% from the chronic value after 7 d and was no longer significant after 22 d. It is possible that the time course of decay is dependent upon the duration of chronic treatment employed. Mezey (1972) treated animals for 2 wk of chronic ethanol treatment in a liquid diet regimen, whereas Wood and Laverty (1979) employed a similar liquid-diet regimen for 5 wk.

Recent work by Thurman and his collaborators has suggested that the adaptive increase in ethanol metabolism in rodents could occur even after a single large dose of ethanol (Thurman et al., 1979, 1983). They have named this phenomenon SIAM, or swift increase in alcohol metabolism. Similar observations were reported by Wilson et al. (1984) in humans. In rodents, this phenomenon is highly strain-dependent and has a strong genetic basis (Thurman et al., 1983). Other investigators (Braggins and Crow, 1981), however, could not find any evidence for SIAM in their work. The contribution of SIAM, if any, to overall acute tolerance requires further exploration.

4.2. Dispositional Tolerance to Barbiturates

4.2.1. General aspects of barbiturate tolerance

It is well known that chronic administration of barbiturates results in the development of tolerance. In general, the characteristics and features of tolerance development to barbiturates are very similar to those for ethanol. The pharmacological end point that is used most often in studies of tolerance to barbiturates is the duration of loss of righting reflex or anesthesia. Although the early investigators were impressed with the remarkable ability of barbiturates to induce tolerance, they did not determine blood or brain levels at awakening in many of their studies to distinguish between dispositional and pharmacodynamic tolerance. Since dispositional tolerance is so pronounced with barbiturates, some investigators believed that tolerance, especially to short-acting barbiturates, was entirely the result of enzyme induction (Remmer, 1969; Stevenson and Turnbull, 1970). In recent years, Okamoto and her colleagues have carried out extensive systematic studies on barbiturate tolerance in cats, and have provided us with a clearer understanding of the time course of development of both dispositional and functional tolerance to various barbiturates. These studies are discussed below.

4.2.2. Absorption, Distribution, and Excretion

The possibility of decreased absorption contributing to the development of tolerance to barbiturates has been explored in both the earlier and recent literature. In experiments on human subjects receiving different doses of phenobarbital for 11 or 22 d, Butler et al., (1954) and Svensmark and Buchthal (1963) measured the plasma concentration of phenobarbital, and compared it with the calculated (i.e., predicted) values obtained by a mathematical formula requiring information about daily doses, the volume of distribution of the drug, and proportion of the drug eliminated daily. Both these authors found excellent agreement between the observed and the calculated plasma concentration over the treatment period studied. Okamoto et al. (1975) have also presented evidence to suggest that the rate of absorption is not altered in tolerant animals compared to controls. These authors made cats tolerant to pentobarbital by twice daily (morning and evening) oral administration of sodium pentobarbital for 5 wk. Comparisons of blood pentobarbital concentrations after the oral morning dose on the last day of chronic treatment with the values obtained on the first day of the similar treatment revealed, if anything, somewhat higher concentrations in the pentobarbital-tolerant animals than in the controls. The peak blood pentobarbital levels were achieved in both groups 1 and 2 h, after administration and there was no significant difference between the two groups in the rate of rise of blood pentobarbital concentration between 15 and 30 min.

Butler et al. (1954) have presented evidence to show that increased excretion of barbiturates does not occur in the chronically treated subject. These authors carried out studies with barbital, phenobarbital, and pentobarbital in dogs and human subjects. For example, in one study with 11 subjects, six of whom were on pentobarbital for several months and the other five of whom were treated for 12–13 d, they found a range of elimination of 11–17% in the former group with a mean of 13.8% and 14–17% with a mean of 18.6% in the latter. These values were not significantly different, and also were similar to the values of approximately 16%/24 h in nontolerant subjects.

The possibility of altered distribution (i.e., a decrease in brain/blood ratio) as a result of chronic treatment with barbiturates was reviewed by Kalant et al. (1971). Since two studies reported a lower brain/blood ratio (Timer et al., 1966; Büch et al., 1969) and the third one (Ebert et al., 1964) found no difference in chronically treated animals than in controls, it was not possible to dismiss the contribution of this factor to tolerance development.

A good deal of information on this topic, however, has become available during the last few years. Various investigators have examined phenobarbital concentrations in brain and other tissues in epileptic patients undergoing temporal lobectomy receiving phenobarbital either singly or in combination with other antiepileptic drugs (Sherwin et al., 1973; Vajda et al., 1974; Houghton et al., 1975). All these studies show a significant relationship between brain and blood, and provide evidence that CSF:plasma or brain:plasma ratios do not differ after acute vs chronic administration. They also suggest that sampling plasma concentrations are useful indicators of brain concentration. Sherwin et al. (1973) further extended the human work by confirming it in rats. The results obtained for the mean ratio of brain to plasma levels in rats was 0.91 ± 0.07, which proved to be almost identical to that found in the human (0.91 ± 0.08). Okamoto and Boisse (1975) confirmed beyond any doubt that the CSF/blood concentration ratio does not change during chronic treatment. These authors examined pentobarbital concentration in blood and CSF in cats on the first (acute) and the last (chronic) day during a 5-wk treatment with twice daily equieffective anesthestic doses of sodium pentobarbital. The rate of rise of pentobarbital and the time to peak concentration (approximately 1 h) after oral dosing was identical in tolerant vs control animals. Comparisons of CSF/blood ratios showed a somewhat higher partitioning of pentobarbital into the CSF in the chronic than in controls at all times studied (15 min to 10 h). The pentobarbital CSF/blood ratios were 0.765 ± 0.019 for the chronic and 0.734 ± 0.021 for the acute group ($p < 0.1$). In contrast to the studies mentioned above, Tagashira et al. (1979) reported results that suggest a decreased entry of phenobarbital into the brain. These authors examined phenobarbital concentrations in serum and brain in rats receiving phenobarbital-containing food *ad libitum* on two different dosage schedules. In the first schedule, food containing phenobarbital (1 and 2 mg/g) was available for 13 d. In the second schedule, a graded incremental dosage schedule of 0.5 and 1.0–4.0 mg/g was provided for 42 d. On both schedules, brain phenobarbital levels did not correlate at all with serum levels. For example, in the fixed-dosage schedule study, serum phenobarbital levels on the first and third day of feeding were approximately 49 and 167 μg/mL compared to the brain phenobarbital levels of 12 and 20 μg/g on d 1 and 3, respectively. In the graded incremental-dosage schedule study, phenobarbital levels in the serum increased markedly with the increased dosage (final level 219 μg/mL), but the level in the brain did not increase above 34 μg/mL. These results are puzzling and difficult to reconcile with other observations discussed earlier.

4.2.3. Metabolism

It is common knowledge that barbiturates induce hepatic microsomal enzymes and increase elimination of drugs. Literally hundreds of publications have documented the ability of barbiturates to induce their own metabolism as well as the metabolism of a large variety of other compounds that are metabolized by the microsomal mixed-function oxygenase system. Many authors have reviewed this material, and the interested reader can refer to these reviews for extensive coverage on these and other topics (*see* Conney, 1967; Remmer, 1970; Gillette, 1971; Okey et al., 1986). We shall concern ourselves here primarily with key differences among the various barbiturates in relation to the mechanisms underlying the development of dispositional tolerance to these drugs.

The long-acting barbiturates such as phenobarbital and barbital are much better inducers of drug metabolism than the short-acting ones such as pentobarbital and secobarbital, simply because they are poorly metabolized by the microsomal enzyme system and their duration of action is much longer. Short-acting barbiturates given in sufficient amount and at sufficiently short intervals produce induction equivalent to that seen with the long-acting barbiturates. It should also be obvious that the induction of metabolism is not restricted to the parent compound being administered, but extends to other barbiturates and to many other drugs metabolized by the mixed-function oxygenase system. Apparently, this induction would be more important and significant for short-acting barbiturates, whose duration of action is determined mainly by the rate at which these drugs are metabolized, than the long-acting (phenobarbital and barbital) or the ultra-short acting (hexobarbital and thiopental) barbiturates. Thus, tolerance to phenobarbital and barbital is primarily attributed to functional tolerance, because only a small proportion of the drug is metabolized and most of it is eliminated by excretion. For example, Ebert et al. (1964) showed that, in rats made tolerant to barbital, the amount of barbital metabolized during 24 h was 6% in the barbital-tolerant animal compared to 3% in control.

Although it has been suggested that tolerance to pentobarbital and other short-acting barbiturates is mainly dispositional and that diminished responses by the CNS play a minor role (Remmer, 1969), other findings (Okamoto et al., 1975; Boisse and Okamoto,

1978a,b) indicate that both types of tolerance (dispositional and functional) occur with short-acting barbiturates. These authors examined tolerance to the impairment of neurological functions produced by repeated pentobarbital administration in cats receiving equieffective anesthetic doses of pentobarbital at 12-h intervals for 5 wk. The dispositional tolerance occurred rapidly and was almost complete within a week (see Fig. 1). Thereafter, there was no significant change in pentobarbital metabolism during the next 4 wk. However, the dosage still had to be increased to maintain maximal CNS depression over these weeks (Fig. 2). This would indicate functional tolerance because the same level of peak CNS depression was seen at a higher blood concentration. Comparison of a long-acting (barbital) and a short-acting (pentobarbital) barbiturate using equieffective doses of these two barbiturates (i.e., equieffective with respect to dose and recovery) in their chronically equivalent procedure showed that functional tolerance developed at the same slow rate for both barbiturates (Boisse and Okamoto, 1978a,b). The average increase in dose from the beginning to the end, however, was significantly less for barbital (1.38-fold) than for pentobarbital (2.31-fold). Since no dispositional component to barbital tolerance was evident, the greater increase in dose for pentobarbital reflects the contribution of dispositional tolerance.

The magnitude of metabolic tolerance differs in different studies. Okamoto et al. (1985) reported in cats an approximately two-fold increase in the rate of pentobarbital elimination. Yamamoto et al. (1977) also found a twofold increase in mice after 10 daily sodium pentobarbital (75 mg/kg) injections. However, if mice were implanted with a 75 mg sc pellet of pentobarbital and tested at 48 h after implantation, these authors found a 4.5-fold increase in pentobarbital disappearance. Many factors such as age, sex, species, dose, frequency of administration, and the duration of action of the barbiturate employed, and so on, have a profound effect on induction of drug metabolism (Conney, 1967; Remmer, 1970; Gillette, 1971). In some studies, the changes in pentobarbital metabolism and P450 content are similar (Okamoto et al., 1985), whereas in other studies, parallel changes in cytochrome P450 and rates of metabolism were not found (see Lemberger and Rubin, 1976).

Okamoto et al. (1985) have also provided data on the loss of dispositional and pharmacodynamic tolerance to pentobarbital after

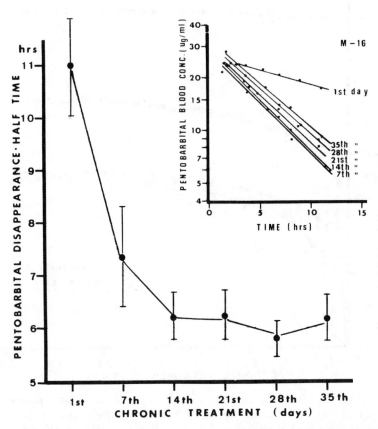

Fig. 1. Pharmacokinetic analysis of pentobarbital disappearance from blood after each successive week of chronic treatment. Inset is an example of the pentobarbital elimination curves obtained weekly during chronic treatment. Abscissa: time in hours after a morning dose of sodium pentobarbital. Ordinate: pentobarbital blood concentration in μg/mL in log scale. Note the first-order kinetic nature of the curves. The main figure shows the relationship between the changes in pentobarbital elimination half-life with the progression of treatment. Abscissa: duration of chronic treatment in days. Ordinate: the half-life of pentobarbital disappearance from blood in hours. The half-life was calculated for each animal by computer by fitting the elimination curves into the first-order decay equation. Each point represents the average of 15 observations. (From Okamoto et al., 1975; reproduced with the permission of the authors and the publisher.)

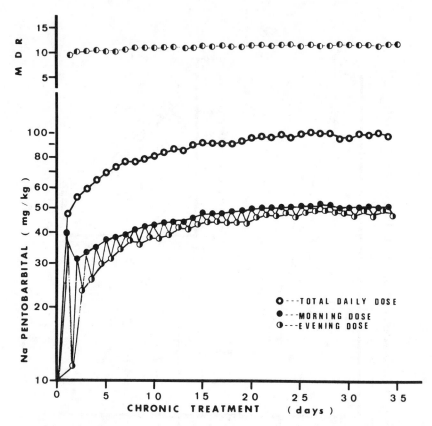

Fig. 2. Average morning, evening, and total daily doses for a group of 60 cats during 35 days of chronic treatment with sodium pentobarbital and their relationship to maximal daily CNS depression (MDR). Abscissa: number of days of chronic treatment. Ordinate at lower half: dose of sodium pentobarbital in mg/kg, administered through the gastric cannula. Closed circles (●) represent the average morning dose for each day. Half-closed circles (◐) indicate the average evening dose each day. Open circles (○) indicate the average total daily dose (morning dose plus evening dose) for each day of chronic treatment. Ordinate at the upper half: the average maximal CNS depression rating (MDR) recorded daily during the chronic treatment. Average S.E. for the daily MDR was 1.35% (range 0.9–1.9%). If plotted, these S.E.s would fall within the diameter of the circles used to plot the MDRs. (From Okamoto et al., 1975; reproduced with the permission of the authors and the publisher.)

termination of chronic treatment. There was a rapid recovery from dispositional tolerance, and by the end of 2 wk after termination of drug administration the $t\frac{1}{2}$ values had returned to the prechronic level. In contrast, functional tolerance was still clearly present at 2 wk, especially to lower levels of CNS depression ratings. One should, however, bear in mind that the recovery results in this study are confounded by the influence of test doses repeated every few days. Ideally, one should use separate groups of animals to test return of initial sensitivity, in order to avoid the contamination of results by repeated test dosing and its unknown effects on the dispositional and functional components of tolerance.

4.3. Dispositional Tolerance to Benzodiazepines

4.3.1. General Aspects of Benzodiazepine Tolerance

Clinical studies have clearly documented the development of tolerance to benzodiazepines (Warner, 1965; Kaplan et al., 1973; Kanto et al., 1974; Hillestad et al., 1974; Korttila et al., 1975; Sellman et al., 1975b; Klotz, 1976; Grundstrom et al., 1978; Pertusson and Lader, 1981; Aranko, 1985). Similarly, tolerance to benzodiazepines has been demonstrated in a variety of animal species, such as mice (Goldberg et al., 1967; File, 1981, 1983a,b), rats (Hoogland et al., 1966; Goldberg et al., 1967; Margules and Stein, 1968; McMillan and Leander, 1978; File, 1982a,b), pigeons (Cesare and McKearney, 1980), and monkeys (Sepinwall et al., 1978). The major emphasis of tolerance demonstration in most of the above-mentioned studies of benzodiazepines has been in relation to depressant effects, and various tests, such as spontaneous motor activity, rotarod performance, avoidance learning tasks, and so on, were employed. Tolerance has also been shown to anticonvulsant (Goldberg et al., 1967; Browne, 1976; Lippa and Regan, 1977; Gastaut and Low, 1979; Feely et al., 1982; File, 1983a,b; Gent and Haigh, 1983) and skeletal muscle relaxant effects of benzodiazepines (Hoogland et al., 1966; Jablonska et al., 1975). There appears to be some disagreement concerning tolerance development to the anxiolytic effects of benzodiazepines (Shearman et al., 1979; Rickels, 1982).

In spite of extensive demonstrations of tolerance, very few studies have examined the time-course kinetics of tolerance acquisition and loss, and the extent and degree of tolerance development. However, some information is available. Sellers (1978) reported that prior benzodiazepine users need threefold higher diazepam doses than nonusers for endoscopy. McMillan and Leander (1978) and Cesare and McKearney (1980) also reported at least a threefold shift to the right of the dose–effect curve for chlordiazepoxide on operant behavior in rats and pigeons, respectively. Since these two investigators examined tolerance at only one time period after the chronic treatment, these studies do not provide any information on the rate of tolerance development. Ryan and Boisse (1983) provided a thorough and systematic investigation pertaining to this issue. These authors used the method of "maximally tolerable" dosing technique and administered chlordiazepoxide intragastrically on a twice-daily basis. They found a fivefold increase in maintenance dose. Tolerance developed more rapidly during the first 9–10 d, but continued to develop thereafter more slowly without an apparent ceiling.

Tolerance to benzodiazepines develops fairly rapidly. Both clinical and animal studies report that the initial sedative effects diminish in intensity and disappear after a few days of treatment. The time course of recovery from tolerance differs in various studies. For example, File (1982) found complete recovery from tolerance to the sedative effects of lorazepam within 48 h of the last dose, whereas Margules and Stein (1968) found that tolerance to the sedative effects of oxazepam persisted even when the drug was withheld for several weeks.

4.3.2. Absorption, Distribution, and Excretion

Very little attention has been given to examination of changes in absorption, distribution, and excretion in benzodiazepine-tolerant vs control subjects. There is only one major study that has tackled this issue directly, although some indirect information on this topic is available from other studies in the literature. Hoogland et al. (1966) compared the tissue distribution and excretion of chlordiazepoxide-2-c^{14} in tolerant (treatment with chlordiazepoxide—100 mg/kg daily for 5 d) and control rats. They found a significant increase in the urinary and fecal excretion in tolerant animals, which excreted 74% of the administered C^{14}, compared to 56% for the controls. Interestingly, the brain:plasma ratio was also significantly lower in the tolerant than in control animals. Kaplan et al. (1973) compared pharmacokinetic profiles for diazepam in human subjects after single intravenous vs oral and chronic oral (10 mg every 24 h for 15 d) administration. Since the simulated chronic diazepam blood level curves with each administration of the drug every 24 h during chronic treatment agreed quite well with the observed blood level data, these authors discounted changes in absorption, distribution, excretion, and metabolism as a result of this chronic treatment. Aranko et al. (1985) compared the rise in serum lorazepam concentrations after an oral challenge dose of 3 mg lorazepam in different groups of patients receiving either high, low, or no benzodiazepine treatment. As expected, the baseline benzodiazepine levels were different in different groups. However, the rise in serum lorazepam concentrations was similar in all groups. Thus, tolerance to benzodiazepines is unlikely to be the result of changes in their absorption and distribution.

4.3.3. Metabolism

Many studies have been undertaken to examine the induction of hepatic microsomal enzymes and the role of metabolic factors in the development of tolerance to benzodiazepines. Induction, no change, and a decrease in benzodiazepine metabolism have all been reported. Several studies have shown that large doses of chlordiazepoxide, diazepam and oxazepam (50–400 mg/kg for 5 d) can induce hepatic microsomal enzymes in animals (Hoogland et al., 1966; Jablonska et al., 1975) and produce an increase in the elimination rate of the drugs. Following high doses of triazolam and lorazepam, tolerance is accompanied by lower plasma concentrations (File 1981, 1982b). Many human studies have also shown results that suggest that metabolic factors are involved in the development of tolerance. Kanto et al. (1974) found lower plasma concentrations of diazepam and nordiazepam after chronic treatment in humans, along with a loss of drug effects 1–6 w later. Sellman et al. (1975b) reported a faster disappearance of diazepam from plasma in patients on chronic diazepam therapy compared to healthy volunteers without previous diazepam treatment. A higher concentration of the diazepam metabolite (N-desmethyldiazepam) was also observed in this study.

Other studies found no changes in the activity of drug-metabolizing enzymes and in the rate of benzodiazepine metabolism with chronic treatment (Solomon et al., 1971; Kaplan et al., 1973; Vorne et al., 1975; Aranko et al., 1985). In two other studies, lower plasma levels were reported after repeated treatment with one particular dose level, but not at others (Christensen, 1973; File, 1982b). It appears that induction of benzodiazepine metabolism is usually observed either with high doses or with a sustained chronic treatment regimen especially in animals, and is seen with some benzodiazepines and not with others (Hoogland et al., 1966, Orme et al., 1972, Valerino et al., 1973; Tanayama et al., 1974; Jablonska et al., 1975).

In other work (Klotz et al., 1976), an increase in the elimination half-life and a decrease in total plasma clearance was observed in both humans and animals that were chronically treated with either diazepam or desmethyldiazepam. These authors suggested that the disposition of diazepam may be regulated by the plasma concentration of its metabolite, desmethyldiazepam, which may act by product inhibition on enzymatic conversion of diazepam to desmethyldiazepam resulting in a slower elimination.

In conclusion, there is no disagreement as to the development of functional tolerance to benzodiazepines, but the significance of dispositional factors in benzodiazepine tolerance is still not clear. It is possible that both functional and dispositional tolerance play a role either concurrently or at different stages of tolerance development. On the basis of observed biphasicity in the rate of increase in the chlordiazepoxide maintenance dose, Ryan and Boisse (1983) suggested that both functional and dispositional components are involved in producing tolerance. Unfortunately, the only dispositional factor that has been studied extensively is induction of hepatic microsomal enzymes and increased elimination. Because of the long half-life of benzodiazepines and many of their metabolites, it is difficult to interpret changes in elimination half-life with chronic use of this class of drugs.

4.4. Dispositional Tolerance to Opiates (Morphine)

4.4.1. General Aspects of Opiate Tolerance

A variety of techniques have been employed to carry out chronic morphine treatment in experimental animals. Common techniques range from multiple sc and ip administration to sc pellet implantation. Depending on the treatment regimen and test system employed, an 8–21-fold difference in the response to morphine between naive and morphinized animals has been reported (Kalant, 1977; Lange et al., 1980a, b, 1983). Tolerance has been thought to occur primarily to the depressant effects of opiates (Seever and Deneau, 1963; Fernandes et al., 1982). Recent work, however, has shown tolerance to the stimulant effects of opiates such as hyperactivity (Vasko and Domino, 1978; Browne and Segal, 1980), convulsions (Frenk et al., 1978; Mucha et al., 1978), and hyperthermia (Mucha et al., 1987).

Although there have been extensive investigations into the characteristics and mechanistics aspects of opiate tolerance, little recent work has been carried out with respect to the dispositional aspect of opiate tolerance. In fact, most of the work on dispositional tolerance of opiates was carried out during the 1950s and 1960s. Detailed information on the pharmacokinetics of opiates in naive and opiate-treated subjects can be found in excellent reviews by Way and Adler (1962), Misra (1972, 1978), and Lemberger and Rubin (1976).

4.4.2. Absorption and Distribution

In several animal species and man, the pharmacokinetics of morphine is best described in terms of a multicompartment model (Hipps et al., 1976; Dahlström and Paalzow, 1978; Murphy and Hug, 1981; Plomp et al., 1981). After a single dose of morphine, the $t\frac{1}{2}$ in plasma is approximately 1.2 h for the first 5–6 h; thereafter, the $t\frac{1}{2}$ increases to about 10 h (Berkowitz et al., 1974; Dahlstrom and Paalzow, 1978; Spector et al., 1978; Plomp et al., 1981). In the brain, morphine levels decline slowly up to 2 h after injection, and then disappear more rapidly subsequently (Spector et al., 1978; Plomp et al, 1981). Morphine has been reported to persist in the brain from 24 h (Mullis et al., 1979) to several weeks (Misra et al., 1971a) after a single injection of morphine. In naive rodents and dogs, maximal brain and plasma morphine concentrations have been reported to occur around 30–40 min after sc or ip administration (Mulé and Woods, 1962; Johannesson and Woods, 1964; Loh et al., 1969; Berkowitz et al., 1974; Patrick et al., 1975; Kim et al., 1986). Within this interval, the plasma/brain morphine ratio is approximately 5/1. This ratio, however, declines to unity at around 90–120 min, but increases subsequently (Mulé and Woods, 1962; Johannesson and Woods, 1964; Loh et al., 1969; Plomp et al., 1981).

Morphine-tolerant animals have been reported to show lower plasma level of opiates than controls at 30–60 min following a challenge dose of morphine (Szerb and McCurdy, 1956; Mulé and Woods, 1964; Loh et al., 1969) or etorphine (Manara et al., 1978; Tavani et al., 1979). Brain concentration of morphine, however, has been reported to be no different from (Woods, 1954; Johannesson and Woods, 1964; Johannesson and Schou, 1963) or lower than that of controls (Szerb and McCurdy, 1956; Mulé and Woods, 1962; Loh et al., 1969). It should be pointed out that, in those studies that reported a lower brain level of morphine (Szerb and McCurdy, 1956; Mulé and Woods, 1962; Loh et al., 1969) or etorphine (Tavani et al., 1979), the differences observed were quite minimal. This is in contrast to other opiates, such as levorphanol (Goldstein et al., 1973) or methadone (Misra et al., 1973), where a much lower brain level of these narcotics has been reported in levorpharnol- or methadone-tolerant animals.

The lower levels of plasma and brain opiates in morphine-tolerant animals observed in the studies mentioned above has been suggested to be a secondary consequence of tolerance development (Manara et al., 1978; Tavani et al., 1979). Acutely, morphine and etorphine have been shown to impair the circulation or reduce blood flow (Tavani et al., 1979; Hurwitz, 1981). In morphine-tolerant animals, such impairment was absent. The lower brain or plasma levels of opiates in morphine-tolerant animals might therefore reflect a better clearance (hepatic and renal) because of more efficient circulation in these animals.

When brain morphine levels were examined during continuous morphine treatment, a different pattern was observed. With sc pellet implantation (Patrick et al., 1975; Inturrisi, 1976) or continuous ip infusion of morphine (Patrick et al., 1978), tolerance to the analgesic effect of morphine developed quite rapidly. With sc morphine pellets (75 mg), a steady-state level of plasma morphine has been shown to occur between 10 and 70 h after implantation (Berkowitz et al., 1974; Inturrisi, 1976; Yoburn et al., 1985). Brain morphine levels that were at maximum values between 4 and 24 h declined quite rapidly to half of their maximum value by 48 h after implantation (Patrick et al., 1975). Similarly, in rats that were infused with escalating doses of morphine to a maximal daily dose of 200 mg/kg/d, brain morphine concentration rose to a maximum level at 24 h after the start of the 200 mg/kg/d dosage, but declined very rapidly subsequently, despite the maintenance of the morphine infusion (Patrick et al., 1978). Although there were some increases in the urinary and fecal excretion of morphine and an increase in the morphine levels in the tissues surrounding the infusion cannulae, these increases were quite minor and cannot explain the large declines in brain morphine levels. The ratio of conjugated to free morphine, however, changed in a fashion parallel to that observed for the decline of brain morphine levels.

The possibility of a change in the permeability of the BBB to morphine induced by chronic morphine administration cannot be ruled out. In earlier work, Kerr (1974) reported that rats rendered tolerant to the lethal effect of morphine by multiple ip injections of high doses of morphine failed to show tolerance to such an effect produced by intracerebral ventricular (icv) injections of morphine. Similarly, recent work by Lange et al., (1980a,b, 1983) also demonstrated that, in mice implanted with morphine pellets, tolerance to the analgesic effects of morphine can be demonstrated by sc, but not icv morphine administration. Cross-tolerance to the analgesic effect of heroin or etorphine was also absent. However, when the pellet is removed, tolerance to the analgesic effect of icv morphine or cross-tolerance to sc heroin and etorphine can be readily demonstrated (Lange et al., 1980a,b, 1983). These authors suggested that the failure to demonstrate tolerance to icv morphine or cross-tolerance to sc etorphine and heroin was the result of possible changes in the BBB, which prevents the access of hydrophilic but not of lipophilic compounds. The tolerance and cross-tolerance after pellet removal was suggested to be a form of ''withdrawal tolerance,'' which reflects a functional change in the opiate systems (Lange et al., 1980a,b, 1983).

Recent work by the same group of investigators (Roerig et al., 1987), however, revealed that tolerance to the respiratory depression induced by icv administration of morphine, heroin, or etorphine developed even though the morphine pellet was not removed. This finding would question the possibility of a general alteration in the BBB, which would prevent the access of these opiates to the brain. It should be pointed out, however, that the $ED_{50}s$ for sc or icv administration of these opiates in inducing respiratory depression were much smaller (5–20-fold) than those required for their analgesic effect. It is possible therefore that, if an alteration in the BBB occurred, such alteration would be less effective in modifying the respiratory depressant effect as compared to the analgesic effect of these opiates.

Since neither brain nor plasma morphine levels were determined in the studies of Lange et al. (1980a,b, 1983) and Roerig et al. (1987), it is difficult to evaluate whether such changes in the BBB exist and the significance of this change with respect to brain morphine levels. It should be pointed out that, in those studies that reported minimal or no change in brain morphine concentration mentioned earlier, there was a time lapse between the last treatment dose and the challenge dose of morphine. On the other hand, in the studies by Patrick et al. (1978) or Lange et al. (1980a,b) brain morphine level or morphine sensitivity was determined during the continuous high treatment dose of morphine. It is possible that the handling of morphine or opiates in the body might be, in part, dependent on the levels of morphine existing in the body.

4.4.3. Metabolism and Excretion

Detailed information concerning the metabolism and excretion of opiates can be found in many excellent reviews (Way and Adler, 1960, 1962; Misra, 1972, 1978; Inturrisi, 1976; Lemberger and Rubin, 1976). In general, N- and O-dealkylation, hydrolysis, and conjugation are the major metabolic pathways for the biotransformation of morphine and its congeners. Glucuronide conjugation is an important pathway for the metabolic elimination of morphine and its surrogates (Way and Adler, 1962; Lemberger and Rubin, 1976). Morphine-3-glucuronide is the main product of this conjugation process. Morphine-6-glucronide has also been reported to be a minor metabolite of morphine in rodent, rabbit, dog, and man (Misra, 1978). This metabolite has been shown to have analgesic activity following icv administration in mice (Hosoya and Oka, 1970; Schultz and Goldstein, 1972). Normorphine is a minor metabolite of morphine in several species through the N-dealkylation process (Misra, 1978). Meperidine, however, is metabolized extensively (about one-third of the administered dose) by the dealkylation reaction (Way and Adler, 1962; Lemberger and Rubin, 1976). Hydrolysis is an important reaction for heroin. It is first deacetylated to 6-monoacetylmorphine (6-MAM), which is then further deacetylated to morphine (Misra, 1978). Both 6-MAM and morphine are thought to account for most of the pharmacological effects of heroin (Lemberger and Rubin, 1976; Misra, 1978). Work by Umans and Inturrisi (1981), however, suggested that the pharmacological effects of heroin are mediated principally by 6-MAM.

In general, it can be said that there is minimal change in the rate of metabolism of opiates, with the exception of methadone, following chronic treatment with opiates (Way and Adler, 1962; Misra, 1978). The activity of the microsomal enzymes responsible for the demethylation of opiates has been reported to be inhibited in rats chronically treated with morphine (Axelrod, 1956; Cochin and Axelrod, 1959). Actually, the demethylated product of morphine has been suggested to play an important role in the analgesic effect of opiates. Consequently, the inhibition of the enzymes responsible for the demethylation following chronic treatment has been speculated as the mechanism for analgesic tolerance to opiates in earlier work. Such a hypothesis, however, was ruled out in later work (*see* review by Lemberger and Rubin, 1976). No consistent evidence for a change in the conjugation process of morphine has been observed. A decrease (Misra, 1978), no change (Way and Adler, 1960), and an increase (Zauder, 1952) in the conjugation of opiates have all been reported. Chronic treatment with methadone, however, resulted in an enhancement of its own metabolism as shown by a shorter half-life of plasma and brain methadone (Misra et al., 1973; Misra, 1978).

Morphine and its surrogates are excreted from the body primarily through the kidney (Way and Adler, 1962). The urinary excretion is generally preceded by biotransformation of morphine or its surrogates to a more polar compound. As mentioned above, since there were no significant changes in the rate of metabolism of opiates following chronic treatment, urinary excretion of opiates can be said not to play a significant role in opiate tolerance (Cochin et al., 1954; Patrick et al., 1978).

4.5. Central Stimulants: Amphetamine and Cocaine

4.5.1. Tolerance and Sensitization to Cocaine and Amphetamine

Amphetamine and cocaine exert a variety of effects both in the periphery and CNS. In humans, tolerance to the subjective effects on mood and appetite has been reported (cf. Kalant et al., 1971). On the other hand, repeated intake has also been shown to produce psychosis (*see* reviews, Kalant, 1966; Segal et al., 1981; Segal and Schuckit, 1983). In experimental animals, tolerance to the hyperthermic and cardiovascular effects of amphetamine has been demonstrated (cf. Caldwell et al., 1980). Tolerance also develops to anorexia and impairment of operant performance induced by cocaine and amphetamine (Lewander, 1978; Woolverton et al., 1978; Poulos et al., 1981; Branch and Dearing, 1982). Such tolerance, however, is usually contingent upon the administration of these drugs prior to exposure to food intake or operant performance, since mere chronic exposure to these drugs is not sufficient to produce tolerance to these effects (*see* review by Demellweek and Goudie, 1983; *see also* chapters by Wolgin and Siegel, this volume). In contrast, the hyperactivity and stereotypy induced by cocaine and amphetamine have been shown to be augmented following repeated administration (*see* review by Post, 1981). Thus, depending on the effects examined and the experimental paradigms employed, tolerance occurs to some effects of cocaine and amphetamine, whereas sensitization manifests to others.

The development of contingent tolerance or sensitization appears to rule out the involvement of dispositional factors (Post,

1981; Demellweek and Goudie, 1983). It should be noted that these studies typically involved only intermittent drug administration, and that dispositional tolerance/sensitization might be obtained with more frequent administration. Moreover, as will be pointed out in Sections 6 and 7, the demonstration of contingent tolerance/ sensitization might not necessarily rule out dispositional factors in such phenomena.

4.5.2. Amphetamine Disposition

The metabolism of amphetamine and its related compounds involves several enzymes and displays marked species variation (for detailed information, *see* Caldwell, 1976, 1980; Lemberger and Rubin, 1976; Caldwell et al., 1980). *p*-Hydroxyamphetamine (POHA) is a product of aromatic hydroxylation of amphetamine. This metabolite can be further hydroxylated in noradrenergic terminals by the enzyme dopamine-β-hydroxylase to *p*-hydroxynorephedrine (PHNOR). POHA and PHNOR had received a lot of attention in earlier work because they were thought to act as false noradrenergic transmitters, and might account for the development of tolerance to the hyperthermic and lipolytic effects of amphetamine and sensitization to its hyperactivity and stereotypic effects [discussion of this topic can be found in the reviews by Caldwell et al. (1980), Demellweek and Goudie (1983)]. Amphetamine can also be subjected to oxidative deamination by liver microsomal enzymes, and phenylacetone and benzoic acid derivatives are the major metabolic products (Caldwell, 1976; Lemberger and Rubin, 1976). In man, monkey, dog, cat, and mouse, approximately 30% of the administered dose of amphetamine is excreted unchanged, whereas only 10% has been reported for rat and guinea pig (Caldwell, 1976; Lemberger and Rubin, 1976). On the other hand, in the rat, 60% of the administered dose is excreted as POHA, whereas only 3–7% was seen in man, dog, and rabbit (Lemberger and Rubin, 1976).

In general, it can be said that the metabolic rate of amphetamine is unchanged following chronic treatment (Lewander, 1968; Ellison et al., 1971; Caldwell, 1976, 1980; Kuhn and Schanberg, 1977, 1978). In man, the half-life of amphetamine has been reported to be unchanged or slightly prolonged in amphetamine patients (Caldwell, 1976, 1980). Total metabolic study suggests an increase in urinary excretion of amphetamine in subjects treated chronically with amphetamine (Caldwell, 1976, 1980). Recent work by Kuhn and Schanberg (1978) suggests two exponential components of amphetamine decay with the half-life for the distribution phase of 0.5–0.9 h and 5–9 h for the elimination phase. A higher amount of amphetamine was observed in plasma at 15–30 min, but not at 1 h or later in rats chronically treated with amphetamine (Kuhn and Schanberg, 1978). These authors suggest that a transient decrease in the metabolism of amphetamine might account for the higher level of plasma and brain amphetamine in the early period following amphetamine administration.

Following chronic administration, brain amphetamine levels produced by a challenge dose have been reported to be higher than (Ellison et al., 1971; Magour et al., 1973; Kuhn and Schanberg, 1977, 1978; Segal et al., 1980) or not different from those in controls (Kuhn and Schanberg, 1977, 1978; McCown and Barret, 1980; Eison et al., 1981; Sparber and Fossom 1984). The higher brain amphetamine levels observed are attributed either to the accumulation of amphetamine from chronic treatment with high treatment doses (Ellison et al., 1971; Magour et al., 1973), or the short interval between the last treatment dose and the test dose of amphetamine. Twenty-four hours after the last treatment dose of amphetamine (5 mg/kg daily for 4 d), brain amphetamine, POHA, and PHNOR were still detected in the brain (Jori et al., 1978). Similarly, accumulation of amphetamine and PHNOR have been shown in rats receiving 5 mg/kg of amphetamine at 12-h intervals for 3 d (Kuhn and Schanberg, 1977, 1978).

Although the accumulation from the chronic treatment regimen might account for the higher brain level of amphetamine observed in some experimental studies, as mentioned above, the possibility of alteration in brain amphetamine distribution following chronic treatment cannot be ruled out. In the studies of Kuhn and Schanberg (1977, 1978), a higher brain level of amphetamine, particularly in the striatal area, was observed in amphetamine-treated rats at 30 min only and not at later times following the challenge dose of amphetamine. An increase in the accumulation of amphetamine in the caudate, nucleus accumbens, and substantia nigra at 30 min following a challenge dose of amphetamine was also observed in rats treated with 1.5 mg/kg ip of amphetamine for 25 d (Eison et al., 1981). Of interest is that, in the same study, no difference in the pattern of amphetamine distribution in the brain was observed in rats treated chronically with amphetamine by sc amphetamine pellets (Eison et al., 1981). Differences in behavior associated with continuous (pellet implant) and intermittent treatment (ip) have been reported (Nelson and Ellison, 1978). How the changes in regional distribution of brain amphetamine or the increase in brain amphetamine levels occurring at early times after amphetamine challenge might be involved in the development of sensitization or tolerance remains to be examined.

4.5.3. Cocaine Disposition

Cocaine is rapidly hydrolyzed by the esterase enzymes, which are present in high levels in plasma and liver, to benzoylecgonine, ecgonine methyl ester, and ecgonine (Montesinos, 1965; cf. Jones, 1984). Cocaine can also undergo microsomal oxidation, yielding norcocaine as a metabolic product (Misra et al., 1974; Inaba et al., 1978). This metabolite has been shown to possess some biological activity like that of cocaine, as evident by its local anesthetic activity (Just and Hoyer, 1977) and its effect on schedule-controlled behavior (Spealman et al., 1979). Although the effects of acute and chronic cocaine treatment on various psychological and behavioral functions have been investigated, few studies have actually determined tissue cocaine levels or its pharmacokinetics (cf. Jones, 1984).

From the available data, it appears that cocaine reaches a peak level in brain or plasma much faster in animals treated chronically with cocaine than in naive ones (Nayak et al., 1976; Ho et al., 1977; Mulé and Misra, 1977; Estevez et al., 1979). In cocaine-treated rats, maximum brain and plasma cocaine levels following an sc administration of 20 mg/kg of cocaine was observed at 2 h compared to 4 h for naive rats. The same phenomenon was also observed following ip cocaine administration in the dog (Mulé and Misra, 1977). The brain/plasma ratio of cocaine also tends to be higher in chronically treated cocaine rats than in controls (Nayak et al., 1976), It should be pointed out that, in the study of Estevez et al. (1979), a higher level of cocaine and norcocaine in brain and plasma of cocaine-treated rats was observed. Furthermore, in this study, the same pattern was also demonstrable even 2 w after the end of cocaine treatment.

In naive rats, the half lives of brain and plasma cocaine following ip or iv administration were 20 and 15 min, respectively (Nayak et al., 1976; Ho et al., 1977). In the dog, a much longer half-life (1.2–1.5 h) for brain and plasma was obtained (Mulé and Misra, 1977). The available data, however, are not sufficient to draw any definite conclusions about the half-life of brain and plasma cocaine in animals chronically treated with cocaine. In the study of Ho et al. (1977), the half-life of cocaine in the brain of cocaine-treated rats was 8.5 min compared to 13 min for naive animals. In contrast, a longer half-life of brain and plasma cocaine in chronically cocaine-treated rats was reported by Nayak et al. (1976). These authors reported that the $t\frac{1}{2}$ of cocaine in the brain and plasma of chronically treated rats was approximately 1.8–2 h and that of the acutely treated animals was 0.8–1 h. The longer half-life observed in both naive and chronically cocaine-treated rats in this study than in other studies was the result of the use of subcutaneous administration. The half-life reported in this study, therefore, might not reflect changes in the elimination of cocaine, but rather a net effect of both absorption and elimination.

Detectable amounts of radioactive cocaine or its metabolites were observed in fat, kidney, and brain 4 w after the termination of cocaine treatment (Mulé and Misra, 1977). In naive animals, no detectable amount of cocaine or its metabolites were observed in the brain, and only small amounts were observed in fat and kidney at 24 h after a single challenge dose of cocaine. Whether or not this persistence of radioactive material reflected mainly the metabolites of cocaine or some unchanged cocaine is not known.

5. Cross-Tolerance and Cross-Sensitization: Dispositional Mechanisms

5.1. Definition

Cross-tolerance is defined as the resistance to a drug imparted by chronic treatment with another drug. It is generally believed that cross-tolerance develops among pharmacologically similar agents, and is more likely to occur between two drugs of the same class rather than between two drugs of different classes. However, this view has been questioned, since there is a lack of clear-cut evidence to support the notion that cross-tolerance always occurs among drugs that have common effects and are thought to work through a common mechanism. In fact, many studies have shown that cross-tolerance does not necessarily occur between two drugs of the same class; moreover, cross-tolerance may occur between two drugs of different classes (*see* Coper, 1978; Khanna and Mayer, 1982).

5.2. Dispositional Cross-Tolerance and Sensitization

Considerable evidence has been presented for cross-tolerance among various barbiturates (Green and Koppanyi, 1944; Gruber and Keyser, 1946; Fraser et al., 1957; Scheinin, 1971), alcohols (LeBlanc and Kalant, 1975; Wood and Laverty, 1978), benzodiazepines (Chan, 1984) and opioids (Coper, 1978), i.e. among similarly acting agents belonging to one particular pharmacological group. Since drugs that belong to one pharmacological group are expected to share many similarities in their acute actions, it is not unusual to expect them to show similar adaptive changes after chronic administration.

This notion of similarities in acute actions and cross-tolerance among drugs that belong to one particular pharmacological group was extended to include drugs that are pharmacologically similar in their actions, but belong to different pharmacological groups. In other words, cross-tolerance was expected not only among various barbiturates or alcohols, and so on, but also between ethanol and many other sedative-hypnotics and anesthetics. Many clinical reports suggested that alcoholics required larger doses of inhalational and intravenous anesthetics than naive individuals (for references, *see* Scheinin, 1971; Khanna and Mayer, 1982; Loft et al., 1982; Newman et al., 1986). Since alcohol, barbiturates, other sedative-hypnotics, and anesthetics are considered to produce their actions through interactions with the neuronal membrane resulting in membrane expansion and distortion, it was generally believed that cross-tolerance among all these depressant drugs was primarily the result of commonality of actions at the CNS level.

Although considerable evidence existed for dispositional cross-tolerance among various barbiturates and among drugs metabolized by the mixed function oxygenase system, the notion of dispositional cross-tolerance between ethanol and other sedative-hypnotics was not entertained because of the widely held view that ethanol is metabolized by an enzyme system that is vastly different from that of barbiturates and other sedative-hypnotic drugs. The suggestion by Lieber and coworkers (Lieber and DeCarli, 1968; Rubin and Lieber, 1968) that ethanol could also be metabolized by an enzyme system in microsomes (they coined the termed MEOS, or microsomal ethanol oxidizing system, for this system) analogous to that for barbiturates and other drugs led to an examination of changes in the microsomal enzyme system resulting from chronic ethanol administration as a possible explanation for cross-tolerance between ethanol and other drugs.

Although the role of the MEOS system in ethanol metabolism is still in dispute (*see* reviews by Khanna and Israel, 1980; Crow, 1985), there appears to be general agreement that chronic ethanol administration causes hypertrophy of the SER and increases microsomal mass, cytochrome P450, and in vitro activities of many drug-metabolizing systems (*see* Lieber, 1975; Khanna et al., 1976; Kalant et al., 1976). It appears that chronic ethanol administration induces a new and unique type of cytochrome P450, which has relatively high affinity for substrates such as aniline, *p*-nitroanisole, benzamphetamine, estradiol, and zoxazolamine, and for certain drugs such as CCl_4, acetaminophen, and cocaine. These findings have been very important in our understanding of the enhanced toxicity of these drugs in alcoholics, because induction of metabolism resulted in increased production of active metabolites that are toxic.

The significance of increased microsomal activity after chronic ethanol administration with respect to in vivo metabolic cross-tolerance to barbiturates, other sedative-hypnotics, and anesthetics is, however, not clear. Some investigators have reported an acceleration of blood clearance of pentobarbital and amobarbital (Ratcliffe, 1969; Misra et al., 1971b; Harris et al., 1983), whereas others found no difference in in-vivo pentobarbital clearance between ethanol-treated rats and controls (Kalant et al., 1970, 1976; Hatfield et al., 1972).

Similarly, there is disagreement among investigators as to whether or not chronic barbiturate treatment confers metabolic cross-tolerance to ethanol. Barbiturates are reported to increase the activity of the MEOS (Lieber and DeCarli, 1970) and to enhance the rate of blood ethanol clearance (Lieber and DeCarli, 1972). However, these findings have been challenged on the basis that marked changes in MEOS activity do not affect ethanol metabolism (Mezey, 1971, 1972; Khanna et al., 1971, 1972). As a result of the questionable role of MEOS in ethanol metabolism in vivo (*see* Khanna and Israel 1980), it is uncertain as to whether metabolic cross-tolerance to ethanol occurs in animals chronically treated with barbiturates and other drugs that alter mixed-function oxidase activity.

Chan (1984) has recently reviewed both the clinical and animal studies on alcohol–benzodiazepine cross-tolerance. Just as Khanna and Mayer (1982) found a lack of well-documented clinical studies on cross-tolerance between ethanol and other sedative-hypnotics, so also did Chan (1984) find that cross-tolerance between ethanol and benzodiazepines is generally assumed, but not well documented. Animal studies on this topic are even fewer (McMillan and Leander, 1978; Cesare and McKearney, 1980; Rosenberg et al., 1983; Lê et al., 1986). These studies do provide clear evidence of cross-tolerance between ethanol and benzodiazepines, but the results on cross-tolerance between pentobarbital and benzodiazepines have been conflicting. Only Lê et al. (1986) examined cross-tolerance in both directions. These authors found that, after chronic ethanol treatment, the extent of tolerance to ethanol and cross-tolerance to benzodiazepines was similar. However, chronic benzodiazepine treatment resulted in much higher levels of tolerance to benzodiazepines than to ethanol. None of these studies have examined the role of dispositional factors in cross-tolerance among these drugs. This area deserves further investigation.

Cross-tolerance between ethanol and other sedative-hypnotics may also result from reduced absorption, altered distribution, or increased excretion. However, this issue has not been seriously examined, and it is not clear whether chronic treatment with either ethanol or sedative-hypnotics alters the other's absorption, excretion, or distribution in such a way that cross-tolerance is enhanced. Since neither ethanol nor barbiturate absorption, excretion, or distribution was significantly influenced by chronic ethanol or barbiturate treatment, respectively, it would seem unlikely that cross-tolerance between ethanol and other sedative-hypnotics could be the result of major alterations in these parameters.

Some studies, however, do suggest minor alterations in the distribution or absorption of barbiturates and benzodiazepines as a result of chronic ethanol treatment. Seidel et al. (1964) reported that chronic ethanol treatment reduced the entry of pentobarbital into the brain of the guinea pig. This work has been disputed by Kalant et al. (1970), who pointed out that the drug levels in the brain were measured at only one time point and that the pentobarbital concentration in the blood did not exhibit the expected decline with time. In contrast to the findings of Seidel et al. (1964), Kalant et al. (1970) did not find a significant difference in distribution between plasma and brain, but observed a slower pentobarbital uptake from the site of injection into the plasma and brain in ethanol-treated rats. This was shown by the lower concentration of pentobarbital in the blood and brain 15 min after injection. It is difficult to assess the contribution of this change to the observed cross-tolerance, since there was no longer a difference in drug levels between treated animals and controls at 30 min after injection. Since the cross-tolerant animals slept for almost 30 min, it is unlikely that the change in blood levels at 15 min could account for the shorter sleep duration.

Although the evidence for changes in distribution as a factor in cross-tolerance is limited, it should not be overlooked. Lower concentrations of diazepam were found in the blood of chronic alcoholics than of controls as early as 15 min after intravenous administration of diazepam (Sellman et al., 1975a,c). Similar results were obtained using ethanol-treated rats that received diazepam intraperitoneally (Sellman et al., 1975a). The differences in concentration were attributed to increased early distribution, since no changes were observed in the concentration of diazepam metabolites or rate of disappearance of diazepam from the blood. The possibility that such changes may play a role in cross-tolerance at least underscores the importance of measuring blood and brain drug levels, even during the early phase of an experiment when absorption and distribution are major factors in determining the disposition of the drug.

Compounding the problem of interpretation of some of these positive findings on dispositional cross-tolerance is the fact that, in some studies, it has been difficult to support the commonly accepted notion of cross-tolerance between ethanol and other sedative-hypnotics. In a review of the literature on this topic, it became clear that this belief is based primarily on clinical impressions, but is supported by rather meager experimental evidence, and there are many studies that fail to support the commonly accepted belief of cross-tolerance between ethanol and other sedative-hypnotics (*see* review by Khanna and Mayer, 1982). For example, Gougos et al., (1986) recently reported that chronic treatment with ethanol by gastric intubation, under conditions offering minimum opportunity for associative processes, although producing clear tolerance to ethanol-induced hypothermia, ataxia, and sleep, produced only a marginal degree of cross-tolerance to these effects of pentobarbital (*see* Fig. 3a,b). Similar findings were reported in another study (Newman et al., 1986). These authors examined three different responses (loss of righting reflex, analgesia, and duration of anesthesia) to ethanol, thiopental, and diazepam at three different dose levels in alcohol-fed rats and their pair-fed controls. Chronic treatment with ethanol attenuated all three responses to ethanol and diazepam (the reduction of diazepam effect was significant only at 20 mg/kg), but did not affect the response to thiopental in any of the three tests at any of the three doses. Although there is more positive than negative evidence concerning cross-tolerance to ethanol in barbiturate-tolerant subjects (Frankel et al., 1977), some investigators were unable to see clear evidence of cross-tolerance to ethanol in animals fed chow containing pentobarbital (Commissaris and Rech, 1981, 1983).

The issue of dispositional cross-tolerance between ethanol and other psychotropic drugs has only been examined recently. It appears that ethanol metabolism is accelerated in animals by methadone and that the metabolism of methadone is enhanced by chronic ethanol treatment (Kreek, 1984). Changes in metabolism, however, were ruled out as an explanation for cross-tolerance between ethanol and morphine (Khanna et al., 1979). Evidence has also been presented that short-term exposure to ethanol enhances cocaine metabolism (Smith et al., 1981). These observations do raise the possibility of cross-tolerance between ethanol and other drugs, and suggest that examination of pharmacokinetic parameters should not be overlooked in any studies of cross-tolerance.

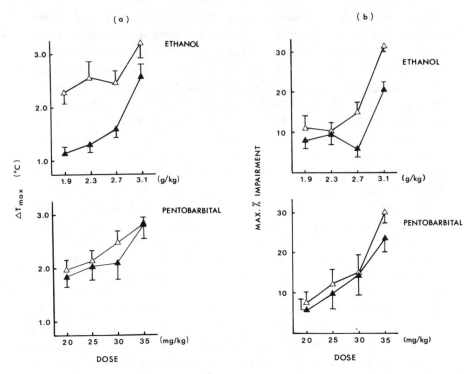

Fig. 3. The hypothermic response (a) and the degree of motor impairment (b) following administration of various doses of ethanol and pentobarbital in rats chronically intubated with ethanol. Ethanol testing was done on d 18 and pentobarbital testing on d 25. Chronic ethanol (▲) equicaloric sucrose (△). Results shown are means ± SEM with n = 8 animals per group at each dose. (From Gougos et al., 1986; reproduced by permission).

6. Variables or Factors that Influence Dispositional Tolerance and Sensitization

6.1. Behavioral Manipulations

6.1.1. Conditioning

In recent years, the role of Pavlovian conditioning in drug tolerance has received a lot of attention. Following the work of Siegel (1975, 1976) on morphine tolerance, numerous investigators have provided evidence for the involvement of Pavlovian conditioning in the development of tolerance and cross-tolerance to a variety of drugs (*see* review by Siegel, 1983 and Siegel, this volume), as well as sensitization to the behavioral effects of cocaine (Hinson and Poulos, 1981). Essentially, these investigators assert that tolerance reflects a special case of classical conditioning of responses that are compensatory to the agonistic actions of the drugs. According to this model, the compensatory responses become associated with the stimuli that accompany drug administration, and these compensatory responses are then evoked upon the presentation of the relevant stimuli. In other words, tolerance should be demonstrable only in the presence of stimuli previously associated with drug administration. For example, saline injection in the environment previously associated with ethanol administration produced a compensatory hyperthermia that has been suggested to account for tolerance to the hypothermic effect of ethanol (Lê et al., 1979; Crowell et al., 1981).

These compensatory responses, however, are not restricted to drug-opposite (responses), but might also involve modification in the pharmacokinetic parameters of the drug. Roffman and Lal (1974) demonstrated that hexobarbital metabolism can be brought under stimulus control. Acute hypoxia resulted in an inhibition of drug metabolism; when mice were exposed to air flow and hypoxia pairing for 6 d, the presentation of air flow produced a reduction in hexobarbital narcosis and an increase in hexobarbital metabolism (Roffman and Lal, 1974). Work by Melchior and Tabakoff (1985) also demonstrated that disposition of ethanol can be brought under stimulus control. At low brain ethanol levels (from 15–90 min after ip injection of 3.5 g/kg), larger volumes of distribution were observed when mice were tested in the environment previously paired with ethanol administration. Moreover, when mice were tested for tolerance to the hypothermic effect of ethanol by icv rather than ip administration, tolerance was observed equally in both the ethanol-linked and saline-linked environments. These authors suggest that the environment-dependent tolerance observed with ip administration might reflect changes in the disposition of ethanol. Although this study is of interest, it is, however, difficult to conceive how the volume of distribution of ethanol can be changed by this process since V_D for ethanol reflects total body water.

Recent work by El-Ghundi et al. (1987, 1989) also showed that the absorption of ethanol was slower and the metabolism lower in rats tested in the environment paired previously with ethanol administration than in rats paired with saline injections (see Fig. 4). No difference in the volume of distribution of ethanol was observed in the ethanol- or saline-linked environment. However, the impairment of ethanol metabolism in the ethanol-linked environment was observed in both ethanol and saline control groups. Therefore, it is likely that such impairment might be the result of stress associated with the transportation of the animals to the special room that constituted the linked ethanol-environment, whereas no stress was produced by the saline-linked environment, which was the animals' home cage. When these rats were tested with pentobarbital (26 mg/kg, ip), a slower absorption, longer half-life, and a lower C_{max} were observed in the ethanol-treated group in the ethanol-linked environment (see Fig. 5). A longer half-life of pentobarbital was also observed in the saline group in the ethanol-linked environment, which again suggests that the impairment of metabolism was the result of other factors, rather than the result of conditioning.

Although the role of conditioning in the development of tolerance to morphine, heroin (Siegel, 1975, 1976, 1983), and amphetamine (Poulos et al., 1981), and of sensitization to cocaine (Hinson and Poulos, 1981) has been demonstrated, none of these studies have examined the possible change in pharmacokinetic parameters of these drugs associated with such manipulation. In light of the changes observed with ethanol and barbiturates, the potential contribution of changes in the disposition of these drugs in conditioned tolerance and sensitization cannot be ruled out.

6.1.2. Stress

In response to stress, many biochemical changes occur in the body, most notably an increase in the peripheral release of catecholamines (Popper et al., 1977; Kvetnansky, 1980). The release of catecholamines would modify blood flow, and as a consequence, the absorption and/or distribution of the drug might be altered. Stress induced by acute aggregated housing has been shown to potentiate amphetamine toxicity (Chance, 1946; Lasagna and McCann, 1957), and a higher brain amphetamine level has been indicated as a possible mechanism (Consolo et al., 1965). Stress can produce analgesia,

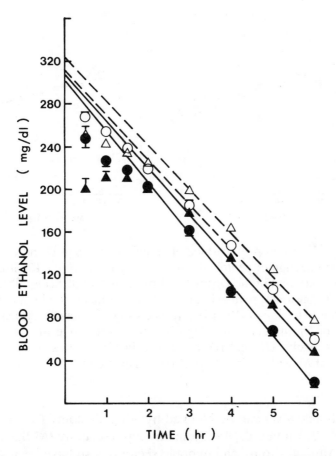

Fig. 4. Blood ethanol elimination curve in E and C groups after an ip injection of ethanol 2.5 g/kg in an environment previously paired with ethanol (DE) or saline (HE) administration. E refers to rats that were given 41 or 47 ethanol injections in DE and the same amount of saline injections in the HE prior to testing in DE or HE, respectively. C refers to rats that were previously injected with saline in either DE or HE. The distinctive environment (DE) was characterized by dim light and a constant background noise. The home environmental (HE) was the colony room where the rats were housed all the time except the test periods in the DE. Results shown are mean ± SEM with N = 21–24 animals per groups.

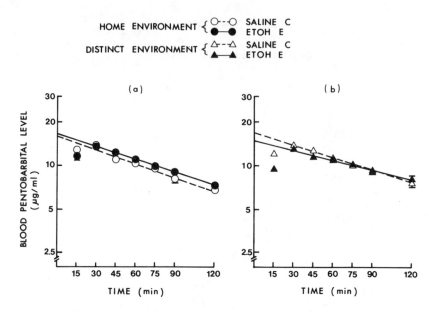

Fig. 5. Blood pentobarbital elimination curves in E and C groups after an ip injection of pentobarbital 25 mg/kg in the Home Environment (HE) (Fig. 5a), and in the Distinctive Environment (DE) (Fig. 5b). E refers to rats that were previously given 51 or 55 ethanol conditioning trials in DE prior to testing in DE or HE, respectively. C refers to rats that were given 51 or 55 saline conditioning trials in DE prior to testing in DE and HE, respectively. DE and HE as defined in Fig. 4. Results shown are mean ± SEM with n = 17–21 animals per group.

and such analgesia can be blocked by adrenalectomy (MacLennan et al., 1982; Lim et al., 1983) and hypophysectomy (Millan et al., 1980). Adrenalectomy and hypophysectomy also have been shown to alter the disposition of morphine (Holaday et al., 1979; Applebaum and Holtzman, 1984).

During the course of chronic treatment by daily injection(s) of the drug, several stress factors such as handling, injection, and change in the volume of body fluids are likely to take place. For example, handling has been shown to enhance the hypothermic effect of ethanol (Peris and Cunningham, 1986). Whether this potentiation of ethanol-induced hypothermia by handling has a dispositional basis still remains to be examined, since no complete pharmacokinetic study of ethanol has been carried out on this topic. The possible stress factors associated with injections and the behavioral procedures employed to assess drug effects are likely to change with repeated injections. In other words, the organism might habituate to these procedures and, as a consequence, might modify the intensity of the stressors. How this habituation might alter the pharmacokinetic parameters of the drug still remains to be investigated. It should be noted that the absorption of a challenge dose of amphetamine in animals chronically treated with amphetamine by pellet implantation has been shown to be different from that in animals treated with daily injections (Eison et al., 1981). Moreover, with repeated injections, the organisms might be able to pick up certain cues, such as removal from their home cage, or weighing, which might allow them to participate or predict the administration of the drug. Such processes, as discussed earlier, have been shown to modify the disposition of the drug.

6.2. Aging

The interaction between drugs and aging has been a subject of interest during the last several years. It is now well documented that senescent organisms are, in general, more responsive to a variety of psychoactive drugs than young or adult ones. For example, senescent rodents have been shown to be more sensitive to the effects of morphine on thermoregulation (McDougal et al., 1980), the hyperactivating effects of amphetamine (Hicks et al., 1980), the anticonvulsant effect of phenobarbital (Kitani et al., 1986), and to a variety of effects of ethanol (*see* review by Wood and Armbrecht, 1982). The development of tolerance to ethanol (Ritzmann and Springer, 1980) and morphine (McDougal et al., 1981) has also been shown to be slower in aging rodents.

The changes in response to drugs in aging animals probably involve both pharmacodynamic and pharmacokinetic factors. In senescent organisms, changes in the function and/or anatomical structure of target tissue decrease receptor numbers (Schocken and Roth, 1977; Vestal et al., 1979). Changes in pharmacokinetic parameters of the drugs, and in many physiological functions that can affect responses, also occur with aging. For example, changes in splanchic blood flow, gastrointestinal motility, and body fat content, as well as the decrease in the activity of hepatic microsomal oxidase systems, have been reported in senescent organisms (for detailed information on the subject, *see* review by Schmucker, 1985).

Acutely, a decrease in the volume of distribution (York, 1983) and in the rate of metabolism of ethanol (Hahn and Burch, 1983) has been demonstrated in aging rats. A clear increase in the rate of metabolism of ethanol occurs following chronic treatment with ethanol in young rats, but occurs minimally in older rats (Israel et al., 1977; Britton et al., 1984). In the case of amphetamine, a longer plasma half-life has been observed in senescent rats (Hicks et al., 1980). However, to our knowledge, except for ethanol, there have been no studies concerned with the possible changes in the pharmacokinetic characteristics of most other psychotropic drugs in aging organisms following chronic drug treatment.

6.3. Nutrition/Diet

Drugs can influence the nutritional status of the organism in a variety of ways. Drugs like amphetamine or opiates can modulate the appetite for food and/or food intake (*see* review by Garattini and Samanin, 1978). Ethanol can impair the absorption and digestion of the intake nutrients (Barona and Lindenbaum, 1977). Diet, on the other hand, can also influence the absorption, metabolism, or binding of the drug to protein. For example, high-fat diets can stimulate or enhance the absorption of fat-soluble drugs. A high-fat meal also produces a substantial increase in plasma free fatty acids, which can compete with the drug for binding to albumin and, therefore, can alter the concentration of free drug in plasma (Spector and Fletcher, 1978). High-protein diet and deficiency in thiamine, iron, and energy have also been shown to modify the activity of microsomal enzymes (*see* reviews by Chabra and Tredger, 1978; Williams, 1978; Anderson et al., 1985; Hathcock, 1985).

Much work on diet or nutrition and drug interaction has emphasized primarily the acute aspects with little emphasis on nutrition and chronic intake of drugs. Composition of the diet, however, has been shown to influence dispositional tolerance to barbiturates. When chronic ethanol treatment was carried out by feeding the rat a liquid diet with high fat content, the metabolism of barbiturates has been suggested to increase (Rubin et al., 1970). Work from our laboratory (Gougos, 1984) has also shown that chronic ethanol treatment alone (by daily gavage in tap water) did not affect pentobarbital metabolism. However, when ethanol was given in a liquid diet with high fat content, the metabolism of pentobarbital was enhanced (*see* Fig. 6). Since this did not occur with the control liquid diet or with ethanol alone an interaction between diet and ethanol treatment is likely to account for such an enhancement of barbiturate metabolism.

In the study of self-administration of psychoactive drugs in experimental animals, food deprivation has been used to facilitate the initiation or enhancement of drug self-administration (Meisch, 1984). Reducing the body weight to 80% of free feeding weight by food restriction, however, has been shown to reduce the absorption of ethanol given either by ip or oral routes in Wistar and Sprague-Dawley rats (Linseman, 1987). Many psychoactive drugs affect food intake acutely. For example, amphetamine produces anorexia. How such anorexic effects will influence the pharmacokinetic parameters of amphetamine during and following chronic treatment with amphetamine remains to be examined.

6.4. Others

A variety of other factors including sex, species and strain can influence the metabolism of psychoactive drugs. It is beyond the scope of this chapter to go into detail about these subjects. Information concerning the influence of these factors on amphetamine, opiate and ethanol metabolism can be found in selected reviews (Kalant, 1971; Caldwell, 1976; Misra, 1978). It is, however, noteworthy that differences in the metabolism of drugs among species or strains has been used as a tool to investigate the role of some metabolic products in tolerance development. For example, the parahydroxylated metabolites of amphetamine were thought to play a critical role in the development of tolerance to amphetamine. However, this hypothesis was rejected, since tolerance was also observed in the guinea pig, which does not produce these metabolites (Caldwell et al., 1980).

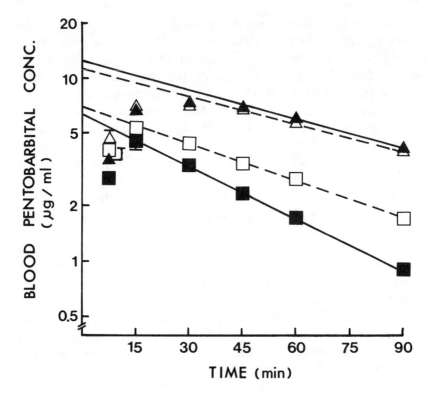

Fig. 6. Blood pentobarbital concentration in rats at various times after ip administration of 2 mg/kg of pentobarbital. Prior to pentobarbital administration, rats were treated for 36 d with either ethanol or sucrose by either gastric intubation or in a liquid diet. Treatment groups: ■, ethanol liquid diet; □, equicaloric sucrose liquid diet; ▲ ethanol gastric intubation; △, equicaloric sucrose gastric intubation; n = 13–15 animals per group. The descending portion of the curve is extrapolated to the ordinate.

7. Summary and Conclusion

Given the extensive body of literature on tolerance and sensitization to psychoactive drugs (Kalant et al., 1971; Post, 1981; Kalant and Lê, 1984), few studies have paid attention to the role of dispositional factors in such phenomena. Most often when dispositional factors were of concern, the emphasis was primarily on the extent of change in the rate of metabolism (biotransformation) following chronic drug treatment. Probably related to this notion, several common assumptions or arguments have been made in the literature to rule out the involvement of dispositional factors in observed tolerance or sensitization. These include:

1. Tolerance/sensitization is observed at an early time following the administration of the challenge dose, at which time minimal metabolism has taken place
2. Tolerance/sensitization occurs to certain effects of the drug but not to others, or that both tolerance and sensitization are manifested to different effects of the drug in the same organisms
3. No change or significant change in the rate of metabolism of the drug under study has been demonstrated
4. Cross-tolerance or cross-sensitization occurs to other drugs that do not show metabolic cross-tolerance or the drug is not subject to metabolism.

Although there is some validity in the arguments or assumptions mentioned above to support the existence of functional tolerance and sensitization, they by no means rule out the contribution or involvement of dispositional factors. In the case of morphine tolerance, for example, although the degree of tolerance is quite extensive and there is minimal change in the rate of morphine metabolism, the involvement of dispositional factors in such tolerance cannot be ruled out. This was illustrated quite clearly from the work of Lange et al. (1980a,b, 1983), which showed a 21-fold tolerance to morphine by sc, but minimal tolerance with icv injection in mice that were implanted with morphine pellets. A significant change of distribution of the drug was clearly indicated from these studies. Changes in regional brain distribution of amphetamine have also been shown to occur in certain conditions following chronic amphetamine administration (*see* Section 4.5.2). Such changes might account for some of the observed concurrent manifestation of both tolerance and sensitization to different effects of the drug in the same organism. For example, an increase in the level of amphetamine in dopamine-rich regions might account for some of the observed sensitization to its hyperactivity or stereotypic effects (Kuhn and Schanberg, 1978; Eison et al., 1981). As a consequence of this, the level of drug in other brain areas might decline and, therefore, might account for some of the tolerance to other critical effects of the drug.

Most often, when the pharmacokinetic parameters of the drug have been determined following chronic drug treatment, little attention has been paid to the absorptive phase of the drug. As evident from the data reviewed above, changes in the rate of absorption of the drug or in time taken to reach peak brain and plasma levels have been observed for ethanol, barbiturates, and cocaine (Nayak et al., 1976; Melchior and Tabakoff, 1985; El-Ghundi et al., 1987). The changes in the absorption of the drug appear to be sensitive to behavioral procedures associated with chronic treatment, such as stress and conditioning (Section 6). The changes in the rate of absorption of the drug were observed not only immediately following the termination of treatment, but also a long time after the cessation of drug treatment. Moreover, chronic treatment with one drug can also modify the absorption of another drug (El-Ghundi et al., 1987).

In summary, chronic treatment with psychotropic drugs can produce many significant changes in their pharmacokinetic parameters ranging from absorption and distribution to metabolism. Some of the changes are not necessarily associated with drug treatment *per se*. Many other variables such as stress, conditioning, and diet involved in the procedure associated with drug treatment are equally important. It is clear that failure to observe changes in pharmacokinetic parameters of the drug in certain experimental paradigms cannot be readily inferred in others. It is also clear that one cannot rule out the involvement of dispositional factors, particularly absorption and distribution, in the observed tolerance and sensitization to various psychoactive drugs in all experimental conditions. Clearly, there is a need for further investigation into the role of dispositional factors in various paradigms of tolerance and sensitization.

Acknowledgment

The authors would like to thank Ms. V. Cabral for the preparation of this chapter. This review was supported by NIAAA Grant No. AA07003-02.

References

Anderson R. E., Pantuck E. J., Conney A. H., and Kappas A. (1985) Nutrient regulation of chemical metabolism in human. *Fed Proc.* **44,** 130–133.

Applebaum B. D. and Holtzman S. G. (1984) Characterization of stress-induced potentiation of opioid effects in the rat. *J. Pharmacol. Exp. Ther.* **231,** 555–585.

Aranko K. (1985) Task-dependent development of cross-tolerance to psychomotor effects of lorazepam in man. *Acta Pharmacol. Toxicol.* **56,** 373–381.

Aranko K., Mattila M. J., Nuutila A., and Pellinen J. (1985) Benzodiazepines, but not anti-depressants or neuroleptics, induce dose-dependent development of tolerance to lorazepam in psychiatric patients. *Acta Psychiat. Scand.* **72,** 436–446.

Axelrod J. (1956) The enzymatic *N*-demethylation of narcotic drugs. *J. Pharmacol. Exp. Ther.* **117,** 322–350.

Barona E. and Lindenbaum J. (1977) Metabolic effects of alcohol on the intestine, in *Metabolic Aspects of Alcoholism* (Lieber C.S., ed.), pp. 81–116. University Park, Baltimore.

Berkowitz B. A., Cerreta K. V., and Spector S. (1974) The influence of physiologic and pharmacologic factors on the disposition of morphine as determined by radioimmunoassay. *J. Pharmacol. Exp. Ther.* **191,** 527–534.

Boisse N. R. and Okamoto M. (1978a) Physical dependence to barbital compared to pentobarbital. II. Tolerance characteristics. *J. Pharmacol. Exp. Ther.* **204,** 507–513.

Boisse N. R. and Okamoto M. (1978b) Physical dependence to barbital compared to pentobarbital. IV. Influence of elimination kinetics. *J. Pharmacol. Exp. Ther.* **204,** 526–540.

Braggins T. J. and Crow K. E. (1981) The effects of high ethanol doses on rates of ethanol oxidation in rats. *Eur. J. Biochem.* **119,** 633–640.

Branch, M. N. and Dearing M. E. (1982) Effects of acute and daily cocaine administration on performance under a delayed-matching-to-sample procedure. *Pharmacol. Biochem. Behav.* **16,** 713–718.

Britton R. S., Videla L. A., Rachamin G., Okuno F., and Israel Y. (1984) Effect of age on metabolic tolerance and hepatomegaly following chronic ethanol administration. *Alcoholism: Clin. Exp. Res.* **8,** 528–534.

Browne T. R. (1976) Clonazepam: A review of a new anti-convulsant drug. *Arch. Neurol.* **33,** 326–332.

Browne R. G. and Segal D. S. (1980) Alterations in beta-endorphin-induced locomotor activity in morphine-tolerant rats. *Neuropharmacol.* **19,** 619–621.

Büch H., Grund W., Buzello W., and Rummel W. (1969) Narkotische Wirksamkeit und Gewebsverteilung der optischen Antipoden des Pentobarbitals bei der Ratte. *Biochem. Pharmacol.* **18,** 1005.

Butler T. C., Mahaffee C., and Waddell W. (1954) Phenobarbital: Studies of elimination, accumulation, tolerance, and dosage schedules. *J. Pharmacol. Exp. Ther.* **111,** 425–435.

Caldwell J. (1976) The metabolism of amphetamines in mammals. *Drug Metab. Rev.* **5,** 219–280.

Caldwell J. (1980) The metabolism of amphetamines and related stimulants in animals and man, in *Amphetamines and Related Stimulants: Chemical, Biological, Clinical, and Sociological Aspects* (Caldwell J., ed.), pp. 29–46, CRC, Florida.

Caldwell J., Croft J. E., and Sever P. S. (1980) Tolerance to the amphetamines: An examination of possible mechanisms, in *Amphetamines and Related Stimulants: Chemical, Biological, Clinical, and Sociological Aspects* (Caldwell J., ed.), pp. 131–146. CRC, Florida.

Carpenter T. M. (1940) The metabolism of alcohol: A review. *Quart. J. Stud. Alc.* **1,** 201–226.

Cesare D. A. and McKearney J. W. (1980) Tolerance to suppressive effects of chlordiazepoxide on operant behavior: Lack of cross tolerance to pentobarbital. *Pharmacol. Biochem. Behav.* **13,** 545–548.

Chabra R. S. and Tredger J. M. (1978) Interactions of drugs and intestinal mucosal endoplasmic reticulum, in *Nutrition and Drug Interrelations* (Hathcock J. N. and Coon J., eds.) pp. 253–277, Academic, New York.

Chan A. W. K. (1984) Effects on combined alcohol and benzodiazepine: A review. *Drug Alc. Depend.* **13,** 315–341.

Chance M. R. A. (1946) Aggregation as a factor influencing the toxicity of sympathomimetic amines in mice. *J. Pharmacol. Exp. Ther.* **87,** 214–219.

Chin J. H. (1979) Ethanol and the blood–brain barrier, in *Biochemistry and Pharmacology of Ethanol, Vol. 2, Chapter 7* (Majchrowicz E. and Noble E. P., eds.), pp. 101–118. Plenum, New York and London.

Christensen J. D. (1973) Tolerance development with chlordiazepoxide in relation to the plasma levels of the parent compound and its main metabolites in mice. *Acta Pharmacol. Toxicol.* **33**, 267–272.

Cochin J. and Axelrod J. (1959) Biochemical and pharmacological changes in the rat following chronic administration of morphine, nalorphine and normorphine. *J. Pharmacol. Exp. Ther.* **125**, 105–110.

Cochin J., Haggart J., Woods L. A., and Seevers M. H. (1954) Plasma levels, urinary and fecal excretion of morphine in non-tolerant and tolerant dogs. *J. Pharmacol. Exp. Ther.* **111**, 74–83.

Commissaris R. L. and Rech R. H. (1981) Tolerance to pentobarbital and ethanol following chronic pentobarbital administration in the rat. *Subst. Alc. Action/Misuse* **2**, 331–339.

Commissaris R. L. and Rech R. H. (1983) Tolerance and cross-tolerance to central nervous system depressants after chronic pentobarbital or chronic methaqualone administration. *Pharmacol. Biochem. Behav.* **18**, 327–331.

Conney A. H. (1967) Pharmacological implications of microsomal enzyme induction. *Pharmacol. Rev.* **19**, 317–366.

Consolo S., Garratini S., and Valzelli L. (1965) Amphetamine toxicity in aggressive mice. *J. Pharm. Pharmacol.* **17**, 53–58.

Coper H. (1978) Cross-tolerance and cross-dependence between different types of addictive drugs, in *The Bases of Addiction* (Fishman J., ed.), pp. 235–256, Dahlem Konferenzen, Berlin.

Crow K. E. (1985) Ethanol metabolism by the liver. *Rev. Drug Metab. Drug Interactions* **5**, 113–158.

Crowell C. R., Hinson R. E., and Siegel S. (1981) The role of conditional drug responses in tolerance to the hypothermic effects of ethanol. *Psychopharmacology* **73**, 51–54.

Dahlström B. E. and Paalzow L. (1978) Pharmacokinetics of morphine in relation to analgesia: Effects of routes of administration, in *Factors Affecting the Action of Narcotics* (Adler M. L., Manara L., and Samarin R., eds.), pp. 233–248, Raven, New York.

Demellweek C. and Goudie A. J. (1983) Behavioral tolerance to amphetamine and other psychostimulants: The case for considering behavioral mechanisms. *Psychopharmacology* **80**, 287–307.

Ebert A. G., Yim G. K. W., and Miya T. S. (1964) Distribution and metabolism of barbital-^{14}C in tolerant and intolerant rats. *Biochem. Pharmacol.* **13**, 1267–1274.

Eison M. S., Ellison G., and Eison A. S. (1981) The regional distribution of amphetamine in rat brain is altered by dosage and by prior exposure to the drug. *J. Pharmacol. Exp. Ther.* **218**, 237–241.

El-Ghundi M., Kalant H., Lê A. D., and Khanna J. M. (1987) The contribution of environmental cues to cross-tolerance between ethanol and pentobarbital. *Alcoholism: Clin. Exp. Res.* **2**, Abst. 63.

El-Ghundi M., Kalant H., Lê A. D., and Khanna J. M. (1989) The contribution of environmental cues to cross-tolerance between ethanol and pentobarbital. *Psychopharmacology* **97**, 194–201.

Ellison T., Okum R., Silver A., and Siegel M. (1971) Metabolic fate of amphetamine in the cat during the development of tolerance. *Arch. Int. Pharmacodyn. Ther.* **190**, 135–149.

Estevez V. S., Ho B. T., and Englert L. F. (1979) Metabolism correlates of cocaine-induced stereotypy in rats. *Pharmacol. Biochem. Behav.* **10**, 267–271.

Feely M., Calvert R., and Gibson J. (1982) Clobazam in catamenial epilepsy. *Lancet* **2**, 71.

Feinman L., Baraona E., Matsuzaki S., Korsten M., and Lieber C. S. (1978) Concentration dependence of ethanol metabolism in vivo in rats and man. *Alcoholism: Clin. Exp. Res.* **2**, 381–385.

Fernandes H., Kluwe S., and Coper H. (1982) Development and loss of tolerance to morphine in the rat. *Psychopharmacology* **78**, 234–238.

File S. E. (1981) Rapid development of tolerance to the sedative effects of lorazepam and triazolam in rats. *Psychopharmacology* **73**, 240–245.

File S. E. (1982a) Development and retention of tolerance to the sedative effects of chlordiazepoxide: Role of apparatus cues. *Eur. J. Pharmacol.* **81**, 637–643.

File S. E. (1982b) Recovery from lorazepam tolerance and the effects of a benzodiazepine antagonist (RO 15-1788) on the level of tolerance. *Psychopharmacology* **77**, 284–288.

File S. E. (1983a) Strain differences in mice in the development of tolerance to the anti-PTZ effects of diazepam. *Neuroscience Letters* **42**, 95–98.

File S. E. (1983b) Tolerance to the anti-pentylenetetrazole effects of diazepam in mouse. *Psychopharmacology* **79**, 284–286.

Fleming R. and Stotz E. (1935) Experimental studies in alcoholism. I. The alcohol content of the blood and cerebrospinal fluid following oral administration in chronic alcoholism and the psychoses. *Arch. Neurol. Psychiat. (Chicago)* **33**, 492–506.

Fleming R. and Stotz E. (1936) Experimental studies in alcoholism. II. The alcohol content of the blood and cerebrospinal fluid following intravenous administration of alcohol in chronic alcoholism and the psychoses. *Arch. Neurol. Psychiat. (Chicago)* **35**, 117–125.

Frankel D., Khanna J. M., LeBlanc A. E., and Kalant H. (1977) Effect of *p*-chlorophenylalanine on the development of cross-tolerance between pentobarbital and ethanol. *Can. J. Physiol. Pharmacol.* **55**, 954–957.

Fraser H. F., Wikler A., Isbell H., and Johnson N. K. (1957) Partial equivalence of chronic alcohol and barbiturate intoxication. *Quart. J. Stud. Alc.* **18**, 541–551.

Frenk H., Urea G., and Liebeskind J. C. (1978) Epileptic properties of leucine- and methionine-enkephalin: Comparison with morphine and reversibility by naloxone. *Brain Res.* **147**, 327–337.

Garattini S. and Samanin R., eds (1978) *Central Mechanisms of Anorectic Drugs*. Raven, New York.

Gastaut H. and Low M. D. (1979) Anti-epileptic properties of clobazam, a 1,5-benzodiazepine, in man. *Epilepsia* **20**, 437–446.

Gent J. P. and Haigh J. R. M. (1983) Development of tolerance to the anti-convulsant effects of clobazam. *Eur. J. Pharmacol.* **94**, 155–158.

Gettler A. O. and Freireich A. W. (1935) The nature of alcohol tolerance. *Amer. J. Surg.* **27**, 237–333.

Gillette J. R. (1971) Factors affecting drug metabolism. *Annals N.Y. Acad. Sci.* **179**, 43–66.

Gilman A. G., Goodman L. S., Rall T. W., and Murad F., eds (1985) *Goodman and Gilman's The Pharmacological Basis of Therapeutics, Seventh Edition*. MacMillan Publishing, A Division of MacMillan, Inc., New York.

Goldberg M. E., Manian A. A., and Efron D. H. (1967) A comparative study of certain pharmacologic responses following acute and chronic administration of chlordiazepoxide. *Life Sci.* **6**, 481–491.

Goldstein A., Judson B. A., and Sheehan P. (1973) Cellular and metabolic tolerance to an opioid narcotic in mouse brain. *Brit. J. Pharmacol.* **47**, 138–140.

Gougos A. (1984) *Studies on Cross-Tolerance Between Ethanol and Barbiturates*. M.Sc. Thesis, University of Toronto, Ontario, Canada.

Gougos A., Khanna J. M., Lê A. D., and Kalant H. (1986) Tolerance to ethanol and cross-tolerance to pentobarbital and barbital. *Pharmacol. Biochem. Behav.* **24**, 801–807.

Green M. W. and Koppanyi T. (1944) Studies on barbiturates. XXVII. Tolerance and cross-tolerance to barbiturates. *Anesthesiology* **5**, 329–340.

Gruber C. M. and Keyser G. F. (1946) A study on the development of tolerance and cross-tolerance to barbiturates in experimental animals. *J. Pharmacol. Exp. Ther.* **86,** 186–196.

Grundstrom R., Holmberg G., and Hansen T. (1978) Degree of sedation obtained with various doses of diazepam and nitrazepam. *Acta Pharmacol. Toxicol.* **43,** 13–18.

Guynn R. W. and Pieklik J. R. (1975) Dependence on dose of acute effects of ethanol on liver metabolism in vivo. *J. Clin. Invest.* **56,** 1411–1419.

Hahn H. K. and Burch R. E. (1983) Impaired ethanol metabolism with advancing age. *Alcoholism: Clin. Exp. Res.* **7,** 299–301.

Harris J. M., Rao S. N., and Okamoto M. (1983) Demonstration of dispositional and functional cross-tolerance between ethanol and pentobarbital. *Alcoholism: Clin. Exp. Res.* **7,** 111.

Hatfield G. K., Miya T. S., and Bousquet W. F. (1972) Ethanol tolerance and ethanol-drug interactions in the rat. *Toxicol. Appl. Pharmacol.* **23,** 459–469.

Hathcock J. N. (1985) Metabolic mechanisms of drug-nutrient interactions. *Fed. Proc.* **44,** 124–129.

Hawkins R. D. and Kalant H. (1972) The metabolism of ethanol and its metabolic effects. *Pharmacol. Rev.* **243,** 67–157.

Hawkins R. D., Kalant H., and Khanna J. M. (1966) Effects of chronic intake of ethanol on rate of ethanol metabolism. *Can. J. Physiol. Pharmacol.* **44,** 241–257.

Hicks P., Strong R., Schoolar J. C., and Samorajski T. (1980) Aging alters amphetamine-induced stereotyped gnawing and neostriatal elimination of amphetamine in mice. *Life Sci.* **27,** 715–722.

Hillestad L., Hansen T., and Melsom H. (1974) Diazepam metabolism in normal man. II. Serum concentration and clinical effects after oral administration and cumulation. *Clin. Pharmacol. Ther.* **16,** 485–489.

Hinson R. E. and Poulos C. X. (1981) Sensitization to the behavioral effects of cocaine: Modification by Pavlovian conditioning. *Pharmacol. Biochem. Behav.* **15,** 559–562.

Hipps P. P., Eveland M. R., Meyer E. R., Sherman W. R., and Cicero T. J. (1976) Mass fragmentography of morphine: Relationships between brain levels and analgesic activity. *J. Pharmacol. Exp. Ther.* **196,** 642–648.

Ho B. T., Taylor D. L., Estevez V. S., Englert L. F., and McKenna M. L. (1977) Behavioral effects of cocaine: Metabolic and neurochemical approach, in *Cocaine and Other Stimulants*

(Ellinwood E. H., Jr. and Kilbey M. M., eds.), pp. 229–240, Plenum, New York.

Holaday J. W., Law P. Y., Loh H. H., and Li C. H. (1979) Adrenal steroids indirectly modulate morphine and β-endorphin effects. *J. Pharmacol. Exp. Ther.* **208,** 176–183.

Hoogland D. R., Miya T. S., and Bousquet W. F. (1966) Metabolism and tolerance studies with chlordiazepoxide-2-^{14}C in the rat. *Toxicol. Appl. Pharmacol.* **9,** 116–123.

Hosoya E. and Oka T. (1970) Studies on morphine glucoronide. *Med. Ctr. J. Univ. Michigan* **36,** 241–246.

Houghton G. W., Richens A., Toseland P. A., Davidson S., and Falconer M. A. (1975) Brain concentrations of phenytoin, phenobarbitone and primidone in epileptic patients. *Eur. J. Clin. Pharmacol.* **9,** 73–78.

Hurwitz A. (1981) Narcotic effects on phenol red disposition in mice. *J. Pharmacol. Exp. Ther.* **216,** 90–94.

Inaba T., Stewart D. J., and Kalow W. (1978) Metabolism of cocaine in man. *Clin. Pharmacol. Ther.* **23,** 547–552.

Inturrisi C. E. (1976) Disposition of narcotics and narcotic antagonists. *Annals N.Y. Acad. Sci.* **218,** 273–287.

Israel Y., Khanna J. M., Kalant H., Stewart D. J., Macdonald J. A., Rachamin G., Wahid S., and Orrego H. (1977) The spontaneously hypertensive rat as a model for studies on metabolic tolerance to ethanol. *Alcoholism: Clin. Exp. Res.* **1,** 39–42.

Jablonska J. K., Knoblock K., Majka J., and Wisniewska-Knypl, J. M. (1975) Stimulatory effects of chlordiazepoxide, diazepam, and oxazepam on the drug-metabolizing enzymes in microsomes. *Toxicology* **5,** 103–111.

Jacobsen E. (1952) The metabolism of ethyl alcohol. *Pharmacol Rev.* **4,** 107–135.

Johannesson T. and Schou J. (1963) Analgesic activity and brain concentration of morphine in tolerant and non-tolerant rats given morphine alone or with neostigmine. *Acta Pharmacol. Toxicol.* **20,** 213–221.

Johannesson T. and Woods L. A. (1964) Analgesic action and brain and plasma levels of morphine and codeine in morphine tolerant, codeine tolerant and non-tolerant rats. *Acta Pharmacol. Toxicol.* **21,** 381–396.

Jones R. T. (1984) The pharmacology of cocaine, in *Cocaine: Pharmacology, Effects, and Treatment of Abuse* (Grabowski J., ed.), pp. 34–53, NIDA Research Monograph 50, Rockville, Maryland.

Jori A., Caccia S., and Dolfini E. (1978) Tolerance to anorectic drugs, in *Central Mechanisms of Anorectic Drugs* (Garattini S. and Samanin A., eds.), pp. 179–190, Raven, New York.

Just W. W. and Hoyer A. J. (1977) The local anesthetic potency of norcocaine, a metabolite of cocaine. *Experientia* **33**, 70–71.

Kalant O. J. (1966) *The Amphetamines, Toxicity and Addiction.* Brookside Monographs No 5 of the Addiction Research Foundation, University of Toronto Press, Toronto.

Kalant H. (1971) Absorption, diffusion, distribution and elimination of ethanol: Effects on biological membranes, in *The Biology of Alcoholism, Vol. 1, Biochemistry* (Kissin B. and Begleiter H., eds.), pp. 1–62, Plenum, New York.

Kalant H. (1977) Comparative aspects of tolerance to, and dependence on, alcohol, barbiturates and opiates, in *Alcohol Intoxication and Withdrawal, Vol. 3B, Advances in Experimental Medicine and Biology* (Gross M. M., ed.), pp. 169–186. Plenum, New York.

Kalant H. and Lê A. D. (1984) Effects of ethanol on thermoregulation. *Pharmacol. Ther.* **23**, 313–364.

Kalant H., Khanna J. M., Lin G. Y., and Chung S. (1976) Ethanol—a direct inducer of drug metabolism. *Biochem. Pharmacol.* **25**, 337–342.

Kalant H., Khanna J. M., and Marshman J. (1970) Effect of chronic intake of ethanol on pentobarbital metabolism. *J. Pharmacol. Exp. Ther.* **175**, 318–324.

Kalant H., Khanna J. M., Seymour F., and Loth J. (1975) Acute alcoholic fatty liver—metabolism or stress. *Biochem. Pharmacol.* **24**, 431–434.

Kalant H., LeBlanc A. E., and Gibbins R. J. (1971) Tolerance to, and dependence on, some non-opiate psychotropic drugs. *Pharmacol Rev.* **23**, 135–191.

Kalant H., Roschlau W. H. E., and Sellers E. M. eds. (1985) *Principles of Medical Pharmacology.* Department of Pharmacology, Faculty of Medicine, University of Toronto, Ontario, Canada.

Kanto J., Iisalo E., Lektinen V., and Salminew J. (1974) The concentrations of diazepam and its metabolites in the plasma after an acute and chronic administration. *Psychopharmacologia (Berl.)* **36**, 123–131.

Kaplan S. A., Jack M. L., Alexander K., and Weinfeld R. E. (1973) Pharmacokinetic profile of diazepam in man following single intravenous and oral and chronic oral administration. *J. Pharmaceut. Sci.* **62**, 1789–1796.

Kerr F. W. L. (1974) Tolerance to morphine and the blood brain/CSF barrier. *Fed. Proc.* **33**, 528.

Khanna J. M. and Israel Y. (1980) Ethanol metabolism, in *Liver and Biliary Tract Physiology. I. International Review of Physiology, Vol. 21* (Javitt N. B., ed.), pp. 275–315, University Park, Baltimore.

Khanna J. M. and Mayer J. M. (1982) An analysis of cross-tolerance among ethanol, other general despressants and opioids. *Subst. Alc. Actions/Misuse* **3**, 243–257.

Khanna J. M., Kalant H., and Lin G. (1972) Significance in vivo of the increase in microsomal ethanol oxidizing system after chronic administration of ethanol, phenobarbital and chlorcyclizine. *Biochem. Pharmacol.* **21**, 2215–2226.

Khanna J. M., Kalant H., Lin G., and Bustos G. O. (1971) Effect of carbon tetrachloride treatment on ethanol metabolism. *Biochem. Pharmacol.* **20**, 3269–3279.

Khanna J. M., Kalant H., Yee Y., Chung S., and Siemens A. J. (1976) Effect of chronic ethanol treatment on metabolism of drugs in vitro and in vivo. *Biochem. Pharmacol.* **25**, 329–335.

Khanna J. M., Lê A. D., Kalant H., and LeBlanc A. E. (1979) Cross-tolerance between ethanol and morphine with respect to their hypothermic effects. *Eur. J. Pharmacol.* **59**, 145–149.

Khanna J. M., Lindros K., Israel Y., and Orrego H. (1977) In vivo metabolism of ethanol at high and low concentrations, in *Alcohol and Aldehyde Metabolizing Systems, Vol. III* (Thurman R. G., Williamson J. R., Drott H. R., and Chance B., eds.), pp. 325–334, Academic, New York.

Kim C., Speisky M. B., and Kalant H. (1986) Simultaneous determination of biogenic amines and morphine in discrete rat brain regions by high-performance liquid chromatography with electrochemical detection. *J. Chromatogr.* **370**, 303–313.

Kitani K., Sato Y., Kanai S., Nokubo M., Ohta M. and Masuda Y. (1986) The effect of age on the adaptation of the brain to anticonvulsant effect of phenobarbital in mice. *Life Sci.* **39**, 483–491.

Klotz U., Antonin K. H., and Bieck P. R. (1976) Comparison of the pharmacokinetics of diazepam after single and sub-chronic doses. *Eur. J. Clin. Pharmacol.* **10**, 121–126.

Korttila K., Mattila M. J., and Linnoila M. (1975) Saturation of tissues with *N*-desmethyldiazepam as a cause for elevated serum levels of this metabolite after repeated administration of diazepam. *Acta Pharmacol. Toxicol.* **36**, 190–192.

Kreek M. J. (1984) Opioid interactions with alcohol, in *Dual Addiction: Pharmacological Issues in the Treatment of Concomitant Alcoholism and Drug Abuse, Vol. 3*, pp. 35–46, The Haworth, New York.

Kuhn C. M. and Schanberg S. M. (1977) Distribution and metabolism of amphetamine in tolerant animals, in *Cocaine and Other Stimulants* (Ellinwood E. H., Jr. and Kilbey M. M., eds.), pp. 161–177, Plenum, New York.

Kuhn C. M. and Schanberg S. M. (1978) Metabolism of amphetamine after acute and chronic administration to the rat. *J. Pharmacol. Exp. Ther.* **207**, 544–554.

Kvetnansky R. (1980) Recent progress in catecholamines under stress, in *Catecholamines and Stress: Recent Advances* (Usdin E., Kvetnansky R., and Kopin I., eds.), pp. 7–19, Elsevier, North Holland.

Lange D. G., Fujimoto J. M., Furman-Lane C. L., and Wang R. I. H. (1980a) Unidirectional non-cross tolerance to etorphine in morphine-tolerant mice and the role of the blood brain barrier. *Toxicol. Appl. Pharmacol.* **54**, 177–186.

Lange D. G., Roerig S. C., Fujimoto J. M. and Busse L. W. (1983) Withdrawal tolerance and unidirectional non-cross-tolerance in narcotic pellet-implanted mice. *J. Pharmacol. Exp. Ther.* **224**, 13–20.

Lange D. G., Roerig S. C., Fujimoto J. M. and Wang R. I. H. (1980b) Absence of cross-tolerance to heroin in morphine-tolerant mice. *Science* **208**, 72–74.

Lasagna L. and McCann W. P. (1957) Effect of "tranquilizing" drugs on amphetamine toxicity in aggregated mice. *Science* **125**, 1241–1242.

Lê A. D., Khanna, J. M., Kalant, H. and Grossi, F. (1986) Tolerance to and cross-tolerance among ethanol, pentobarbital and chlordiazepoxide. *Pharmacol. Biochem. Behav.* **24**, 93–98.

Lê A. D., Poulos C. X., and Cappell H. (1979) Conditioned tolerance to the hypothermic effect of ethyl alcohol. *Science* **206**, 1109–1110.

LeBlanc A. E. and Kalant H. (1975) Ethanol-induced cross-tolerance to several homologous alcohols in the rat. *Toxicol. Appl. Pharmacol.* **32**, 123–128.

Lemberger L. and Rubin A. C. (1976) *Physiologic Disposition of Drugs of Abuse*, (Monographs in Pharmacology and Physiology, Vol. 1, Vesell E. S. and Garattini S., eds), Spectrum Public Inc., New York.

Levy J. (1935) Contribution a l'étude de l'accoutumance experimentale aux poisons. III. Alcoolisme experimentale l'accoutumance a l'alcohol peut-elle etre consideree comme une consequence de l'hyposensibilité cellulaire? *Bull. Soc. Chem. Biol.* **17,** 47–59.

Lewander T. (1968) Urinary excretion and tissue levels of catecholamines during chronic amphetamine intoxication. *Psychopharmacologia (Berl.)* **13,** 394–407.

Lewander T. (1978) Experimental studies on anorexigenic drugs: Tolerance, cross-tolerance and dependence, in ̄ *Central Mechanisms of Anoretic Drugs* (Garratini S. and Samanin R., eds.), pp. 343–356, Raven, New York.

Lieber C. S. (1975) Alcohol and the liver. Transition from metabolic adaptation to tissue injury and cirrhosis, in *Alcoholic Liver Pathology* (Khanna J. M., Israel Y., and Kalant H., eds.), pp. 171–188. Addiction Research Foundation of Ontario, Canada.

Lieber C. S. (1977) Metabolism of ethanol, in *Metabolic Aspects of Alcoholism* (Lieber C. S., ed.), pp. 1–29, University Park, Baltimore.

Lieber C. S. and DeCarli L. M. (1968) Ethanol oxidation by hepatic microsomes: Adaptive increase after ethanol feeding. *Science* **162,** 917–918.

Lieber C. S. and DeCarli L. M. (1970) Effect of drug administration on the activity of the hepatic microsomal ethanol oxidizing system. *Life Sci.* **9,** 267–273.

Lieber C. S. and DeCarli L. M. (!972) The role of hepatic microsomal ethanol oxidizing system (MEOS) for ethanol metabolism in vivo. *J. Pharmacol. Exp. Ther.* **181,** 279–288.

Lim A. T. W., Oei T. P., and Funder J. W. (1983) Prolonged foot-shock induced analgesia: Glucocorticoids and non-pituitary opioids are involved. *Neuroendocrinology* **37,** 48–51.

Linseman M. A. (1987) Alcohol consumption in free-feeding rats: Procedural, genetic and pharmacokinetic factors. *Psychopharmacology* **92,** 254–261.

Lippa A. S. and Regan B. (1977) Additional studies on the importance of glycine and GABA in mediating the actions of benzodiazepines. *Life Sci.* **21,** 1779–1784.

Loft S., Jensen V., Rorsgaard S., and Dyrberg V. (1982) Influence of moderate alcohol intake on thiopental anesthesia. *Acta Anaesth. Scand.* **26,** 22–26.

Loh H. H., Shen F. H., and Way E. L. (1969) Inhibition of morphine tolerance and physical dependence development and brain serotonin synthesis by cycloheximide. *Biochem. Pharmacol.* **18,** 2711–2721.

MacLennan A. J., Drucan R. C., Hyson R. L., Mater S. F., Madden J., and Barchas J. (1982) Corticosterone: A critical factor in an opioid form of stress induced analgesia. *Science* **215,** 1530–1532.

Magour S., Coper H., and Fahndrich C. (1973) Effect of chronic intoxication with amphetamine on its concentration in liver and brain on ^{14}C leucine incorporation into microsomal and cytoplasmic proteins of rat liver. *J. Pharm. Pharmacol.* **26,** 105–108.

Manara L., Aldinio C., Cerletti C., Coccia P., Luini A., and Serra G. (1978) In vivo tissue levels and subcellular distribution of opiates with reference to pharmacological action, in *Factors Affecting the Action of Narcotics* (Adler M. L., Manara L., and Samanin R., eds.), pp. 271–296, Raven, New York.

Margules D. L. and Stein L. (1968) Increase of 'anti-anxiety' activity and tolerance of behavioral depression during chronic administration of oxazepam. *Psychopharmacologia (Berl.)* **13,** 74–80.

McCown T. J. and Barrett R. J. (1980) Development of tolerance to the rewarding effects of self-administered S(+) – amphetamine. *Pharmacol. Biochem. Behav.* **12,** 137–141.

McDougal J. N., Marques P. R., and Burks T. F. (1980) Age-related changes in body temperature responses to morphine in rats. *Life Sci.* **27,** 2679–2685.

McDougal J. N., Marques P. R., and Burks T. F. (1981) Reduced tolerance to morphine thermoregulatory effects in senescent rats. *Life Sci.* **28,** 137–145.

McMillan D. E. and Leander J. D. (1978) Chronic chlordiazepoxide and pentobarbital interactions on punished and unpunished behavior. *J. Pharmacol. Exp. Ther.* **207,** 515–520.

Meier G. W., Hatfield J. L., and Foshee D. P. (1963) Genetic and behavioral aspects of pharmacologically induced arousal. *Psychopharmacologia (Berl.)* **4,** 81–90.

Meisch R. A. (1984) Alcohol self-administration by experimental animals, in *Research Advances in Alcohol and Drug Problems* (Smart R. G., Glaser F. B., Israel Y., Kalant H., Popham R. E., and Schmidt W., eds.), Vol. 8, pp. 23–45, Plenum, New York and London.

Melchior C. L. and Tabakoff B. (1985) Features of environment-dependent tolerance to ethanol. *Psychopharmacology* **87,** 94–100.

Mendelson J. H. and Mello N. K. (1964) Metabolism of C^{14}-ethanol and behavioral adaptation of alcoholics during experimentaly induced intoxication. *Trans. Amer. Neurol. Assoc.* **89,** 133–135.

Mendelson J. H., Mello, N. K., Corbett C., and Ballard R. (1965) Puromycin inhibition of ethanol ingestion and liver alcohol dehydrogenase activity in the rat. *J. Psychiat. Res.* **3**, 133–143.

Mezey E. (1971) Effect of pentobarbital administration on ethanol oxidizing enzymes and on rates of ethanol degradation. *Biochem. Pharmacol.* **20**, 508–510.

Mezey E. (1972) Duration of the enhanced activity of the microsomal ethanol-oxidizing enzyme system and rate of ethanol degradation in ethanol-fed rats after withdrawal. *Biochem. Pharmacol.* **21**, 137–142.

Millan M. J., Przewlocki R., and Herz A. (1980) A non-β-endorphinergic adrenohypophyseal mechanism is essential for an analgetic response to stress. *Pain* **8**, 343–353.

Mirsky A., Piker P., Rosenbaum M., and Ledever H. (1941) Adaptation of the central nervous system to varying concentrations of alcohol in the blood. *Quart. J. Stud. Alc.* **2**, 35–45.

Misra A. L. (1972) Disposition and metabolism of drugs of dependence, in *Chemical and Biological Aspects of Drug Dependence* (Mulé S. J. and Brill H., eds.), pp. 219–276, The Chemical Rubber Company, Cleveland, Ohio.

Misra A. (1978) Metabolism of opiates, in *Factors Affecting the Action of Narcotics* (Adler M. L., Manara L., and Samanin R, eds.), pp. 297–343, Raven, New York.

Misra A. L., Mitchell C. L., and Woods L. A. (1971a) Persistence of morphine in central nervous system of rats after a single injection and its bearing on tolerance. *Nature* **232**, 48–50.

Misra A. L., Mulé S. J., Bloch R., and Vadlamani N. L. (1973) Physiological disposition and metabolism of levo-methadone-1-^{3}H in nontolerant and tolerant rats. *J. Pharmacol. Exp. Ther.* **185**, 287–299.

Misra A. L., Nayak P. K., Patel N. L., and Mulé S. J. E. (1974) Identification of norcocaine as a metabolite of [^{3}H]cocaine in rat brain. *Experientia* **30**, 1312–1314.

Misra P. S., Lefèvre A., Ishii H., Rubin E., and Lieber C. S. (1971b) Increase of ethanol, meprobamate and pentobarbital metabolism after chronic ethanol administration in man and in rats. *Amer. J. Med.* **51**, 346–351.

Moisset B. and Welch B. L. (1973) Effects of *d*-amphetamine upon open field behavior in two inbred strains of mice. *Experientia* **29**, 625–626.

Montesinos F. (1965) Metabolism of cocaine. *Bull. Narc. (U.N.)* **17**, 11–17.

Mucha R. F., Kalant H., and Kim C. (1987) Tolerance to hyperthermia produced by morphine in rat. *Psychopharmacology* **92**, 452–458.

Mucha R. F., Niesink R., and Kalant H. (1978) Tolerance to morphine analgesia and immobility measured in rats by changes in log-dose-response curves. *Life Sci.* **23**, 357–364.

Mulé S. J. and Misra A. L. (1977) Cocaine: Distribution and metabolism in animals, in *Cocaine and Other Stimulants* (Ellinwood E. H., Jr. and Kilbey M. M., eds.), pp. 215–228, Plenum, New York.

Mulé S. J. and Woods L. A. (1962) Distribution of N-C^{14} methyl labeled morphine: I. In central nervous system of nontolerant and tolerant dogs. *J. Pharmacol. Exp. Ther.* **136**, 232–241.

Mullis K. B., Perry D. C., Finn A. M., Stafford B., and Sadeé W. (1979) Morphine persistence in rat brain and serum after single doses. *J. Pharmacol. Exp. Ther.* **208**, 228–231.

Murphy H. R. and Hug C. C. (1981) Pharmacokinetics of intravenous morphine in patients anesthetized with enflurane-nitrous oxide. *Anesthesiology* **54**, 187–192.

Nayak P. N., Misra A. L., and Mulé S. J. (1976) Physiological disposition and biotransformation of [^{3}H] cocaine in acutely and chronically treated rats. *J. Pharmacol. Exp. Ther.* **196**, 556–569.

Nelson L. R. and Ellison G. (1978) Enhanced stereotypies after repeated injections but not continuous amphetamines. *Neuropharmacology* **17**, 1081–1084.

Newman H. W. and Lehman A. J. (1938) Nature of acquired tolerance to alcohol. *J. Pharmacol. Exp. Ther.* **62**, 301–306.

Newman H. W., Wilson R. H. L., and Newman E. J. (1952) Direct determination of maximal daily metabolism of alcohol. *Science* **116**, 328–329.

Newman L. M., Curran M. A., and Becker G. L. (1986) Effects of chronic alcohol intake on anesthetic responses to diazepam and thiopental in rats. *Anesthesiology* **65**, 196–200.

Okamoto M. and Boisse N. R. (1975) Effect of chronic pentobarbital treatment on blood-CSF kinetics. *Eur. J. Pharmacol.* **33**, 205–209.

Okamoto M., Rao S., Reyes J., and Rifkind A. B. (1985) Recovery from dispositional and pharmacodynamic tolerance after chronic pentobarbital treatment. *J. Pharmacol Exp. Ther.* **235**, 26–31.

Okamoto M., Rosenberg H. C, and Boisse N. R. (1975) Tolerance characteristics produced during the maximally tolerable chronic pentobarbital dosing in the cat. *J. Pharmacol. Exp. Ther.* **192,** 555–569.

Okey A. B., Roberts E. A., Harper P. A., and Denison M. S. (1986) Induction of drug-metabolizing enzymes: Mechanisms and consequences. *Clin. Biochem.* **19,** 132–141.

Orme M., Breckenridge A., and Brooks R. V. (1972) Interactions of benzodiazepines with warfarin. *Brit. Med. J.* **3,** 611–614.

Patrick G. A., Dewey, W. L., Huger F. P., Daves E. D., and Harris. L. S. (1978) Disposition of morphine in chronically infused rats: Relationship to antinociception and tolerance. *J. Pharmacol. Exp. Ther.* **205,** 556–562.

Patrick G. A., Dewey W. L., Spaulding T. C., and Harris. L. S. (1975) Relationship of brain morphine levels to analgesic activity in acutely treated mice and rats, and in pellet-implanted mice. *J. Pharmacol. Exp. Ther.* **193,** 876–883.

Peris J. and Cunningham C. L. (1986) Handling-induced enhancement of alcohol's acute physiological effects. *Life Sci.* **38,** 273–279.

Petursson H. and Lader M. H. (1981) Benzodiazepine dependence. *Brit. J. Addict.* **76,** 133–145.

Pikkarainen P. M. and Lieber C. S. (1980) Concentration dependency of ethanol elimination rates in baboons: Effect of chronic alcohol consumption. *Alcoholism: Clin. Exp. Res.* **4,** 40–43.

Plomp G. J. J., Maes R. A. A., and van Ree J. M. (1981) Disposition of morphine in rat brain: Relationship to biological activity. *J. Pharmacol. Exp. Ther.* **217,** 181–188.

Popper C. W., Chiuch C. C., and Kopin J. J. (1977) Plasma catecholamine concentrations in unanesthetized rats during sleep, wakefulness, immobilization and after decapitation. *J. Pharmacol. Exp. Ther.* **202,** 144–148.

Post R. M. (1981) Central stimulants. Clinical and experimental evidence on tolerance and sensitization, in *Research Advances in Alcohol and Drug Problems, Vol. 6* (Israel Y., Glaser F. B., Kalant H., Popham R. E., Schmidt W., and Smart R. J., eds.), pp. 1–65, Plenum, New York.

Poulos C. X., Wilkinson D. A., and Cappell H. (1981) Homeostatic regulation and Pavlovian conditioning in tolerance to amphetamine-induced anorexia. *J. Comp. Physiol. Psychol.* **95,** 735–746.

Ratcliffe F. (1969) The effect of chronic ethanol administration on the responses to the amylobarbitone sodium in the rat. *Life Sci.* **8,** 1051–1061.

Remmer H. (1969) Tolerance to barbiturates by increased breakdown, in *Scientific Basis of Drug Dependence* (Steinberg H., ed.), pp. 111–128, Grune & Stratton, Inc., New York.

Remmer H. (1970) The role of the liver in drug metabolism. *Clin. J. Med.* **49**, 617–629.

Rickels K. (1982) Benzodiazepines in the treatment of anxiety. *Amer. J. Psychother.* **36**, 358–370.

Ritzmann R. F. and Springer A. C. (1980) Age-differences in brain sensitivity and tolerance to ethanol in mice. *Age* **3**, 15–17.

Roerig S. C., Fujimoto J. H., and Lange D. G. (1987) Development of tolerance to respiratory depression in morphine- and etorphine-pellet implanted mice. *Brain Res.* **400**, 278–284.

Roffman M. and Lal H. (1974) Stimulus control of hexobarbital narcosis and metabolism in mice. *J. Pharmacol. Exp. Ther.* **191**, 358–369.

Rosenberg H. C., Smith S., and Chiu, T. H. (1983) Benzodiazepine-specific and non-specific tolerance following chronic flurazepam treatment. *Life Sci.* **32**, 279–285.

Rubin E. and Lieber C. S. (1968) Hepatic microsomal enzymes in man and rat—Induction and inhibition by ethanol. *Science* **162**, 690–691.

Rubin E., Bacchin P., Gang H., and Lieber C. S. (1970) Induction and inhibition of hepatic microsomal and mitochondrial enzymes by ethanol. *Lab. Invest.* **22**, 569–580.

Ryan G. P. and Boisse N. R. (1983) Experimental induction of benzodiazepine tolerance and physical dependence. *J. Pharmacol. Exp. Ther.* **226**, 100–107.

Scheinin B. (1971) *The Cross-Tolerance Between Ethanol and General Anesthetics: An Experimental Study on Rats.* Ph.D. Thesis, University of Turku, Finland.

Schmucker D. L. (1985) Aging and drug disposition: An update. *Pharmacol. Rev.* **37**, 133–148.

Schultz R. and Goldstein A. (1972) Inactivity of narcotic glucuronides as analgesics and on guinea-pig ileum. *J. Pharmacol. Exp. Ther.* **183**, 404–410.

Schweisheimer W. (1913) Der Alcoholgehalt des Blutes unter verschiedenen Bedingungen. *Deuts. Arch. Klin. Med.* **109**, 271–313.

Seever M. H. and Deneau G. A. (1963) Physiological aspects of tolerance and physical dependence, in *Physiological Pharmacology* (Root W. S. and Hofmann F. G., eds.), pp. 565–571, Academic Press, New York.

Segal D. S. and Schuckit M. A. (1983) Animal models of stimulant-induced psychosis, in *Stimulants: Neurochemical, Behavioral, and Clinical Perspectives* (Creese I., ed.), pp. 131–167, Raven, New York.

Segal D. S., Geyer M. A., and Schuckit M. A. (1981) Stimulant-induced psychosis: An evaluation of animal models, in *Essays in Neurochemistry and Pharmacology, Vol. 4* (Youdim M. B. H., Lovenberg W., Sharman D. F., and Lagnado J. R., eds.), pp. 95–129, John Wiley and Sons, London.

Segal D. S., Weinberger S. B., Cahill J., and McCunney S. J. (1980) Multiple daily amphetamine administration: Behavioral and neurochemical alterations. *Science* **207,** 904–907.

Seidel G., Streller I., and Soehring K. (1964) Der Einfluss subchronischer Alkohologaben auf die Pentobarbitalaufnahme des Meerschweinchenhirns. *Naunyn-Schmiedeberg's Arch. Pharmakol. Exp. Pathol.* **247,** 312–313.

Sellers E. M. (1978) Addictive drugs: Disposition, tolerance and dependence interrelationships. *Drug Metab. Rev.* **8,** 5–11.

Sellman R., Kanto J., Raijola E., and Pekkarinen A. (1975a) Human and animal study on elimination from plasma and metabolism of diazepam after chronic alcohol intake. *Acta. Pharmacol. Toxicol.* **36,** 33–38.

Sellman R., Kanto J., Raijola E., and Pekkarinen A. (1975b) Induction effects of diazepam on its own metabolism. *Acta Pharmacol. Toxicol.* **37,** 345–351.

Sellman R., Pekkarinen A., Kangas L., and Raijola E. (1975c) Reduced concentrations of plasma diazepam in chronic alcoholic patients following an oral administration of diazepam. *Acta Pharmacol. Toxicol.* **36,** 25–32.

Sepinwall J., Grodsky F. S., and Cook L. (1978) Conflict behavior in the squirrel monkey: Effects of chlordiazepoxide, diazepam, and *N*-desmethyldiazepam. *J. Pharmacol. Exp. Ther.* **204,** 88–102.

Shearman G. T., Miksic S., and Lal H. (1979) Lack of tolerance development to benzodiazepines in antagonism of the Pentylenetetrazole discriminative stimulus. *Pharmacol. Biochem. Behav.* **10,** 795–797.

Sherwin A. L., Eisen A. A., and Christine D. (1973) Anticonvulsant drugs in human epileptogenic brain. *Neurology* **29,** 73–77.

Shigeta Y., Nomura F., Iida S., Leo M. A., Felder M. R., and Lieber C. S. (1984) Ethanol metabolism in vivo by the microsomal ethanol oxidizing system in deermice lacking alcohol dehydrogenase (ADH). *Biochem. Pharmacol.* **33,** 807–814.

Shigeta Y., Nomura F., Leo M. A., Iida S., Felder M. R., and Lieber C. S. (1983) Alcohol dehydrogenase (ADH) independent ethanol metabolism in deermice lacking ADH. *Pharmacol. Biochem. Behav.* **18(Suppl. 1)**, 195–199.

Shocken D. and Roth G. (1977) Reduced beta-adrenergic receptor concentration in aging man. *Nature (London)* **267**, 858–865.

Siegel S. (1975) Evidence from rats that morphine tolerance is a learned response. *J. Comp. Physiol. Psychol.* **89**, 498–506.

Siegel S. (1976) Morphine analgesic tolerance: Its situation specificity supports a Pavlovian conditioning model. *Science* **193**, 323–325.

Siegel S. (1983) Classical conditioning, drug tolerance, and drug dependence, in *Research Advances in Alcohol and Drug Problems* (Smart R. G., Glaser F. B., Israel Y., Kalant H., Popham R. E., and Schmidt W., eds.), pp. 207–246, Plenum, New York.

Smith A. C., Freeman R. W., and Harbison R. D. (1981) Ethanol enhancement of cocaine-induced hepatoxicity. *Biochem. Pharmacol.* **30**, 453–458.

Solomon H. M., Barakat, M. J., and Ashley C. J. (1971) Mechanisms of drug interaction. *J. Amer. Med. Assoc.* **216**, 1977–1999.

Sparber S. B. and Fossom L. H. (1984) Amphetamine cumulation and tolerance development: Concurrent and opposing phenomena. *Pharmacol. Biochem. Behav.* **20**, 415–424.

Spealman R. D., Goldberg S. R., and Kelleher R. T. (1979) Effects of norcocaine and some norcocaine derivatives on schedule-controlled behavior of pigeons and squirrel monkeys. *J. Pharmacol. Exp. Ther.* **210**, 196–205.

Spector A. A. and Fletcher J. E. (1978) Nutritional effects on drug-protein binding, in *Nutrition and Drug Interrelations* (Hathcock J. N. and Coon J., eds.). pp. 447–473, Academic, New York.

Spector S., Ngai S. H., Hempstead J., and Berkowitz B. A. (1978) Pharmacokinetics of naloxone in rats, in *Factors Affecting the Action of Narcotics* (Adler M. L., Manara L., and Samanin R., eds.), pp. 249–256, Raven, New York.

Stevenson I. H. and Turnbull M. J. (1970) The sensitivity of the brain to barbiturate during chronic administration and withdrawal of barbitone sodium in the rat. *Brit. J. Pharmacol.* **39**, 325–33.

Svensmark O. and Buchthal F. (1963) Dosage of phenytoin and phenobarbital in children. *Danish Med. Bull.* **10**, 234–235.

Szerb J. C. and McCurdy D. H. (1956) Concentration of morphine in blood and brain after intravenous injection of morphine in nontolerant, tolerant and neostigmine-treated rats. *J. Pharmacol Exp. Ther.* **118,** 446–450.

Tagashira E., Izumi T., and Yanaura S. (1979) Experimental dependence on barbiturates. *Psychopharmacology* **60,** 111–116.

Tanayama S., Momose S., and Takagaki E. (1974) Metabolism of 8-chloro-6-phenyl-4H-S-triazolo[4,3-a]-[1,4]benzodiazepin (D-40TA), a new central depressant. III. Metabolism and tolerance studies in rats. *Xenobiotica* **4,** 57–65.

Tavani A., Luini A., and Manara L. (1979) Time course of etorphine levels in tissues of opiate tolerant and nontolerant rats. *J. Pharmacol. Exp. Ther.* **211,** 140–144.

Thurman R. G., Bradford B. U., and Glassman E. (1983) The swift increase in alcohol metabolism (SIAM) in four inbred strains of mice. *Pharmacol. Biochem. Behav.* **18(Suppl. 1),** 171–175.

Thurman R. G., McKenna W. R., Brentzel H. J., Jr. and Hesse S. (1975) Significant pathways of hepatic ethanol metabolism. *Fed. Proc.* **34,** 2075–2081.

Thurman R. G., Yuki T., Bleyman M. A., and Wendell G. (1979) The adaptive increase in ethanol metabolism due to pretreatment with ethanol: A rapid phenomenon. *Drug Alc. Depend.* **4,** 119–129.

Timar M., Licurici V., and Lazarescu M. (1966) Incorporation du barbital C^{14} chez les rats dans les conditions de l'accoutumance à certain barbituriques. *Med. Pharmacol. Exp.* **14,** 24–30.

Tobon F. and Mezey E. (1971) Effect of ethanol administration on hepatic ethanol and drug metabolizing enzymes and on rates of ethanol degradation. *J. Lab. Clin. Med.* **77,** 110–121.

Troshina A. E. (1957) O wekhanizmakh privykaniya organisma K alkogolyu (On the mechanism of habituation of the organism to alcohol). *Sb. Nauch. Tr. Ryazan. Med. Inst.* **4,** 1. (Abstract in Quart. J. Stud. Alc. **20,** 783–784, 1959).

Umans J. G. and Inturrisi C. E. (1981) Pharmacodynamics of subcutaneously administered *D*-acetylmorphine, 6-acetylmorphine and morphine in mice. *J. Pharmacol. Exp. Ther.* **218,** 409–415.

Valerino D. M., Vesell E. S., Johnson A. O., and Aurori K. C. (1973) Effects of various centrally active drugs on hepatic microsomal enzymes: A comparative study. *Drug Metab. Disp.* **1,** 669–678.

Vajda F., Williams F. M., Davidson S., Falconer M. A., and Breckrenridge A. (1974) Human brain, cerbrospinal fluid, and plasma concentrations of diphenylhydantoin and phenobarbital. *Clin. Pharmacol. Ther.* **15,** 597–603.

Vasko M. R. and Domino E. F. (1978) Tolerance development to the biphasic effects of morphine on locomotor activity and brain acetylcholine in the rat. *J. Pharmacol. Exp. Ther.* **207,** 848–858.

Vestal R., Wood A., and Shand D. (1979) Reduced beta-adrenoceptor sensitivity in the elderly. *Clin. Pharmacol. Ther.* **26,** 818–886.

Vorne M. S., Puolakka J. O., and Idanpaan-Heikkila J. E. (1975) Diazepam, ethanol and drug metabolizing enzymes in rat liver. *Arch. Int. Pharmacodyn. Ther.* **216,** 280–287.

Wallgren H. (1970) Absorption, diffusion, distribution and elimination of ethanol. Effect on biological membranes, in *International Encyclopedia of Pharmacology and Therapeutics, Section 20: Alcohols and Derivatives, Vol. 1,* (Trémolières J, ed.), pp. 161–193. Pergamon, Oxford.

Wallgren H. and Barry H., III., eds. (1970) *Actions of Alcohol.* Elsevier, Amsterdam.

Warner R. S. (1965) Management of the office patient with anxiety and depression. *Psychosomatics* **6,** 347–351.

Way E. L. and Adler T. K. (1960) The pharmacologic implications of the fate of morphine and its surrogates. *Pharmacol. Rev.* **12,** 383–445.

Way E. L. and Adler T. K. (1962) The biological disposition of morphine and its surrogates. *World Health Organization,* pp. 32–35. Geneva, Switzerland.

Widmark E. M. P. (1932) Die theoretischen Grundlagen und die praktische Verwendbarkeit der gerichtlich-medizinischen Alkoholbestimmung, in *Forstchritte der naturwissenschaftlichen Forschung, Heft 11,* Abderhalden E., ed.), Urbain und Schwartzenberg, Berlin.

Wilkinson P. K. (1980) Pharmacokinetics of ethanol: A review. *Alcoholism: Clin. Exp. Res.* **4,** 6–21.

Williams R. T. (1978) Nutrients in drug detoxication reactions, in *Nutrition and Drug Interrelations* (Hathcock J. N. and Coon J., eds.), pp. 303–318, Academic, New York.

Wilson J. R., Erwin V. G., and McClearn G. E. (1984) Effects of ethanol. I. Acute metabolic tolerance and ethnic differences. *Alcoholism: Clin. Exp. Res.* **8,** 226–232.

Wood J. and Laverty R. (1978) Cross-tolerance between ethanol and other anesthetic agents. *Proc. Univ. Otago Med. Sch.* **56,** 108–109.

Wood J. and Laverty R. (1979) Effect of depletion of brain catecholamines on ethanol tolerance and dependence. *Eur. J. Pharmacol.* **58,** 285–293.

Wood W. G. and Armbrecht H. J. (1982) Behavioral effects of ethanol in animals: Age differences and age changes. *Alcoholism: Clin. Exp. Res.* **6**, 3–12.

Woods, L. A. (1954) Distribution and fate of morphine in non-tolerant and tolerant dogs and rats. *J. Pharmacol. Exp. Ther.* **112**, 158–175.

Woolverton W. L., Kandel D., and Schuster C. R. (1978) Tolerance and cross-tolerance to cocaine and d-amphetamine. *J. Pharmacol. Exp. Ther.* **205**, 525–535.

Yamamoto I., Ho I. K., and Loh H. H. (1977) Acceleration of pentobarbital metabolism in tolerant mice induced by pentobarbital pellet implantation. *Life Sci.* **20**, 1353–1362.

Yoburn B. C., Chen J., Huang T., and Inturrisi C. E. (1985) Pharmacokinetics and pharmacodynamics of subcutaneous morphine pellets in the rat. *J. Pharmacol. Exp. Ther.* **235**, 282–286.

York J. L. (1983) Increased responsiveness to ethanol with advancing age in rats. *Pharmacol. Biochem. Behav.* **19**, 687–691.

Zauder H. L. (1952) The effect of prolonged morphine administration on the in vivo and in vitro conjugation of morphine by rats. *J. Pharmacol. Exp. Ther.* **104**, 11–19.

Dopamine Receptor Changes During Chronic Drug Administration

Andrew J. Greenshaw, Glen B. Baker, and Thomas B. Wishart

1. The Parametric Estimation of Neural Receptors: Basic Principles and Clinical Relevance

Brain–behavior relationships remain poorly understood in terms of neurotransmitter function. This contentious statement applies particularly to the analysis of effects of long-term drug administration. In this area, particular problems are those of the relative paucity of systematic neurochemical data and the increasing number of identified neuroactive substances. The analysis of brain–behavior relationships may be approached from various levels, and the most prominent of these currently is that of receptor function. The reason for this current focus is, most probably, that the neural receptor represents a convenient reference point from which the overall activity of neurotransmitter systems may be viewed. In terms of long-term changes in the activity of neural systems, in keeping with the classical observations of tissue-bath pharmacology, changes in neural activity are often evident from direct measurement of receptor activity, or indeed, from other functional receptor analyses (vide infra) that are not evident from such measures as neurotransmitter turnover.

The present chapter outlines general principles and emerging concepts in this field. This is not intended to be a review of all studies in the literature, but an overview and critical appraisal of representative work in this area. Emphasis is placed here on the relevance of this area of research to the field of biological psychiatry, in accord with the main orientation of the authors.

The problems of definition of receptors for neurotransmitters are beyond the scope of this chapter. Suffice it to say that a binding site cannot necessarily be equated with a receptor, unless it can be shown that interaction with this site by the proposed neurotransmitter in vivo results in a relevant physiological or biochemical response. Readers unfamiliar with the binding site vs receptor controversy may consult a number of comprehensive reviews (e.g., Laduron, 1988.) For the present purpose, the terms "receptor" and "binding site" are used interchangeably. For many aspects of chronic drug action, it is evident that our understanding of the significance of receptor changes is simply correlative.

It is perhaps useful to reiterate some fundamentals of receptor analysis before examining the current status of this area. Receptors are, of course, target sites for the action of neuroactive substances and of classical hormones. The target sites (or binding sites) of interest in the present context are putative neurotransmitter binding sites or specific binding sites for drugs. In some cases, these will be equivalent, as in the case of antipsychotic drugs, which usually bind well to dopamine receptors. In other cases, they are not simply equivalent as in the case of imipramine binding sites, which may represent 5-hydroxytryptamine uptake or transport sites on the neural membrane (Langer, 1987).

There are numerous approaches to the study of receptors. Directly, they may be investigated by radiochemical methods in which a compound (or ligand) is labeled with a radioactive marker, usually a beta-emitter such as tritium or, in cases where greater specific activity is necessary, a gamma-emitter such as ^{125}Iodine (*see* Boulton et al., 1986). The presence of the radiolabeled ligand after incubation with the tissue preparation, followed by a suitable washing procedure, may be quantified and yields estimates of binding-site characteristics. The mathematics involved are an application and extension of the Michaelis-Menton approach so widely used in enzymology (*see* Iversen et al., 1983; Boulton et al., 1986). With these procedures, estimates of the number of binding sites (B_{max}) and of their affinity for the radiolabeled ligand (K_d) may be calculated. Displacement studies are routinely conducted to determine the relative affinity of different ligands for a binding site (expressed as IC_{50}: the concentration of the competing ligand required to inhibit 50% of binding of the reference ligand). It is apparent that changes in binding-site activity may be viewed at two levels. These sites may undergo a change in their abundance or "number" (B_{max}). Similarly, they may be altered in terms of their propensity to form complexes with ligands and, thereby, undergo a change in their "affinity" (K_d). Although it is possible that either or both of these parameters may be altered by chronic exposure to drugs, the most usual consequence of chronic drug administration is a change in B_{max} (Boulton et al., 1986).

It is important to note that K_d estimates may be raised by the presence of drugs in tissue from drug-treated subjects. This does not represent an adaptation of the receptor, but is simply a function of competition for sites. This consideration is important for evaluating many data in the literature. Binding sites undergo biophysical changes when they are activated by ligands (agonists) or inactivated by ligands (antagonists). Typically, these changes give rise to altered levels of biochemical, or more specifically for brain, neurochemical activity. Such events in neural tissue either result in changes in ion permeability (e.g., with respect to chloride channels) or changes in phosphate metabolism (e.g., with cyclic adenosine monophosphate or phosphoinositide metabolism). Typically, these changes in binding sites are measured in vitro or ex vivo. Recent advances have, however, made it possible to monitor binding activity in vivo using specialized imaging techniques such as Positron Emission Tomography (PET), as described below.

Binding activity may be estimated indirectly by means of functional analysis using techniques ranging from cell-culture studies through classical tissue bath preparations to whole animal tests. Although the latter are usually conducted in the preclinical context, various clinical indices are also useful (e.g., prolactin secretion in relation to dopamine D_2 receptors). Various strategies will be described in context in the following sections of this chapter.

2. Emergent Changes in Receptor Number and Affinity

In accord with the effects of altered ligand availability on binding sites outlined above, it is often possible to predict changes in binding that are induced by chronic drug administration. Such predictable effects are either direct or are indirectly mediated by an identifiable metabolic change. In the former case, whether the drug is an active ligand at neurotransmitter receptors or at "drug-specific" binding sites, adaptive changes are attributable to the direct interaction of the drug molecule with the binding site in question. In the latter situation, the drug molecule influences mechanisms governing availability of the endogenous ligand. These effects are usually manifest in terms of changes in the uptake, release, or catabolism of the endogenous ligand in the neural milieu. It is essential to understand fully the neuropharmacological profile of drug action, including mechanisms such as feedback control of neurotransmitter synthesis, in attempts to predict or understand receptor changes.

The following sections describe changes in dopamine receptor activity induced by chronic drug administration. It is important to realize that, as noted above, although changes in B_{max} are most usual, K_d values are altered by some drug treatments (*see* Section 6.2). Radiochemical assays of receptor binding provide information allowing a distinction of these two aspects of binding activity. Functional analyses involving electrophysiological or overt behavioral responses only provide a "global" estimate of the "sensitivity" of the integral binding system. Although there is a great need for characterizing overall functional responses, the importance of describing in detail the mechanism of altered responsiveness (vis-à-vis K_d and B_{max} changes) must be recognized. This consideration has obvious importance for our understanding of mechanisms of tolerance and sensitization to drug effects at the receptor level, as in the case of antidepressants (*see* chapter by Willner in this book; but *see also* Section 5.1).

In terms of definition, the word "chronic" is probably far too liberally applied in the analysis of drug effects. It is the opinion of the authors that there can be no simple resolution of the imprecision of this term, considering the present state of the literature. There is obviously a vast discrepancy in the use of "chronic" in the clinical literature relative to that in preclinical studies or basic scientific studies involving nonhuman species. Chronic ranges from typically about 21 d in the laboratory relative to several months or years in the clinic, although some investigators (*see* Section 6.2) have assessed drug action for over a year in laboratory animals. In the present context, the adjectives "chronic" or "long-term" are used generally, and specific details of the duration of drug administration are given wherever it is deemed appropriate.

3. Dopamine Function in Biological Psychiatry

Increases in our understanding of the mechanisms of action of neuroleptic drugs, introduced in the 1950s for the treatment of psychoses (Delay et al., 1952; Hamon et al., 1952), established the importance of dopamine function in the field of biological psychiatry. These drugs, which are used to relieve the florid symptoms of schizophrenia, are now thought to produce their therapeutic effects through blockade of dopamine D_2 receptors (Seeman, 1987). Parallel developments in the analysis of Parkinson's disease, with the observation that dopamine replacement therapy (with L-dopa and carbidopa) was clinically efficacious (Hornykiewicz, 1973), further underscored the functional importance of dopamine changes. These developments were accompanied by the implication of dopamine dysfunction in movement disorders such as tardive dyskinesia consequent to neuroleptic drug therapy (Klawans, 1973), and additionally by proposals that dopamine systems may play a role in the regulation of learning, motivation, and reinforcement (Beninger, 1983). In view of these considerations, the importance of understanding the effects of prolonged drug administration on dopamine receptor function with an analysis of behavioral sequelae is self-evident.

4. Dopamine Receptor Subtypes

In the mid-1970s the now established concept of two dopamine receptor subtypes was first proposed (Cools and Van Rossum, 1976): D_1 dopamine receptors mediating excitation and D_2 dopamine receptors mediating inhibition. D_1 receptors are linked clearly to adenylate cyclase, their stimulation leading to accumulation of cyclic-AMP. D_2 receptors do not stimulate adenylate cyclase. Indeed, it is now established that the D_2 receptor sybtype actually inhibits cyclic-AMP formation in the pars anterior and pars intermedia of the pituitary gland, and in the striatum. Recent evidence has, however, demonstrated that the D_2-mediated inhibition of prolactin release in the pars anterior of the pituitary is independent of the adenylate cyclase system (Memo et al., 1986). It has been proposed that the D_2 response in this region may be mediated by second messenger system involving phosphoinositide (PI) metabolism (Simmond and Strange, 1985). Support for an involvement of PI systems in D_2 function comes from a report of decreased inositol triphosphate levels in rat striatal slices after D_2 stimulation (Pizzi et al., 1987). The dopamine receptor subtype story has been complicated by the fact that each receptor may exist in both high- and low-affinity states, the high-affinity state having an approximately 1000 times greater binding affinity for dopamine than the low-affinity state. The terms D_3 and D_4 were originally used to describe the D_1 (high) and D_2 (high) states, respectively (Seeman, 1987). The terms D_3 and D_4 have now been generally disregarded, and it is conventional (for general assay purposes) to convert all high-affinity sites to their low-affinity state by virtue of their sensitivity in this respect to temperature, guanine nucleotides, and sodium ions. The technical problems of high/low-affinity binding and dopamine receptors have been discussed by Seeman (1987). It is of interest to note that the brain D_1 and D_2 binding sites appear to be functionally equivalent to the vascular DA-1 and DA-2 binding sites for dopamine (*see* Seeman, 1987).

Evidence from various sources suggests an involvement of guanine nucleotides, particularly GTP, in the coupling of the D_1 receptor to adenylate cyclase. Such effects appear to be agonist-specific. For example, Gpp(NH)p inhibits adenylate cyclase and markedly increases K_d values of the agonists [^{3}H]-apomorphine and [^{3}H]-ADTN without any effect on B_{max}; such effects are not evident with respect to the binding of antagonists such as [^{3}H]-haloperidol. The involvement of guanine nucleotides in dopamine receptor binding has been discussed recently by Mishra (1986). A clear understanding of the linkages of these receptor systems to second messengers must, however, await further studies.

The importance of mechanisms of receptor activation and consequent molecular events must not be underestimated in the present context. Although the relevance of such phenomena to drug tolerance or sensitization is currently unclear, a full understanding will almost certainly be crucial for a comprehensive analysis of the consequences of adaptive changes in receptor function (e.g., *see* Olianas and Onali, 1987).

5. Effects of Chronic Dopamine Agonists

The effects of chronic administration of dopamine agonists on receptor function are both complex and controversial. These studies are important in relation to a number of clinical areas mentioned earlier, including dopamine replacement therapy in Parkinson's disease, the etiology of psychoses (particularly with respect to amphetamine psychosis: *see* Connell, 1958), and drug abuse.

5.1. Chronic Amphetamine Administration

The psychomotor stimulant drug amphetamine acts principally on catecholamine systems, primarily in terms of dopamine release (Fischer and Cho, 1979) and may therefore be classified as an indirect agonist. Chronic administration of this drug of abuse leads to two distinct behavioral syndromes, depending on its pattern of administration (Robinson and Becker, 1986).

5.1.1. Continuous Administration

With rats, in response to continuous administration (i.e., defined as typically *multiple daily injections* or *continuous infusion*) of amphetamine, there is an initial period of hyperreactivity, followed by a decrease in behavior. After 4–5 d a pattern of bizarre grooming and biting emerges, associated with limb-flicks and wet-dog shakes. Although at high doses amphetamine and related compounds induce neurotoxic reactions in the brain that are not restricted to dopamine systems (*see* Robinson and Becker, 1986), it is evident that low-dose continuous administration of amphetamine leads to specific alterations in dopamine function. These low-dose effects are characterized by decreases in striatal dopamine concen-

trations and associated tyrosine hydroxylase activity with a concomitant loss of striatal dopamine receptors (*see* Ellison and Eison,
1983, for a review). It is proposed that amphetamine toxicity is induced by the formation of neurotoxic compounds such as 6-hy-
droxydopamine (6-OHDA) in the brain (Schmidt et al., 1985;
Seiden and Vosmer, 1984), but the mechanisms are not yet clearly
characterized.

5.1.2. Intermittent Administration

The behavioral consequences of intermittent administration (i.e.,
single daily injections) of amphetamine are markedly different from
those of continuous administration. There is a lack of tolerance to
the initial effects of this drug on motor activity. Furthermore, the
initial stimulant effects on behavior are increased by repeated drug
administration. The dependence of this phenomenon on various test
conditions and drug regimens has been discussed at length by
Robinson and Becker (1986). These authors have pointed out that
the most salient features of the phenomenon are:

1. Sensitization may be induced by a single low dose of
   amphetamine
2. Sensitization is potentiated by intermittent administration of the drug
3. The phenomenon persists for a period of months following withdrawal of amphetamine.

In relation to the prominent dopamine-releasing actions of amphetamine and the involvement of the nigrostriatal dopamine system in motor activity, a great deal of work has been directed to the
analysis of nigrostriatal dopamine function in this context. An early
proposal was that the phenomenon represented an expression of
postsynaptic dopamine receptor supersensitivity. Experimental evidence has been obtained by indirect measurements (e.g., measuring
behavioral responses to a dopamine receptor agonists such as apomorphine) (Klawans and Margolin, 1975) or by direct measures of
dopamine receptor binding (Klawans et al., 1979). Generally, however, studies of this kind do not support a simple dopamine receptor
supersensitivity hypothesis. In fact, although the behavioral experiments with agonists are in accord with the receptor supersensitivity
hypothesis, binding studies generally report either no change in

dopamine receptors or a decrease in their number (B_{max}). Studies to date have almost exclusively been conducted with D_2 ligands or ligands with high affinites for both D_1 and D_2 receptors, and most often exclusively with male rats. These data are summarized in Table 1 taken from a recent excellent and extensive review by Robinson and Becker (1986).

It is possible that behavioral sensitization to amphetamine is the result of a change in presynaptic dopamine receptors. The concept of presynaptic receptors (autoreceptors) controlling release and synthesis of dopamine is well accepted. The assessment of their function and adaptive responses is, however, rather complex. The interpretation of receptor changes in terminal projections in studies using lesions of dopamine-containing neurons is made difficult by adaptive changes in postsynaptic receptors (increase in receptor number on the postsynaptic membrane obscures the analysis of presynaptic decreases). Similarly, changes in binding after lesions of cell bodies in areas of dopamine terminal innervation reveal nothing about presynaptic binding. Binding studies have, however, revealed that D_2 autoreceptors are present on dopamine cell bodies in the substantia nigra (Filloux et al., 1987). There is currently no evidence that presynaptic dopamine receptors represent a unique class of binding site, although indirect electrophysiological (White and Wang, 1984) and neuropharmacological measures have demonstrated autoreceptor-mediated changes (*see* Roth, 1984, for an extensive review).

At the present stage, it is only possible to speculate that the small decrease in D_2 binding that has been reported after chronic intermittent amphetamine administration may reflect a decrease in the number of presynaptic binding sites. Robinson and Becker (1986) have extensively reviewed the evidence for functional changes in autoreceptors in this context. These authors conclude that "although electrophysiological studies have supported the dopamine autoreceptor subsensitivity hypothesis, biochemical/pharmacological studies designed to test the same hypothesis have not. . . ." Indeed Robinson and Becker (1986) have identified discrepancies between the electrophysiological data currently available and the phenomenon of behavioral sensitization. Among the main points of contention is the observation that there are major discrepancies between the drug regimens necessary for behavioral and electrophysiological manifestation of sensitization. Further-

Table 1
The Effect of Amphetamine Sensitization on Striatal Dopamine Receptor Binding

Reference	Species	Sex	AMPH[7,8]	Injection schedule[9]	Withdrawal period	Ligand	Competitor	Binding
Akiyama et al., 1982a	Rats	M	M	4 mk/d × 14 d	7 d → 4 mk[3]	[3H]spiperone	Spiperone	Down
Akiyama et al., 1982b	Rats	M	M	4 mk/d × 14 d	7 d	[3H]spiperone	Butaclamol	Down
Daiguji Meltzer, 1982	Rats	M	D	5 → 15 mk 2/d × 20d[2]	17–20 h	[3H]spiroperidol	ADTN	Down
Hitzemann et al., 1980	Rats	F	D	6 mk 2 ×/d × 1–4 d	16–20 h	[3H]spiroperidol	Butaclamol or sulperide	Down
Howlett and Nahorski, 1979	Rats	M	D	5 → 15 mk 2×/d × 20 d[2]	17–20 h	[3H]spiroperidol	Butaclamol or dopamine	Down
Howlett and Nahorski, 1978	Rats	M	D	5 → 15 mk 2 ×/d × 20d[2]	17–20 h	[3H]spiperone	?	Down
Kaneno and Shimazono, 1981	Rats	M	M	6 mk/d × 14 d	10 d	[3H]spiroperidol	In vivo[4]	Down
Muller and Seeman, 1979	Rats	M	?	10 mk/d × 14 d (oral)	1 d	[3H]apomorphine	Apomorphine	Down
Riffee et al., 1982	Mice	M	D	4 mk/d × 14 d	3 d	[3H]spiroperidol	Apomorphine or butaclamol	Down
Robertson, 1982	Rats	M	D	5–10 mk 2 ×/d × 21 d	1 d	[3H]spiroperidol	Domperidone	Down
Robertson, 1983	Rats	M	D	10 mk 2 ×/d × 21 d	24–36 h	[3H]spiroperidol	Domperidone	Down
Akiyama et al., 1982b	Rats	M	M	4 mk/d × 14 d	7 d	[3H]spiperone	ADTN[6]	NC[5]
Algeri et al., 1980	Rats	M	D	10 mk/d × 7 d	1 d	[3H]haloperidol	Haloperidol	NC[5]
Burt et al., 1977	Rats	?	D	5 mk/d × 3 wk	5–7 d	[3H]haloperidol	Dopamine	NC[5]
Howlett and Nahorski, 1979	Rats	M	D	5 → 15 mk 2 ×/d × 4d[2]	17–20 h	[3H]spiroperidol	Butaclamol or dopamine	NC[5]
Howlett and Nahorski, 1978	Rats	M	D	5 → 15 mk 2 ×/d × 4d[2]	17–20 h	[3H]spiperone	?	NC[5]
Jackson et al., 1981	Rats	M	D	5 mk/d × 25 d	7 d	[3H]spiperone	Butaclamol	NC[5]
Muller and Seeman, 1979	Rats	?	?	10 mk/d × 14 d (oral)	1 d	[3H]spiperone	Pimozide	NC[5]
Owen et al., 1981	Vervet	M/F	D	4 → 12 mk/d × 35 d	1 d?	[3H]spiperone	Butaclamol	NC[5]
Riffee et al., 1982	Mice	M	D	4 mk/d × 14 d	1 or 5 d	[3H]spiroperidol	Apomorphine or butaclamol	NC[5]
Borison et al., 1978	Rats	M	D	3.75 mk/d × 5 wk	5 d	[3H]dopamine	Butaclamol	Up
Klawans et al., 1979	Guinea pigs	M	D	5 mk/d × 4 wk	7 d	[3H]dopamine	Apomorphine or butaclamol	Up
Robertson, 1979	Rats	M	D	10 mk 2 ×/d × 21 d	24–36 h	[3H]ADTN	Dopamine	Up
Robertson, 1983	Rats	M	D	5 mk/d × 22 d	2 d	[3H]spiroperidol	?	Up

[1]From Robinson and Becker, 1986.
[2]Also added 25 → 75 mg/ml to drinking water.
[3]Given 4 mg/kg of AMPH 1 h before kill.
[4]Cerebellum used to estimate nonspecific binding.
[5]NC = no change.
[6]ADNT = 2-amino-6,7-dihydroxy-1,2,3,4-tetrahydronaphthalene.
[7]M = methylamphetamine.
[8]D = D-amphetamine.
[9]mk = milligrams/kilograms.

more, dopamine release induced by amphetamine may not, in fact, be mediated by dopamine autoreceptors (Langer and Arbilla, 1984).

At the present time, behavioral sensitization to amphetamine is not well understood (Demellweek and Goudie, 1983). Although it seems clear that alterations in dopamine release are important, there is simply no convincing evidence for alterations in dopamine binding as a critical factor. It is interesting to note that, although dopamine mechanisms are purportedly involved in sensitization to the behavioral effects of chronic opiate administration, changes in dopamine binding are not observed under these conditions (Kalivas, 1985). The problem of attempting to relate tolerance and sensitization of psychoactive drugs to changes in receptor function is a complex one. This is well-illustrated by the variety of behavioral factors that may modify such phenomena (*see* Demellweek and Goudie, 1983). Although changes in behaviors are always accompanied by changes in neurotransmitter function (the fundamental tenet of contemporary behavioral neurobiology), the latter may be manifest as various changes in neurotransmitter release, uptake or receptor density and/or affinity. Furthermore, the interaction between different neural systems is clearly important in the context of long-term drug administration (e.g., *see* Rupniak et al., 1986a, 1987). The net influence of such changes must eventually be accounted for in any complete analysis of tolerance and sensitization. Given the current incomplete status of this research area, it is not surprising that these behavioral phenomena have been described as "enigmatic" (Demellweek and Goudie, 1983). The reader is referred to Kilbey and Sannerud (1985) for an overview of behavioral studies in relation to models of tolerance and sensitization.

5.2. Chronic Administration of Direct Dopamine Agonists

In contrast to the status of studies with chronic administration of amphetamine, relatively few behavioral studies have been conducted involving chronic administration of direct dopamine receptor agonists or of the dopamine precursor L-dopa. Pycock et al., (1982) reported that L-dopa administration (orally) to rats induced a transient enhancement of spontaneous motor activity, which declined to control levels between 3 and 6 mo. The effects of acute challenges with dopamine agonists and antagonists were assessed in these animals. At 6 mo after initiation of L-dopa administration, behavioral responses (sterotypy ratings) to these acute challenges were significantly greater than in animals receiving chronic vehicle treatment. These effects were accompanied by specific increases in D_2 dopamine receptor binding (B_{max} characterized with [^{3}H]-spiperone). In this study, the effects of bromocriptine on behavior were not increased by chronic L-dopa treatment. In another study with L-dopa, the effects of oral administration of this compound in combination with carbidopa, or of pergolide (a preferential D_2 agonist) administered intraperitoneally were assessed (Reches et al., 1984). Both [^{3}H]-spiperone binding and rotational responses to apomorphine in rats with unilateral 6-OHDA lesions of the substantia nigra were measured in this study. After 14 d of drug treatment (two divided daily doses in the case of each drug), a decrease in B_{max} was observed in both intact and denervated striata. The extent of this decrease in D_2 binding was equivalent irrespective of 6-OHDA lesioning. The significance of the apomorphine responses in this study is impossible to assess because of the vast differences between control turning counts for the control groups for the L-dopa and pergolide treatments, respectively.

Globus et al. (1982) examined the effects of chronic treatment with bromocriptine on both apomorphine-induced sterotypy and D_2 binding in rats. Interestingly, in relation to the amphetamine studies discussed earlier, bromocriptine induced an increase in the behavioral response to apomorphine without influencing [^{3}H]-spiperone binding. These results are in contradiction to those of Quick and Iversen (1978), who reported decreased D_2 binding after chronic bromocriptine administration under similar conditions.

More recently Traub et al. (1985) have, however, reported results similar to those of Quik and Iversen (1978). In this latter study, dopamine agonist-induced rotations in rats with unilateral 6-OHDA lesions of the substantia nigra (SN) were used to assess behavioral supersensitivity. Agonist-induced rotation increased significantly over a 3-wk period of bromocriptine administration. Furthermore, bromocriptine reversed the increase in dopamine binding induced by the 6-OHDA lesions. The enhanced behavioral sensitivity lasted up to 3 wk following the last dose of bromocriptine. Chronic bromocriptine, however, failed to affect dopamine agonist-induced rotation in animals with unilateral kainate lesions of the striatum, and did not affect behavioral sensitivity to apomorphine in intact animals. These data indicate that the changes in behavioral sensitivity to dopamine agonists induced by bromocriptine may be independent of changes in D_2 receptor binding.

This conclusion is, although perfectly logical, intuitively unappealing. Impressive correlations exist for D_2 density and behavioral responses; and potencies of dopamine agonists in the 6-OHDA-induced rotation model correlate highly with D_2 receptor affinities (*see* Seeman, 1980). Nevertheless, as Traub et al. (1985) have remarked, behavioral supersensitivity to apomorphine precedes measurable changes in D_2 binding (Hyttel, 1979; Staunton et al., 1981). Traub et al. (1985) have suggested that, at least in some cases, effects of chronic administration of dopamine agonists or altered responsiveness to such compounds may be the result of alterations in systems other than those primarily concerning dopamine. Some evidence for this proposal has been provided by Pizzolato et al. (1985), who report increased local glucose utilization in many nondopamine containing systems, during stereotypy induced by bromocriptine. A certain amount of caution is, however, necessary in interpreting lesion-induced changes in dopamine receptor binding. Some authors claim that 6-OHDA lesions of the SN are ineffective for increasing striatal D_2 binding (e.g., Staunton et al., 1982). The discrepancy between this proposal and the many other directly contradictory reports of increased B_{max} for D_2 sites under the same conditions (e.g., Breese et al., 1987) is alarming. Nevertheless, a recent paper by Savasta et al. (1987) indicates that the upregulation of D_2 sites following lesions of the SN may be selective to the ventro- and dorsolateral regions of this structure. It is possible that tissue sample differences in terms of striatal topography or effective extent of SN lesions may provide an answer to the enigmatic state of the data describing 6-OHDA-induced changes in dopamine receptor binding. These considerations currently obscure to some extent our level of understanding of chronic drug effects on dopamine receptors.

It is important to note that, in the above studies, drugs were administered once daily with the exception of the Pycock et al. (1982) investigation, in which L-dopa was available in the diet.

Doses of bromocriptine used were similar in all studies. It is unlikely that the differences in reported effects on D_2 receptors are the result of simple differences in drug administration or dose regimens. Single or repeated daily injections of apomorphine are reported to reduce the ability or subsequent low-dose apomorphine challenges to inhibit spontaneous locomotor activity (Bernardi and Scavone, 1985). In this study, similar effects were observed after repeated administration of amphetamine. However, although repeated apomorphine increased striatal levels of the dopamine metabolites DOPAC and HVA, these effects were not observed with amphetamine. Bernardi and Scavone (1985) interpreted these effects in terms of the development of autoreceptor subsensitivity in response to chronic apomorphine, but not amphetamine, administration.

As with amphetamine, it has been recently observed that continuous infusions of the D_2 agonist $(+)$-4-propyl-9-hydroxynaphthoxazine (PHNO) by osmotic minipumps led to the development of tolerance to the stimulant action of this drug on locomotor activity (Martin-Iverson et al., 1987, 1988). Intermittent administration of PHNO results in behavioral sensitization, as is the case with amphetamine (Martin-Iverson et al., 1987, 1988). These effects are, however, dependent on circadian factors. Intermittent administration of PHNO results in sensitization in both phases of the light/dark cycle. The tolerance to continuous administration of PHNO only occurred in the light phase of the cycle: motor stimulant effects during the dark phase underwent sensitization. It is of considerable interest that the tolerance to PHNO was reversed by administration during PHNO treatment of the selective D_1 agonist SKF 38933 (Martin-Iverson et al., 1987, 1988). Although dopamine receptor density (Marzullo and Friedhoff, 1982) is reported to exhibit circadian changes, the relationship of this phenomenon to effects of chronic drug administration is unexplored.

The importance of D_1 receptor activation for the expression of D_2 agonist effects (Walters et al., 1987; Carlson et al., 1987) is supported by restoration of behavioral effects of bromocriptine by dopamine-releasing agents or by SKF 38933 (Jackson and Jenkins, 1985; Jackson and Hashizume, 1986), and by the observation that SKF 38933 may potentiate behavioral effects of the selective D_2 agonist RU 24213 (Mashurano and Waddington, 1986). The interactive effects of D_1 and D_2 agonists have already been recognized as potentially important in the treatment of Parkinson's disease (Waddington, 1986; Robertson and Robertson, 1986). In this respect, the possible necessity of D_1 activation for restoration of dopamine function is highly important. The relevance of the findings discussed above will be considered later (*see* Section 7.1.2).

5.3. Effects of Chronic Cocaine

In relation to interactions of drugs of abuse with dopamine receptor systems, it has recently been demonstrated that chronic cocaine administration increases and decreases D_2 binding in rat brain in a regionally selective manner (Goeders and Kuhar, 1987). In this study, daily injections of cocaine for 15 d decreased the B_{max} for [^{3}H]sulpiride binding in the striatum. In contrast, D_2 binding in nucleus accumbens was actually increased by this treatment. No changes in K_d were observed. Other investigators have reported increases in striatal D_2 binding (using [^{3}H]spiroperidol) following acute or chronic cocaine administration (Trulson and Ulissey, 1987; Taylor et al., 1979). Goeders and Kuhar (1987) have attributed these differential results to possible methodological differences.

Goeders and Kuhar (1987) have proposed that the observed differential changes in mesolimbic (nucleus accumbens) and nigrostriatal (caudate nucleus) dopamine binding may help to explain the time course of changes in behavioral actions of cocaine. These authors suggest that "the tolerance to effects of cocaine in the striatum and the gradual sensitization of receptors in the nucleus accumbens may be related to the progressive attenuation of locomotor activity and the augmentation of restrictive sterotyped behavior" reported in their study (Goeders and Kuhar, 1987).

As cocaine self-administration is selectively blocked by the D_2 receptor antagonist pimozide, but not by the D_1 antagonist SCH 23390 (Woolverton, 1986), the time course of changes in responsiveness of cocaine to D_2 blockade may be of interest in the analysis of cocaine self-administration. Indeed, it is interesting to note that administration of the D_2 agonist bromocriptine may reverse cocaine craving in humans (Dackis et al., 1987).

6. Effects of Chronic Dopamine Antagonists

It has been demonstrated in several species of animals that chronic administration of a variety of neuroleptic drugs leads to increased and enhanced behavioral responses to dopamine agonists such as apomorphine and amphetamine, and that this enhanced sensitivity to dopamine stimulation persists for a period of several weeks after withdrawal of neuroleptic treatment. It has been proposed that such responses are the result of increases in receptor number following chronic DA receptor blockade. The results of representative studies are presented in summary form in Table 2.

6.1. Short-Term Antagonist Administration

For studies involving short-term neuroleptic administration for periods of up to 3 w, the general consensus is that dopamine receptor binding is increased. In general, binding changes associated with these effects are reported to be changes in B_{max} without alterations in K_d. Changes in K_d that have sometimes been reported (Ebstein et al., 1979; Owen et al., 1980; Murugaiah et al., 1982; Rupniak et al., 1984, 1985) are discussed in Section 6.2. The relative effects of antagonists on D_1 and D_2 binding are dependent on the affinity of these compounds for either receptor subtype. Memo et al. (1987b) have demonstrated elegantly that chronic selective blockade of D_1 receptors or D_2 receptors, using SCH 23390 or sulpiride, respectively, leads to increased sensitivity to receptor-mediated changes in adenyl-cyclase activity (stimulation in the case of D_1 activation and inhibition in the case of D_2 activation using selective agonists). These effects were accompanied by increased sensitivity to the behavioral effects of methylphenidate. Interestingly, this supersensitivity could only be blocked by combined D_1/D_2 antagonism. Neuroleptic-induced changes in D_1 receptor binding have been reported for both the striatum and nucleus accumbens and, although changes in behavioral sensitivity to drugs that act at D_1 sites (Johansson et al., 1987; Sanger, 1987; Nakajima, 1986; Molloy and Waddington, 1987) have not been assessed after chronic administration of antagonists, the Memo et al. (1987a,b) studies indicate that this is an interesting area for future studies.

It is particularly interesting to note that receptor changes are not evident in the case of the frontal cortex (Memo et al., 1987b; Bacopoulos, 1979, 1981), a region that is claimed to be devoid of presynaptic (auto)receptors for dopamine (Roth, 1984, but *see* Timmermans and Thoolen, 1987, for a critical appraisal). The significance of this phenomenon will be discussed in section 6.3

A peculiar finding has been the observation that the combination of chronic antagonist treatment and 6-0HDA lesions of the SN leads to greater increases in both D_2 binding and behavioral sensitivity to apomorphine than does either treatment alone (Staunton et al., 1982; *see* Fig. 1). This effect was proposed to result from potentiation of postsynaptic supersensitivity phenomena, in view of the terminal loss that occurs following 6-0HDA. The proposal that supersensitivity with haloperidol alone may not have been maximal was discounted by the assessment of high haloperidol dose effects, which were not as effective as the combined treatment. Binding was measured in the nucleus accumbens, an area exhibiting dopamine autoreceptor activity. The clearest significance of this study is that receptor changes induced either by lesion-induced reductions in dopamine or by receptor blockade, respectively, may not be viewed as functionally equivalent. This consideration extends the earlier caveat concerning the assessment of effects of lesions in this context.

6.2. Long-Term Antagonist Administration

Ebstein et al. (1979) have reported increases in D_2 binding following 21 and 70 d of access to a haloperidol-containing diet. Rats were killed 4 d following withdrawal from haloperidol. In this study, the K_d was decreased after 70 d of drug treatment. Long-term changes in D_2 binding were also assessed by Owen et al. (1980). In this study, animals had access to haloperidol in their drinking water for 9 mo with or without a subsequent withdrawal period of 7–10 d. Sensitivity to behavioral stimulant effects of apomorphine was reduced in rats maintained on haloperidol, whereas increased sensitivity was observed following withdrawal. The B_{max} for [^{3}H]-spiperone was increased regardless of whether or not these subjects were in withdrawal. The K_d was, however, reported to in-

Table 2
Effects on Chronic DA Antagonists on DA Binding and Receptor Function

| Reference | Drug/Schedule | Effect on Binding B_{max} unless specified | | Effect on cAMP | Effect on behavioral responsiveness to agonists |
		D_1	D_2		
Liskowksy and Potter, 1987	Haloperidol 21 d, continuous daily	0	↑ in striatal and frontal cortex	?	?
Hytell and Christensen, 1983	12 d oral daily dose cisflupenthixol	?	?	?	Supersensitivity to methylphenidate: only blocked by combined D_1 and D_2 antagonism
Memo et al., 1987a	Haloperidol 21 d increasing doses injected?	?	?	Striatum and nucleus accumbens ↑ of both D_1 and D_2 effects	?
Mackenzie and Zigmond, 1985	Fluphenazine and Haloperidol 21 d daily sc	0	↑ in striatum	?	?
Memo et al., 1987b	(−)Sulpiride 21 d daily ip SCH 23390 (× 3 daily sc)	?	?	SN striatum, pituitary and N Acc changes not in FC or hippocampus → of both D_1 and D_2 receptors	?
Mackenzie and Zigmond, 1984	Haloperidol 54 d sc daily	Assessment of high and low affinity states†	↑ striatum	?	?
Fleminger et al., 1983	Haloperidol 21 d daily ip	0	↑ striatum, nucleus accumbens and olfactory tubercles	?	Increased responsiveness to low and high dose apomorphine
	cis Flupenthixol	↑ striatum only	?	↑ striatum only	
Burt et al., 1976	21 d Haloperidol Fluphenazine sc daily	?	↑ significant D_2, [^{3}H]haloperidol used 0.2–1.4 nM, striatum	?	?

Reference	Treatment				
Bacopaulos, 1981	21 d Fluphenazine sc daily	?	↑ striatum no change in frontal cortex	?	?
VonVoigtlander et al., 1975	14 d oral diets Haloperidol and thioridazine	?	?	0 striatum	Increased responsiveness to amphetamine
Heal et al., 1976	4–5 d chlorpromazine haloperidol	?	?	0 striatum	Increased responsiveness to L-DOPA plus tranylcypromine
Porceddu et al., 1985	pimozide daily ip 12 d SCH 23390 (sc × 3 daily) Haloperidol (sc × 3 daily) Sulpiride (ip × 3 daily)	↑ slight ↑ 0	? ? ?	?	?
Creese and Chen, 1985	21 d SCH 23390 (ip daily)	↑	striatum	?	?
Vaccheri et al., 1987	21 d SCH 23390 (sc daily)	?	striatum	?	Increased response to high dose apomorphine. No change in response to low dose apomorphine. Effect persisted for at least 77 d after withdrawal.
Staunton et al., 1981	Haloperidol 14–15 d sc daily (alone and with 6-OHDA lesions)	?	↑ Potentiation of effect by 6-OHDA Nucleus accumbens	?	Increased responsiveness to high dose apomorphine. Potentiated by 6-OHDA lesions.

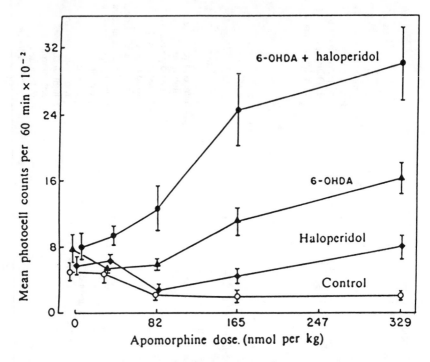

Fig. 1. Effect of various doses of apomorphine-HC1 on locomotor activity. Note that the dose 329 nmol/kg corresponds to 0.1 mg/kg apomorphine-HC1. Values are the cumulative number of photocell beam interruptions (mean ± SEM) for the 60-min period following apomorphine administration. From Staunton et al. (1982) with permission.

crease in rats maintained on the drug. No effects on K_d were evident after withdrawal. These K_d effects are clearly not consistent with those of the Ebstein et al. (1979) study. Murugaiah et al. (1982) investigated effects of continuous access to trifluoperazine (in drinking water) in rats for 12 or 14 mo. After the first 3 mo, D_2 binding (B_{max}) progressively increased in drug-treated animals compared to age-matched controls. In animals withdrawn from the drug at 12 mo, the B_{max} declined after a further 3 mo, reaching control levels by 6 mo. Trifluoperazine induced a transient increase in K_d in this study, evident after 3 mo of drug administration and returning to control values by 6 mo of continuous drug access. Between 6 and 12 mo, the K_d was observed to rise again, returning to control levels 1 mo after discontinuation of drug treatment (*see* Fig. 2). Alterations in behavioral sensitivity to apomorphine and

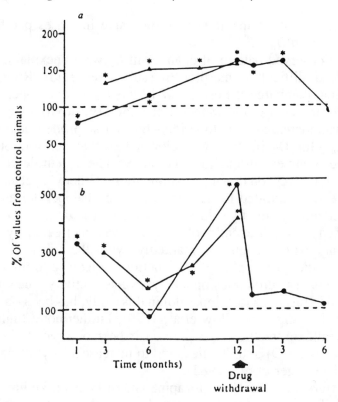

Fig. 2. Comparison of the alterations in dopamine ($10^{-4}M$)-specific [^{3}H]-spiperone (O.125–4.0 nM) binding to rat striatal preparations produced by the administration of trifluoperazine dihydrochloride continuously for up to 12 mo in two distinct studies. a, Receptor numbers (B_{max}); b, dissociation constant (K_d). The experimental groups were as follows: Expt. 1. (●), male Wistar rats received trifluoperazine dihydrochloride (2.5–3.5 mg/kg/d) for up to 12 mo when the drug was withdrawn. Expt. 2. (▲): male Wistar rats received trifluoperazine dihydrochloride (4.5–5.6 mg/kg/d) for up to 12 mo. In each case, a group of age-matched control animals from the same supplier received distilled water alone and was maintained alongside the drug-treated animals. At intervals during the course of the experiments, striatal preparations from drug-treated and control animals were examined for dopamine ($10^{-4}M$)-specific [^{3}H]-spiperone binding. For comparison of the dissociation constant (K_d) and receptor numbers (B_{max}) in different animal groups at different time intervals, values in drug-treated animals are expressed as a percentage of K_d and B_{max} values in tissue from control animals obtained on the same day and assayed in parallel throughout the experiment. From Murugaiah et al. (1982) with permission.

trifluoperazine were manifest as progressive increases paralleling the changes in B_{max}.

The observed pattern of changes in K_d were unexplained and are, indeed, difficult to interpret. In a subsequent study, Rupniak et al. (1986a) compared the effects of 12 mo access to haloperidol or sulpiride (in the drinking water). Behavioral sensitivity to apomorphine was increased in haloperidol- but not sulpiride-treated rats. The B_{max} for D_2 binding was elevated by haloperidol, but not sulpiride treatment throughout the study. These neuroleptics had differential effects on dopamine stimulation of adenyl cyclase. This response was inhibited for the first month by haloperidol and was increased after 12 mo by sulpiride treatment. The B_{max} for D_1 sites, as defined by [^{3}H]flupenthixol binding, was unaffected by either treatment, which contrasts markedly with the adenyl cyclase changes indicative of altered D_1 function. Tolerance to the inhibitory effects of neuroleptics on adenyl cyclase activity, such as that observed in response to haloperidol in this study, has been observed by other investigators (Clow et al, 1979). Furthermore, high-dose cis-flupenthixol treatment has previously been observed to increase B_{max} values for D_1 and D_2 sites, and to increase adenyl cyclase activity (Fleminger et al., 1983).

Differential effects of clozapine and of haloperidol have been reported by Rupniak et al. (1985). In this study, rats were given access to these drugs in their drinking water for up to 12 mo. Stereotyped behavior induced by apomorphine was inhibited during 12 mo of haloperidol treatment, but after this 12-mo period, this behavioral response was enhanced. Clozapine did not affect behavioral responses to apomorphine in this study. D_2 binding was enhanced throughout the 12 mo of haloperidol treatment; clozapine had no effect on the B_{max} of D_2 sites. In contrast, clozapine increased the B_{max} of D_1 sites after 9–12 mo of treatment, although adenyl cyclase activity was unaffected by this drug. Haloperidol inhibited adenyl cyclase for the first month of treatment. It has been suggested that differential effects on receptor function (vis D_1 and D_2 subtypes) of typical neuroleptics (e.g., haloperidol) and atypical compounds such as sulpiride and clozapine may be relevant for the reported low incidence of tardive dyskinesia induced by atypical compounds (Rupniak et al., 1985), but see Section 7.1.3. below.

6.3. Analyses of Presynaptic Receptor Changes Induced by Chronic Dopamine Antagonists

It is evident that, although tolerance occurs to initial motoric effects of neuroleptics in the clinical context, such effects are rarely observed in relation to antipsychotic effects (but *see* Chouinard and Jones, 1980, and Table 3). It has been proposed that the lack of tolerance to antipsychotic effects may be related to regional differences in brain dopamine function induced by chronic administration of these drugs (Laduron, 1988).

Of particular interest in this context is the apparent lack of autoreceptors for dopamine projections to certain areas of the cortex. The evidence for this regional difference and region specific effects of drugs on presynaptic dopamine receptor function have been reviewed by Roth (1984) and by Timmermans and Thooley (1987).

As indicated earlier (Section 5.1.2), binding studies have been largely uninformative in the analysis of presynaptic dopamine receptor function. Indirect techniques have been useful but, as Roth (1984) has pointed out, only the following techniques have proved particularly useful in assessing effects of chronic drug treatment.

1. Blocking impulse flow in dopamine neurons with γ-butyrolactone and using agonist-induced reversal of increased dopamine levels or tyrosine hydroxylation as an index of presynaptic receptor function.
2. Driving dopamine neurons with supramaximal electrical stimulation and monitoring agonist-induced reversal of the stimulation-induced increase in tyrosine hydroxylation as an index of presynaptic receptor activation (Ross, 1979).
3. Iontophoretic application of dopamine and dopamine agonists, and the assessment of changes in the activity of identified dopamine neurons (Skirboll and Bunney, 1979).

In the first two techniques, levels of dopamine, L-dopa, and the metabolites DOPAC and HVA are monitored as indexes of re-

Table 3
Possible Tolerance to Antipsychotic Effects
of Fluphenazine Esters

Year of treatment	Dose, mg/2 wk	Annual dose increase, %
1	50 ± 14	—
2	59 ± 10	18
3	157 ± 33	166
4	234 ± 69	49

'Ten patients of mean age 34 y were drawn from an outpatient clinic population of 300 schizophrenics treated for at least 3 yr with fluphenazine (decanoate or enanthate) injections (doses are means ± SEM). Serum prolactin was elevated in all cases (70 ± 11 ng/mL), and 80% had mild or moderate tardive dyskinesia, the severity of which corresponded only weakly with the increasing dosage requirements. The data are adapted from Chouinard and Jones, *Am. J. Psychiatry* 137:16, 1980. The findings suggest that tolerance may occur to the antipsychotic effects of some neuroleptics in a minority of patients after several years of continuous treatment; they do not prove that new, iatrogenic dysfunction was added. From Baldessarini, 1985, with permission.

ceptor function. The data in Fig. 3 illustrate the lack of tolerance to changes in DOPAC in prefrontal and cingulate cortex following chronic administration of haloperidol (28 d in drinking water) to rats (Galloway and Roth, 1983).

Tolerance to neuroleptic-induced increases in concentrations of dopamine metabolites has recently been ascribed to changes in their basal levels following chronic neuroleptic treatment (Finlay et al., 1987). Such an influence is proposed to be a result of decreased basal activity of dopamine-containing neurons following repeated exposure to neuroleptics. The complexity of mechanisms of tolerance to effects of neuroleptics on dopamine metabolites is illustrated by analyses of depolarization block following repeated neuroleptic administration (Grace and Bunney, 1986). In summary, dramatic decreases in the number of spontaneously active dopamine neurons in the substantia nigra have been reported following repeated haloperidol administration (21 d). Intracellular recordings indicate depolarization block in these cells, presumably as a result of compensatory excitatory feedback drive of nigral cell bodies. This effect could be overcome by hyperpolarization of the

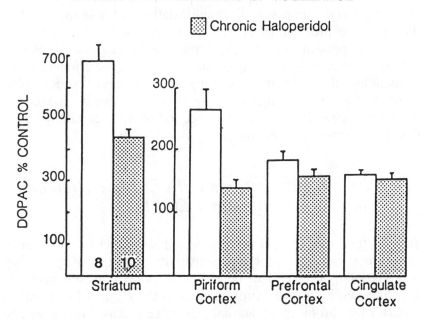

Fig. 3. Effects of acute and chronic haloperidol treatment on DOPAC levels in various brain regions. Haloperidol was chronically administered to rats in drinking water for 28 d. Haloperidol (1 mg/kg) or vehicle challenge was administered 60 min before decapitation. The *bars* on the columns represent the SEM, and the *numbers* in the columns indicate the number of rats. From Galloway and Roth, 1983, with permission.

neural membrane by autoreceptor-activating doses of apomorphine. This phenomenon is consistent with lower basal levels of dopamine metabolites observed by Finlay et al. (1987) following acute withdrawal of repeated haloperidol administration. These observations are of considerable potential for understanding the relative changes in autoreceptor function and dopamine cell firing *per se* induced by repeated drug administration.

Important questions remain concerning the specificity of tolerance to effects of repeated neuroleptic administration on dopamine cell-firing and metabolite levels. These are principally: (1) To what extent is tolerance to neuroleptic effects on metabolite levels the result of changes in presynaptic dopamine receptor function? (2) Do apparently region-specific manifestations of tolerance reflect differential expression of autoreceptor function?

At the present time, no clear conclusions are possible. Nevertheless, further understanding of these issues will be critical for understanding the relevance of long-term changes in dopamine autoreceptors for pharmacotherapy of schizophrenia, movement disorders such as Parkinson's disease, and possibly certain forms of depression (*see* later sections).

6.4. Lithium Interactions with Chronic Dopamine Antagonist Effects

Administration of lithium is reported to prevent behavioral expression of dopamine receptor supersensitivity (*see* review by Bunney and Garland, 1983). The decrease in behavioral sensitivity is accompanied by a reversal of changes in the B_{max} of D_2 receptors induced by chronic drug administration (e.g., *see* Gianutsos and Friedman, 1987; also Gallager et. al., 1988; Perr et al., 1978). It is interesting to speculate that lithium effects on phosphoinositide systems that are implicated in D_2 receptor function (Pizzi et al., 1987) are related to this phenomenon. Klawans (1973) has proposed that lithium may be efficacious in preventing the development of neuroleptic-induced tardive dyskinesia. Gianutsos and Friedman (1987) and Friedman et al. (1983) have, however, suggested that such effects may be related more directly to the reversal of changes in muscarinic binding induced by neuroleptics such as fluphenazine. Changes in various receptor populations other than dopamine receptor subtypes have been reported following chronic dopamine antagonist treatment. These include changes in [3H]-flunitrazepam and [3H]QNB binding. A detailed discussion of this material is beyond the scope of the present chapter, but the reader is referred to recent studies by Rupniak and colleagues (1986a, 1987) for a current discussion. The relevance of changes in dopamine receptor binding vs changes in other neural systems in the mechanisms of drug action (both clinically therapeutic and detrimental) is discussed briefly in Section 7 of this chapter.

6.5. Effects of Chronic Antidepressant Treatment on Dopamine Receptor Binding

Recent observations that behavioral responses to dopamine agonists are altered following chronic antidepressant treatment implicated changes in dopamine receptors in the therapeutic actions of these drugs (Spyraki and Fibiger, 1981; Maj et al., 1984; Maj and Wedzony, 1985). As Klimek and Nielsen (1987) have pointed out, chronic administration of antidepressants induces subsensitivity of presynaptic striatal dopamine receptors (Serra et al., 1979; Antelman and Chiodo, 1981; Maj et al., 1985; Nielsen, 1986). It is of interest in this context that Klimek and Nielsen (1987) have reported decreases in B_{max} for D_1 binding, although no changes occurred in B_{max} for D_2 binding after 14 d of administration of various antidepressants. The implications of these results are discussed by Willner (this volume), and will not be considered here.

7. Effects of Chronic Drug Administration on Dopamine Receptors in Humans

The dopamine hypothesis of schizophrenia relates overactivity of mesocorticolimbic dopamine-containing pathways to psychotic symptoms (Matthysse, 1973; Van Rossum, 1967). This hypothesis has received substantive support from the fact that dopamine receptor antagonists are well established as antipsychotic drugs. Seeman (1987) has pointed out that the clinical observations that psychotic symptoms are decreased in schizophrenics who develop Parkinson's disease and that Parkinsonian patients apparently do not develop schizophrenia also support this hypothesis. Parkinson's disease is characterized by motoric impairment that is principally related to loss of striatal dopamine-containing nerve terminals (Hornykiewicz, 1973). The basic line of evidence for a role of dopamine receptors in schizophrenia has been fully outlined in a recent article by Seeman (1987). In the clinical context, the study of dopamine receptor changes is important for several distinct reasons.

1. Long-term changes in response to chronic administration of dopamine antagonists may reflect alter-

ations in dopamine receptor function. Such receptor changes may reveal useful information related to both the etiology and treatment of psychosis.

2. Treatment with antipsychotic drugs quite frequently results in the emergence of disorders of motor function such as tardive dyskinesia (Kane et al., 1986). The extent to which such side effects may relate to changes in dopamine receptor function is again relevant to treatment and a clearer understanding of etiology.

3. Parkinson's disease is a progressive illness that is treated by means of strategies directed to restoration of brain dopamine levels or receptor activation *per se* through agonist drug administration. The changes in dopamine receptor function over the course of the illness *per se* and in response to drug therapy may help elucidate the reasons for the variable efficacy of drug therapy over time and lead to more effective treatment regimens.

7.1. Postmortem and In Vivo Binding Studies

7.1.1 Schizophrenia

Seeman (1987) has suggested that prolonged blockade of dopamine receptors by neuroleptics, elevating D_2 receptors by about 30%, results in a compensatory reduction in dopamine turnover. This is manifested by a fall in plasma or CSF HVA concentrations in patients, which generally parallels clinical improvement (Pickar et al., 1986). As outlined earlier (Section 6.3), this reduction may be because of the depolarization block induced by neuroleptics (Grace and Bunney, 1986) or, as Seeman (1987) suggests, by a shift of D_2 receptors into a nonfunctional low-affinity state. In relation to dopamine receptor subtypes, it is pertinent to note that, although neuroleptics may bind to both D_1 and D_2 sites, predominantly they exhibit highest affinity for D_2 sites. Furthermore, D_2 sites in both the nucleus accumbens and striatum exhibit equivalent affinities (Reynolds et al., 1982; Richelson and Nelson, 1984; Seeman and Ulpian, 1983). Also in relation to neuroendocrine markers for dopamine receptor activation, neuroleptic potencies for elevating plasma prolactin also generally correlate with dissociation constants for D_2 binding in the anterior pituitary gland (Meltzer and Stahl, 1976). Both D_1 and D_2 binding have now been assessed in postmortem human brain. There are numerous technical issues related to high- vs low-affinity states and binding. These are beyond the scope of the present chapter, but see Seeman (1987) and Section 4.

Generally, it appears that the density (B_{max}) of D_1 (high-affinity) receptors is unchanged in striata of schizophrenics (Seeman, 1987). Conflicting data exist concerning the degree of adenyl cyclase stimulation induced by dopamine in brain tissue from schizophrenics. Memo et al. (1983) report an increase in this functional measure of D_1 activity, in contrast to the observations of Carenzi et al. (1975). From the available data, it appears that D2 binding is increased in schizophrenic brain as a function of chronic neuroleptic treatment. Owen et al. (1978) have reported an increase in [³H]-spiperone binding in the nucleus accumbens, caudate nucleus, and putamen of 19 schizophrenic patients. This effect was observed in five patients free of medication for one year prior to death. These researchers have proposed that the receptor changes are disorder- rather than drug-related. Mackay et al. (1980) reported data that conflict with the results of Owen et al. (1978), claiming that increased binding of [³H]-spiperone in caudate nucleus and nucleus accumbens of schizophrenics is really the result of drug treatment and is not seen in brains of drug-free patients. Increases in K_d in this study were attributed to treatment drug competition with the ligand in the assay system (*see* Section 2). An increase in "fixed concentration" [³H]-flupenthixol binding only in brains of drug-free schizophrenics has been measured by Cross et al. (1981). Scatchard analysis revealed increased B_{max} and K_d relative to controls in brains of drug-treated patients; in drug-free patients' brains, only B_{max} increased.

A study by Lee et al. (1978) indicates that apomorphine binding is unchanged in schizophrenic brain. [³H]-Haloperidol and [³H]-spiperone binding was increased. The role of clinical drug treatment is unclear from this study.

Seeman et al. (1984) have examined the densities and affinity of D_2 receptors in 81 controls and 59 schizophrenics from four countries. A bimodal distribution for B_{max} was observed in the caudate nucleus and putamen and in the nucleus accumbens, each mode representing a change of (a) 125% of control B_{max} and (b) 230% of control B_{max}. These researchers have argued that the results are not attributable to different drug histories, since K_d was elevated to a similar extent in both groups, indicating similar neuroleptic dose histories. The distinction between drug-treated and drug-free schizophrenics is not possible in this study because of too few drug-free patients.

Kornhuber et al. (1989) have measured [^{3}H]-spiperone binding in caudate putamen from 27 schizophrenic and 27 control. B_{max} and K_d increased only in patients receiving neuroleptics within 3 mo prior to death. After at least 3 mo of withdrawal, B_{max} was actually lower than, but not statistically different from, control values. K_d normalized by 2 wk following drug withdrawal. Predominantly positive schizophrenic symptoms were not related to higher B_{max} values relative to predominantly negative symptoms. Also, there was no relationship between binding parameters and tardive dyskinesia or extrapyramidal symptoms.

The status of laterality in dopamine systems in relation to schizophrenia is unclear (Reynolds et al., 1987). There is no neurochemical evidence for altered dopamine levels in striatum, although a strong unilateral effect is reported for the amygdala. Crawley et al. (1986) have reported higher D_2 binding in the right striatum, which was undetectable in the left, but the history of medication of patients prior to death was not reported. In the Reynolds et al. (1987) study, the right putamen exhibited a 19% higher B_{max} for D_2 sites than the left in controls. The increased B_{max} for D_2 sites was accentuated in tissue from schizophrenic patients. The relationship of these binding changes to drug treatment is unknown.

Kornhuber et al.'s (1989) study indicates that the changes in K_d rapidly adapt to control values following withdrawal of neuroleptics, whereas B_{max} declines at least to control levels over a longer period. This indicates that the K_d changes are perhaps artifactual, reflecting competition between the administered drug and the ligand in the assay system. The observation of no relationship to typology of symptoms or to movement disorders in this study is also particularly interesting. The bimodal pattern of receptor (B_{max}) increases observed by Seeman and colleagues indicates that, perhaps, responsiveness to an antipsychotic drug at the receptor level may characterize some underlying dynamic aspect of altered receptor function in schizophrenia. This difference is, so far, undetectable in drug-free post-mortem brain tissue.

Positron emission tomography (PET) has been successfully used to measure D_2 binding in humans (Wagner et al., 1983; Wong et al., 1984; Sedvall et al., 1986a,c). In the original studies [11C]-methylspiperone was used to label D_2 sites. Recent studies with [11C]-raclopride and [11C]-piquindone have the advantage of more specificity since methylspiperone also labeled the 5-HT$_2$ sites (Farde et al., 1985, 1986; Sedvall et al., 1986b). More recently, the labeled D_1 antagonist [11C]-SCH23390 has been used for human brain receptor imaging (Sedvall et al., 1986c). The results of the few PET studies available to date are complicated by various pharmacokinetic and pharmacodynamic factors that influence the rapid establishment of a pseudoequilibrium between bound and free ligand and the availability of receptors (which may be occupied in certain cases by active metabolites of the ligand of choice). The reader is referred to Farde et al. (1987) and to Perlmutter and Raichle (1986) for a clear overview of such considerations. At the present time, as with post-mortem studies, conflicting data have been generated by PET analysis. Farde et al. (1986) and DeLisi et al. (1986) report that D_2 densities in the striata of schizophrenics never treated with antipsychotics were equivalent to control values. In drug-free schizophrenics (>6 mo drug free), Crawley et al. (1986) reported an 11% increase in [77Br]-bromospiperone binding in the striatal region. Seeman has pointed out that this effect may have been an underestimation relative to the deficiency of D_2 occupation with this gamma-scintigraphy approach (Seeman, 1987). In drug-treated patients, however, a 27% increase in D_2 binding, characteristic of postmortem studies, was observed by Wong et al. (1986a,b); nevertheless, an increase in D_2 binding was also observed in drug-free schizophrenics by these researchers. The drug-free changes may be related to the use of [11C]-methylspiperone, which is not as selective for the D_2 site as is [11C]-raclopride.

Farde et al. (1987) reported that D_2 ligand receptor occupancy was markedly reduced following treatment with the D_1/D_2 antagonist *cis*-flupenthixol decanoate or the more selective D_2 antagonist sulpiride, whereas D_1 occupancy was reduced only to a minor degree. In this study, the use of [11C]-SCH23390 as a D_1 label in addition to [11C]-raclopride as a D_2 label indicates that comparative studies of D_1/D_2 receptor characteristics are now feasible in vivo.

Furthermore, they indicate that, at clinical doses, even cis-flupen-thixol (which has significant affinity for the D_1 site) acts mainly via D_2 receptors in vivo. As Seeman (1987) has recently pointed out, the available data are insufficient to yield a clear picture of in vivo changes in dopamine receptor function in relation to the treatment of schizophrenia.

7.1.2. Parkinsonism

In postmortem brain tissues from Parkinsonian patients, various changes in D_2 receptor characteristics have been reported. The observation that increases in D_2 binding (B_{max}) may occur in striata from patients never exposed to L-dopa suggests D_2 supersensitivity as a feature of the disease process (Lee et al., 1978; Rinne, 1981, 1982; Rinne et al., 1979, 1980, 1983). Reductions in D_2 binding have been measured in brain tissue from Parkinsonian patients treated with L-dopa (Lee et al., 1978; Olsen et al., 1980; Reisine et al., 1977; Rinne et al., 1981; Yahr, 1984). In contrast, other studies have reported either no change (Quik et al., 1979; Riederer and Jellinger, 1982, 1983; Riederer et al., 1978; Seemann et al., 1984) or a slight increase in B_{max} for such tissues (Bokobza et al., 1983). Such data are complicated by factors such as the duration of both the disease process and drug therapy, as well as by the age of the patients at the time of death (Guttman et al., 1986). Recently, these factors have been evaluated in a study by Guttman et al. (1986). These researchers report that changes in D_2 density in caudate nucleus and putamen as a function of these factors do not represent a critical determinant of the efficacy of L-dopa maintenance therapy. Thus, dopamine "receptor dropout" may not be responsible for diminished efficacy of dopamine replacement therapy over the time course of the illness (Fabbrini et al, 1987; Nutt, 1987). In this study, K_d was also apparently not related to these factors. In relation to D_1 receptor function, striatal dopamine-dependent adenyl cyclase activity has exhibited various changes (increases, e.g., Nagatsu et al., 1978; and decreases, e.g.; Shibuya, 1979). In the striatum, D_1 densities have been reported to increase (Raisman et al., 1985; Rinne et al., 1985) or remain unchanged in Parkinsonian brain (Pimoule et al., 1985). An interesting observation is that [^{3}H]-flupenthixol binding may be reduced in the substantia nigra of Parkinsonian patients who were not treated with L-dopa (Rinne et al., 1985). The significance of this latter study is unclear because in vivo studies indicate little activity of this D_1/D_2 antagonist at D_1 sites in the human brain (*see* Section 7.1.1), and yet according to Guttman et al. (1986), at least for the striatum, D_2 binding appears to be unchanged.

Cash et al. (1987) have evaluated postmortem D_1 and D_2 receptor densities in the caudate, putamen, and substantia nigra of Parkinsonian patients. In this study, DARPP-32 (dopamine and adenosine 3':5'-monophosphate-regulated phosphoprotein-^{32}K) concentrations were also measured. DARPP-32 is believed to represent the intracellular messenger for D_1 receptors, being localized in the same structures as D_1 sites (Hemmings et al. 1987). In this study, D_2 binding was defined with [^{3}H]-spiperone and D_1 binding with [^{3}H]-SCH23390. D_2 density was unchanged in the putamen. D_1 density, although unchanged in the putamen and the substantia nigra pars reticulata, was decreased by 28% in the substantia nigra pars compacta. Furthermore, DARPP-32 was decreased in each region. The decrease in the concentration of this putative index of D_1 activity was 45% in the putamen, 66% in the substantia nigra pars reticulata, and 79% in the pars compacta. These researchers suggest that these D_1 changes may reflect degeneration of pallidonigral GABA-containing neurons, and possibly also a loss of D_1 sites on the dopamine-containing neurons that arise in the pars compacta. These results are potentially of great importance in view of the proposed dependence of dopamine function on the balance of activation of D_1 and D_2 binding sites (*see* Section 5.2).

The importance of D_1 activation in the treatment of Parkinson's disease has been proposed on the basis of apparently greater therapeutic efficacy of bromocriptine (D_2 agonist) and L-dopa therapy relative to L-dopa alone (Parkes et al., 1976; Lieberman et al., 1976). The rational for this is based on the higher affinity of D_1 sites for dopamine (as a metabolic product of L-dopa). As Robertson and Robertson (1986, 1987) have proposed, coactivation of D_1 sites preferentially by dopamine, combined with more D_2-selective effects of bromocriptine, may lead to greater therapeutic effects. An alternative hypothesis has been proposed by Goldstein et al. (1985), who suggest that bromocriptine has a lower "intrinsic activity" at the dopamine receptor than dopamine. However, once dopamine has activated the receptor, bromocriptine displaces dopamine and induces an increased response. This may be viewed as a "priming" hypothesis. Such issues await empirical resolution. Nevertheless, it is apparent that increased D_1 receptor activation may not be advantageous either alone or in combination with intravenous L-dopa for the treatment of Parkinsonism (Braun et al., 1987). See Clark and White (1987) for a recent review of D_1 dopamine receptor function.

7.1.3. Tardive Dyskinesia

Although the incidence of tardive dyskinesia is clearly related to neuroleptic drug administration, the role of dopamine neurotransmission and of dopamine receptors *per se* in this phenomenon remains unclear (Rupniak et al., 1986b; Kane et al., 1986). The dopamine receptor hypothesis of tardive dyskinesia is based on the possible relationship between dopamine receptor antagonism, and subsequent increases in dopamine receptor density, and the emergence of tardive dyskinesias (Goetz and Klawans, 1982). Recently, this hypothesis has been criticized by Fibiger and Lloyd (1984). These researchers have outlined a number of inconsistencies in the literature, which are summarized as follows: (1) There is a discrepancy between the time course of emergence of tardive dyskinesia (typically after years of neuroleptic treatment) and changes in dopamine receptor density (within weeks of initial administration of neuroleptics). (2) Tardive dyskinesia persists for long periods following withdrawal of neuroleptics, whereas receptor density rapidly reverts to control levels after drug withdrawal. Similarly, Fibiger and Lloyd (1984) have pointed out that "spontaneous behavioral changes rarely emerge as a result of changes in dopamine receptor binding induced by neuroleptic drugs over a period of two to four weeks with laboratory animals." (*See* Dewey and Fibiger, 1983; Clow et al., 1980; but *see* Pittman et al., 1984). Furthermore, although increased dopamine density is generally observed in postmortem analyses of brains of patients who received neuroleptic treatment (equivalent results being obtained from laboratory animal studies), only a subpopulation of such patients exhibit tardive dyskinesia. A further limitation of the hypothesis is that it does not account for the prevalence of oral dyskinesias as a manifestation of the disease (Marsden et al., 1975). This indicates a selective and unexplained vulnerability of neuronal systems controlling oral musculature in the manifestation of the disorder.

It remains possible that neuroleptic-induced proliferation of dopamine receptors may be a necessary but not sufficient condition for tardive dyskinesia (Jenner and Marsden, 1986; Fibiger and Lloyd, 1986: Waddington, 1986). Nevertheless, other reports focus on possible neuroleptic-induced destruction of a subpopulation of GABA-containing neurons in the striatum (*see* Fibiger and Lloyd, 1984 for a review) as a major factor in this group of disorders.

7.2. Central Vs Peripheral Receptor Estimations— The Problem of Assessing Long-Term Changes in Receptor Function in the Clinical Context

As Siever and his colleagues (1984) have pointed out, the assessment of human receptor *function* is, in practice, quite difficult. The study of peripheral receptors is, notwithstanding the usual ethical and practical difficulties of clinical studies, quite straightforward. Peripheral tissues are readily available. Such analyses entail, however, necessary assumptions concerning both (1) the equivalence of brain and peripheral binding sites and (2) the relative equivalence of drug action in brain and periphery.

Functional assessments have been carried out with peripherally active pharmacological challenges. Evidence for central mediation of some peripheral indices of pharmacologic challenges indicates the potential of this approach for understanding central receptor changes in humans. Both prolactin and growth hormone have been widely used to assess the functional state of in vivo dopamine receptors (Sachar, 1978; Meltzer et al., 1978).

Neuroendocrine markers, such as prolactin secretion, have the advantage of potentially revealing the actual status of central receptor function in humans (Sachar, 1978; see Fig. 4). Despite the advantages of relative accessibility and physiological relevance, such procedures suffer from the disadvantage that hormonal responses are almost invariably governed by multiple neural receptor systems. This means, of course, that the inferences concerning receptor function using this approach are less specific. In general, and from the limited data currently available, the effects of drug treatments on measures of human receptor function parallel those of preclinical studies involving nonhuman species.

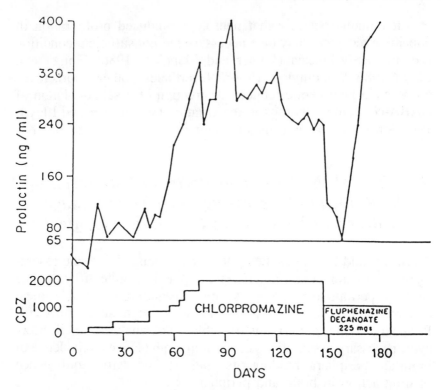

Fig. 4. Serum prolactin levels following treatment with chlorpromazine for approximately 5 mo, followed by fluphenazine decanoate approximately 75 mg/wk. Note the rapid decline and subsequent restoration of prolactin levels in response to the shift from chlorpromazine to fluphenazine treatment. From Meltzer et al. (1978) with permission.

In relation to neuroleptic drug treatment, most human receptor analyses to date relate to D_2 receptor function (Seeman, 1987). Particular emphasis has been placed on prolactin secretion (which is elevated by D_2 receptor antagonists) (Meltzer et al., 1978). Of great interest is the issue of D_1 vs D_2 dopamine receptor activity, as described earlier. As Seeman and Grigoriadis (1987) have pointed out, although there is no correlation between the dissociation constants of neuroleptics at D_1 binding sites and clinical doses, the interaction, or link, between D_1 and D_2 receptors remains of great potential importance. This is particularly evident considering the recent work of Martin-Iverson and his colleagues (Martin-Iverson et al., 1987, 1988). It will be necessary to develop a means of assessing relative D_1/D_2 receptor changes in humans in vivo to elucidate the importance of their interaction in the context of long-term drug administration, as already discussed (Section 7.1.1). This may now be feasible with PET, although it must be recognized that efficient neuroendocrine analysis, if possible, is economically far more feasible.

8. Summary

This chapter has outlined some basic features of dopamine receptor adaptation in the context of psychotherapeutic drug action. The vast majority of data in this area are represented by preclinical studies. For clinical studies, indirect indices of receptor function and binding studies with peripheral tissues predominate. Despite the limited availability of human data, results of clinical studies largely parallel those with nonhuman species.

For neuroleptic drug use, it is apparent that to understand the basis of adverse reactions, such as tardive dyskinesia and supersensitivity psychosis, it is necessary to explore both changes in D_1 and D_2 receptors and to assess the interactions of neuroleptics with nondopamine receptor systems (Rupniak et al., 1987). Our understanding of the functions of D_1 and D_2 binding sites is still in its infancy at the preclinical level. Despite this, it is already evident that the D_1/D_2 link may have important clinical implications. Circadian influences on dopamine function are of particular interest at the present time and should be explored directly in terms of radioligand binding.

In the human context, PET studies offer great advances in our potential understanding of long-term effects of drugs. This approach may be advantageous both in terms of potential accuracy of measuring functional changes and (notwithstanding economic factors) accessibility of brain in vivo. A concluding caveat must be stated relating to the issue of in vivo (and ex vivo: as in autoradiography) imaging vs in vitro binding studies. In vivo studies, by their very nature, have an implicit set of assumptions concerning the measurement of binding under nonequilibrium conditions. This situation contrasts markedly against the relatively stable situation with in vitro assays. These issues have been discussed by

Perlmutter and Raichle (1986), which is recommended to readers who may be unfamiliar with this problem. Receptor analyses offer an extremely powerful approach in our attempt to understand effects of long-term drug administration. Nevertheless, it is an approach that realistically must be viewed conservatively. Indeed, comparisons of in vitro with both in/ex vivo studies and functional analyses are necessary for the development of a clear perspective.

Acknowledgments

The authors wish to acknowledge the ongoing support of the Alberta Heritage Foundation (AJG is a Heritage Medical Research Scholar) the Alberta Mental Health Advisory Council and the Medical Research Council of Canada during the period of preparation of this manuscript. The assistance of D. J. McManus and D. D. Mousseau is greatly appreciated. Ms. H. Stelte provided excellent secretarial support; to her we owe special thanks in this endeavor.

References

Akiyama K., Sato M., Kashihara K., and Otsuki S. (1982a) Lasting changes in high affinity ^{3}H-spiperone binding to the rat striatum and mesolimbic area after chronic methamphetamine administration: evaluation of dopaminergic and serotonergic receptor components. *Biol. Psychiatry* **17**, 1389–1402.

Akiyama K., Sato M., and Otsuki S. (1982b) Increased ^{3}H-spiperone binding sites in mesolimbic area related to methamphetamine-induced behavioral hypersensitivity. *Biol. Psychiatry* **17**, 223–231.

Algeri S., Brunello N., and Vantini G. (1980) Different adaptive responses by rat striatal dopamine synthetic and receptor mechanisms after repeated treatment with *D*-amphetamine, methylphenidate and nomifensine. *Pharmacol Res. Commun.* **12**, 675–681.

Antelman S. M. and Chiodo L. A. (1981) Dopamine autoreceptor subsensitivity: a mechanism common to the treatment of depression and the induction of amphetamine psychosis? *Biol. Psychiat.* **16**, 717.

Bacopoulos N. G. (1981) Biochemical mechanism of tolerance to neuroleptic drugs; regional differences in rat brain. *Eur. J. Pharmacol.* **70,** 585–586.

Bacopoulos N. G., Spokes E. G., Bird E. D., and Roth R. H. (1979) Antipsychotic drug action in schizophrenic patients: Effect on cortical dopamine metabolism after long-term treatment. *Science* **205,** 1405–1407.

Baldessarini R. J. (1985) *Chemotherapy in Psychiatry.* Harvard University Press, Cambridge, Mass.

Beninger R. J. (1983) The role of dopamine in locomotor activity and learning. *Brain Res. Rev.* **6,** 173–196.

Bernardi M. M. and Scavone C. (1985) Differential biochemical and behavioral effects of single and chronic administration of amphetamine and apomorphine. *Gen. Pharmac.* **16,** 407–410.

Bokobza G., Ruberg M., Scatton B., *et al.* (1983) ^{3}H-Spiperone binding, dopamine and HVA concentrations in Parkinson's disease and supranuclear palsy. *Eur. J. Pharmacol.* **99,** 167–175.

Borison R. L., Hitri A., Klawans H. L., and Diamond B. I. (1978) A New Animal Model for Schizophrenia: Behavioral and Receptor Binding Studies, in *Catecholamines: Basic and Clinical Frontiers.* (Usdin E., Kopin I. J., and Barchas J., eds.), Pergamon Press, New York, pp. 719–721.

Boulton A. A., Baker G. B., and Hrdina P. D. (eds.) (1986) *Neuromethods, Vol. 4, Receptor Binding.* Humana Press, Clifton, N. J.

Braun A., Fabbrini G., Mouradian M. M., Serrati C., Barone P., and Chase T. N. (1987) Selective D-1 dopamine receptor agonist treatment of Parkinson's disease. *J. Neural. Transm.* **68,** 41–50.

Breese G. R., Duncan G. E., Napier T. C., Bondy S. C., Iorio L. C., and Mueller R. A. (1987) 6-Hydroxydopamine treatments enhance behavioral responses to intracerebral microinjection of D_1- and D_2-dopamine agonists into nucleus accumbens and striatum without changing dopamine antagonist binding. *J. Pharmacol. Exp. Ther.* **240,** 167–176.

Bunney W. E. and Garland B. L. (1983) Possible receptor effects of chronic lithium administration. *Neuropharmacology* **22B,** 367–372.

Burt D. R., Creese I., and Snyder S. H. (1977) Antischizophrenic drugs: chronic treatment elevates dopamine receptor binding in brain. *Science* **196,** 326–328.

Carenzi A., Gillin J. C., Guidotti A., Schwartz M. A., Trabucchi M., and Wyatt R. J. (1975) Dopamine-sensitive adenyl cyclase in human caudate nucleus. *Arch. Gen. Psychiatry* **32,** 1056–1059.

Carlson J. H., Bergstrom D. A., and Walters J. R. (1987) Stimulation of both D1 and D2 dopamine receptors appears necessary for full expression of postsynaptic effects of dopamine agonists: a neurophysiological study. *Brain Res.* **400,** 205–218.

Cash R., Raisman R., Ploska A., and Agid Y. (1987) Dopamine D-1 receptor and cyclic AMP-dependent phosphorylation in Parkinson's disease. *J. Neurochem.* **49,** 1075–1083.

Chouinard G. and Jones B. D. (1980) Neuroleptic-induced supersensitivity psychosis: clinical and pharmacological characteristics. *Am. J. Psychiatry* **137,** 16–31.

Clark D. and White F. J. (1987) Review: D1 dopamine receptor—the search for a function: a critical evaluation of the D1/D2 dopamine receptor classification and its functional implications. *Synape* **1,** 347–388.

Clow A., Jenner P., Theodorou A., and Marsden C. D. (1979) Striatal dopamine receptors become supersensitive while rats are given trifluoperazine for six months. *Nature* **278,** 59–61.

Clow A., Theodora A., Jenner P., and Marsden C. D. (1980) Cerebral dopamine function in rats following withdrawal from one year of continuous neuroleptic administration. *Eur. J. Pharmacol.* **63,** 145–157.

Connell P. H. (1958) *Amphetamine Psychosis. Chapman and Hill, London.*

Cools A. R. and Van Rossum J. M. (1976) Excitation-mediating and inhibition-mediating dopamine-receptors: A new concept towards a better understanding of electrophysiological, biochemical, pharmacological, functional and clinical data. *Psychopharmacologia (Berlin)* **45,** 243–254.

Crawley J. C. W., Owens D. G. C., Crow T. J., Johnson, A. C., Oldland S. R. D., Owen, F., Poulter M., Smith T., Veall N., and Zanelli G. D. (1986) Dopamine D2 receptors in schizophrenia studies *in vivo. Lancet* **ii,** 224–225.

Creese I. and Chen A. (1985) Selective D-1 dopamine receptor increase following chronic treatment with SCH 23390. *Eur. J. Pharmacol.* **109,** 127–128.

Cross A. J., Crow T. J., and Owen F. (1981) ^{3}H-Flupenthixol binding in post-mortem brains of schizophrenics: evidence for a selective increase in dopamine D2 receptors. *Psychopharmacology* **74**, 122–124.

Dackis C. A., Gold M. S., Sweeney D. R., Byron Jr. J. P., and Climko R. (1987) Single-dose bromocriptine reverses cocaine craving. *Psychiatry Res.* **20**, 261–264.

Daiguji M. and Meltzer H. Y. (1982) Effect of chronic phencyclidine or *D*-amphetamine treatment on [^{3}H]-spiroperidol binding in rat caudate-putamen and nucleus accumbens. *Eur. J. Pharmacol.* **85**, 339–342.

Delay J., Diniker P., and Harl J. -M. (1952) Traitement des etats d'excitation et d'agitation par une methode medicamenteuse derivee de l'hibernotherapie. *Ann. Med. Psychol.* **110(pt.2)**, 267–273.

DeLisi L. E., Crow T. J., and Hirsch S. R. (1986) The third biannual winter workshop on schizophrenia *Arch. Gen. Psychiat.* **43**, 706–711.

DeLisi L. E., Holcomb H. H., Cohen R. M., Pickar D., Carpenter W., Morihisa J. M., King A. C., Kessler R., and Buchsbaum M. S. (1985) Positron emission tomography in schizophrenic patients with and without neuroleptic medication. *J. Cerebr. Blood Flow and Metab.* **5**, 201–206.

Demellweek C. and Goudie A. J. (1983) Behavioural tolerance to amphetamine and other psychomotor stimulants: a case for considering behavioural mechanisms. *Psychopharmacology* **80**, 287–307.

Dewey K. J. and Fibiger H. C. (1983) The effects of dose and duration of chronic posogide administration on dopamine receptor supersensitivity. *Naunyn-Schmied. Arch. Pharmacol.* **322**, 261–270.

Ebstein R. P., Pickholz D., and Belmaker R. H. (1979) Dopamine receptor changes after long-term haloperidol treatment in rats. *J. Pharm. Pharmacol.* **31**, 558.

Ellison G. D. and Eison M. A. (1983) Continuous amphetamine intoxication: an animal model of the acute psychotic episode. *Psychol. Med.* **13**, 751–762.

Fabbrini G., Juncos J., Mouradian M. M., Serrati C., and Chase T. N. (1987) Levodopa pharmacokinetic mechanisms and motor fluctuations in Parkinson's disease. *Ann. Neurol.* **21**, 370–376.

Farde L., Ehrin E., Eriksson L., Greitz T., Hall H., Hedstrom C., Litton J. -E., and Sedvall G. (1985) Substituted benzamides as ligands for visualization of dopamine receptor binding in the human brain by positron emission tomography. *Proc. Natl. Acad. Sci. USA* **82**, 3863–3867.

Farde L., Hall H., Ehrin E., and Sedvall G. (1986) Quantitative analysis of D2 dopamine receptor binding in the living human brain by PET. *Science* **231**, 258–261.

Farde L., Halldin C., Stone-Elander S., and Sedvall G. (1987) PET analysis of human dopamine receptor subtypes using ^{11}C-SCH 23390 and ^{11}C-raclopride. *Psychopharmacology* **92**, 278–284.

Fibiger H. C. and Lloyd K. G. (1986) Reply from Fibiger and Lloyd. *TINS*, **9**, 260.

Fibiger H. C. and Lloyd K. G. (1984) Neurobiological substrates of tardive dyskinesia: the GABA hypothesis. *TINS* **7**, 462–464.

Filloux F. M., Wamsley J. K., and Dawson T. M. (1987) Dopamine D2 auto- and postsynaptic receptors in the nigrostriatal system of the rat brain: localization by quantitative autoradiography with [^{3}H]sulpiride. *Eur. J. Pharmacol.* **138**, 61–68.

Finlay J. M., Jakubovic A., Fu D. S., and Fibiger H. C. (1987) Tolerance to haloperidol-induced increases in dopamine metabolites: fact or artifact? *European J. Pharmacol.* **137**, 117–121.

Fischer J. F. and Cho A. K. (1979) Chemical release of dopamine from striatal homogenates: evidence for an exchange-diffusion model. *J. Pharmacol. Exp. Ther.* **208**, 203–209.

Fleminger S., Rupniak N. M. J., Hall M. D., Jenner P., and Marsden C. D. (1983) Changes in apomorphine-induced sterotype as a result of subacute neuroleptic treatment correlates with increased D-2 receptors, but not with increases in D-1 receptors. Biochem. Pharmacol. **32**, 2921–2927.

Friedman E., Gianutsos G., and Kuster J. (1983) Chronic fluphenazine and closzapine elicit opposite changes in brain muscarinic receptor binding: implications for understanding tardive dyskinesia. *J. Pharmacol. Exp. Ther.* **226**, 7–12.

Gallager D. W., Pert A., and Bunney W. E. (1988) Haloperidol-induced presynaptic dopamine supersensitivity is blocked by chronic lithium. *Nature* **273**, 309–312.

Galloway M. P. and Roth R. H. (1983) Biochemical response of dopamine neurons to atypical neuroleptics is not dependent on autoreceptors. *Soc. For. Neurosci. Abs.* **9**, 1003 (291.18).

Gianutsos G. and Friedman E. (1987) Prevention of fluphenazine-induced changes in dopaminergic and muscarinic receptors by lithium. *Pharmacol. Biochem. Behav.* **26**, 635–637.

Globus M., Bannet J., Lerer B., and Behmaker R. (1982) The effect of chronic bromocriptine and L-Dopa on spiperone binding and apomorphine induced sterotype. *Psychopharmacology* **78**, 81.

Goeders N. E. and Kuhar M. J. (1987) Chronic cocaine administration induces opposite changes in dopamine receptors in the striatum and nucleus accumbens. *Alcohol and Drug Res.* **7,** 207–216.

Goetz C. G. and Klawans H. L. (1982), in *Butterworth's International Medical Reviews. Neurology 2: Movement Disorders.* (Marsden C. D. and Fahn S., eds.), Butterworth Scientific, London, pp. 263–276.

Goldstein M., Lieberman A., and Meller E. (1985) A possible molecular mechanism for the antiparkinson action of bromocriptine in combination with levodopa. *Trends Pharmacol. Sci.* **6,** 436–437.

Grace A. A. and Bunney B. S. (1986) Induction of depolarization block in midbrain dopamine neurons by repeated administration of haloperidol: analysis using *in vivo* intracellular recording. *J. Pharmacol. Exp. Ther.* **238,** 1092–1100.

Greenshaw A. J., Sanger D. J., and Nguyen T. V. (1988) Animal models for assessing antidepressants, neuroleptic and anxiolytic drug action, in *Neuromethods: Analysis of Psychiatric Drugs.* (Boulton A. A., Baker G. B., and Coutts R. T., eds.)

Guttman M., Seeman P., Reynolds G. P., Riederer P., Jellinger K., Tourtellotte W. W. (1986) Dopamine D2 receptor density remains constant in treated Parkinson's disease. *Ann. Neurol.* **19,** 487–492.

Hamon J., Paraire J., and Velluz J. (1952) Remarques sur l'action du 4560 R. P. sur l'agitation maniaque. *Ann. Med. Psychol.* **110,** 331–335.

Heal D. J., Green A. R., Boullin D. J., and Grahame-Smith D. G. (1976) Single and repeated administration of neuroleptic drugs to rats: effects on striatal dopamine-sensitive adenylate cyclase and locomotor activity produced by tranylcypromine and L-tryptophan or L-Dopa. *Psychopharmacology* **49,** 287–300.

Hemmings Jr. H. C., Walaas S. I., Ouimet C. C., and Greengard P. (1987) Dopaminergic regulation of protein phosphorylation in the striatum: DARPP-32. *TINS* **10,** 377–383.

Hitzemann R., Wu J., Hom D., and Loh H. (1980) Brian locations controlling the behavioral effects of chronic amphetamine intoxication. *Psychopharmacology* **72,** 93–101.

Hornykiewicz O. (1973) Parkinson's disease: from brain homogenate to treatment. *Fed. Proc.* **32,** 183–190.

Howlett D. R. and Nahorski S. R. (1979) Acute and chronic amphetamine treatments modulate striatal dopamine receptor binding sites. *Brain Res.* **161,** 173–178.

Howlett D. R. and Nahorski S. R. (1978) Effect of acute and chronic amphetamine administration on β-adrenoceptors and dopamine receptors in rat corpus striatum and limbic forebrain. *Br. J. Pharmacol.* **64,** 411P.

Hyttel J. (1979) Neurochemical parameters in the hyperresponsive phase after a single dose of neuroleptics to mice. *J. Neurochem.* **33,** 229.

Hyttel J. and Christensen A. V. (1983) Biochemical and pharmacological differentiation of neuroleptic effect on dopamine D-1 and D-2 receptors. *J. Neural. Transm.* **18,** 157–164.

Iversen L. L., Iversen S. D., and Snyder S. H. (1983) (eds.) *Handbook of Psychopharmacology, vol. 17, Biochemical Studies on CNS Receptors.* Plenum P, New York.

Jackson D. M. and Hashizume M. (1986) Bromocriptine induces marked locomotor stimulation in dopamine-depleted mice when D1 dopamine receptors are stimulated with SKF 38393. *Psychopharmacol.* **90,** 147–149.

Jackson D. M. and Jenkins O. F. (1985) Hypothesis: bromocriptine lacks intrinsic dopamine receptor stimulating properties. *J. Neural. Transm.* **62,** 219–230.

Jackson D. M., Bailey R. C., Christie M. J., Crisp E. A., and Skerritt J. H. (1981) Long-term d-amphetamine in rats: lack of a change in postsynaptic dopamine receptor sensitivity. *Psychopharmacology* **73,** 276–280.

Jenner P. and Marsden C. D. (1986) Is the dopamine hypothesis of tardive dyskinesia completely wrong? *TINS*, 256.

Johansson P., Levin E., Gunne L., and Ellison G. (1987) Opposite effects of a D1 and a D2 agonist on oral movements in rats. *Eur. J. Pharmacol.* **134,** 83–88.

Kalivas P. W. (1985) Sensitization to repeated enkephalin adminsistration into the ventral tegmental area of the rat. II. Involvement of the mesolimbic dopamine system. *J. Pharmacol. Exp. Ther.* **235,** 544–550.

Kane J. M., Woerner M., Lieberman J. A., and Kinon B. J. (1986) Tardive dyskinesia and drugs. *Drug Dev. Res.* **9,** 41–51.

Kaneno S. and Shimazono Y. (1981) Decreased in vivo [^{3}H]spiroperidol binding in rat brain after repeated methamphetamine administration. *Eur. J. Pharmacol.* **72,** 101–105.

Kibey M. M. and Sannerud C. A. (1985) Models of Tolerance: Do They Predict Sensitization to The Effects of Psychomotor Stimulants, in *Behavioural Pharmacology: The Current Status. Neurology and Neurobiology, 13.* (Seiden L. S. and Balster R. L., eds.), Alan R. Liss, Inc., New York, NY, pp. 295–321.

Klawans H. L. (1973) The pharmacology of extrapyramidal movement disorders, in *Monographs on Neural Science*. (Cohen M. M., ed.), Karger, Basel, pp. 1–137.

Klawans H. L. and Margolin D. I. (1975) Amphetamine-induced dopaminergic hypersensitivity in guinea pigs. *Arch. Gen. Psychiatry* **32**, 725–732.

Klawans H. L., Hitri A., Carvey P. M., Nausieda P. A., and Weiner W. J. (1979) Effect of chronic dopaminergic agonist on striatal membrane dopamine binding. *Adv. Neurol.* **24**, 217–224.

Klimek V. and Nielsen M. (1987) Chronic treatment with antidepressants decreases the number of [^{3}H]SCH 23390 binding sites in the rat striatum and limbic system. *Eur. J. Pharm.* **139**, 163–169.

Kornhuber J., Riederer P., Reynolds G. P., Beckmann H., Jellinger K., and Gabriel E. (1988) ^{3}H-Spiperone binding sites in post-mortem brains from schizophrenic patients. Relationship to neuroleptic drug treatment, abnormal movements and positive symptoms. *J. Neural. Trans.* (in press)

Laduron P. M. (1988) Dopamine receptors and neuroleptic drugs, in *Neuromethods* Vol. 12 (Boulton A. A., Baker G. B., and Juorio A. V., eds.), Humana Press, NJ.

Langer S. Z. (1987) The imipramine binding site in depression. *ISI Atlas of Science: Pharmacology* **1**, 143–146.

Langer S. Z. and Arbilla S. (1984) The amphetamine paradox in dopaminergic neurotransmission. *Trends Pharmacol. Sci.* **9**, 387–390.

Lee T., Seeman P., Tourtellotte W. W., Farley I. J., and Hornykiewicz O. (1978) Binding of ^{3}H-neuroleptics and ^{3}H-apomorphine in schizophrenic brains. *Nature* **274**, 897–900.

Lieberman A., Kupersmith M., Estey E., and Goldstein M. (1976) Treatment of Parkinson's disease with bromocriptine. *N. Engl. J. Med.* **296**, 1400–1404.

Liskowsky D. R. and Potter L. T. (1987) Dopamine D2 receptors in the striatum and frontal cortex following chronic administration of haloperidol. *Neuropharmacol.* **26**, 481–483.

Mackay A. V. P., Bird E. D., Spokes E. G., Rossor M., Iversen L. L., Creese I., and Snyder S. H. (1980) Dopamine receptors and schizophrenia: drug effect or illness? *Lancet* **ii**, 915–916.

Mackenzie R. G. and Zigmond M. J. (1985) Chronic neuroleptic treatment increases D-2 but not D-1 receptors in rat stratum. *Eur. J. Pharmac.* **113**, 159–165.

Mackenzie R. G. and Zigmond M. J. (1984) High- and low-affinity states of striatal D2 receptors are not affected by 6-hydroxydopamine or chornic haloperidol treatment. *J. Neuro-chem.*

Maj. J. Rogz Z., Skuza G., and Sowinska H. (1984) Repeated treatment with antidepressant drugs increases the behavioural response to apomorphine. *J. Neural. Transm.* **60**, 273.

Maj J., Rogoz Z., Skuza G., and Sowinska H. (1985) The effect of repeated treatment with antidepressant drugs on the action of *d*-amphetamine and apomorphine in rats *Neuropharmacology* **85**, 133.

Marsden C. D., Tarsy D., and Baldessarini R. J. (1975) Spontaneous and Drug-Induced Movement Disorders in Psychotic Patients, in *Psychiatric Aspects of Neurologic Disease*. (Benson D. F. and Blumer D., eds.), Grune & Stratton, New York, pp. 219–266.

Martin-Iverson M. T., Stahl S. M., and Iversen S. D. (1988) Chronic administration of a selective dopamine D_2 agonist: Factors determining behavioral tolerance and sensitization *Psychopharmacology*. **95**, 534–539.

Martin-Iverson M. T., Stahl S. M., and Iversen S. D. (1987) Factors determining the behavioural consequences of continuous treatment with 4-propyl-9-hydroxynaphthoxazine, a selective dopamine D_2 agonist, in *Parkinson's Disease: Current Clinical and Experimental Approaches* (Clifford Rose F, ed.), Libby, London.

Marzullo G. and Friedhoff A. J. (1982) Apomorphine-induced pecking in young chicks: diurnal cycles. *Soc. Neurosci. Abstr.* **8**, 394.

Mashurano M. and Waddington J. L. (1986) Stereotyped behavior in response to the selective D2 dopamine receptor agonist RU 24213 is enhanced by pretreatment with the selective D1 agonsit SKF 38393. *Neuropharmacol.* **25**, 947–949.

Matthysse S. (1973) Antipsychotic drug actions: a clue to the neuropathology of schizophrenia? *Fed. Proc.* **32**, 200–205.

Meltzer H. Y. and Stahl S. M. (1976) The dopamine hypothesis of schizophrenia: a review. *Schizophr. Bull.* **2**, 19–76.

Meltzer H. Y., Goode D. J., and Fang V. S. (1978) The effect of psychotropic drugs on endocrine function. I. Neuroleptics, precursors and agonists, in *Psychopharmacology A Generation of Progress*. (Lipton M. A., DiMascio A., and Killam K. F., eds.), Raven, New York.

Memo M., Castelletti L., Maissale C., Valerio A., Carruba M. O., and Spano P. F. (1986) Dopamine inhibition of prolactin release and calcium influx induced by neurotensin in anterior pituitary is independent of cyclic AMP systems. *J. Neurochem.* **47**, 1689.

Memo M., Kleinman J. E., and Hanbauer I. (1983) Coupling of dopamine D1 recognition sites with adenylate cyclase in nuclei accumbens and caudatus of schizophrenics. *Science* **221**, 1304–1307.

Memo M., Pizzi M., Missale C., Carruba M. O., and Spano P. F. (1987a) Modification of the function of D1 and D2 dopamine receptors in striatum and nucleus accumbens of rats chronically treated with haloperidol. *Neuropharmacol.* **26**, 477–480.

Memo M., Pizzi M., Nisoli E., Missale C., Carruba M. O., and Spano P. (1987b) Repeated administration of (-)sulpiride and SCH 23390 differentially up-regulate D-1 and D-2 dopamine receptor function in rat mesostriatal areas but not in cortical-limbic brain regions. *Eur. J. Pharmacol.* **138**, 45–51.

Mishra R. K. (1986) Central Nervous System Dopamine Receptors, in *Neuromethods*, vol. 4. (Boulton A. A., Baker G. B., and Hrdina P. D., eds.), Humana Press, Clifton, N. J.

Molloy A. G. and Waddington J. L. (1987) Assessment of grooming and other behavioural responses to the D-1 dopamine receptor agonist SK & F 38393 and its R- and S-enantiomers in the intact adult rat. *Psychopharmacology* **92**, 164–168.

Muller P. and Seeman P. (1979) Presynaptic subsensitivy as a possible basis for sensitization by long-term dopamine mimetics. *Eur. J. Pharmocal.* **55**, 149–157.

Muller P. and Seeman P. (1978) Dopaminergic supersensitivity after neuroleptics: time course and specificity. *Psychopharmacology (Berlin)* **60**, 1–11.

Murugaiah K., Theodorou A., Mann S., Clow A., Jenner P., and Marsden C. D. (1982) Chronic continuous administration of neuroleptic drugs alters cerebral dopamine receptors and increases spontaneous dopaminergic action in the striatum. *Nature* **296**, 570–572.

Nagatsu T., Kanamor T., Kato T., Iizuka R., and Narabayashi H. (1978) Dopamine-stimulated adenylate cyclase activity in the human brain: changes in Parkinsonism. *Biochem. Med.* **19**, 360–365.

Nakajima S. (1986) Suppression of operant responding in the rat by dopamine D1 receptor blockage with SCh 23390. *Physiol. Psychol.* **14**, 111–114.

Nielsen J. A. (1986) Effects of chronic antidepressant treatment on nigrostriatal and mesolimbic dopamine autoreceptors in the rat. *Neurochem. Int.* **9**, 61–67.

Nutt J. G. (1987) On-off phenomenon: relation to levodopa

pharmacokinetics and pharmacodynamics. *Ann. Neurol.* **22,** 535–540.

Olianas M. C. and Onali P. (1987) Supersensitivity of striatal D2 dopamine receptors mediating inhibition of adenylate cyclase and stimulation of guanosine triphosphatase following chronic administration of haloperidol in mice. *Neurosci. Lett.* **78,** 349–354.

Olsen R., Reisine T., and Yamamura H. (1980) Neurotransmitter receptors—biochemistry and alterations in neuropsychiatric disorders. *Life Sci.* **27,** 801–808.

Owen F., Baker H. F., Ridley R. M., Cross A. J., and Crow T. J. (1981) Effect of chronic amphetamine administration on central dopaminergic mechanisms in the vervet. *Psychopharmacology* **74,** 213–216.

Owen F., Cross A. J., Crow T. J., Longden A., Poulter M., and Riley G. J. (1978) Increased dopamine-receptor sensitivity in schizophrenia. *Lancet* **ii,** 223–226.

Owen F., Cross A. J., Waddington J. L., Poulter M., Gamble S. J., and Corw T. J. (1980) Dopamine-mediated behaviour and ^{3}H-spiperone binding to striatal membranes in rats after nine months haloperidol administration. *Life Sci.* **26,** 55–59.

Parkes J. D., Marsden C. D., Donaldson I., Galea-DeBono A., Walters J., Kennedy G., and Asselman P. (1976) Bromocriptine treatment in Parkinson's disease. *J. Neurol. Neurosurg. Psychiatry* **39,** 184–193.

Perlmutter J. S. and Raichle M. E. (1986) *In vitro* or *in vivo* receptor binding: where does the truth lie? *Ann. Neurol.* **19,** 384–385.

Pert A., Rosenblatt J. E., Sivit C., Pert C. B., and Bunney W. E. (1978) Long-term treatment with lithium prevents the development of dopamine receptor supersensitivity. *Science* **201,** 171–173.

Pickar D., Lavarca R., Doran A. R., Wolkowitz O. M., Roy A., Breier A., Linnoila M., and Paul S. M. (1986) Longitudinal measurement of plasma homovanillic acid levels in schizophrenic patients. *Arch Gen. Psychiatry* **43,** 669–676.

Pimoule C., Schoemaker H., Reynolds G. P., and Langer S. Z. (1985) [^{3}H]SCH 23390 labeled D1 dopamine receptors are unchanged in schizophrenia and Parkinson's disease. *Eur. J. Pharmacol.* **114,** 235–237.

Pittman K. J., Jakubovic A., and Fibiger H. C. (1984) The effects of chronic lithium on behavioral and biochemical indices of dopamine receptor supersensitivity in the rat. *Psychopharmacology* **82,** 371–377.

Pizzi M., D-Agostini F., Da Prada M., Spano P. F., and Haefely W. E.

(1987) Dopamine D2 receptor stimulation decreases the inosital trisphosphate level of rat striatal slices. *Eur. J. Pharmacol.* **136,** 263–264.

Pizzolato G., Soncrant T. T., and Rapoport S. I. (1985) Time-course and regional distribution of the metabolic effects of bromocriptine in rat brain. *Brain Res.* **341,** 303–312.

Porceddu M. L., Ongini E., and Biggio G. (1985) [³H]SCH 23390 binding sites increase after chronic blockade of D-1 dopamine receptors. *Eur. J. Pharmacol.* **118,** 367–370.

Pycock C. J., Dawburn D., and O'Shaughnessy C. (1982) Behavioural and biochemical changes following chronic administration of L-dopa to rats. *Eur. J. Pharmacol.* **79,** 201.

Quik M. and Iversen L. L. (1978) Subsensitivity of the rat striatal dopaminergic system after treatment with bromocriptine: effects on [³H]spiperone binding and dopamine-stimulated cyclic AMP formation. *Naunyn-Schmiedeberg's Arch. Pharmacol.* **304,** 141–145.

Quik M., Spokes E., MacKay A., and Bannister R. (1979) Alterations in ³H-spiperone binding in human caudate nucleus, substantia nigra and frontal cortex in the Shy-Drager syndrome and Parkinson's disease. *J. Neurol. Sci.* **43,** 429–437.

Raisman R., Cash R., Ruberg M., Javoy-Agid F., and Agid Y. (1985) Binding of ³H-SCH 23390 to D1 receptors in the putamen of control and parkinsonian subjects. *Eur. J. Pharmacol.* **113,** 467–468.

Reches A., Wagner H. R., Jackson-Lewis V., Yablonskaya-Alter E., and Fahn S. (1984) Chronic levodopa or pergolide administration induces down-regulation of dopamine receptors in denervated striatum. *Neurology* **34,** 1208–1212.

Reisine T., Fields J., and Yamamura H. (1977) Neurotransmitter receptor alterations in Parkinson's disease. *Life Sci.* **21,** 335–344.

Reynolds G. P., Czudek C., Bzowej N., and Seeman P. (1987) Dopamine receptor asymmetry in schizophrenia. *Lancet* **i,** 979.

Reynolds G. P., Dowey L., Rossor M. N., and Iversen L. L. (1982) Thioridazine is not specific for limbic dopamine receptors. *Lancet* **2,** 499.

Richelson E. and Nelson A. (1984) Antagonism by neuroleptics of neurotransmitter receptors of normal human brain in vitro. *Eur J. Pharmacol.* **103,** 197–204.

Riederer P. and Jellinger K. (1983) Morphologie und Pathobiochemie der Parkinson-Krankeit, in *Pathophysiologie, Klinik und Therapie des Parkinsonismus.* (Ganshirt H., ed.), Basel, Roche, pp. 49–60.

Riederer P. and Jellinger K. (1982) Rezeptorfunktion bei Parkinson-Krankeit, Morbus Alzheimer, seniler Demenz und Schizophrenie, in

Psychopathologie des Parkinson-Syndroms. (Fischer P. A., ed.), Basel, Roche, pp. 71–81.

Riederer P., Rausch W. D., Birkmayer W., Jellinger K., and Danielczyk W. (1978) Dopamine-sensitive adenylate cyclase activity in the caudate nucleus and adrenal medulla in Parkinson's disease and in liver cirrhosis. *J. Neural Transm. (Suppl.)* **14,** 153–161.

Riffee W. H., Wilcox R. E., Vaugh D. M., and Smith R. V. (1982) Dopamine receptor sensitivity after chronic dopamine agonists. *Psychopharmacology* **77,** 146–149.

Rinne J. O., Rinne J. K., Laakso K., Lonnberg P., and Rinne U. K. (1985) Dopamine D-1 receptors in the parkinsonian brain. *Brain Res.* **359,** 306–310.

Rinne U. (1982) Brain Neurotransmitter Receptors in Parkinson's Disease, in *Movement Disorders.* (Marsden C. and Fahn S., eds.), Butterworth, Stoneham, MA, pp. 59–074.

Rinne U. (1981) Treatment of Parkinson's disease: problems withna progressing disease. *J. Neural Transm.* **51,** 161–174.

Rinne U., Koskinen V., and Lonnberg P. (1980) Neurotransmitter Receptors in the Parkinsonian Brian, in *Parkinson's Disease—Current Progress, Problems and Management.* (Rinne U., Klingler M., and Stamm G., eds.), Elsevier/North-Holland Biomedical Press, New York, pp. 93–107.

Rinne U., Lonnberg P., and Koskinen V. (1981) Dopamine receptors in the parkinsonian brain. *J. Neural Transm.* **51,** 97–106.

Rinne U., Rinne J. O., Rinne J. K., *et al.* (1983) Brain receptors changes in Parkinson's disease in relation to the disease process and treatment. *J. Neural Transm. (Suppl.)* **18,** 279–286.

Rinne U., Sonninen V., Laaksonen H. (1979) Responses of brain neurochemistry to levodopa treatment in Parkinson's disease. *Adv. Neurol.* **24,** 259–274.

Robertson H. A. (1983) Chronic *D*-amphetamine and phencyclidine: effects on dopamine agonist and antagonist binding sites in the extrapyramidal and mesolimbic systems. *Brain Res.* **267,** 179–182.

Robertson H. A. (1982) Chronic phencyclidine, like amphetamine, produces a decrease in [^{3}]spiroperidol binding in rat striatum. *Eur. J. Pharmacol.* **78,** 363–365.

Robertson H. A. (1979) Effect of chronic *D*-amphetamine or β-phenylethylamine on dopamine binding in rat striatum and limbic system. *Soc. Neurosci. Abstr.* **5,** 570.

Robertson H. A. and Robertson G. S. (1987) Combined L-DOPA and bromocriptine therapy for Parkinson's disease: a proposed mechanism of action. *Clin. Neuropharmacol.* **10**, 384–387.

Robertson H. A. and Robertson G. S. (1986) The antiparkinson action of bromocriptine in combination with levodopa. *Trends Pharmacol. Sci.* **7**, 224–225.

Robinson T. and Becker B. (1986) Enduring changes in brain and behavior produced by chronic amphetamine administration: a review and evaluation of animal models of amphetamine psychosis. *Brain Res. Rev.* **11**, 157–198.

Roth R. H. (1984) CNS dopamine autoreceptors: distribution, pharmacology, and function. *Ann. N. Y. Acad. Sci.* **430**, 27–53.

Rupniak N. M. J., Briggs R. S., Petersen M. M., Mann S., Reavill C., Jenner P., and Marsden C. D. (1986a) Differential alterations in striatal acetylcholine function in rats during 12 months' continuous administration of haloperidol, sulpiride, or clozapine. *Clin. Neuropharmac.* **9**, 282–292.

Rupniak N. M. J., Hall M. D., Kelly E., Fleminger S., Kilpatrick G., Jenner P., and Marsden C. D. (1985) Mesolimbic dopamine function is not altered during continuous chronic treatment of rats with typical or atypical neuroleptic drugs. *J. Neural. Transm.* **62**, 249–266.

Rupniak N. M. J., Jenner P., and Marsden C. D. (1986b) Acute dystonia induced by neuroleptic drugs. *Psychopharmacology* **88**, 403–419.

Rupniak N. M. J., Kilpatrick G., Hall M. D., Jenner P., and Marsden C. D. (1984) Differential alterations in striatal dopamine receptor sensitivity induced by repeated administration of clinically equivalent doses of haloperidol, sulpiride or clozapine in rats. *Psychopharmacology* **84**, 512–519.

Rupniak N. M. J., Prestwich S. A., Horton R. W., Jenner P., and Marsden C. D. (1987) Alterations in cerebral glutamic acid decarboxylase and ^{3}H-flunitrazepam binding during continuous treatment of rats for up to 1 year with haloperidol, sulpiride or clozapine. *J. Neural. Transm.* **68**, 113–125.

Sachar E. J. (1978) Neuroendocrine responses to psychotropic drugs, in, *Psychopharmacology: A Generation of Progress.* (Lipton M. A., DiMascio A., and Killam K. F., eds.), Raven, New York.

Sanger D. J. (1987) The actions of SCH 23390, a D1 receptor antagonist, on operant and avoidance behavior in rats. *Pharmacol. Biochem. Behav.* **26**, 509–513.

Savasta M., Dubois A., Feuerstein C., Manier M., and Scatton B. (1987) Denervation supersensitivity of striatal D2 dopamine receptors is restricted to the ventro-and dorsolateral regions of the striatum. *Neurosci. Lett.* **74,** 180–186.

Schmidt C. J., Ritter J. K., Sonsalla P. K., Hanson G. R., and Gibb, D. W. (1985) Role of dopamine in the neurotoxic effects of methamphetamine. *J. Pharmacol. Exp. Ther.* **233,** 539–544.

Sedvall G., Farde L., Persson A., and Wiesel F. -A. (1986a) Imaging of neurotransmitter receptors in the living human brain. *Arch. Gen. Psychiatry* **43,** 995–1005.

Sedvall G., Farde L., Stone-Elander S., and Halldin C. (1986b) Dopamine D1-Receptor Binding in the Living Human Brain, in *Neurobiology of Central D-1-Dopamine Receptors.* (Breese G. R., and Creese I., eds.), Plenum, New York, pp. 119–124.

Sedvall G., Ehrin E., and Farde L., (1986c) Steroselective binding of ^{11}C-labelled piquindone (Ro 22-1319) to dopamine-D2 receptors in the living human brain. *Human Psychopharmacol* **2,** 23–30.

Seeman P. (1987) Dopamine receptors and the dopamine hypothesis of schizophrenia. *Synapse* **1,** 133–152.

Seeman P. (1980) Brain dopamine receptors. *Pharmacol. Rev.* **32,** 229–313.

Seeman P. and Grigoriadis D. (1987) Dopamine receptors in brain and periphery. *Neurochem. Int.* **10,** 1–25.

Seeman P. and Ulpian C. (1983) Neuroleptics have identical potencies in human brain limbic and putamen regions. *Eur. J. Pharmacol.* **15,** 145–148.

Seeman P., Ulpian C., Bergeron C., Riederer P., Jellinger K., Gabriel E., Reynolds G. P., and Tourtellotte W. W. (1984a) Bimodal distribution of dopamine receptor densities in brains of schizophrenics. *Science* **225,** 728–731.

Seemann D., Danielczyk W., and Ogris E. (1984) Dopaminergic Agonists—Effects on Multiple Receptor Sites in Parkinson's Disease, in *Recent Research in Neurology.* (Callaghan N. and Galvin R., eds.), Pitman, London, pp. 49–60.

Seiden L. S. and Vosmer G. (1984) Formation of 6-hydroxydopamine in caudate nucleus of the rat brain after a single large dose of methylamphetamine. *Pharmacol. Biochem. Behav.* **21,** 29–31.

Serra G., Srgiolas A., Klimek V., Fadda F., and Gessa G. L. (1979) Chronic treatment with antidepressants prevents the inhibitory effect of small doses of apomorphine on dopamine synthesis and motor activity. *Life Sci.* **25,** 415.

Shibuya M. (1979) Dopamine-sensitive adenylate cyclase activity in the striatum in Parkinson's disease. *J. Neural Transm.* **44,** 287–295.

Siever L. J., Uhde T. W., and Murphy D. L. (1984) Strategies for assessment of noradrenergic receptor function in patients with affective disorders, in *Neurobiology of Mood Disorders*. (Post R. M. and Ballenger J. L., eds.), Williams and Wilkins, Baltimore.

Simmond S. H. and Strange P. G. (1985) Inhibition of inositol phospholipid breakdown by D2 dopamine receptors in dissociated bovine anterior pituitary cells. *Neurosci. Lett.* **60,** 267.

Skirboll L. R. and Bunney B. S. (1979) The effects of acute and chronic haloperidol treatment on spontaneously firing neurons in the caudate nucleus of the rat. *Life Sci.* **25,** 1419–1434.

Spyraki C. and Fibiger H. C. (1981) Behavioural evidence for supersensitivity of postsynaptic dopamine receptors in the mesolimbic system after chronic administration of desipramine. *European J. Pharmacol.* **74,** 195.

Staunton D. A., Magistretti P. J., Koob G. F., Shoemaker W. J., and Bloom F. E. (1982) Dopaminergic supersensitivity induced by denervation and chronic receptor blockade is additive. *Nature* **299,** 72–74.

Staunton D. A., Wolfe B. B., Groves P. M., and Molinoff P. B. (1981) Dopamine receptor changes following destruction of the nigrostriatal pathway: lack of a relationship to rotational behaviour. *Brain Res.* **211,** 315.

Taylor D. L., Ho D. T., and Fagin J. D. (1979) Increased dopamine receptor binding in rat brain by repeated cocaine injection. *Commun. Psychopharmacol.* **3,** 137–142.

Timmermans P. B. M. W. M. and Thooley M. J. M. C. (1987) Autoreceptors in the central nervous system. *Med. Res. Rev.* **7,** 307–332.

Traub M., Wagner H. R., Hassan M., Jackson-Lewis V., and Fahn S. (1985) The effects of chronic bromocriptine treatment on behaviour and dopamine receptor binding in the rat striatum. *Eur. J. Pharmacol.* **118,** 147–154.

Trulson M. E. and Ulissey M. J. (1987) Chronic cocaine administration decreases dopamine synthesis rate and increases [^{3}H]spiroperidol binding in rat brain. *Brain Res. Bull.* **19,** 35–38.

Vaccheri A., Dall'Olio R., Gandolfi O., Roncada P., and Montanaro N. (1987) Enhanced stereotyped response to apomorphine after chronic D-1 blockade with SCH 23390. *Psychopharmacology* **91,** 394–396.

Van Rossum J. M. (1967) The Significance of Dopamine-Receptor Blockade for the Action of Neuroleptic Drugs, in *Proceedings of the 5th Collegium Internationale Neuropsychopharmacologicum*. (Brill

H., Cole J. O., Deniker P., Hippius H., and Bradley P. B., eds), pp. 321–329.

vonVoigtlander P. F., Losey E. G., and Triezenberg H. J. (1975) Increased sensitivity to dopaminergic agents after chronic neuroleptic treatment. *J. Pharmacol. Exp. Ther.* **193**, 88–94.

Waddington J. L. (1986a) Biomocriptine, selective D2 dopamine receptor agonists and D1 dopaminergic tone. *Trends Pharmacol. Sci.*, 223–224.

Waddington J. L. (1986b) Reply from Waddington. *TINS*, **9**, 260.

Wagner H. N., Burns D., Dannals F. R., Wong D. F., Langstrom B., Duelfer T., Frost J. J., Ravert H. T., Links J. M., Rosenbloom S. H., Lukas S. E., Kramer A. V., and Kuhar M. J. (1983) Imaging dopamine receptors in the human brain by positron emission tomography. *Science* **221**, 1264–1266.

Walters J. R., Bergstrom D. A., Carlson J. H., Chase T. N., and Braun A. R. (1987) D1 dopamine receptor activation required for postsynaptic expression of D2 agonist effects. *Science* **236**, 719–722.

White F. J., and Wang R. Y. (1984) Electrophysiological evidence for A10 dopamine autoreceptor subsensitivity following chronic *d*-amphetamine treatment. *Brain Res.* **309**, 283–292.

Wong D. F., Gjedde A., Wagner Jr. H. N., Dannals R. F., Douglass K. H., Links J. M., and Kuhar M. J. (1986a) Quantification of neuroreceptors in the living human brain. II. Inhibition studies of receptor density and affinity. *J. Cereb. Blood Flow Metab.* **6**, 147–153.

Wong D. F., Wagner Jr. H. N., Dannals R. F., Links J. M., Frost J. J., Ravert H. T., Wilson A. A., Rosenbaum A. E., Gjedde A., Douglass K. H., Petronis J. D., Folstein M. F., Toung J. K. T., Burns H. D., and Kuhar M. J. (1984) Effects of age on dopamine and serotonin receptors measured by positron tomography in the living human brain. *Science* **226**, 1393–1396.

Wong D. F., Wagner Jr. H. N., Tune L. E., Dannals R. F., Pearlson G. D., Links J. M., Tamminga C. A., Broussolle E. P., Ravert H. T., Wilson A. A., Toung J. K. T., Malat J., Williams J. A., O'Tuama L. A., Snyder S. H., Kuhar M. J., and Gjedde A. (1986b) Positron emission tomography reveals elevated D2 dopamine receptors in drug-naive schizophrenics. *Science* **234**, 1558–1563.

Woolverton W. L. (1986) Effects of a D1 and a D2 dopamine antagonist on the self-administration of cocaine and piribedil by rhesus monkeys. *Pharmacol. Biochem. Behav.* **24**, 531–535.

Yahr M. (1984) Limitation on Long-term use of anti-Parkinson drugs. *Can. J. Neurol. Sci.* **11**, 191–194.

Sensitization
to the Actions of
Antidepressant Drugs

Paul Willner

1. Why Are Antidepressants Slow to Act?

The slow onset of action of antidepressants is legendary: all clinicians know that no improvement will be apparent for at least two weeks, and that progress will be gradual thereafter. Clearly, some slow process of sensitization is at work. However, as with all legends, the truth of this one should be approached with caution: the mere fact of a slow onset does not, in itself, prove that sensitization is occurring.

For the purposes of this review, sensitization is taken to mean that, for a given dose of drug, the size of the response increases over time where all other relevant factors remain constant. One relevant factor that clearly does not remain constant over time is the plasma drug concentration, which builds up gradually with repeated dosing, eventually reaching a steady state. This consideration leads to an important distinction between "true" and "pseudo-sensitization" (Fig. 1). Let us suppose that, on a first administration, a low drug dose, D_L, gives rise to a low tissue con-

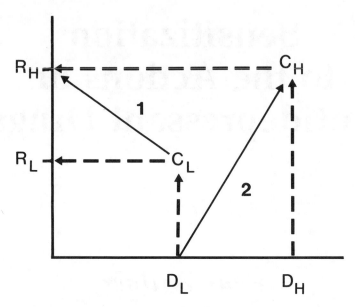

Fig. 1. Hypothetical relationships between low (L) or high (H) drug dose (D), tissue concentration (C) and response (R). Dotted lines show effects seen on the first drug administration: $D_L - C_L - R_L$ and $D_H - C_H - R_H$. Solid lines show effects of a low dose potentiated by repeated treatment: Case 1 (true sensitization) $D_L - C_L - R_H$; Case 2 (dispositional or pseudo-sensitization) $D_L - C_H - R_H$.

centration at the site of action, C_L, resulting in a low response, R_L, whereas a higher dose, D_H, gives rise to a higher tissue concentration, C_H, and a higher response, R_H. Now suppose that, on a second administration, we observe potentiation of the response to the lower drug dose ($D_L - R_H$). If (Case 1) the tissue concentration is the same as before, then true sensitization has occurred ($D_L - C_L - R_H$). However (Case 2), it is also possible that the low dose on this occasion gives rise to a high tissue concentration, with no alteration in the relationship between tissue concentration and response ($D_L - C_H - R_H$). This situation can arise in one of two ways. It may be that physiological alterations caused by the first drug administration are responsible: for example, the suppression of drug metabolism or an increased permeability of the blood–brain barrier. By analogy to the terminology of drug tolerance, such effects might be called "dispositional sensitization" (*see* chapter by Lê and Khanna, this volume). However, a low drug dose might also cause high tissue concentrations in the absence of physiological changes of any kind. All that is required for this "pseudo-sensitization" is that, at the time of the second administration, the first dose has not yet been fully cleared from the body. It is clearly of some importance to distinguish between true sensitization and an artifactual increase in drug effect that simply reflects the kinetics of the approach to steady-state drug tissue levels.

In the case of tricyclic antidepressants, the achievement of steady-state plasma levels takes anything between one and four weeks of treatment (Peet and Coppen, 1979). The significance of these changes in effective dosage for the clinical action of tricyclics is still unclear. As the approach to steady state is exponential, the greatest increases in plasma concentration take place within the first week, and it seems unlikely that the small increases thereafter contribute greatly to the gradual clinical improvement, particularly in view of the very wide spread of plasma levels between individuals taking the same dose (Peet and Coppen, 1979). However, the buildup of dose within the first few days of treatment almost certainly does contribute to the delayed onset of clinical effects (Frazer et al., 1985; Pollock et al., 1985).

In addition to changes in drug plasma concentrations, a second factor that does not remain constant over time is clinical state. In most clinical studies, improvement is roughly linear over some four to six weeks. Six weeks of treatment has a far more beneficial effect than two weeks, but can this be considered to be sensitization, when the improvement week by week is roughly constant? The only satisfactory way of answering this question would be a clinical trial of antidepressants in mild depression, in which hitherto untreated patients were compared to patients recovering under drug therapy from a severe depression. In view of the extreme difficulty of trying to match these two groups for clinical symptomatology, it is most unlikely that such a trial will ever be carried out. However, since antidepressants tend to be relatively ineffective in mild depressions, we might anticipate that, if it were, there might be a better outcome in the partially treated group, which would demonstrate that sensitization had occurred.

The nature of depression provides, potentially, a third approach to explaining the slow action of antidepressants without invoking the notion of sensitization. Depression is not simply a mood disorder, but a wholesale reorganization of thoughts and feelings about the world and the depressed person's place in it (Beck,

1967; Willner, 1985b). These habitual modes of thought may persist after the removal of the circumstances in which they arose. Furthermore, depressed people behave in ways that tend to exacerbate their condition: they lose jobs, drive away friends, and place a heavy strain on close relationships (Lewinsohn, 1974). These secondary effects of depression frequently serve as self-fulfilling prophecies that fully justify being depressed. In cases where this kind of social disintegration has occurred, drug treatment does not provide the necessary reconstruction. The weight of the task of reconstituting an acceptable social environment may be sufficiently heavy, following a potentially successful course of treatment, to mask an underlying improvement.

It is impossible, at present, to assess the significance of these psychological and social factors in the gradual recovery from depression over weeks of treatment. What is clear, however, is that they are relatively unimportant for understanding the delayed onset of drug action, because the delayed onset of action of antidepressants is something of a myth. One source of the belief that antidepressants are slow to act is the gradual buildup towards adequate plasma blood levels, which is caused in part by the kinetics of the steady state, and in part by the common clinical practice of beginning treatment at a low dose, and increasing the dose over the course of one or two weeks. However, some antidepressants—for example, amoxapine (Fujimori, 1981) or salbutamol (Lecrubier et al., 1980)—can usually be relied on to produce improvement within the first week; even with tricyclics, improvement can be detected almost from the very beginning of treatment, provided that the drug is administered at a sufficiently high dose (Frazer et al., 1985; Pollock et al., 1985). A second factor is that many patients show an early temporary placebo response, which tends to mask a real and lasting improvement in drug-treated patients. A third problem is that many antidepressants, tricyclics in particular, have a number of unpleasant side effects that contribute adversely to global measures of psychological well-being. Finally, the design of many trials simply fails to include a sensitive early assessment (Frazer et al., 1985; Haskell et al., 1975). Nevertheless, in appropriately conducted studies, although there are considerable differences between individual patients, the overall picture is one of gradual improvement from the outset of antidepressant therapy (e.g., Bech et al., 1984).

2. Slow Changes in Monoamine Receptor Function

Although it is possible that the gradual recovery under antidepressant treatment might be explicable without there being any true pharmacological sensitization, it is known that, on chronic administration, antidepressants do, in fact, alter the properties of a number of monoamine receptor systems. This section will describe these effects, and assess whether they have characteristics appropriate to explain a clinical action that has a gradual but rapid onset, and increases in size over a period of several weeks.

2.1. Beta-Receptor "Downregulation"

The best-known and best-established effect of chronic antidepressant treatment is a decrease in the functional responsiveness of beta-adrenergic receptors. Vetulani and Sulser were the first to demonstrate that a variety of antidepressants, including tricyclics, monoamine oxidase inhibitors (MAOI), and electroconvulsive shock (ECS), all decreased the ability of noradrenaline (NA) to stimulate the synthesis of cyclic-AMP (cAMP) in brain slices (Vetulani and Sulser, 1975; Vetulani et al., 1976). This initial observation has been extensively confirmed and extended to a wide range of atypical antidepressants (Sulser et al., 1983). Subsequently, it was shown in radioligand binding studies that antidepressants also decrease the number of beta-receptors (Banerjee et al., 1977). This effect, also, has been shown to occur in response to a wide range of chemically diverse antidepressants (*see* Willner, 1985b), but not in response to a variety of nonantidepressants (Sellinger-Barnette et al., 1980; Wise and Halliday, 1985).

It is generally assumed that the "downregulation" of beta-receptors is a form of adaptive regulation, secondary to a persistent increase in receptor stimulation brought about by the blockade of NA reuptake or, in the case of some atypical antidepressants (e.g., mianserin), an increase in NA release (*see* Willner, 1985b). If overstimulation of beta receptors is prevented, by blocking the receptors with propranolol (Okada et al., 1982; Wolfe et al., 1978) or by destroying NA terminals with the neurotoxin 6-hydroxydopamine (6-OHDA) (Janowsky et al., 1982; Wolfe et al., 1978), then "downregulation" does not occur. Less obviously, the reduction of beta-receptor binding by antidepressants also appears to require active 5-HT neurons: antidepressants fail to reduce beta-receptor binding after inhibition of 5-HT synthesis (Brunello et al., 1985a, b; Manier et al., 1985) or neurotoxic lesions of 5-HT terminals (Brunello et al., 1982; Nimgaonkar et al., 1985; Sulser et al., 1983). However, it is important to be aware that these effects of 5-HT are only observed in receptor-binding studies. Abolishing 5-HT transmission does not appear to alter the reduction of beta-receptor function (cAMP production) by antidepressants (Manier et al., 1985; Gandolfi et al., 1984).

Effects of antidepressants on beta-receptor number and function increase in size over the course of two to four weeks of treatment. Changes may be apparent within the first two days (Frazer et al., 1985), but it is unusual for there to be any detectable change within the first 4–5 days of treatment. The similarity between this time course and the time course of the clinical action of antidepressants has led many workers to consider beta-receptor "downregulation" to be a prime candidate for their mechanism of action (Sulser, 1978). However, there are a number of compelling arguments against this hypothesis.

First, although the majority of antidepressant drugs do clearly "downregulate" beta-receptors, a growing number apparently do not. Antidepressants reported not to alter the properties of beta-receptors include 5-HT uptake inhibitors (alaproclate, citalopram, fluoxetine, and zimelidine), catecholamine uptake inhibitors (maprotiline, nomifensine, and buproprion), mianserin, and trimipramine (Ask et al., 1986; Barbaccia et al., 1986; Ferris and Beaman, 1983; Garcha et al., 1985; Hall et al., 1984; Hauser et al., 1985; Hytell et al., 1984). These drugs may not be totally ineffective: some studies have reported that zimelidine, mianserin, and buproprion did in fact reduce beta-receptor binding, after prolonged administration of high doses (Hall et al., 1984; Gandolfi et al., 1983). Similarly, decreases in NA-stimulated cAMP production have been observed with mianserin, maprotiline, and buproprion, after administration at high doses or using particularly sensitive methods (Mishra et al., 1980; Gandolfi et al., 1983, 1984; Schoffelmeer et al., 1984). What is clear, however, is that the generality of beta-receptor "downregulation" effects is not nearly so easy to demonstrate as it may have appeared from the early work. Negative results have been reported even with the archetypal "downregulator," imipramine (Antkiewicz-Michaluk, 1985), and the effect of one of the most frequently prescribed antidepressants, amitriptyline, is notoriously weak (Wise and Halliday, 1985).

A second problem lies in the fact that "downregulation" is primarily a response to the overstimulation of beta receptors by elevated intrasynaptic concentrations of NA (Wolfe et al., 1978; Delina-Stula et al., 1982). The hypothesis that antidepressants act by "downregulating" beta receptors, therefore, has the unfortunate corollary that the acute and chronic effects of antidepressants should be in opposite directions. However, there is little clinical evidence to suggest that antidepressants initially make patients worse, before making them better (except to the extent that some antidepressants have unpleasant side effects). Similarly, in animal models of depression, the changes that take place during chronic treatment are generally in the size of the response, rather than its direction (*see* Section 3, below)

Even at beta-adrenergic synapses, there is no good evidence that the direction of the response reverses with chronic treatment. The important variable is not simply the responsiveness of beta receptors, but rather, transmission across the synapse, which is a joint function of receptor responsiveness and intrasynaptic transmitter concentration. The net outcome of antidepressant effects on these two variables is difficult to measure, but a few studies have attempted to do so. Two studies have measured postsynaptic responses to electrical stimulation of the locus ceruleus. One of these measured the cAMP response, which was unaffected by chronic treatment with three of the six drugs tested, slightly reduced by two drugs at high doses, and only clearly reduced by one drug, desmethylimipramine (DMI) (Korf et al., 1979). The second study measured electrophysiological responses in hippocampal cells, which were unaffected by chronic DMI treatment (Huang, 1979). The response of hippocampal cells to iontophoretic application of NA was also unaffected by chronic treatment with a variety of antidepressants (De Montigny and Aghajanian, 1978; De Montigny et al., 1981; Gallager and Bunney, 1979) [though a decrease in the response to iontophoretically applied NA has been observed in cingulate cortex (Olpe and Schellenberg, 1980)].

A number of studies have used the pineal gland as a model system: in this structure, a beta-adrenergic input stimulates the synthesis of cAMP, leading to secretion of the hormone melatonin.

Most pineal studies have examined effects of DMI, which is usually found either to increase beta-mediated functions after chronic treatment or to have no effect (Weiss B. et al., 1982; Cowen et al., 1983). There have been two studies of plasma melatonin levels in human subjects, during three weeks of treatment with DMI: one found a persistent elevation in depressed patients (Thompson et al., 1983); the other found an initial elevation, followed by a return to predrug levels, in normal volunteers (Cowen et al., 1985). All in all, the best estimate from these and other studies (*see* Willner, 1985b) is that beta-receptor "downregulation" is a homeostatic response that serves only to restore transmission across beta-adrenergic synapses towards the level obtained prior to the commencement of antidepressant treatment.

A fourth problem concerns the status of beta receptors in depressed patients. Central beta-receptor function in depression has not been studied. There is some evidence that peripheral (lymphocytic) beta receptors might actually be "downregulated" in depressed patients prior to treatment (Extein et al., 1979; Pandey et al., 1979), though problems in interpreting studies of peripheral NA function are considerable (*see* Willner, 1985b). What is clear, however, is that many depressed patients have an extremely high level of pituitary-adrenal activity (Carroll, 1982), and that both cortisol and ACTH "downregulate" beta receptors (Mobley and Sulser, 1980; Kendall et al., 1982). In animals treated with a combination of ACTH and imipramine, the reduction in beta-receptor function was actually no greater than in animals treated with ACTH alone (Kendall et al., 1982). This result casts serious doubt on whether antidepressants do "downregulate" beta receptors in the brain of a depressed person (as opposed to a normal rat).

Finally, there is evidence that runs directly counter to the hypothesis. If antidepressants work by "downregulating" beta receptors, then beta-blockers should have antidepressant properties and beta-receptor agonists should induce depression. Exactly the reverse is the case. There is no evidence that beta-blockers have antidepressant activity, but depression is frequently reported as a side effect of their use in the control of hypertension (Petrie et al., 1982). Conversely, salbutamol, an agonist at both subtypes of beta receptor (Leclerc et al., 1981), is a rapidly acting antidepressant (Lecrubier et al, 1980). In keeping with these observations, a comparison of the potency of various antidepressants shows that the more potently a drug "downregulates" beta receptors, the less potent it is clinically (Willner, 1984a, 1985b).

There seems to be no escaping the conclusion that, if the "downregulation" of beta receptors is relevant to the clinical action of antidepressants, it is more likely to be as an unwanted side effect than as their mechanism of action.

2.2. Desensitization of Alpha$_2$ Receptors

The effects of antidepressants on alpha$_2$ adrenoceptors have also attracted considerable attention. Alpha$_2$ receptors are usually studied using clonidine, which at high doses, stimulates both alpha$_2$ and alpha$_1$ receptors, but at low doses appears to be a selective agonist at the alpha$_2$ receptor (Starke et al., 1977). Autoreceptors on NA cell bodies and axon terminals are of the alpha$_2$ type. Consequently, stimulation of alpha$_2$ receptors inhibits the firing of NA cells in the locus ceruleus, and reduces the release of NA in terminal regions. This has a number of physiological and behavioral consequences that can be used to measure alpha$_2$-receptor activation, including hypothermia (Von Voightlander et al., 1978) and sedation (Delina-Stula et al., 1979), though there is evidence (Nassif et al., 1983) that clonidine-induced sedation may also involve postsynaptic alpha$_2$ receptors, which in the cerebral cortex are in the majority.

The effects of antidepressants on alpha$_2$ receptor function are somewhat variable. In general, the tricyclics that potently inhibit NA uptake and strongly stimulate alpha$_2$ receptors (imipramine, DMI) are almost always found to desensitize alpha$_2$ receptors after chronic treatment (*see* Willner, 1985b, Ch. 13). Some other antidepressants, amitriptyline and mianserin in particular, in addition to enhancing the intrasynaptic concentration of NA, are also potent alpha-receptor antagonists, which tends to neutralize their NA-enhancing effect (Brown et al., 1980). Chronic administration of amitriptyline is often found not to decrease alpha$_2$ function, and an increase in alpha$_2$ function is sometimes observed with mianserin (e.g., Sugrue, 1982). However, the effects of amitriptyline and mianserin appear to be dose-dependent: decreases in alpha$_2$ function have been reported following low doses of both drugs (Heal et al., 1983; Kostowski and Malatynska, 1983; Von

Voightlander et al., 1978). Reduced alpha$_2$-receptor function has also been observed following repeated administration of a number of other atypical antidepressants, including iprindole, nomifensine, opipramol, trimipramine, and viloxazine (Von Voightlander et al., 1978), though the very specific 5-HT uptake inhibitor citalopram (Hytell et al., 1984) does not appear to desensitize alpha$_2$ receptors (Plaznik and Kostowski, 1985). There may be some regional specificity in these effects, as various tricyclic and atypical antidepressants failed to desensitize alpha$_2$ autoreceptors on cortical NA terminals, at doses that did "downregulate" beta receptors (Schoffelmeer et al., 1984). However, the effects are not confined to cell body autoreceptors, since DMI, at least, has been found to desensitize alpha$_2$ receptors in the hippocampus (Plaznik et al., 1984; Plaznik and Kostowski, 1985).

Unlike beta-receptor "downregulation," desensitization of alpha$_2$ receptors has been clearly demonstrated in depressed patients undergoing treatment with antidepressants. In these studies, the status of central alpha$_2$ receptors was assessed by the ability of clonidine to reduce blood pressure or plasma MHPG (the major central metabolite of NA), and to induce sedation. Subsensitivity of alpha$_2$ receptors has been observed following two to four weeks of treatment with the tricyclics amitriptyline and DMI, or the MAOI clorgyline (Charney et al., 1983; Glass et al., 1982; Mavroidis et al., 1984; Siever et al., 1981). However, as in the majority of animal studies, mianserin was ineffective in depressed patients (Charney and Heninger, 1983).

The relatively clear picture of a functional desensitization of alpha$_2$ receptors by the majority of antidepressants is not found in the receptor binding literature. To take the most obvious example, imipramine and DMI, which are almost universally found to desensitize alpha$_2$ receptors in functional studies, have variously been found to decrease, to increase, and not to alter alpha$_2$-receptor binding (see Willner, 1985b, Ch. 13). One important reason for this discrepancy may be the postsynaptic location of the majority of the receptors sampled in binding studies (Morris et al., 1981). In addition to these studies of brain alpha$_2$ receptors, alpha$_2$-receptor binding has also been measured in blood platelets taken from depressed patients. An early study reported a decrease during antidepressant treatment (Garcia-Sevilla et al., 1981), but a number of more recent reports have failed to confirm this effect (e.g., Pimoule et al., 1983). An effect in blood platelets is, in any case, of doubtful relevance. In general, studies of alpha$_2$-receptor binding have added little of substance to the functional literature.

What is the functional significance of alpha$_2$-receptor desensitization? Tricyclic antidepressants, as well as many atypical antidepressants, are potent NA uptake inhibitors. However, the stimulation of alpha$_2$ autoreceptors causes a reduction of firing in NA cells, which is proportional to, and commensurate with, the blockade of NA uptake (Quinaux et al., 1982). Nonetheless, electrophysiological recording from cells postsynaptic to NA terminals shows that, in general, the net outcome is a clear enhancement of NA transmission (Menkes and Aghajanian, 1981; Menkes et al., 1980; Sangdee and Frantz, 1979). Tertiary tricylics, such as amitriptyline, combine a relatively weak uptake blockage with relatively potent antagonism of postsynaptic alpha$_1$ receptors. Even so, amitriptyline also enhances NA transmission at low doses, but reduces transmission at higher doses (Menkes and Aghajanian, 1981; Menkes et al., 1980; Sangdee and Frantz, 1979); it is at low doses that amitriptyline desensitizes alpha$_2$ receptors (Kostowski and Malatynska, 1983; Von Voightlander et al., 1978). Following chronic antidepressant treatment, there is no change in the blockade of NA uptake by tricyclics (Schildkraut, 1975). Desensitization of alpha$_2$ receptors, therefore, has the effect of amplifying the acute increase in NA transmission, by mitigating the decrease in presynaptic NA activity.

Nevertheless, there is still some question as to its potential clinical relevance, since antidepressants appear to desensitize alpha$_2$ receptors far more rapidly than they affect other receptor systems. Although most studies have tended to use two or three weeks as the period of chronic drug treatment, the two studies that demonstrated the most general effects on alpha$_2$ receptors (effective drugs included amitriptyline and mianserin) used only four days of treatment (Kostowski and Malatynska, 1983; Von Voightlander et al., 1978). Indeed, in the case of DMI, desensitization of alpha$_2$ receptors in the hippocampus has been demonstrated after only two days of treatment (Plaznik et al., 1984; Plaznik and Kostowski, 1985). These results suggest that any beneficial effects of alpha$_2$ -receptor desensitization might come into play very early in treatment, perhaps even during the period in which a rising plasma drug concentration is the major factor responsible for changes in drug effect. This, in turn, implies that alpha$_2$-receptor desensitization contributes little to the slow improvement over several weeks of antidepressant treatment.

2.3. Sensitization of Alpha$_1$ Receptors

In contrast to the "downregulation" of beta receptors, it is usually observed that chronic antidepressant administration causes an increase in the responsiveness of postsynaptic alpha$_1$ receptors. Also, unlike the alpha$_2$ receptor, the change in alpha$_1$-receptor function is usually seen only after chronic treatment for at least 10 d.

These effects have been observed in both behavioral and electrophysiological studies. Increases in electrophysiological responsiveness to alpha$_1$-receptor stimulation have been observed in a variety of subcortical brain areas: facial motor nucleus (Menkes et al., 1980), lateral geniculate nucleus (Menkes and Aghajanian, 1981), and amygdala (Wang and Aghajanian, 1980). Paradigms used in behavioral studies include locomotor activation by a high dose of clonidine (Maj et al., 1979a), clonidine-induced aggression (Maj et al., 1982), spinal flexor reflex activity (Maj et al., 1983), and the potentiation of the acoustic startle response by intrathecal application of the specific alpha$_1$-receptor agonist phenylephrine (Menkes et al., 1983b). Sensitization has been observed in all of these paradigms, following chronic treatment with a range of typical and atypical antidepressants. The list of effective agents includes amitriptyline and mianserin, which on acute administration are alpha$_1$-receptor antagonists. One study failed to observe potentiation by antidepressants of the locomotor stimulant effect of intraventricular phenylephrine (Heal, 1984). However, others have observed that chronic treatment with DMI did potentiate the locomotor stimulant effect when phenylephrine was administered to the hippocampus (Plaznik and Kostowski, 1985; Plaznik et al., 1984).

Although the sensitization of alpha$_1$ receptors is a property shared by a wide range of antidepressants, results with specific 5-HT uptake inhibitors have been somewhat variable. Fluoxetine was ineffective in electrophysiological studies (Menkes et al., 1980; Wang and Aghajanian, 1980) and did not enhance clonidine-induced aggression (Maj et al., 1982). On the other hand, apomorphine-induced aggression, which also appears to by mediated by alpha$_1$ receptors (Maj et al., 1979b), was enhanced by chronic zimelidine (Maj et al., 1979b). Citalopram was ineffective in the clonidine-induced aggression test (Maj et al., 1982), but did enhance the locomotor stimulant effect of phenylephrine administration to the hippocampus (Plaznik and Kostowski, 1985). However, this effect of phenylephrine was also present on subacute (2-d) treatment (Plaznik and Kostowski, 1985). In a clinical study, alaproclate has been found to increase the pupillary response to phenylephrine, and again, this effect was present after a brief period (1 wk) of treatment (Thompson and Checkley, 1985). Thus, although in some cases 5-HT uptake inhibitors may enhance responses to alpha$_1$-receptor stimulation, a time-dependent sensitization does not appear to occur with these drugs.

In general, chronic antidepressant treatment has not usually been reported to increase alpha$_1$-receptor binding (Peroutka and Snyder, 1980; *see* Willner, 1985b, Ch.13, for review), though some studies have succeeded in observing increased binding of prazosin, the most specific alpha$_1$-receptor ligand currently available (Campbell and McKernan, 1982; Maj et al., 1983; Vetulani and Antikiewicz-Michaluk, 1985). However, there is evidence that the negative results reported in most studies might reflect an inappropriate use of antagonist ligands: in two studies, antidepressants have been found to increase the affinity of alpha$_1$ receptors for the receptor agonist phenylephrine (Menkes et al., 1983a; Hong et al., 1986).

From these results, it seems that a change in the binding characteristics of alpha$_1$ receptors might explain their increased functional responsiveness. However, it should be noted that the effects of antidepressants on alpha$_1$ receptors cannot easily be explained by the principle of homeostasis (unlike the effects on alpha$_2$ and beta receptors). Indeed, the mechanism by which antidepressants sensitize alpha$_1$ receptors is at present something of a mystery. The only clue lies in the observation that a transient blockade of alpha$_1$ receptors may be a more general property of antidepressants than is usually realized; in which case, it might, after all, be possible to explain alpha$_1$-receptor sensitization as a homeostatic response (Vetulani and Antkiewicz-Michaluk, 1985).

2.4. Sensitization of Dopamine Receptors

In addition to increasing transmission through alpha$_1$ receptors, antidepressants appear to have a similar effect on transmission through DA synapses, particularly in the mesolimbic system. Two mechanisms have been described that potentially contribute to effects of this kind, a desensitization of presynaptic inhibitory autoreceptors and a sensitization of postsynaptic responses.

The desensitization of DA autoreceptors by antidepressants, first described by Serra and colleagues in 1979 (Serra et al., 1979), has subsequently proved extremely elusive, with roughly equal numbers of positive (e.g., Chiodo and Antelman, 1980) and negative (e.g. MacNiell and Gower, 1982) reports (*see* Willner, 1983, 1985b, Ch.10, for review). The problem here seems to be that desensitization of cell body autoreceptors may indeed be reliably observed, but only following withdrawal from chronic antidepressant treatment (Scavone et al., 1986; Towell et al., 1986). As DA autoreceptor desensitization is only apparent after the termination of antidepressant administration, it seems most unlikely that this effect contributes significantly to the clinical and behavioral actions of antidepressants.

Postsynaptic responsiveness of the DA systems has been studied using the DA receptor agonist apomorphine and the indirectly acting agonist amphetamine; it is assumed that the locomotor stimulant effect of moderate doses of these compounds is mediated by the mesolimbic DA system, and the stereotyped behavior seen at higher doses is mediated by the nigro-striatal system (Moore and Kelly, 1978). These responses are unaffected by acute antidepressant treatment, but responses to mesolimbic stimulation are enhanced by chronic treatment.

The enhancement of mesolimbic DA function is clearest in the case of ECS, which has been the subject of numerous studies (e.g., Modigh, 1979; Green and Deakin, 1980; *see* also Willner, 1983, 1985b, Ch.10). Studies of antidepressant drugs report similar effects, though they vary somewhat in their account of which drugs are effective and which are not. With one exception (Arnt et al., 1984), it has usually been found that mesolimbic function was enhanced by tricyclic (imipramine, DMI) and atypical (mianserin, iprindole) antidepressants (Maj et al., 1984a,b; Martin-Iverson et al., 1983; Serra et al., 1979; Spyraki and Fibiger, 1981; Willner and Montgomery, 1981). Imipramine has also been found to enhance the locomotor stimulant effect of apomorphine administration to the hippocampus (Smialowski and Maj, 1985). The effect of amitriptyline appears to increase with the length of the withdrawal period (Maj et al., 1984a, b): this time dependence may explain why amitriptyline clearly enhanced mesolimbic function with a 3-d withdrawal period (Martin-Iverson et al., 1983), but not at 2 or 24 h (Arnt et al., 1984); nevertheless, clear effects of amitriptyline were apparent within 2 h of the final drug administration in some studies (Maj et al., 1984a, b). Negative results have been reported with the 5-HT uptake inhibitors chlorimipramine (Spyraki et al., 1985), zimelidine and fluoxetine (Martin-Iverson et al., 1983), using a 3–4 day withdrawal period. However, 5-HT uptake inhibitors were clearly effective in another study, which used shorter withdrawal periods of 2 and 24 h (Arnt et al., 1984). It has also been reported that following chronic (14 d) but not acute treatment, a variety of antidepressants, including the very selective 5-HT uptake inhibitor citalopram, enhanced the locomotor hyperactivity elicited by quinpirole, a selective agonist at the D_2 subtype of DA receptor (Skuza et al., 1988).

In contrast to the relatively clear picture of increased functional responsiveness in the mesolimbic DA system following chronic antidepressant treatment, studies in which stimulant-induced stereotyped behavior have been used to assess nigro-striatal DA function are more equivocal. Nigro-striatal function is clearly potentiated by ECS (e.g., Modigh, 1979; *see* also Willner, 1983, 1985b, Ch. 10) or by MAOIs (Campbell et al., 1985). With some exceptions, however (e.g., Willner et al., 1984), enhancement of nigro-striatal function is not usually found with tricyclic antidepressants (Delina-Stula and Vassout, 1979; Maj et al., 1979a; Spyraki and Fibiger, 1981).

Studies of DA-receptor binding are of little help in understanding the mechanisms underlying the effects of antidepressants on DA function. Numerous studies of DA-receptor binding in the striatum have failed to observe any alteration after chronic antidepressant treatment (*see* Willner, 1983, 1985b); these include studies in which functional responsiveness in the nigro-striatal system was clearly enhanced (e.g. Campbell et al., 1985). A similar constancy of DA-receptor binding in the nucleus accumbens has also been reported (Martin-Iverson et al., 1983), which again is in clear conflict with the increased functional responsiveness. However, after chronic treatment, both imipramine and mianserin have been reported to increase the affinity of the agonist quinpirole at D_2 receptor sites in the nucleus accumbens, but not in the striatum (Klimek, 1988).

In one study, antidepressants that have negligible affinity for muscarinic receptors did not potentiate DA-dependent behavior, but chronic administration of an anticholinergic drug did do so, suggesting that the sensitization of DA receptors may depend upon the anticholinergic properties of antidepressants (Martin-Iverson et al., 1983). However, the drugs that were ineffective in this study did potentiate DA-dependent behaviors in other studies (Arnt et al., 1984; Maj et al., 1984a, b); consequently, this promising hypothesis must be abandoned. Destruction of central NA neurons has been reported to abolish the facilitation of DA function by ECS (Green and Deakin, 1980). However, NA depletion did not prevent the facilitation of DA function by DMI (Martin-Iverson et al., 1983), so although the effect of ECS may be mediated indirectly through a primary action on NA neurons, the effect of DMI appears to be independent of NA. As in the case of alpha$_1$ receptors, the mechanisms by which antidepressant drugs sensitize DA receptors remain obscure.

2.5. Sensitization of 5-HT Receptors?

Unlike the relatively clear consensus over the effects of antidepressants, chronically administered, on NA and DA function, their effects on 5-HT are more controversial. Broadly speaking, electrophysiological studies show clear evidence that antidepressants enhance 5-HT function; behavioral studies show a similar effect, though much less clearly, and receptor binding studies are thoroughly confusing.

In electrophysiological studies, antidepressants appear to enhance 5-HT function by two distinct mechanisms. One group of drugs, which includes the specific 5-HT uptake inhibitors (fluoxetine, indalpine, zimelidine, and citalopram) and the atypical antidepressant trazodone are found to desensitize 5-HT autoreceptors, causing a gradual increase, over the course of around three weeks, in the firing rate of 5-HT neurons, and in the release of 5-HT (Blier and De Montigny, 1983; Blier et al., 1984b; Dowdall and De Montigny, 1984; Chaput et al., 1984). Tricyclic antidepressants are not usually found to desensitize 5-HT autoreceptors (Blier and De Montigny, 1980; Svensson, 1980), though a recent study did report such an effect (Goodwin et al., 1985). However, tricyclics and some atypical antidepressants, including iprindole, mianserin, and the benzodiazepine derivative adinazolam, are found to increase the sensitivity of postsynaptic 5-HT receptors, as measured by the response to iontophoretic application of 5-HT, or to the 5-HT agonist 5-MeODMT (Blier et al., 1984b; Gallager and Bunney, 1979; De Montigny and Aghajanian, 1978; Menkes and Aghajanian, 1981; Menkes et al., 1980; Turmel and De Montigny, 1984; Wang and Aghajanian, 1980). As a result of these diverse actions, drugs in both classes have been found to increase the postsynaptic response to electrical stimulation of 5-HT pathways (Blier and De Montigny, 1983; Wang and Aghajanian, 1980). A similar enhancement of the response to stimulation of the afferent 5-HT pathway has also been observed with the MAOI clorgyline, even though this drug actually decreased the postsynaptic response to iontophoretically applied 5-HT (De Montigny and Blier, 1984; Blier et al, 1984a).

Most studies of 5-HT-receptor binding recognize at least two subtypes of 5-HT binding site, the $5\text{-}HT_1$ site labeled by 5-HT, and the $5\text{-}HT_2$ site labeled by spiroperidol and ketanserin (Leysen, 1984). Although some studies have indicated that antidepressants may induce complex changes in binding to $5\text{-}HT_1$ receptors (Fuxe et al., 1983), the overwhelming impression from numerous other studies is that antidepressants do not alter $5\text{-}HT_1$-receptor binding (e.g., Lucki and Frazer, 1982: *see* Willner, 1985a,b, Ch.17, for review). Consistent with this conclusion are the observations that the effects on locomotor activity of the specific $5\text{-}HT_1$-receptor agonists RU-24949 and TFMPP were not altered by chronic administration of DMI and amitriptyline, respectively (Green et al., 1984; Frazer et al., 1985). The MAOIs may behave differently, since a decrease in $5\text{-}HT_1$-receptor binding and function has usually been observed with these drugs (Lucki and Frazer, 1982; Frazer et al., 1985). However, these results should be treated with caution: in electrophysiological studies, MAOIs decreased postsynaptic sensitivity to iontophoretic application of 5-HT, but nevertheless, increased the response to stimulation of the afferent 5-HT pathway (Blier et al., 1984a).

Unlike the 5-HT_1 receptor, the 5-HT_2 receptor does respond to chronic antidepressant treatment. There is one report of the tricyclic DMI increasing 5-HT_2-receptor binding (Green et al., 1983), but this result is exceptional. Other studies are almost unanimous in reporting that virtually all antidepressant drugs decrease binding to 5-HT_2 receptors (e.g., Kellar et al., 1981: *see* Willner, 1985a;b, Ch.17 for review). However, the mechanism of this effect is obscure, and its significance even more so. Citalopram, the most potent and selective antagonist of 5-HT uptake, is found not to reduce 5-HT_2-receptor binding, indicating that the effect, when present, is probably not caused by persistent exposure to high intrasynaptic concentrations of 5-HT (Hytell et al., 1984). More problematic by far is that ECS has the opposite effect: ECS increases 5-HT_2-receptor binding (Green et al., 1983; Kellar et al., 1981; Vetulani et al., 1981). As the effects of ECS on 5-HT function are broadly similar to those of antidepressant drugs (*see* Willner, 1985a,b), changes in 5-HT_2-receptor binding seem to be rather unreliable.

In studies of behavioral responses to 5-HT_2-receptor stimulation (head twitches and the "5-HT syndrome"), ECS clearly and reliably enhances 5-HT function, in keeping with its effects on 5-HT_2 receptors (e.g., Green et al., 1983: see also Willner, 1985a;b, Ch.17). However, the effects of antidepressant drugs on 5-HT_2-receptor binding and function cannot be so easily reconciled. The 5-HT uptake inhibitor citalopram, for example, decreased the headtwitch response to 5-HT-receptor agonists, after a dose regime that had no effect on binding to 5-HT_2 receptors (Arnt et al., 1984); a similar decrease in functional sensitivity has been reported with other 5-HT uptake inhibitors (Fuxe et al., 1981; Stolz et al., 1983). On the other hand, tricyclics and other atypical antidepressants are usually found to increase functional measures, after dosage regimes that decrease 5-HT_2-receptor binding (*see* Willner, 1985a,b). These effects of tricyclics appear to be time dependent: increased 5-HT function is seen after a prolonged period of withdrawal (48 h), but decreases are more common after brief withdrawal periods (2 h) (e.g., Friedman and Dallob, 1979; Stolz et al., 1983). After withdrawal periods of 12–24 h, which arguably are of the greatest clinical relevance, the majority of well-controlled studies seem to show an increase in 5-HT-receptor function, but at present, there is no clear consensus on this issue (*see* Willner, 1985a,b for a fuller discussion).

It is, in any case, difficult to draw definitive conclusions from studies of postsynaptic receptor function, given that antidepressants also cause presynaptic changes in the activity of 5-HT neurons, and increase intrasynaptic concentrations of 5-HT. As yet, there have been no behavioral studies that successfully assess the effect of antidepressants on transmission across 5-HT synapses, as opposed to the sensitivity of 5-HT receptors. The best guess at present, based primarily on the electrophysiological evidence, is that the net outcome of chronic antidepressant treatment is to increase 5-HT transmission. Further studies are clearly needed in this area.

3. Sensitization to the Behavioral Effects of Antidepressants

The effects so far described represent not so much sensitization to antidepressants, as sensitization by antidepressants of responses to monoamine receptor agonists. For many years, it was only by interacting antidepressants with other types of drug that interesting behavioral changes could be elicited. In most behavioral tests, antidepressants have nonspecific sedative actions (Tucker and File, 1986), with the exception of buproprion, which is a stimulant (Nielsen et al., 1986). On chronic administration, one study has reported the development of tolerance to the sedative effect (Cuomo et al., 1983), but more usually, sedation remains constant with chronic treatment (Tucker and File, 1986; Vogel et al., 1986). There are scattered reports in the literature of chronic antidepressant treatment increasing intracranial self-stimulation rates (Aulakh et al., 1983; Fibiger and Phillips, 1981), and increasing aggressive behavior (*see* Vogel et al., 1986); and the stimulant effect of buproprion increases over time (Nielsen et al., 1986). In general, however, the behavioral effects of antidepressant drugs, applied acutely or chronically, have appeared unremarkable.

This situation has altered dramatically with the development in recent years of animal models of depression, a range of abnormal behaviors in which the behavior is "normalized" by antidepressants (Willner, 1984b, 1985b). In animal models of depression, antidepressants can usually (though not always) be shown to "work" on acute administration, but their effects are clearly potentiated by subchronic or chronic administration. This section will examine the physiological basis of some of these behavioral abnormalities and of their reversal by antidepressants, and the extent to which changes in monoamine receptor function can explain the sensitizing effect of chronic antidepressant treatment.

3.1. The "Behavioral Despair" Model

The Porsolt swimming test (Porsolt et al., 1977), also known as the "behavioral despair" model, has been by far the most popular paradigm for studies attempting to elucidate physiological mechanisms. In this test, rats or mice are forced to swim in a confined space; after an initial frenzied attempt to escape, the animal eventually adopts an immobile posture. Antidepressants are found to delay the onset of immobility; this behavioral activation is in marked contrast to their more usual sedative effect, as observed, for example, in the open field. The original experiments using this procedure employed a subacute dosage regime, in which antidepressants were administered, at relatively high doses, on three occasions in the 24 hours preceding the behavioral test (Porsolt et al., 1977, 1979). It was subsequently shown that a small antidepressant effect may be present after a single treatment at a high dose (Kitada et al., 1981; Zebrowska-Lupina, 1980), but large effects are observed after chronic treatment (typically two weeks) at lower doses (Araki et al., 1985; Kitada et al., 1981; Kostowski, 1985; Nomura et al., 1984).

Three important studies have demonstrated that potentiation of the antidepressant response by repeated treatment does actually involve true sensitization. In the original report of potentiation by chronic treatment, it was shown that chronic administration of low doses produced larger "antidepressant" effects than a single high dose, but achieved lower brain concentrations. The most impressive effect was seen with DMI, for which the suppression of immobility was four times greater, after a chronic dosage regime known to result in less than half the brain concentration (Kitada et al., 1981). These results have recently been confirmed by a study in which a single large dose of DMI did not significantly suppress immobility; chronic treatment at half the acute dose did suppress immobility, and achieved substantially lower brain levels of DMI (Mancinelli et al., 1987). Similar findings have also been reported for a subacute dosage regime, in which DMI was administered five times at intervals corresponding to the half-life of the drug (5 h in rats and 70 min in mice). The ''antidepressant'' effect of the subacute schedule was substantially greater, in both species, than a single injection of twice the dose; brain concentrations were the same in the two cases (Poncelet et al., 1986).

A number of different neurotransmitters have been implicated in the physiological control of immobility in the ''behavioral despair'' model. Immobility is prevented by anticholinergic drugs (Browne, 1979), but potentiated in animals with cholinergic receptor supersensitivity (Overstreet et al., 1986), suggesting that cholinergic mechanisms promote immobility. DA mechanisms have the opposite effect. The onset of immobility is delayed by DA receptor stimulants (apomorphine and bromocriptine) or uptake inhibitors (mazindol and nomifensine) (Porsolt et al., 1979), by electrical stimulation of DA cell bodies in the ventral tegmental area, or by infusions of DA or apomorphine into the nucleus accumbens (Plaznik et al., 1985a,b). Conversely, immobility is potentiated by DA receptor antagonists (Porsolt et al., 1979; Kitada et al., 1983). Certain atypical neuroleptics have been found to reduce immobility (Browne, 1979; Gorka and Janus, 1985). However, these effects were only observed at low doses, and probably reflect a preferential blockade of DA autoreceptors (Costall et al., 1980), and consequent increase in DA function.

Immobility may also be prevented by treatments that increase NA function. Such treatments include $alpha_2$-receptor antagonists (Porsolt et al., 1979; Zebrowska-Lupina, 1980), NA uptake inhibitors (Porsolt, 1981), electrical stimulation of the locus ceruleus, and administration of NA to the hippocampus or nucleus accumbens (Plaznik and Kostowski, 1985; Plaznik et al., 1985a,b). The selective $alpha_1$-receptor agonist phenylephrine also reversed immobility, when administered systemically (Porsolt et al., 1979), or into the cerebral ventricles (Kitada et al., 1983), hippocampus (Plaznik and Kostowski, 1985), or nucleus accumbens (Plaznik et al., 1985a,b). Conversely, immobility is potentiated by two treatments that reduce the activity of NA neurons; $alpha_2$-receptor agonists (Parale and Kulkarni, 1986) and neonatal 6-OHDA treatment (Willner and Thompson, unpublished).

It is therefore possible that the activating effect of antidepressants in the "behavioral despair" test could be mediated by their potentiating effects on DA and/or NA function. The evidence supports an involvement of both of these systems. Early studies of the effects of DA-receptor antagonists were somewhat equivocal (Borsini et al., 1981; Kitada et al., 1983). More recently, however, a series of studies by Samanin and colleagues found that low doses of DA receptor antagonists, which had no effect when administered alone, blocked the effects of subacute treatment with DMI (Pulvirenti and Samanin, 1986) or of chronic treatment (7 or 21 d) with a range of tricyclic (DMI and amitriptyline) or atypical (mianserin, iprindole, nomifensine, and amineptine) antidepressants (Berettera et al., 1986; Borsini et al., 1984, 1985a,b; Pulvirenti and Samanin, 1986). A claim that the effects of DMI were antagonized by the selective D_2-receptor antagonist sulpiride, but not by the nonspecific D_1/D_2 antagonist haloperidol (Borsini et al., 1984), was not substantiated in a later study, in which both of these drugs were found to be effective (Pulvirenti and Samanin, 1986; *see* Fig. 2). The mesolimbic DA projection is implicated in these effects by the observation that sulpiride blocked the activation of behavior by chronic DMI or imipramine when applied directly to the nucleus accumbens (Cervo and Samanin, 1987). In one study that appears to argue against an involvement of DA, the antiimmobility effect of DMI was found to be unaffected by 6-OHDA lesions of the mesolimbic DA projection (Plaznik et al., 1985a). However, rather than ruling out a role for DA, this study may simply confirm that antidepressants potentiate postsynaptic responses to DA, but do not alter the activity of presynaptic DA neurons (*see* Section 2.4).

By contrast, antidepressants enhance NA function by both presynaptic (Section 2.2) and postsynaptic (Section 2.3) mechanisms. In keeping with this observation, it is found that the antiimmobility action of antidepressants can be antagonized either by presynaptic reduction of NA activity or by postsynaptic antagonism of alpha$_1$ receptors. One presynaptic manipulation that has been used is the destruction of NA neurons, by administration of the neurotoxin 6-OHDA either to the locus ceruleus (Plaznik et al., 1985a; Kostowski, 1985) or to the medial amygdala (Araki et

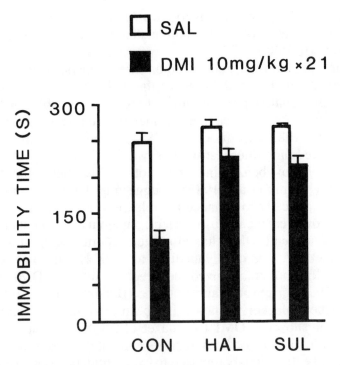

Fig. 2. Reversal by haloperidol (0.5 mg/kg) or sulpiride (100 mg/kg) of the activating effect of DMI (10 mg/kg for 21 d: black bars) in the "behavioral despair" test. (Drawn from data of Pulvirenti and Samanin, 1986).

al., 1985): in both cases, the effects of both acute and chronic treatment with DMI were abolished. Stimulation of $alpha_2$ receptors is a second method used to reduce NA activity: low doses of clonidine have been found to antagonize the action of a variety of antidepressants (though the size of this effect varies somewhat between studies) (Zebrowska-Lupina, 1980; Parale and Kulkarni, 1986; Poncelet et al., 1986). Similarly, blockade of postsynaptic $alpha_1$ receptors by the antagonist phenoxybenzamine has also been found to reduce the effect of subacute antidepressant treatment (Kitada et al., 1983; Borsini et al, 1981). In studies by the Italian group, the more specific $alpha_1$ antagonist prazosin failed to alter the antiimmobility effect of chronic treatment with DMI, amitriptyline, or iprindole (Berettera et al., 1986; Borsini et al., 1981, 1985a; Pulvirenti and Samanin, 1986). However, it is likely that the explanation of this discrepancy is simply that the dose of prazosin used (3 mg/kg) was too low: a higher dose (5 mg/kg) did block the effects of DMI, administered acutely or chronically (Kostowski, 1985).

Investigations of beta receptors are so inconsistent that their involvement in this model seems rather unlikely. The beta-receptor agonist isoprenaline has been found to reverse immobility when administered to the hippocampus or nucleus accumbens (Plaznik and Kostowski, 1985; Plaznik et al., 1985a,b); the therapeutic effect of NA injected into these structures was prevented by the alpha-blocker phenoxybenzamine, but not by the beta-blocker propranolol (Plaznik et al., 1985b), suggesting that isoprenaline may act presynaptically to enhance the release of NA (Misu and Kubo, 1983). However, the opposite effect, antagonism of the therapeutic effect of subacute DMI, has also been reported with isoprenaline (Kitada et al., 1983; Miyauchi et al., 1984). In some studies, propranolol, too, antagonized the effects of subacute DMI (Borsini et al., 1981) or chronic amitriptyline (Borsini et al., 1985a), but another beta-blocker, atenolol. has been reported to potentiate the effects of subacute DMI (Kitada et al., 1983; Miyauchi et al., 1984). In other words, both agonists and antagonists at beta receptors have been variously reported to exert antidepressant effects and also to antagonize the effects of antidepressants. In other studies, propranolol did not alter the effects of chronic iprindole (Berettera et al., 1986) or DMI (Pulvirenti and Samanin, 1986). The beta-receptor agonist salbutamol, an effective antidepressant (Lecrubier et al., 1980), does not demonstrate antidepressant activity in this model (Porsolt et al., 1979).

There are similar inconsistencies in the literature concerning the possible involvement of 5-HT in "behavioral despair." Both agonists and antagonists of 5-HT have sometimes been found to prevent immobility (e.g., Porsolt et al., 1979), and there are also reports of 5-HT-mediated increases in immobility (e.g., Gorka et al., 1979). More usually, however, systemic administration of drugs acting predominantly on 5-HT systems has been found not to influence immobility (Porsolt et al., 1979: see also Willner, 1985b Ch.13). Similarly, no effects were observed when 5-HT agonists were administered to the nucleus accumbens (Plaznik et al., 1985a,b). The effects of subacute treatment with DMI were unaffected by 5-HT agonists or antagonists (Borsini et al., 1981; Kitada et al., 1983), and neurotoxic lesions of 5-HT neurons did not alter the therapeutic effects of imipramine or DMI, administered acutely or chronically (Araki et al., 1985) in one study, and subacutely in another (Poncelet et al., 1986). On balance, it seems that manipulations of 5-HT have little effect on behavior in the "behavioral despair" model.

Finally, mention should be made of the endogenous opiate system. The opiate antagonist naloxone has been found to block the effect of acute tricylic treatment in the behavioral despair test, though not of MAOIs and atypical antidepressants (Devoize et al, 1984). However, this effect would probably disappear on chronic treatment: DMI has been found to potentiate morphine analgesia on acute administration, but to antagonize morphine after chronic administration (O'Neill and Valentino, 1986).

In sum, the actions of antidepressants in the "behavioral despair" model appear to depend primarily on an increase in transmission through alpha-$_1$ and/or DA receptors. Although the relative importance placed on these two systems varies between different laboratories, it seems likely that both systems are involved, irrespective of the treatment regime.

3.2. The Learned Helplessness Model

The term "learned helplessness" refers to a pattern of behavioral changes that are seen following exposure to uncontrollable aversive events, typically, electric shocks. The most significant of these effects is a subsequent failure to act, even in circumstances where control may be effectively reestablished. The performance deficit is not, however, observed in animals subjected to controllable shocks (Seligman, 1975; Maier and Seligman, 1976). In addition to their passivity in the face of stress, "helpless" animals show a variety of other behavioral changes, including decreased locomotor activity, exploratory behavior and aggression, poor performance in rewarded tasks, early waking, and loss of appetite and weight. Indeed, the range of symptomatic parallels between learned helplessness and depression is so striking that it has been suggested that "helpless" animals meet DSM-III criteria for major depressive disorder (Weiss J. et al., 1982). The learned helplessness model of depression has attracted considerable attention, though for the most part, this has been directed at its psychological interpretation, rather than its physiological basis (*see* Maier and Seligman, 1976; Willner, 1985b)

Like "behavioral despair," learned helplessness appears to involve, primarily, alpha-adrenergic, dopaminergic, and cholinergic mechanisms. Learned helplessness may be simulated pharmacologically either by drugs that reduce DA and/or NA function (Anisman et al., 1979a,b; Weiss et al., 1975), or by drugs that increase cholinergic function (Anisman et al., 1981a). Conversely, the debilitating effect of uncontrollable stress on later performance may be reversed by the DA agonist apomorphine, by clonidine, which at high doses is an $alpha_1$-receptor agonist (Anisman et al., 1979a,b, 1980), or by anticholinergic drugs (Anisman et al., 1979b, 1981a). It has also been reported that learned helplessness was reversed by beta-adrenergic agonists (Martin et al., 1986a); however, in an earlier study, the beta-receptor antagonist propranolol failed to exacerbate the effects of uncontrollable shock (Anisman et al., 1981b). Studies of 5-HT mechanisms are also inconsistent: 5-HT agonists, for example, have been reported to induce helplessness, to reverse helplessness, and to be without effect (Brown et al., 1982; Sherman and Petty, 1980; Anisman et al., 1979a).

As in the "behavioral despair" model, antidepressants are clearly effective in reversing learned helplessness. Acute antidepressant treatment has sometimes been found to work (Kametani et al., 1983; Martin et al., 1986a). More usually, however, chronic treatment is required (Sherman et al., 1979, 1982), the effect developing over three to four days of treatment (Petty et al., 1982; Martin et al., 1986a). In one study, similar effects were found using a single high dose or repeated low doses (Martin et al., 1986a). In addition to causing deficits in aversively motivated tasks, inescapable shock is also found to suppress intracranial self-stimulation. This effect was prevented by chronic treatment with an antidepressant (DMI), but not by acute treatment (Zacharko et al., 1984).

In this model, however, the potentiation of antidepressant effects by chronic treatment appears to depend solely on increases in drug tissue levels, rather than sensitization. Reversal of learned helplessness has been demonstrated using acute administration of antidepressants to frontal cortex (Sherman and Petty, 1980). Imipramine concentrations in this region, after systemic administration, were highly correlated with level of performance. Performance was not, however, affected by the duration of treatment, over and above the increase in drug concentration (Petty et al., 1982). The question of whether a true sensitization may occur on more prolonged treatment has not yet been addressed.

Only two studies have examined the mechanism of antidepressant action in the learned helplessness model. Neurotoxic lesions of the ascending 5-HT projections were found to have no effect either on the inception of learned helplessness, or on its reversal by DMI, imipramine, chlorimipramine, or nialamide (Soubrie et al., 1986). Both alpha$_1$- and beta-adrenergic antagonists were found to abolish the therapeutic effect of DMI and chlorimipramine, administered for 3–5 d; the effects of DA-receptor antagonists were not studied (Martin et al., 1986b). The effects of antidepressants were also blocked by the opiate antagonist naloxone (Martin et al., 1986b); the significance of this observation is difficult to assess, given that opiate antagonists also prevent many of the behavioral consequences of inescapable shock (Maier, 1984). This study suggests that both alpha$_1$- and beta-adrenergic receptors may be involved in the acute effect of antidepressants on learned helplessness. However, it remains to be established whether the two receptor systems are equally implicated following chronic antidepressant administration, remembering that over the course of two to three weeks of treatment the responsiveness of alpha$_1$ receptors rises, but the responsiveness of beta receptors falls.

3.3. The Chronic Mild Stress Model

Animals exposed to chronic unpredictable stress have been shown to display a number of behavioral abnormalities, including a failure to increase their fluid intake when sucrose was added to their drinking water (Katz, 1982). Recently, a version of this model has been developed in which very mild stressors are employed, such as periods of food and water deprivation, small temperature reductions, white noise of moderate intensity, changes of cage mates, and so on. Over a period of weeks of exposure to the stress regime, rats gradually reduced their consumption of a highly preferred sucrose solution. Antidepressant treatment did not increase sucrose consumption in nonstressed animals, but following the reduction of sucrose intake by stress, normal behavior was restored by chronic treatment with DMI or imipramine. In the early stages, antidepressant treatment did not affect sucrose intake in some experiments, but actually reduced it in others; this, together with the prolonged time course of the onset of antidepressant effects (at least two and up to four weeks), strongly suggests a true sensitization effect (Willner, et al. 1987, 1988; Muscat et al., 1988). The therapeutic action of imipramine in this model appears to be mediated by an increase in DA receptor responsiveness; a low dose of the DA receptor antagonist pimozide selectively abolished the recovery of sucrose consumption in imipramine-treated chronically stressed animals (Willner et al., 1988; *see* Fig. 3). The therapeutic effects of DMI and amitriptyline in this model were similarly reversed by either SCH-23390 or sulpiride, selective antagonists at the D_1 and D_2 subtypes of DA receptor (unpublished data).

3.4. Other Models

In the "behavioral despair," learned helplessness, and chronic mild stress models, the effect of antidepressants is to activate behavior in a passive animal. There are also a number of models in which antidepressants suppress abnormal behavior in an active animal. One such model is "muricide" (mouse killing), which is seen in a minority of "normal" rats, but in a high proportion following social isolation or a variety of insults to the brain (Albert and Walsh, 1982; Eichelman, 1979). A second model is the olfactory bulbectomized rat, which displays hyperactivity, deficient passive avoidance learning, and certain hormonal changes (Cairncross et al., 1979; Jesberger and Richardson, 1985). The close relationship of these two models is shown by the fact that olfactory bulbectomy is one of the procedures that can induce muricide (Albert and Walsh, 1982).

Though somewhat unreliable (and ethically objectionable), drug-induced suppression of muricide is one of the oldest tests for antidepressant activity: most antidepressants suppress muricide, at relatively high doses, on acute systemic administration. They are also effective if administered to the medial amygdala (Horovitz, 1965; Ueki, 1982). Chronic treatment potentiated the antimuri-

WEEK 10 □ VEHICLE ▨ PIMOZIDE

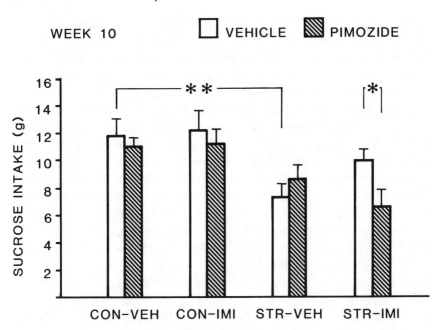

Fig. 3. Consumption of a 1% sucrose solution by rats subjected for 10 wk to chronic mild stress and/or treated with imipramine (6 mg/kg/d) from wk 5 onwards. In week 10 only, half of each group were pretreated with pimozide (0.15 mg/kg: hatched bars), which specifically reduced intake in the STR-IMI group. *, $p < 0.05$; **, $p < 0.01$. (From Willner et al., 1988.)

cidal effect of imipramine, DMI, or amitriptyline (but not chlorimipramine) (Enna et al., 1981; Shibata et al., 1984): the inhibition of muricide was substantially greater 24 h after a course of chronic treatment than 1 h after an acute dose (Shibata et al., 1984), though the chronic regime is known to produce lower drug concentrations (Vetulani et al., 1976). Chronic treatment also potentiated the antimuricidal effect of DMI administered to the medial amygdala (Shibata et al., 1984). These observations suggest that potentiation of the antidepressant effect is caused by sensitization rather than by changes in drug concentration. The situation in relation to olfactory bulbectomized animals is slightly different, and more straightforward, in that chronic treatment (5–10 d) is usually necessary before antidepressant effects are observed (Lloyd et al., 1982; Noreika et al., 1981; Jesberger and Richardson, 1985). Nevertheless, antidepressants appear to be effective acutely if administered to the medial amygdala, and certain antidepressants are active on acute systemic administration (Broekkamp et al., 1980; Garrigou et al., 1981). Out of the spectrum of effects of olfactory bulbectomy, the passive avoidance deficit responds most specifically to antidepressants (Cairncross et al., 1979).

One series of experiments has suggested that muricide, and its reversal by antidepressants, may depend upon noradrenergic mechanisms (Shibata et al., 1984; Iwasaki et al., 1986). However, most of the literature points to 5-HT as being the more important physiological substrate for this behavior. A large number of studies have shown that reducing 5-HT activity, by blocking 5-HT synthesis or by chemical or electrical lesions to the raphé nuclei, will induce muricide in a high proportion of rats that do not already show this behavior, or potentiate muricide in those that do (Albert and Walsh, 1982; Eichelman, 1979). Conversely, muricide is suppressed if 5-HT activity is raised by systemic administration of 5-HT precursors, receptor agonists, or uptake blockers, or by intraventricular injections of 5-HT (Albert and Walsh, 1982; Eichelman, 1979). Small depletions of 5-HT, which did not by themselves induce muricide, have been shown to block the antimuricidal effect of tricyclic antidepressants, administered acutely (Eisenstein et al., 1982; Marks et al., 1978). It has not yet been demonstrated that 5-HT is also involved in the potentiating effect of chronic administration.

Behavior in the olfactory bulbectomy model also appears to depend primarily on 5-HT. One widely quoted study has suggested a role for NA, based upon the observation that the behaviorally active isomer of mianserin increased NA concentrations in the amygdala, whereas the behaviorally inactive isomer did not (Janscar and Leonard, 1984). However, these data now appear to be somewhat misleading, as a later study using an analog of mianserin failed to demonstrate a correlation between NA levels and behavioral activity (O'Connor and Leonard, 1986). A role for 5-HT was first suggested by the observation that the effects of bulbectomy may be mimicked by 5-HT lesions, and reversed by 5-HT agonists administered systemically, or by 5-HT (but not NA) administered to the medial amygdala (Broekkamp et al., 1980; Cairncross et al., 1979; Garrigou et al., 1981; Lloyd et al., 1982). 5-HT uptake inhibitor antidepressants (fluoxetine and zimelidine) are active in this model on acute administration, whereas tricylics require subchronic treatment (Broekkamp et al., 1980; Joly and Sanger, 1986). The 5-HT receptor antagonist metergoline blocks the therapeutic effect of 5-HT uptake inhibitors administered systemically (Broekkamp et al., 1980; Joly and Sanger, 1986) and of acute antidepressant administration to the medial amygdala (Garrigou et al., 1981). Metergoline also blocked the effects of imipramine and mianserin administered chronically (Broekkamp et al., 1980; *see* Fig. 4), suggesting that sensitization to the effects of these drugs is mediated by an increase in 5-HT function.

4. Some Methodological Comments

It seems appropriate to conclude by drawing together some of the methodological issues that have arisen in the course of this review. First and foremost is the need to demonstrate that sensitization has actually occurred. There are two circumstances in which sensitization is a likely explanation for a time-dependent increase in the effect of an antidepressant. The first is when the effect of chronic treatment is greater than the effect of acute treatment at any dose: this situation appears to apply in the olfactory bulbectomy model, and to some of the effects of antidepressants on monoamine receptor function. The second is when the effect continues to increase over a prolonged period. Antidepressant concentrations build up in the rat brain over some 3–7 d (though this period may be longer or shorter depending on the half-life of the drug). As a rule of thumb, therefore, sensitization is a likely explanation of effects occurring in the second week or later.

In other circumstances, it is essential that sensitization be demonstrated, rather than assumed. In the clinical situation, changes in brain drug concentrations appear to play a major role in the delayed onset of action, at least within the first week of treatment; the subsequent slow improvement probably involves processes of physiological sensitization, but nevertheless, a variety of cognitive and social factors have not been definitively excluded. Evidence of true sensitization, as distinct from rising plasma drug concentrations, has been adduced in three animal models of depression, "behavioral despair," chronic mild stress, and olfactory bulbectomy. However, the slow onset of drug action in the learned helplessness model apparently does depend upon changes in drug

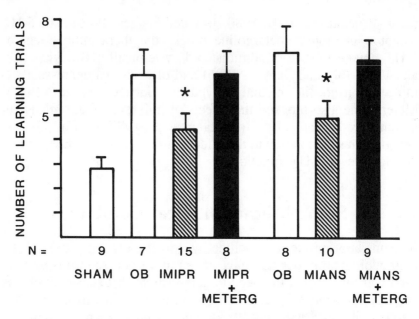

Fig. 4. Blockade by metergoline (10 mg/kg) of the improvement of passive avoidance learning in olfactory bulbectomized rats by imipramine (10 mg/kg for 5 d) or mianserin (10 mg/kg twice daily for 10 d). *, $p < 0.05$ relative either to untreated OB rats or to antidepressant + metergoline. (Adapted from Broekkamp et al, 1980.)

concentration (Petty et al., 1982): a salutary reminder of the importance of keeping an open mind.

The timing of experimental tests is a second area that requires particular attention. Many studies of chronic antidepressant administration have allowed a period of drug washout before testing for biochemical or functional changes. There are often good reasons for doing so, in order to avoid interference by the drug in a biochemical assay or to avoid pharmacokinetic interactions in a functional study. However, the use of a washout period introduces the possibility that the observed changes are withdrawal effects. This does, in fact, seem to be the explanation of the controversial desensitization of DA autoreceptors by antidepressants (Scavone et al., 1986; Towell et al., 1986), and the question of withdrawal effects arises constantly in the behavioral literature on antidepressants and 5-HT (Willner, 1985a).

A third problem concerns the techniques used to measure physiological changes, and their interpretation. In studies of the effects of chronic antidepressant treatment on monoamine receptors, discrepancies between effects on receptor binding and on functional measures are the rule, rather than the exception. This may in part reflect the predominant use of antagonists as ligands in binding studies; there is some evidence that in the case of alpha$_1$ (Menkes et al., 1983a) and dopamine (De Ceballos et al., 1985) receptors, more congruent results may be obtained if agonist ligands are used. Whatever the reason, it seems most unwise to infer functional changes from binding data.

Even where functional measures are used, it is essential to establish whether tolerance or sensitization in the response to a receptor agonist reflects a similar change in transmission across the synapse: a change in postsynaptic receptor function might be offset by a presynaptic change in the opposite direction. In the case of alpha$_1$ receptors, the increase in postsynaptic responsiveness is probably additive with an increase in presynaptic activity; on the other hand, the same increase in presynaptic activity acts counter to the "downregulation" of beta receptors. There are two approaches to assessing the net effect of antidepressants on synaptic transmission. The first is to attempt to infer the net outcome by enumerating the many effects of antidepressants on the individual components of synaptic activity (*see* Willner, 1985a,b). The second, which is more empirical but perhaps less direct, is to measure the integrated synaptic activity by changes in the output of the system, as reflected in an appropriate functional response. The demonstration that antidepressants potentiate the behavioral effects of alpha$_1$- and DA-receptor agonists tells us only about postsynaptic responsiveness in these systems. However, the fact that the therapeutic action of antidepressants in the "behavioral despair" test is blocked by alpha$_1$- and/or DA-receptor antagonists provides strong evidence that antidepressants do, in fact, increase the level of transmission through alpha$_1$ and DA synapses.

A final methodological point, frequently overlooked, is the importance of distinguishing between actions and mechanisms of action. Describing the effects of antidepressants on monoamine receptors is not the same thing as explaining their clinical actions. Indeed, it seem likely that one of the best established actions of antidepressants—the "downregulation" of beta receptors—does not contribute to their therapeutic effect. However, data obtained from animal models of depression support the relevance of certain other receptor changes. In the "behavioral despair" model, behavioral sensitization following chronic antidepressant treatment appears to be mediated primarily by sensitization of postsynaptic alpha$_1$ receptors and/or dopamine receptors, though the precise extent to which each of these systems contributes remains to be determined. In the olfactory bulbectomy model, a single study (Broekkamp et al., 1980) suggests that the sensitization of 5-HT receptors is responsible for behavioral sensitization. It is striking that, in both of these models, the same brain systems appear to be involved in the acute actions of antidepressants as in their potentiation by chronic treatment. Clinical studies have produced a certain amount of evidence that antidepressants can alter receptor function in the human brain. However, there is a resounding silence in the literature when it comes to establishing the relevance of particular receptor changes to the therapeutic effects. In animal models, pharmacological blockade of antidepressant action has proved to be the most fruitful line of investigation for establishing relevance. Remarkably, the literature on pharmacological blockade of antidepressant action in a clinical context appears to be confined to two papers, which were published more than ten years ago and deal with a total of seven patients, two treated with imipramine and five treated with the MAOI tranylcypromine. In these patients, antidepressant effects were blocked by the 5-HT synthesis inhibitor PCPA, but not (in the two imipramine-treated patients) by the catecholamine synthesis inhibitor AMPT (Shopsin et al., 1975, 1976). There is a pressing need for these studies to be replicated and extended, using other antidepressants and specific receptor antagonists. Considering the level of interest in establishing the mechanisms of clinical action of antidepressant drugs, it is astonishing that this research strategy has not been widely adopted.

Increasingly, studies of sensitization are contributing to an understanding of the actions of antidepressants. However, what has antidepressant research taught us about the processes underlying drug-induced sensitization? One lesson is that sensitization may be rather difficult to demonstrate: it cannot be assumed merely from the fact that a response increases with repeated treatment, and an increase in one response may be offset by a corresponding decrease in another. The second, and somewhat more positive, contribution of antidepressant research is the provision of a methodology for establishing the physiological mechanisms of behavioral sensitization: the use of pharmacological antagonists to reverse sensitized responses has considerable potential to define the neuropharmacological systems, and also the anatomical locations, in which behaviorally relevant physiological changes are occurring. Nevertheless, having identified some of the systems that are functionally sensitized by repeated antidepressant treatment, it is important to note that we are at present almost totally devoid of insights into the underlying processes. Homeostatic models are of very limited value in this context. To take one obvious example, the principle of homeostasis is no help at all in understanding how antidepressants sensitize postsynaptic DA receptors; homeostasis cannot even account for all of the phenomena of beta-receptor desensitization. A theory of sensitization will require the formulation of additional explanatory principles; they may well be forthcoming from the elucidation, in the not too distant future, of the processes underlying antidepressant sensitization effects.

References

Albert D. J. and Walsh M. L. (1982) The inhibitory modulation of agonistic behaviour in the rat brain: A review. *Neurosci. Biobehav. Rev.* **6,** 125–143.

Anisman H., Irwin J., and Sklar L. S. (1979a) Deficits of escape performance following catecholamine depletion: implications for behavioural deficits induced by uncontrollable stress. *Psychopharmacol.* **64,** 163–170.

Anisman H., Remington G., and Sklar L. S. (1979b) Effects of inescapable shock on subsequent escape performance: catecholaminergic and cholinergic mediation of response initiation and maintenance. *Psychopharmacol.* **61,** 107–124.

Anisman H., Suissa A., and Sklar L. S. (1980) Escape deficits induced by uncontrollable stress: Antagonism by dopamine and noradrenaline agonists. *Behav. Neur. Biol.* **28,** 34–47.

Anisman H., Glazier S. J., and Sklar L. S. (1981a) Cholinergic influences on escape deficits produced by uncontrollable stress. *Psychopharmacol.* **74,** 81–87.

Anisman H., Ritch M., and Sklar L. S. (1981b) Noradrenergic and dopaminergic interactions in escape behavior: Analysis of uncontrollable stress effects. *Psychopharmacol.* **74,** 263–268.

Antkiewicz-Michaluk L. (1985) Action of antidepressant neuroleptics chlorprothixene and levomepromazine on the central noradrenergic system: comparison with other antidepressants. *Pol. J. Pharmacol. Pharm.* **37,** 673–683.

Araki H., Kazuaki K., Uchiyama Y., and Aihara H. (1985) Involvement of amygdaloid catecholaminergic mechanism in suppressive effects of desipramine and imipramine on duration of immobility in rats forced to swim. *Eur. J. Pharm.* **113,** 313–318.

Arnt J., Hytell J., and Overo K. F. (1984) Prolonged treatment with the specific 5-HT-uptake inhibitor citalopram: Effect on dopaminergic and serotonergic functions. *Pol. J. Pharm. Pharmacol.* **36,** 221–230.

Ask A.-L., Fowler C. J., Hall. H., Kelder D., Ross S. B. and Saaf J. (1986) Cortical beta- and alpha$_2$-adrenoceptor binding, hypothalamic noradrenaline and pineal melatonin concentrations measured at different times of the day after repeated treatment of rats with imipramine, zimelidine, alaproclate and amiflamine. *Acta Pharmacol. Toxicol.* **58,** 16–24.

Aulakh C. S., Cohen R. M., Pradhan S. N., and Murphy D. L. (1983) Self-stimulation responses are altered following long-term but not short-term treatment with clorgyline. *Brain Res.* **270,** 383–386.

Banerjee S. P., Kung L. S., Riggi S. J., and Chanda S. K. (1977) Development of beta-adrenergic subsensitivity by antidepressants. *Nature* **268,** 455–456.

Barbaccia M. L., Ravizza L., and Costa E. (1986) Maprotiline: An antidepressant with an unusual pharmacological profile. *J. Pharmacol. Exp. Ther.* **236,** 307–312.

Bech P., Allerup P., Reisby N., and Gram L. F. (1984) Assessment of symptom change from improvement curves on the Hamilton depression scale in trials with antidepressants. *Psychopharmacol.* **84,** 276–281.

Beck A. T. (1967) *Depression: Clinical, Experimental and Theoretical Aspects.* Harper & Row, New York.

Berettera C., Invernizzi R., Pulvirenti L., and Samanin R. (1986) Chronic treatment with iprindole reduces immobility of rats in the behavioural 'despair' test by activating dopaminergic mechanisms in the brain. *J. Pharm. Pharmacol.* **38,** 313–315.

Blier P. and De Montigny C. (1980) Effect of chronic tricyclic antidepressant treatment on the serotonergic autoreceptor: A microiontophoretic study in the rat. *Naunyn-Schmiedebergs Arch. Pharmacol.* **314,** 123–128.

Blier, P. and De Montigny, C. (1983) Electrophysiological investigations on the effects of repeated zimelidine administration on serotonergic neurotransmission. *J. Neurosci.* **3**, 1270–1278.

Blier P., De Montigny C., and Azzaro A. J. (1984a) Modification of serotonergic and noradrenergic neurotransmission by long-term administration of monoamine-oxidase inhibitors. II. Responsiveness of postsynaptic neurons. *Soc. Neurosci. Abstr.* **10,** 16.

Blier P., De Montigny C., and Tardif D. (1984b) Effects of two antidepressant drugs mianserin and indalpine on the serotonergic system: Single-cell studies in the rat. *Psychopharmacol.* **84,** 242–249.

Borsini F., Bendotti G., Velkov V., Rech R., and Samanin R. (1981) Immobility test: effects of 5-hydroxytryptaminergic drugs and role of catecholamines in the activity of some antidepressants. *J. Pharm. Pharmacol.* **33,** 33–37.

Borsini F., Nowakowska E., and Samanin R. (1984) Effect of repeated treatment with desipramine in the behavioural 'despair' test in rats: Antagonism by 'atypical' but not 'classical' neuroleptics or antiadrenergic drugs. *Life Sci.* **34,** 1171–1176.

Borsini F., Nowakowska E., Pulvirenti L., and Samanin R. (1985a) Repeated treatment with amitriptyline reduces immobility in the behavioural 'despair' test by activating dopaminergic and beta-adrenergic mechanisms. *J. Pharm. Pharmacol.* **37,** 137–138.

Borsini F., Pulvirenti L., and Samanin R. (1985b) Evidence of dopamine involvement in the effect of repeated treatment with various antidepressants in the behavioural 'despair' test in rats. *Eur. J. Pharmacol.* **110,** 253–256.

Broekkamp C. L., Garrigou D., and Lloyd K. G. (1980) Serotonin-mimetic and antidepressant drugs on passive avoidance learning by olfactory bulbectomized rats. *Pharmacol. Biochem. Behav.* **13,** 643–646.

Brown J., Doxey J. C., and Handley S. (1980) Effects of alpha-adrenoceptor agonists and antagonists and of antidepressant drugs on pre- and postsynaptic alpha-adrenoceptors. *Eur. J. Pharmacol.* **67,** 33–40.

Brown L., Rosellini R. A., Samuels O. B., and Riley E. P. (1982) Evidence for a serotonergic mechanism of the learned helplessness phenomenon. *Pharmacol. Biochem. Behav.* **17,** 877–883.

Browne R. G. (1979) Effects of antidepressants and cholinergics in a mouse 'behavioral despair' test. *Eur. J. Pharmacol.* **58,** 331–334.

Brunello N., Barbaccia M. L., Chuang D.-M, and Costa E. (1982)

Down-regulation of beta-adrenergic receptors following repeated desmethylimipramine injections: Permissive role of serotonergic axons. *Neuropharmacology* **21**, 1145–1149.

Brunello, N., Mocchetti, I., Volterra, A., Cuomo, V. & Racagni, G. (1985a) Serotonergic modulation of cortical rat noradrenergic system in the mechanism of action of antidepressant drugs. *Psychopharmacol. Bull.* **21**, 379–384.

Brunello N., Volterra A., Cagiano R., Ianeri G. C., Cuomo V., and Racagni G. (1985b) Biochemical and behavioral changes in rats after prolonged treatment with desipramine: Interaction with *p*-chlorophenylalanine. *Naunyn-Schmiedebergs Arch. Pharmacol.* **331**, 20–22.

Cairncross K. D., Cox B., Forster C., and Wren A. F. (1979) Olfactory projection systems, drugs and behaviour: A review. *Psychoneuroendocrinology* **4**, 253–272.

Campbell I. C. and McKernan, R. M. (1982) Central and peripheral changes in alpha-adrenoceptors in the rat in response to chronic antidepressant drug administration, in *New Vistas in Depression*. (Langer S. Z., Takahashi R, Segawa T, and Briley M, eds.) Pergamon, New York, pp. 153–160.

Campbell I. C., Durcan M. J., Cohen R. M., Pickar D., Chugani D., and Murphy D. L. (1985) Chronic clorgyline and pargyline increase apomorphine-induced stereotypy in the rat. *Pharmacol. Biochem. Behav.* **23**, 921–925.

Carroll B. J. (1982) The dexamethsone suppression test for melancholia. *Brit. J. Psychiatr.* **142**, 292–304.

Cervo L. and Samanin R. (1987) Evidence that dopamine mechanisms in the nucleus accumbens are selectively involved in the effect of desipramine and imipramine in the forced swimming test. *Abstr. 6th. Int. Catecholamine Symp.*, Jerusalem, p. 94.

Chaput Y., Blier P., and De Montigny C. (1984) The effects of acute and chronic administration of citalopram on serotonergic neurotransmission: Electrophysiological studies in the rat. *Soc. Neurosci. Abstr.* **10**, 17.

Charney D. S. and Heninger G. R. (1983) Monoamine receptor sensitivity and depression: Clinical studies of antidepressant effects on serotonin and noradrenergic function. *Psychopharmacol. Bull.* **19**, 490–495.

Charney D. S., Heninger G. R. and Sternberg D. E. (1983) Alpha$_2$ adrenergic receptor sensitivity and the mechanism of action of antidepressant therapy: The effect of long-term amitriptyline treatment. *Brit. J. Psychiatr.* **142**, 265–275.

Chiodo L. A. and Antelman S. M. (1980) Repeated tricyclics induce a progressive dopamine autoreceptor subsensitivity independent of daily drug treatment. *Nature* **287**, 451–454.

Costall B., Hui S. -C, and Naylor R. C. (1980) Differential actions of substituted benzamides on pre-and postsynaptic receptor mechanisms in the nucleus accumbens. *J. Pharm. Pharmacol.* **32**, 329–332.

Cowen P. J., Fraser S., Grahame-Smith D. G., Green A. R., and Stanford C. (1983) The effect of chronic antidepressant treatment on beta-adrenoceptor function of the rat pineal. *Brit. J. Pharmacol.* **78**, 89–96.

Cowen P. J., Green A. R., Grahame-Smith D. G., and Braddock L. E. (1985) Plasma melatonin during desmethylimipramine treatment: Evidence for changes in noradrenergic transmission. *Brit. J. Clin. Pharmacol.* **19**, 799–805.

Cuomo V., Cagiano R., Brunello N., Fumagalli R., and Racagni G. (1983) Behavioural changes after acute and chronic administration of typical and atypical antidepressants in rats: Interactions with reserpine. *Neurosci. Lett.* **40**, 315–319.

De Ceballos M. L., Benedi A., De Felipe C., and Del Rio J. (1985) Prenatal exposure of rats to antidepressants enhances agonist affinity of brain dopamine receptors and dopamine-mediated behaviour. *Eur. J. Pharmacol.* **116**, 257–266.

De Montigny C. and Aghajanian G. K. (1978) Tricyclic antidepressants: Long-term treatment increases responsiveness of rat forebrain neurons to serotonin. *Science* **202**, 1303–1306.

De Montigny C. and Blier P. (1984) Modification of serotonergic and noradrenergic neurotransmission by long-term administration of monoamine oxidase inhibitors. I. Activity of presynaptic neurons. *Soc. Neurosci. Abstr.* **10**, 16.

De Montigny C., Blier P., Caille G., and Kouassi E. (1981) Pre- and postsynaptic effects of zimelidine and norzimelidine on the serotonergic system: Single cell studies in the rat. *Acta Psychiatr. Scand.* **63, Suppl. 290**, 79–90.

Delina-Stula A. and Vassout A. (1979) Modulation of dopamine-mediated behavioral responses by antidepressants: Effects of single and repeated treatment. *Eur. J. Pharmacol.* **58**, 443–451.

Delina-Stula A., Baumann P., and Buch, O. (1979) Depression of exploratory activity by clonidine in rats as a model for the detection of pre- and postsynaptic central noradrenergic receptor selectivity of alpha-adrenolytic drugs. *Naunyn-Schmiedebergs Arch. Pharmacol.* **307**, 115–122.

Delina-Stula A., Hauser K., Baumann P., Olpe-H. -R., Waldmeier P., and Sorni A. (1982) Stereospecificity of behavioral and biochemical responses to oxaprotiline—A new antidepressant, in *Typical and Atypical Antidepressants: Molecular Mechanisms,* (Costa E. and Racagni G., eds.) Raven, New York, pp. 265–275.

Devoize J. -L., Rigal F., Eschalier A., Trolese J. -F., and Renoux M. (1984) Influence of naloxone on antidepressant drug effects in the forced swimming test in mice. *Psychopharmacol.* **84,** 71–75.

Dowdall M. and De Montigny C. (1984) Pre- and postsynaptic effects of trazodone on serotonin neurotransmission: Single cell studies in the rat. *Soc. Neurosci. Abstr.* **10,** 17.

Eichelman B. (1979) Role of biogenic amines in aggressive behaviour, in *Psychopharmacology of Aggression.* (Sandler M. ed.) Raven, New York, pp. 61–93.

Eisenstein N., Iorio, L. C., and Clody D. E. (1982) Role of serotonin in the blockade of muricidal behaviour by tricyclic antidepressants. *Pharmacol. Biochem. Behav.* **17,** 847–849.

Enna S. J., Mann E., Kendall D., and Stancer G. M. (1981) Effect of chronic antidepressant administration on brain neurotransmitter receptor binding, in *Antidepressants: Neurochemical, Behavioural and Clinical Perspectives.* (Enna S. J., Malick J. B., and Richelson, E., eds.) Raven, New York, pp. 91–105.

Extein I., Tallman J., Smith C. C., and Goodwin F. K. (1979) Changes in lymphocyte beta-adrenergic receptors in depression and mania. *Psychiatr. Res.* **1,** 191–197.

Ferris R. M. and Beaman O. J. (1983) Buproprion: A new antidepressant drug, the mechanism of action of which is not associated with down-regulation of postsynaptic beta-adrenergic, serotonergic (5-HT$_2$), alpha$_2$-adrenergic, imipramine and dopaminergic receptors in brain. *Neuropharmacology* **22,** 1257–1267.

Fibiger H. C. and Phillips A. G. (1981) Increased intracranial self-stimulation in rats after long-term administration of desipramine. *Science* **214,** 683–684.

Frazer A., Lucki I., and Sills M. (1985) Alterations in monoamine-containing neuronal function due to administration of antidepressants repeatedly to rats. *Acta Pharmacol. Toxicol.* **56, Suppl. 1,** 21–34.

Friedman E. and Dollob A. (1979) Enhanced serotonin receptor activity after chronic treatment with imipramine or amitriptyline. *Commun. Psychopharmacol.* **3,** 89–92.

Fujimori M. (1981) Amoxapine, a new antidepressant: Overview, in *New*

Vistas in Depression. (Langer S. Z., Takahashi R., Segawa T., and Briley M., eds.) Pergamon, New York, pp. 287–294.

Fuxe K., Ogren S. -O., Agnati L. F., Benfenati F., Fredholm B., Andersson K., Zini I., and Eneroth P. (1983) Chronic antidepressant treatment and central 5-HT synapses. *Neuropharmacology* **22,** 389–400.

Fuxe K., Ogren S. -O., Agnati L. F., Eneroth P., Holm A. C., and Andersson K. (1981) Long-term treatment with zimelidine leads to a reduction in 5-hydroxytryptamine neurotransmission within the central nervous system of the mouse and rat. *Neurosci. Lett.* **21,** 57–62.

Gallager D. W. and Bunney W. E. (1979) Failure of chronic lithium treatment to block tricyclic antidepressant-induced 5-HT supersensitivity. *Naunyn-Schmiedebergs Arch. Pharmacol.* **307,** 129–133.

Gandolfi O., Barbaccia M. L., Chuang D. M., and Costa E. (1983) Daily buproprion injections for 3 weeks attenuate the NE-stimulation of adenylate cyclase and the number of beta-adrenergic recognition sites in rat frontal cortex. *Neuropharmacology* **22,** 927–929.

Gandolfi O., Barbaccia M. L., and Costa E. (1984) Comparison of iprindole, imipramine and mianserin action on brain serotonergic and beta-adrenergic receptors. *J. Pharmacol. Exp. Ther.* **229,** 782–786.

Garcha G., Smokcum R. W. J., Stephenson J. D., and Weeramanthri T. B. (1985) Effects of some atypical antidepressants on beta-adrenoceptor binding and adenylate cyclase activity in the rat forebrain. *Eur. J. Pharmacol.* **108,** 1–7.

Garcia-Sevilla J. A., Zis A. P., Hollingsworth P. J., Greden J. F., and Smith C. B. (1981) Platelet alpha$_2$-adrenergic receptors in major depressive disorder. *Arch. Gen. Psychiatr.* **38,** 1327–1333.

Garrigou D., Broekkamp C. L., and Lloyd K. G. (1981) Involvement of the amygdala in the effect of antidepressants on the passive avoidance deficit in bulbectomized rats. *Psychopharmacol.* **74,** 66–70.

Glass I. B., Checkley S. A., Shur E., and Dawling S. (1982) The effect of desipramine upon central adrenergic function in depressed patients. *Brit. J. Pharmacol.* **141,** 372–376.

Goodwin G. M., De Souza R. J., and Green A. R. (1985) Presynaptic serotonin-mediated response in mice attenuated by antidepressant drugs and electroconvulsive shock. *Nature* **317,** 531–533.

Gorka Z. and Janus K. (1985) Effects of neuroleptics displaying antidepressant activity on behavior of rats in the forced swimming test. *Pharmacol Biochem. Behav.* **23,** 203–206.

Gorka Z., Wojtasik E., Kwiatek H., and Maj J. (1979) Action of serotonin-mimetics in the behavioral despair test in rats. *Commun. Psychopharmacol.* **3**, 133–136.

Green A. R. and Deakin J. F. W. (1980) Brain noradrenaline depletion prevents ECS-induced enhancement of serotonin- and dopamine-mediated behaviour. *Nature* **285**, 232–233.

Green A. R., Guy A. P. and Gardner C. R. (1984) The behavioural effects of RU24969, a suggested 5-HT1 receptor agonist in rodents and the effect on the behaviour of treatment with antidepressants. *Neuropharmacology* **23**, 655–661.

Green A. R., Heal D. J., Johnson P., Laurence B. E., and Nimgaonkar V. L. (1983) Antidepressant treatments: effects in rodents on dose-response curves of 5-hydroxytryptamine- and dopamine-mediated behaviours and 5-HT$_2$ receptor number in frontal cortex. *Brit. J. Pharmacol.* **80**, 377–385.

Hall H., Ross S. B., and Sallemark M. (1984) Effect of destruction of central noradrenergic and serotonergic nerve terminals by systemic neurotoxins on the long-term effects of antidepressants on beta-adrenoceptors and 5-HT$_2$ binding sites in the rat cerebral cortex. *J. Neural Transm.* **59**, 9–23.

Haskell D. S., DiMaschio A., and Prusoff B. (1975) Rapidity of symptom reduction in patients treated for depression. *J. Nerv. Ment. Dis.* **160**, 24–33.

Hauser K., Olpe H. -R., and Jones R. S. G. (1985) Trimipramine: A tricyclic antidepressant exerting atypical actions on the central noradrenergic system. *Eur. J. Pharmacol.* **111**, 23–30.

Heal D. J. (1984). Phenylephrine-induced activity in mice as a model of central alpha$_1$-adrenoceptor function: Effects of acute and repeated administration of antidepressant drugs and electroconvulsive shock. *Neuropharmacology* **23**, 1241–1251.

Heal D. J., Lister S., Smith S. L., Davies C. L., Molyneux S. G., and Green A. R. (1983) The effects of acute and repeated administration of various antidepressant drugs on clonidine-induced hypoactivity in mice and rats. *Neuropharmacology* **22**, 983–992.

Hong K. W., Rhim B. Y., and Lee W. S. (1986) Enhancement of central and peripheral alpha$_1$-adrenoceptor sensitivity and reduction of peripheral alpha$_2$-adrenoceptor sensitivity following chronic imipramine treatment in rats. *Eur. J. Pharmacol.* **120**, 275–283.

Horovitz Z. P. (1965) Selective block of rat mouse killing by antidepressants. *Life Sci.* **4**, 1909–1912.

Huang Y. H. (1979) Chronic desipramine treatment increases activity of noradrenergic postsynaptic cells. *Life Sci.* **25**, 739–746.

Hytell J., Overo K. F., and Arnt J. (1984) Biochemical effects and drug levels in rats after long-term treatment with the specific 5-HT-uptake inhibitor, citalopram. *Psychopharmacol.* **83**, 20–27.

Iwasaki K., Fujiwara M., Shibata S., and Ueki S. (1986) Changes in brain catecholamine levels following olfactory bulbectomy and the effect of acute and chronic administration of desipramine in rats. *Pharmacol. Biochem. Behav.* **24**, 1715–1719.

Jancsar S. M. and Leonard B. E. (1984) The effect of (+ / −) mianserin and its enantiomers on the behavioural hyperactivity of the olfactory bulbectomized rat. *Neuropharmacology* **23**, 1065–1070.

Janowsky A. J., Steranka L. R., Gillespie D. D., and Sulser F. (1982) Role of neuronal signal input in the down-regulation of central noradrenergic receptor function by antidepressant drugs. *J. Neurochem.* **39**, 290–292.

Jesberger J. A. and Richardson J. S. (1985) Animal models of depression: Parallels and correlates to severe depression in humans. *Biol. Psychiat.* **20**, 764–784.

Joly D. and Sanger D. J. (1986) The effects of fluoxetine and zimelidine on the behavior of olfactory bulbectomized rats. *Pharmacol. Biochem. Behav.* **24**, 199–204.

Kametani H., Nomura S., and Shimizu J. (1983) The reversal effect of antidepressants on the escape deficit induced by inescapable shock in rats. *Psychopharmacol.* **80**, 206–208.

Katz R. J. (1982) Animal model of depression: Pharmacological sensitivity of a hedonic deficit. *Pharmacol. Biochem. Behav.* **16**, 965–968.

Kellar K. J., Cascio C. S., Butler J. A., and Kurtzke R. N. (1981) Differential effects of electroconvulsive shock and antidepressant drugs on serotonin-2 receptors in rat brain. *Eur. J. Pharmacol.* **69**, 515–518.

Kendall D. A., Duman R., Slopis J., and Enna S. J. (1982) Influence of adrenocorticotropic hormone and yohimbine on antidepressant-induced declines in rat brain neurotransmitter receptor binding and function. *J. Pharmacol. Exp. Ther.* **222**, 566–571.

Kitada Y., Miyauchi T., Kanazawa Y., Naḳamichi H., and Satoh S. (1983) Involvement of alpha- and beta$_1$-adrenergic mechanisms in the immobility-reducing action of desipramine in the forced swimming test. *Neuropharmacology* **22**, 1055–1060.

Kitada Y., Miyauchi T., Satoh A., and Satoh S. (1981) Effects of antidepressants in the forced swimming test. *Eur. J. Pharmacol.* **72**, 145–152.

Klimek V. (1988) Chronic treatment with antidepressants enhances agonist affinity for rat brain D-2 receptors. (Abstr.) Presented at First

Polish-Swedish Symposium on Structure and Function in Neuropharmacology, Warsaw, May 1988.

Korf J., Sebens J. B., and Postrema F. (1979) Cyclic AMP in rat cerebral cortex after stimulation of the locus coeruleus: Decrease by antidepressant drugs. *Eur. J. Pharmacol.* **59,** 23–30.

Kostowski W. (1985) Possible relationship of the locus coeruleus hippocampal noradrenergic neurons to depression and mode of action of antidepressant drugs. *Pol. J. Pharmacol. Pharm.* **37,** 727–743.

Kostowski W. and Malatynska E. (1983) Antagonism of behavioral depression produced by clonidine in the mongolian gerbil: A potential screening test for antidepressant drugs. *Psychopharmacol.* **79,** 203–208.

Leclerc B., Rouot B., Velly J., and Schwartz J. (1981) Beta-adrenergic receptor subtypes. *Trends Pharmacol. Sci.* **17,** 18–20.

Lecrubier Y., Puech A. J., Jouvent R., Simon P., and Widlocher D. (1980) A beta adrenergic stimulant (salbutamol) versus clomipramine in depression: A controlled study. *Br. J. Psychiat.* **136,** 354–358.

Lewinsohn P. M. (1974) A behavioral approach to depression, in *The Psychology of Depression: Current Theory and Research.* (Friedman R. J. and Katz M. M., eds.) Winston/Wiley, New York, pp. 157–185.

Leysen J. (1984) Problems in in vitro receptor binding studies and identification and role of serotonin receptor sites. *Neuropharmacology* **23,** 247–254.

Lloyd K. G., Garrigou D., and Broekkamp C. L. E. (1982) The action of monoaminergic, cholinergic and gabaergic compounds in the olfactory bulbectomized rat model of depression, in *New Vistas in Depression.* (Langer S. Z., Takahashi R., Segawa T., and Briley M. eds.) Pergamon, New York, pp. 179–186.

Lucki I. and Frazer A. (1982) Prevention of the serotonin syndrome in rats by repeated administration of monoamine oxidase inhibitors but not tricyclic antidepressants. *Psychopharmacol.* **77,** 205–211.

MacNiell D. A. and Gower M. (1982) Do antidepressants induce dopamine autoreceptor subsensitivity? *Nature* **298,** 302.

Maier S. F. (1984) Learned helplessness and animal models of depression. *Prog. Neuropsychopharmacol. Biol. Psychiatr.* **8,** 435–446.

Maier S. F. and Seligman M. E. P. (1976) Learned helplessness: Theory and evidence. *J. Exp. Psychol: General* **1,** 3–46.

Maj J., Gorka Z., Melzacka M., Rawlow A., and Pilc A. (1983) Chronic treatment with imipramine: Further functional evidence for the enhanced noradrenergic transmission in flexor reflex activity. *Naunyn-Schmiedebergs Arch. Pharmacol.* **322**, 256–260.

Maj J., Mogilnicka E., and Klimek V. (1979a) The effect of repeated administration of antidepressant drugs on the responsiveness of rats to catecholamine agonists. *J. Neural Transm.* **44**, 221–235.

Maj J., Mogilnicka E., and Kordecka A. (1979b) Chronic treatment with antidepressant drugs: Potentiation of apomorphine-induced aggressive behaviour in rats. *Neurosci. Lett.* **13**, 337–341.

Maj J., Rogoz Z., Skuza G., and Sowinska H. (1982) Effects of chronic treatment with antidepressants on aggressiveness induced by clonidine in mice. *J. Neural Transm.* **55**, 19–25.

Maj. J., Rogoz Z., Skuza G., and Sowinska H. (1984a) Repeated treatment with antidepressant drugs potentiates the locomotor response to (+)-amphetamine. *J. Pharm. Pharmacol.* **36**, 127–130.

Maj J., Rogoz Z., Skuza G., and Sowinska H. (1984b) Repeated treatment with antidepressant drugs increases the behavioural response to apomorphine. *J. Neural Transm.* **60**, 273–282.

Mancinelli A., D'Aranno V., Borsini F., and Meli A. (1987) Lack of relationship between effect of desipramine on forced swimming test and brain levels of desipramine or its demethylated metabolite in rats. *Psychopharmacol.* **92**, 441–443.

Manier D. H., Gillespie D. D., Steranka L. R., and Sulser F. (1985) A pivotal role for serotonin (5HT) in the regulation of beta adrenoceptors by antidepressants: reversibility of the action of parachlorophenylalanine by 5-hydroxytryptophan. *Experientia* **40**, 1223–1227.

Marks P. C., O'Brien M., and Paxinos G. (1978) Chlorimipramine inhibition of muricide: The role of the ascending 5-HT projection. *Brain Res.* **149**, 270–273.

Martin P., Soubrie P., and Simon P. (1986a) Shuttle-box deficits induced by inescapable shocks in rats: reversal by the beta-adrenoreceptor stimulants clenbuterol and salbutamol. *Pharmacol. Biochem. Behav.* **24**, 177–181.

Martin P., Soubrie P., and Simon P. (1986b) Noradrenergic and opioid mediation of tricyclic-induced reversal of escape deficits caused by inescapable shock pretreatment in rats. *Psychopharmacol.* **90**, 90–94.

Martin-Iverson M., Leclere J. -F., and Fibiger H. C. (1983) Cholinergic-dopaminergic interactions and the mechanisms of action of antidepressants. *Eur. J. Pharmacol.* **94**, 193–201.

Mavroidis M. L., Kanter D. R., Greenblum D. N., and Garver D. L. (1984) Adrenergic-receptor desensitization and course of clinical improvement with desipramine treatment. *Psychopharmacol.* **83**, 295–296.

Menkes D. B. and Aghajanian G. K. (1981) Alpha$_1$-adrenoceptor-mediated responses in the lateral geniculate nucleus are enhanced by chronic antidepressant treatment. *Eur. J. Pharmacol.* **74**, 27–35.

Menkes D. B. and Aghajanian G. K., and McCall R. B. (1980) Chronic antidepressant treatment enhances alpha-adrenergic and serotonergic responses in the facial nucleus. *Life Sci.* **27**, 45–55.

Menkes D. B., Aghajanian G. K., and Gallagher D. W. (1983a) Chronic antidepressant treatment enhances agonist affinity of brain alpha$_1$-adrenoceptors. *Eur. J. Pharmacol.* **87**, 35–41.

Menkes D. B., Kehne J. H., Gallager D. W., Aghajanian G. K., and Davis M. (1983b) Functional sensitivity of CNS alpha$_1$-adrenoceptors following chronic antidepressant treatment. *Life Sci.* **33**, 181–188.

Mishra R., Janowsky A., and Sulser F. (1980) Action of mianserin and zimelidine on the norepinephrine receptor coupled adenylate cyclase system in brain: Subsensitivity without reduction in beta-adrenergic receptor binding. *Neuropharmacology* **19**, 983–987.

Misu Y. and Kubo T. (1983) Presynaptic beta-adrenoceptors. *Trends Pharmacol. Sci.* **4**, 506–508.

Miyauchi T., Kitada Y., Nakamichi H., and Satoh S. (1984) Beta-adrenoceptor mediated inhibition of behavioral action of desipramine and of central noradrenergic activity in forced swimming rats. *Life Sci.* **35**, 543–551.

Mobley P. L. and Sulser F. (1980) Adrenal steroids affect norepinephrine sensitive adenylate cyclase system in rat limbic forebrain. *Eur. J. Pharmacol.* **65**, 321–323.

Modigh K. (1979) Long lasting effects of ECT on monoaminergic mechanisms, in *Neuropsychopharmacology*. (Saletu B., Berner P., and Hollister L., eds.) Pergamon, Oxford, pp. 11–20.

Moore K. E. and Kelly P. H. (1978) Biochemical pharmacology of mesolimbic and mesocortical dopaminergic neurons, in *Psychopharmacology: A Generation of Progress*. (Lipton M. A., DiMaschio A., and Killam K. F., eds.) Raven, New York, pp. 221–234.

Morris M. J., Elghozi J. -L., Dausse J. -P, and Meyer P. (1981) Alpha$_1$ and alpha$_2$ adrenoceptors in rat cerebral cortex: Effect of frontal lobotomy. *Naunyn-Schmiedebergs Arch. Pharmacol.* **316**, 42–44.

Muscat R., Towell A., and Willner P. (1988) Changes in dopamine autoreceptor sensitivity in an animal model of depression. *Psychopharmacol.* **94**, 545–550.

Nassif S., Kempf E., Cardo B., and Velley L. (1983) Neurochemical lesion of the locus coeruleus of the rat does not suppress the sedative effect of clonidine. *Eur. J. Pharmacol.* **91**, 69–76.

Nielsen J. A., Shannon N. J., Bero L., and Moore K. E. (1986) Effects of acute and chronic bupropion on locomotor activity and dopaminergic neurons. *Pharmacol. Biochem. Behav.* **24**, 795–799.

Nimgaonkar V. L., Goodwin G. M., Davies C. L., and Green A. R. (1985) Down-regulation of beta-adrenoceptors in rat cortex by repeated administration of desipramine, electroconvulsive shock and clenbuterol requires 5-HT neurones but not 5-HT. *Neuropharmacology* **24**, 279–283.

Nomura S., Shimizu J., Ueki N., Sakaida S., and Nakazawa T. (1984) The activation of mice's behavior and reduction of brain beta-adrenergic receptor binding following repeated administration of antidepressant drugs. *Jpn. J. Psychopharmacol.* **4**, 237–241.

Noreika L., Pastor G., and Liebman J. (1981) Delayed emergence of antidepressant efficacy following withdrawal in olfactory bulbectomized rats. *Pharmacol. Biochem. Behav.* **15**, 393–398.

O'Connor W. T. and Leonard B. E. (1986) Effect of chronic administration of the 6-aza analogue of mianserin (ORG. 3770) and its enantiomers on the behaviour and changes in noradrenaline metabolism of olfactory-bulbectomized rats in the "open field" apparatus. *Neuropharmacology* **25**, 267–270.

O'Neill K. A. and Valentino D. (1986) Chronic desipramine attenuates morphine analgesia. *Pharmacol. Biochem. Behav.* **24**, 155–158.

Okada F., Manier D. H., Janowsky A. J., Steranka L. R., and Sulser F. (1982) Role of aminergic neuronal input in the down-regulation by desipramine (DMI) of the norepinephrine (NE) coupled adenylate cyclase system in rat cortex. *Soc. Neurosci. Abstr.* **8**, 659.

Olpe H. R. and Schellenberg A. (1980) Reduced sensitivity of neurons to noradrenaline after chronic treatment with antidepressant drugs. *Eur. J. Pharmacol.* **63**, 7–13.

Overstreet D. H., Janowsky D. S., Gillin J. C., Shiromani P. J., and Sutin E. L. (1986) Stress-induced immobility in rats with cholinergic supersensitivity. *Biol. Psychiatr.* **21**, 657–664.

Pandey G. N., Dyksen M. W., Garver D. L. and Davis J. M. (1979) Beta-adrenergic receptor function in affective illness. *Am. J. Psychiatr.* **136**, 675–678.

Parale M. P. and Kulkarni S. K. (1986) Clonidine-induced behavioural despair in mice: Reversal by antidepressants. *Psychopharmacol.* **89**, 171–174.

Peet M. and Coppen A. (1979) The pharmacokinetics of antidepressant drugs: relevance to their therapeutic effects, in *Psychopharmacology of Depression.* (Paykel E. S. and Coppen A., eds.) Oxford University Press, Oxford, pp. 91–107.

Peroutka S. J. and Snyder S. H. (1980) Long term antidepressant treatment decreases spiroperidol-labelled serotonin receptor binding. *Science* **210**, 88–90.

Petrie W. M., Maffucci R. J. and Woolsley R. L. (1982) Propranolol and depression. *Am. J. Psychiatr.* **139**, 92–94.

Petty F., Saquitne J. L., and Sherman A. D. (1982) Tricyclic antidepressant drug action correlates with its tissue levels in anterior neocortex. *Neuropharmacology* **21**, 475–477.

Pimoule C., Briley M. S., Gay C., Loo H., Sechter D., Zarifian E., Raisman R., and Langer S. A. (1983) ^{3}H-rauwolscine binding in platelets from depressed patients and healthy volunteers. *Psychopharmacol.* **79**, 308–312.

Plaznik A. and Kostowski W. (1985) Modification of behavioral response to intra-hippocampal injections of noradrenaline and adrenoceptor agonists by chronic treatment with desipramine and citalopram: Functional aspects of adaptive receptor changes. *Eur. J. Pharmacol.* **117**, 245–252.

Plaznik A., Danysz W., and Kostowski W. (1984) Behavioral evidence for alpha$_1$-adrenoceptor up- and alpha$_2$-adrenoceptor down-regulation in the rat hippocampus after chronic desipramine treatment. *Eur. J. Pharmacol.* **101**, 305–306.

Plaznik A., Danysz W., and Kostowski W. (1985a) Mesolimbic noradrenaline but not dopamine is responsible for organization of rat behavior in the forced swim test and an anti-immobilizing effect of desipramine. *Pol. J. Pharmacol. Pharm.* **37**, 347–357.

Plaznik A., Danysz W., and Kostowski W. (1985b) A stimulatory effect of intraaccumbens injections of noradrenaline on the behavior of rats in the forced swim test. *Psychopharmacol.* **87**, 119–123.

Pollock B. G., Perel J. M., and Shostak M. (1985) Rapid achievement of antidepressant effect with intravenous chlorimipramine. *N. Engl. J. Med.* **312**, 1130.

Poncelet M., Gaudel G., Danti S., Soubrie P., and Simon P. (1986) Acute versus repeated administration of desipramine in rats and mice: Relationships between brain concentrations and reduction of immobility in the swimming test. *Psychopharmacol.* **90**, 139–141.

Porsolt R. D. (1981) Behavioral despair, in *Antidepressants: Neurochemical, Behavioral and Clinical Perspectives.* (Enna S. J., Malick J. B., and Richelson E. eds.) Raven, New York, pp. 121–139.

Porsolt R. D., Bertin A., Blavet N., Deniel M., and Jalfre M. (1979) Immobility induced by forced swimming in rats: Effects of agents which modify central catecholamine and serotonin activity. *Eur. J. Pharmacol.* **57,** 201–210.

Porsolt R. D., LePichon M., and Jalfre M. (1977) Depression: A new animal model sensitive to antidepressant treatments. *Nature* **266,** 730–732.

Pulvirenti L. and Samanin R. (1986) Antagonism by dopamine, but not noradrenaline receptor blockers of the anti-immobility activity of desipramine after different treatment schedules in the rat. *Pharmacol. Res. Comm.* **18,** 73–80.

Quinaux N., Scuvee-Moreau J., and Dresse A. (1982) Inhibition of in vitro and ex vivo uptake of noradrenaline and 5-hydroxytryptamine by five antidepressants: Correlation with reduction of spontaneous firing rate of central monoaminergic neurons. *Naunyn-Schmiedebergs Arch. Pharmacol.* **319,** 66–70.

Sangdee R. and Frantz D. N. (1979) Enhancement of central norepinephrine and 5-hydroxytryptamine transmission by tricyclic antidepressants: A comparison. *Psychopharmacol.* **62,** 9–16.

Scavone C., Aizenstein M. L., De Lucia R., and Da Silva Planeta C. (1986) Chronic imipramine administration reduces apomorphine inhibitory effects. *Eur. J. Pharmacol.* **132,** 263–267.

Schildkraut J. J. (1975) Norepinephrine metabolism after short- and long-term administration of tricyclic antidepressant drugs and electroconvulsive shock, in *Neurobiological Mechanisms of Adaptation and Behavior.* (Mandell A. J., ed.) Raven, New York, pp. 137–153.

Schoffelmeer A. N. M., Hoornemen E. M. D., Sminia P., and Mulder A. H. (1984) Presynaptic alpha- and postsynaptic beta-adrenoceptor sensitivity in slices of rat neocortex after chronic treatment with various antidepressant drugs. *Neuropharmacology* **23,** 115–119.

Seligman M. E. P. (1975) *Helplessness: On Depression, Development and Death.* Freeman, San Francisco.

Sellinger-Barnette M. D., Mendels J., and Frazer A. (1980) The effects of psychoactive drugs on beta-adrenergic binding sites in rat brain. *Neuropharmacology* **19,** 447–454.

Serra G., Argiolas A., Klimek V., Fadda F., and Gessa G. L. (1979) Chronic treatment with antidepressants prevents the inhibitory effect of small doses of apomorphine on dopamine synthesis and motor activity. *Life Sci.* **25**, 415–424.

Sherman A. D. and Petty F. (1980) Neurochemical basis of the action of antidepressants on learned helplessness. *Behav. Neural Biol.* **30**, 119–134.

Sherman A. D., Allers G. L., Petty F., and Henn F. A. (1979) A neuropharmacologically relevant animal model of depression. *Neuropharmacology* **18**, 891–893.

Sherman A. D., Saquitne J. L., and Petty F. (1982) Specificity of the learned helplessness model of depression. *Pharmacol. Biochem. Behav.* **16**, 449–454.

Shibata S., Nakanishi H., Watanabe S., and Ueki S. (1984) Effects of chronic administration of antidepressants on mouse-killing behavior (muricide) in olfactory bulbectomized rats. *Pharmacol. Biochem. Behav.* **21**, 225–230.

Shopsin B., Friedman E., and Gershon S. (1976) Parachlorophenylalanine reversal of tranylcypromine effects in depressed patients. *Arch. Gen. Psychiatr.* **33**, 811–819.

Shopsin B., Friedman E., Goldstein M., and Gershon S. (1975) The use of synthesis inhibitors in defining a role for biogenic amines during imipramine treatment in depressed patients. *Psychopharmacol. Commun.* **1**, 239–249.

Skuza G., Rogoz Z., and Maj J. (1988) Antidepressants given repeatedly increase the behavioural effect of dopamine D-2 agonist. (Abstr.) Presented at European Behavioural Pharmacology Society, Athens, September 1988.

Siever L. J., Cohen R. M., and Murphy D. L. (1981) Antidepressants and alpha$_2$-adrenergic autoreceptor desensitization. *Am. J. Psychiatr.* **138**, 681–682.

Smialowski A. and Maj J. (1985) Repeated treatment with imipramine potentiates the locomotor effect of apomorphine administered into the hippocampus in rats. *Psychopharmacol.* **86**, 468–471.

Soubrie P., Martin P., El Mestikawy S., Thiebot M. H., Simon P., and Hamon M. (1986) The lesion of serotonergic neurons does not prevent antidepressant-induced reversal of escape failures produced by inescapable shocks in rats. *Pharmacol. Biochem. Behav.* **25**, 1–6.

Spyraki C. and Fibiger H. C. (1981) Behavioral evidence for supersensitivity of postsynaptic dopamine receptors in the mesolimbic system after chronic administration of desipramine. *Eur. J. Pharmacol.* **74**, 195–206.

Spyraki C., Papadoulou Z., Kourkoubas A., and Varonos D. (1985) Chlorimipramine, electroconvulsive shock and combination thereof:

Differential effects of chronic treatment on apomorphine-induced behaviours and on striatal and mesocortical dopamine turnover. *Naunyn-Schmiedeberg's Arch. Pharmacol.* **329,** 128–134.

Starke K., Taube H. D., and Borowski E. (1977) Presynaptic receptor systems in catecholaminergic transmission. *Biochem. Pharmacol.* **26,** 259–268.

Stolz J. F., Marsden C. A., and Middlemiss D. N. (1983) Effect of chronic antidepressant treatment and subsequent withdrawal on [^{3}H]-5-hydroxytryptamine and [^{3}H]-spiperone binding in rat frontal cortex and serotonin receptor mediated behaviour. *Psychopharmacol.* **80,** 150–155.

Sugrue M. F. (1982) Effect of chronic antidepressants on rat brain alpha$_2$-adrenoceptor sensitivity, in *Typical and Atypical Antidepressants: Molecular Mechanisms.* (Racagni G. and Costa E., eds.) Raven, New York, pp. 55–62.

Sulser F. (1978) New perspectives on the mode of action of antidepressant drugs. *Trends Pharmacol. Sci.* **1,** 92–94.

Sulser F., Janowsky A. J., Okada F., Manier D. H., and Mobley P. L. (1983) Regulation of recognition and action function of the norepinephrine (NE) receptor-coupled adenylate cyclase system in brain: Implications for therapy of depression. *Neuropharmacology* **22,** 425–431.

Svensson T. H. (1980) Effect of chronic treatment with tricyclic antidepressant drugs on identified brain noradrenergic and serotonergic neurons. *Acta Psychiatr. Scand.* **61, (Suppl. 280),** 121–131.

Thompson C., and Checkley S. A. (1985) The effects of alaproclate on the pupillary responses to tyramine, phenylephrine and pilocarpine in depressed patients. *Psychopharmacol.* **85,** 65–88.

Thompson C., Checkley S. A., Corn T., Franey C., and Arendt J. (1983) Down-regulation of pineal beta-adrenoceptors in depressed patients treated with desipramine? *Lancet* **1,** 1101.

Towell A., Muscat R., and Willner P. (1986) Behavioural evidence for autoreceptor subsensitivity in the mesolimbic dopamine system during withdrawal from antidepressant drugs. *Psychopharmacol.* **90,** 64–71.

Tucker J. C. and File S. E. (1986) A review of the effects of tricyclic and atypical antidepressants on spontaneous motor activity in rodents. *Neurosci. Biobehav. Rev.* **10,** 115–121.

Turmel A. and De Montigny C. (1984) Sensitization of rat forebrain neurons to serotonin by adinazolam, an antidepressant triazolobenzodiazepine. *Eur. J. Pharmacol.* **99,** 241–244.

Ueki S. (1982) Mouse-killing behaviour (muricide) in the rat and the effect of antidepressants, in *New Vistas in Depression*. (Langer S. Z., Takahashi R., Segawa T., and Briley M. eds.) Pergamon, New York, pp. 187–194.

Vetulani J. and Antikiewicz-Michaluk L. (1985) Alpha-adrenergic receptor changes during antidepressant treatment. *Acta Pharmacol. Toxicol.* **56, Suppl. 1,** 55–65.

Vetulani J. and Sulser F. (1975) Action of various antidepressant treatments induces reactivity of noradrenergic cyclic AMP generating system in limbic forebrain. *Nature* **257,** 495–496.

Vetulani J., Lebrecht U., and Nowak J. Z. (1981) Enhancement of responsiveness of the central serotonergic system and serotonin-2 receptor density in rat frontal cortex by electroconvulsive shock treatment. *Eur. J. Pharmacol.* **76,** 81–85.

Vetulani J., Stawarz R. J., Dingell J. V., and Sulser F. (1976) A possible common mechanism of action of antidepressant treatments. *Naunyn-Schmiedebergs Arch. Pharmacol.* **293,** 109–114.

Vogel G. W., Minter K., and Woolwine B. (1986) Effects of chronically administered antidepressant drugs on animal behavior. *Physiol Behav.* **36,** 659–666.

Von Voightlander P. F., Triezenberg H. J., and Losey E. G. (1978) Interaction between clonidine and antidepressant drugs: A method for identifying antidepressant-like agents. *Neuropharmacology* **17,** 375–381.

Wang R. Y. and Aghajanian, G. K. (1980) Enhanced sensitivity of amygdaloid neurons to serotonin and norepinephrine after chronic antidepressant treatment. *Commun. Psychopharmacol.* **4,** 83–90.

Weiss B., Heydorn W., and Frazer A. (1982) Modulation of the beta-adrenergic receptor-adenylate cyclase system following acute and repeated treatment with antidepressants, in *Typical and Atypical Antidepressants: Molecular Mechanisms*. (Costa E. and Racagni G. eds.) Raven, New York, pp. 37–53.

Weiss J. M., Bailey W. H., Goodman P. A., Hoffman L. J., Ambrose M. J., Salman S., and Charry J. M. (1982) A model for neurochemical study of depression, in *Behavioural Models and the Analysis of Drug Action*. (Spiegelstein M. Y. and Levy A., eds.) pp. 195–223. Elsevier, Amsterdam.

Weiss J. M., Glazer H. I., Pohorecky L. A., Brick J., and Miller N. E. (1975) Effects of chronic exposure to stressors on avoidance-escape behaviour and on brain norepinephrine. *Psychosom. Med.* **37,** 522–534.

Willner P. (1983) Dopamine and depression: A review of recent evidence. III. The effects of antidepressant treatments. *Brain Res. Rev.* **6**, 237–246.

Willner P. (1984a) The ability of drugs to desensitize beta-adrenergic receptors is not correlated with their clinical potency. *J. Affect. Disord.* **7**, 53–58.

Willner P. (1984b) The validity of animal models of depression. *Psychopharmacol.* **83**, 1–16.

Willner P. (1985a) Antidepressants and serotonergic neurotransmission: an integrative review. *Psychopharmacol.* **85**, 387–404.

Willner P. (1985b) *Depression: A Psychobiological Synthesis.* Wiley, New York.

Willner P. and Montgomery A. (1981) Behavioural changes during withdrawal from the tricyclic antidepressant desmethylimipramine (DMI) I. Interactions with amphetamine. *Psychopharmacol.* **75**, 54–59.

Willner P., Towell A., and Montgomery A. (1984) Changes in amphetamine-induced anorexia and stereotypy during chronic treatment with antidepressant drugs. *Eur. J. Pharmacol.* **98**, 397–406.

Willner P., Towell A., Sampson D, Sophokleous S., and Muscat R. (1987) Reduction of sucrose preference by chronic mild unpredictable stress, and its restoration by a tricyclic antidepressant. *Psychopharmacol.* **93**, 358–364.

Willner P., Muscat R., Sampson D, and Towell A. (1988) Dopamine and antidepressant drugs: The antagonist challenge strategy, in *Progress in Catecholamine Research.* (Belmaker H., Sandler M., and Dahlstrom A. eds.) **Vol. 3.** Alan R. Liss, New York, in press.

Wise H. and Halliday C. A. (1985) Why is amitriptyline much weaker than desipramine at decreasing beta-adrenoceptor numbers? *Eur. J. Pharmacol.* **110**, 137–141.

Wolfe D. B., Harden T. K., Sporn J. R., and Molinoff P. B. (1978) Presynaptic modulation of beta adrenergic receptors in rat cerebral cortex after treatment with antidepressants. *J. Pharmacol. Exp. Ther.* **207**, 446–457.

Zacharko R. M., Bowers W. J., and Anisman H. (1984) Responding for brain stimulation: Stress and desmethylimipramine. *Prog. Neuro-Psychopharmacol. Biol. Psychiat.* **8**, 601–606.

Zebrowska-Lupina I. (1980) Presynaptic alpha-adrenoceptors and the action of tricyclic antidepressant drugs in behavioural despair in rats. *Psychopharmacol.* **71**, 169–172.

Adaptation in Neuronal Calcium Channels as a Common Basis for Physical Dependence on Central Depressant Drugs

John M. Littleton
Hilary J. Little

1. Introduction

There are many ways of approaching the study of drug tolerance and dependence, as exemplified in the various chapters in this book. What we have attempted in this chapter is to take one of the simplest properties of neurones, that is, their electrical excitability, to consider how this might be modified by central depressant drugs and how neurones might therefore adapt to the presence of such drugs. We have then reviewed the neurochemical and behavioral evidence that supports the concept that such adaptation might lead to tolerance and physical dependence. We have, therefore, deliberately oversimplified the situation in an attempt to provide a framework on which more complex explanations can be built. We believe this is justified, not least because it has led to the wide-ranging and testable hypothesis embodied in the title.

Dependence on central depressant drugs, including alcohol, barbiturates, benzodiazepines, and opiates, is arguably the most important mental health problem in the world. By most estimates, the size of the problem is growing, and there is no clearly successful therapeutic approach. It is vitally important that we reach some understanding of the mechanism(s) by which tolerance and dependence on these drugs develops in order that we may interfere, pharmacologically or psychologically, in this progression.

1.1. Decremental and Oppositional Tolerance

Most mechanistic explanations of physical dependence (e.g., Collier, 1965) assume that this is simply an extension of tolerance. Thus, tolerance is seen as an adaptive response normalizing function in the presence of the drug. Removal of the drug exposes this adaptation, which, being no longer opposed by the presence of the drug, causes the withdrawal syndrome. This approach to tolerance and physical dependence, first promulgated by Himmelsbach in 1942, can be seen as an obvious oversimplification if one considers the mechanisms that could produce tolerance. Other chapters in this book discuss tolerance in terms of altered disposition of drugs (Chapter by Lê and Khanna) or in terms of learned behavior (Chapters by Siegel and Wolgin). In the case of "dispositional" tolerance, this is due to reduced access of the drug to its site of action, e.g., the increased rate of elimination of barbiturates after repeated administration lessening the effects of these drugs on the brain. In "learned" or "context-specific" tolerance, the animal modifies its behavior or physiology to lessen the effect of the drug in the specific test situation. In the case of dispositional tolerance, it is difficult to see how the mechanism could cause a withdrawal syndrome on removal of the drug. Alterations in the pharmacokinetic disposition of drugs would not be expected to produce, for example, central nervous system hyperreactivity on removal of the depressant drug. In other words, some mechanisms involved in tolerance are associated with a *decrease* in the effects of the drugs, rather than the institution of a mechanism *opposing* the actions of the drug. This type of tolerance has previously been called "decremental" adaptation, whereas that which can produce a withdrawal syndrome has been called "oppositional" adaptation (Littleton, 1983). All these kinds of adaptation can coexist and interact, but conceivably only one mechanism out of many has a causal relationship to physical dependence (*see* Fig. 1).

Context-specific tolerance need not always be of the decremental type. In this form of tolerance, the response of the animal to the drug is dependent on the test environment. Tolerance is seen when the tests are made in the surroundings in which the animal has received the chronic drug treatment, but not, or to a lesser degree, in a novel environment (Lê et al., 1979). The phenomenon may be due, at least in part, to an adaptive response of the organism in a direction opposite to that of the effect caused by the drug. Siegel (1975) demonstrated hyperalgesia after saline injection in rats tolerant to the analgesic action of morphine. Similarly, when rats were given saline injections in an environment in which they had developed tolerance to the hypothermic effects of ethanol injections, they exhibited hyperthermia (Mansfield and Cunningham, 1980). Since its original demonstration, the phenomenon of context-specific tolerance has been extensively studied in the whole animal (Siegel, 1975; Goudie and Demellweek, 1986; *see* review by Siegel, this volume), but the elucidation of its neuronal and molecular basis is clearly going to be difficult. Changes in neuronal activity must occur at some level during the adaptive response, and these may involve alterations in ion channels such as we describe in this chapter. Such alterations may be transitory in the case of context-specific tolerance, induced by the environment, and more stable when caused by long-term drug administration, but they could still have the same neuronal basis. Interaction between the cues provided by the environment and the drug administration may produce a greater adaptive response than that produced by the drug alone. We suggest that one important mechanism of "oppositional" adaptation at the neuronal level is an alteration in ion channel activity. This activity may be altered, in addition, by factors

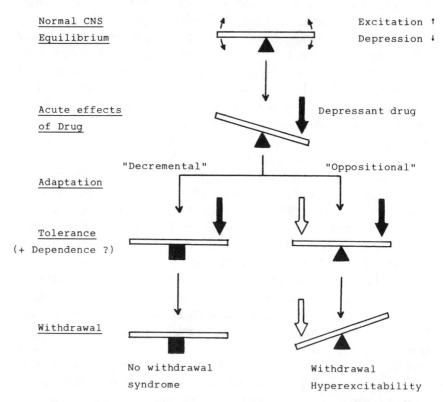

Fig. 1. *The concepts of "decremental" and "oppositional" adaptive mechanisms.* The level of excitation of central neurons is seen as an equilibrium between excitatory and depressant influences. In normal circumstances, these are relatively well balanced. The presence of a CNS depressant drug disturbs this balance towards depression and, thus, produces a reduction in central excitability acutely. Adaptive mechanisms of two kinds are then set in motion. In one "decremental," adaptation produces a lessening of the drug effect. There are several possible mechanisms for this: in the diagram, an alteration in the capacity of the system to be disturbed has been shown. The other type of adaptation, "oppositional" adaptation, implies the institution of an opposing excitatory influence to counteract the presence of the drug. Although both mechanisms can restore the level of CNS excitation to normal (i.e., cause tolerance), only "oppositional" adaptation leads to hyperexcitability on removal of the drug (i.e., causes physical dependence).

factors other than the presence of the drug, and our conceptual framework extends to include these factors.

Since we are attempting to construct a simple model for physical dependence, let us now consider the situation at the level of individual neurones, rather than the whole organism. Here, too, there is a distinction between "decremental" and "oppositional" adaptation. For example, the exclusion of ethanol from its putative site of action in neuronal membranes (Rottenberg et al., 1981) is an example of dispositional tolerance and can be classed as decremental adaptation. On the other hand, functions such as neurotransmitter release can show a mixture of decremental and oppositional changes. Ethanol added in vitro to brain preparations from control animals depresses depolarization-evoked release of some transmitters, and this is lost in preparations taken from animals made tolerant to ethanol in vivo, although release in the absence of ethanol did not differ between the two groups, i.e., decremental adaptation (Clark et al., 1977; Lynch and Littleton, 1983). However, in the case of other transmitters, depolarization-evoked release is greater in the preparations taken from ethanol-tolerant animals (Clark et al., 1977; Lynch and Littleton, 1983). Even at the level of single neurones, therefore, it is likely that several mechanisms, both decremental and oppositional, underlie the tolerance to a single drug. These mechanisms may involve, for example, receptors (Collier, 1965), neuronal enzymes (Goldstein and Goldstein, 1961), second-messenger systems (Collier and Francis, 1975), or ion channels (this chapter). Conceivably, only one of these mechanisms is the oppositional adaptation that has a causal relationship to physical dependence.

This modification of the original postulate by Himmelsbach (1942) to incorporate multiple mechanisms of tolerance has several important consequences. First, there may clearly be situations in which the manifestations of one mechanism of tolerance can be prevented without preventing the manifestation of physical dependence (as a withdrawal syndrome). Thus, Ritzmann and Tabakoff (1976) showed that treatment of mice with 6-hydroxydopamine prevented the manifestation of tolerance to the hypothermic effect of ethanol, but did not reduce the severity of the withdrawal syndrome. The presence of multiple mechanisms for tolerance also has implications for the demonstration of cross-tolerance and cross-dependence between different drugs. Since these phenomena are central to the hypothesis discussed in this chapter, these implications will now be considered.

1.2. Cross-Tolerance and Cross-Dependence

Central depressant drugs all, by definition, depress electrical activity in the neurons of the central nervous system. Therefore, at a very basic level, one might expect that they should evoke a similar adaptive mechanism, and this implies cross-tolerance. On the other hand, the mechanisms by which these drugs reduce electrical activity in neurons and the neuronal pathways that are affected probably differ considerably. On this basis, one would predict that more sophisticated mechanisms of adaptation would differ between the various classes of central depressants. This, in turn, implies that cross-tolerance would not exist. In our view, both the above predictions are correct. Whether cross-tolerance can be demonstrated or not depends on the drug effect studied. If it is one in which the tolerance is mainly due to our putative basic adaptive mechanism, then some cross-tolerance will be observed. If it is an effect in which tolerance is due to a more sophisticated mechanism, specific for the original drug used to produce the tolerance, then no cross-tolerance will be observed. Therefore, an absence of cross-tolerance between two groups of central depressants in some behavioral or psychological tests cannot be taken as proof that the drugs do not share any common adaptive mechanism. It indicates only that the adaptive mechanisms tested to date are unlikely to be shared.

Cross-tolerance between drugs of different pharmacological groups has been demonstrated in some situations, but is not always found (*see* review by Khanna and Meyer, 1982). Cross-dependence, however, which is the ability of a drug to prevent the withdrawal syndrome caused by another type of compound, is well established, both clinically and experimentally. Barbiturates, for example, suppress the alcohol withdrawal syndrome, but cannot always be shown to produce cross-tolerance to ethanol (*see* chapter by Lê and Khanna). The cross-tolerance reported in some instances between ethanol and morphine, such as that to their hypothermic effects (Khanna et al., 1979), may have been related to the functional disturbance produced by the drugs, or may have been influenced by environmental cues. However, morphine has been shown to reduce ethanol drinking and to significantly decrease the ethanol withdrawal syndrome in rats (Sinclair et al., 1973; Blum et al., 1977). Cross-tolerance has also been found between these compounds in vitro (Meyer et al., 1980).

The discussion above has all been related to adaptive mechanisms that could produce cross-tolerance, and both decremental and oppositional adaptive mechanisms are potentially able to do this. However, in the case of cross-dependence, we would expect the important mechanism of adaptation to be of the oppositional type only. Here it is necessary to say a few words about the experimental conditions used to investigate tolerance and dependence and the kinds of adaptation that these are likely to evoke. Tolerance to many drugs may easily and conveniently be produced by single daily injections of the drug over a period of a week or two. In such a regime, concentrations of the drug in the brain almost certainly show marked peaks and troughs. These conditions seem unlikely to produce an oppositional form of tolerance, since there is no continuous pressure to adapt in this way. In order to minimize functional disturbance by the drug when concentrations are fluctuating, a constant opposing force would not be useful (*see* Littleton, 1983). It seems more likely that an adaptive mechanism is activated that simply produces a reduction in the effect of the drug, i.e., decremental adaptation. This is presumably why such regimes, though capable of producing tolerance, rarely cause physical dependence (e.g., Goldstein, 1972). On the other hand, experimental paradigms in which high concentrations of a drug are maintained in the brain for several days are capable of inducing physical dependence. In these circumstances, a response opposing the continuous presence of the drug is an appropriate adaptive mechanism. These comments are meant to show that the mechanisms that underlie tolerance, as evoked by the first type of experiment, designed to study tolerance and cross-tolerance, may not be the same as those induced by the second type of experiment, designed to investigate dependence and cross-dependence.

1.3. Aims of the Present Chapter

This chapter seeks a common adaptive mechanism for central depressant drugs that might underlie physical dependence. This implies cross-dependence, which is the ability of one drug to suppress the physical symptoms of withdrawal from another. Unfortunately, cross-dependence is a poor indication of a common adaptive mechanism, since any state of neuronal hyperactivity, including a withdrawal syndrome, is likely to be suppressed by any depressant drug. Nevertheless, there are some basic similarities between the physical syndromes of withdrawal for all the central depressant drugs discussed here; certainly, all exhibit neuronal hyperexcitability in the central nervous system. The opiates might initially be thought to show a different pattern from those of the other groups of compounds described in this chapter (ethanol, barbiturates, benzodiazepines), but they do share some properties, and there is now good evidence that their effects on ion channels, and the adaptive responses of these, are similar, although the pathways involved may differ. If the physical syndrome of withdrawal from central nervous system depressants does share a common mechanism, then, from our previous comments, we believe this is likely to be by some basic oppositional adaptive mechanism in neurons. Since this mechanism need not be in identical neuronal pathways for the different groups of central depressants, then the withdrawal syndromes may differ subtly and cross-dependence may not be complete.

The approach we have taken is first to review the mechanisms by which each class of central depressant drugs inhibits neuronal excitability and to establish which reported adaptations in neurons are likely to overcome these. We have suggested that a final common path for the inhibition of neuronal activity by central depressants is the inhibition of calcium entry into neurons through voltage-operated calcium channels. The adaptation that seems most likely to underlie physical dependence after chronic administration of these drugs (i.e., a basic oppositional adaptation) is an increase in a particular subtype of neuronal voltage-operated Ca^{2+} channels. These channels are sensitive to inhibition by the "Ca^{2+} antagonist" drugs. We have next considered the implications of this hypothesis and the extent to which these are borne out by experimental evidence. In particular, we have considered potential interactions between Ca^{2+} antagonist drugs, the acute effects of central depressant drugs, and the development of tolerance and physical dependence on these. Our conclusion is that these Ca^{2+} antagonists, previously thought to be of very limited interest in the central nervous system, offer an exciting new tool for the experimental and therapeutic manipulation of tolerance and physical dependence on central depressant drugs.

2. Mechanism of Action of Central Depressants

2.1. Alcohols and General Anesthetics

2.1.1. Introduction

General anesthetics are a group of compounds of very diverse structures and are usually assumed to include the aliphatic alcohols. Since they have no obvious chemical similarities, a common mechanism of action based on physical properties has been sought. This approach rapidly led to the realization that anesthetic potency within the group is very closely related to hydrophobicity (Meyer, 1901; Overton, 1901). This in turn suggests that the partitioning of alcohols and anesthetics into some hydrophobic area is important in their central depressant action. Since most of the hydrophobic material in the brain is within the lipids of cell membranes, most workers have concentrated on an action within neuronal membranes as a potential anesthetic mechanism. However, no evidence excludes actions at other hydrophobic sites, e.g., hydrophobic regions of cytosolic proteins. Although the later stages of the actions of general anesthetics are almost certainly on the ion channels involved in neuronal transmission, the primary site may be on the membrane lipid, the hydrophobic areas of proteins, or (more likely) the lipid/protein interface. Since most evidence of direct relevance to neurons emphasizes the role of depressant drugs on the function of the cell membrane, we will concentrate on these effects. Also, since the emphasis in this chapter is on tolerance and dependence, and since the majority of work in this area has been on ethanol rather than other alcohols or anesthetics, we too will concentrate on work on this drug.

The function of neurons is entirely dependent on the normal operation of membrane proteins. In particular, conduction of the electrical impulse depends on the opening of voltage-operated Na^+ channel proteins, and the transduction of the electrical impulse into release of transmitter depends on the opening of voltage-operated Ca^{2+} channel proteins in the nerve terminal membrane. Released transmitter molecules then interact with receptor proteins in the postsynaptic membrane, which, usually by modulating ion channel opening, alter the electrical excitability of the next neuron in the pathway.

This very simple model of neuronal activity suggests four ways in which alcohols and anesthetics could inhibit neuronal function. They could either act presynaptically to inhibit conduction by interfering with Na^+ channel opening or to inhibit transmitter release by preventing Ca^{2+} channel opening. Alternatively, they could act postsynaptically to inhibit excitatory impulses or to potentiate inhibitory responses. Many excitatory influences on neurons are thought to be conducted down dendrites to the cell body via voltage-operated Ca^{2+} channels (Schwartzkroin and Slawsky, 1977), so that these influences would also be inhibited if ethanol prevented the opening of Ca^{2+} channels. A large proportion of inhibitory synapses in mammalian brain probably utilize γ-aminobutyric acid (GABA) as a transmitter (Schon and Iversen, 1974), and since the mechanism of GABA inhibition of neuronal excitability is to open Cl^- channels in the neuronal membrane (Curtis and Johnston, 1974), depressant drugs would be required to potentiate the opening of this ion channel protein in the latter case. There is evidence that all these processes can be affected by nonspecific CNS depressants in a way that leads to inhibition of neuronal excitability.

2.1.2. Effects on Na^+ Channels

Early electrophysiological studies using ethanol on invertebrate neurons showed inhibition of action potentials and inhibition of transient movements of Na^+ (Armstrong and Burnstock, 1964; Moore et al., 1964). These experiments required high concentrations of ethanol, but effects on Na^+ movement in electrically stimulated mammalian brain slices were seen at 105 mM ethanol, a concentration reached in states of severe intoxication (Bergmann et al., 1974). Inhibition of Na^+ movement into synaptosomes stimulated with batrachotoxin or veratridine is also inhibited by ethanol at high concentrations (Harris and Bruno, 1985; Mullin and Hunt, 1985). In general, although it is clear that ethanol can affect nerve conduction by an inhibitory effect on Na^+ channels, it is likely that this plays a role only in very severe ethanol intoxication. There is no evidence that any change in Na^+ channels that might oppose the effect of ethanol (and therefore make neurons hyperexcitable) occurs after chronic ethanol administration. Indeed, a reduction in Na^+ uptake into synaptosomes has been reported after the chronic administration of ethanol in vivo, in the absence of ethanol in vitro (Harris and Bruno, 1985).

2.1.3. Effect on Ca^{2+} Channels

The depolarization-induced release of neurotransmitters has been reported by several groups to be inhibited by ethanol (e.g., Clark et al., 1977; Lynch and Littleton, 1983). It can occur at concentrations that are intoxicating but not lethal, and several lines of evidence suggest it to be a consequence of inhibition of Ca^{2+} flux into nerve terminals (Harris and Hood, 1980; Leslie et al., 1983). The chronic administration of ethanol to laboratory animals has been reported to increase sensitivity to extracellular Ca^{2+} of neurotransmitter release from brain slices (Lynch and Littleton, 1983), suggesting that some increase in flux through Ca^{2+} channels may have occurred. No evidence for this potentially adaptive mechanism is seen, however, if Ca^{2+} flux or neurotransmitter release is studied in synaptosomal preparations from ethanol-treated rodents (Leslie et al., 1983). It is likely that the difference in results is a consequence of the different preparations used. Brain slice preparations contain both nerve terminals and cell bodies, and depolarization releases transmitter from both (Suetake et al., 1981). The results may therefore be interpretable as indicating an increase in Ca^{2+} channels on nerve cell bodies, rather than on nerve terminals. Recent developments in our understanding of Ca^{2+} channel types and their location make this more likely.

Patch clamp analysis of Ca^{2+} channels on cultured neurons indicate that there are three separate types of channels, distinguished by the duration with which they open on depolarization (Nowycky et al., 1985). These have been termed "T" (transient), "L" (long), and "N" (neither) channels. Further experiments using cell cultures support the concept that "L" channels are located mainly on cell bodies, and that "N" channels are mainly on nerve terminals and are coupled to the release of conventional neurotransmitters (*see* Miller, 1987 and illustration in Fig. 2). "T" channels may be responsible for electrical activity in the cell body and axon hillock. Clearly, neurons are very complex systems, and brain preparations are not ideal to establish whether a change of one or the other type of channel occurs in response to ethanol. For this reason, simpler systems of neuronal origin have been used to establish Ca^{2+} channel characteristics after chronic exposure to ethanol. To date, most important information has come from investigation of adrenal chromaffin cell cultures.

Adrenal chromaffin cells originate in the neural crest and migrate away from the central nervous system. They represent partially differentiated neurons, but without axons and nerve terminals. In culture, they display spontaneous action potentials, the frequency of which can be increased pharmacologically (nicotinic acetylcholine receptor activation in bovine cells). Release of stored adrenaline by exocytosis follows an increase in action potential generation or can be caused by prolonged depolarization of the cell membrane induced by increasing K^+ concentration. It is likely that the cell membranes contain both "N" and "L" type Ca^{2+} channels, but there may be a difference in the extent to which these are coupled to adrenaline release by different stimuli. Thus, release induced by K^+ depolarization is very sensitive to inhibition by the dihydropyridine (DHP) Ca^{2+} channel "antagonists," which selectively reduce Ca^{2+} flux through "L" channels, whereas release induced by nicotinic receptor stimulation is insensitive to these drugs (Boarder et al., 1986). The adrenal cell is therefore a good model for the neuronal cell body, and dihydropyridine sensitive Ca^{2+} flux and adrenaline release on depolarization provide indices of the functional activity of "L" type Ca^{2+} channels in the membrane. Radiolabeled dihydropyridine drugs bind to "L" type Ca^{2+} channels, so binding indices can be used to define the number of these channels in a culture.

When cells of adrenal origin (PC12 cells) are cultured in the presence of ethanol (200 mM for 6 d), there is a 100% increase in the number of [^{3}H]-DHP binding sites on the cell membrane (Messing et al., 1986). The K^+ depolarization-induced Ca^{2+} flux is also enhanced in cultures grown in ethanol. In primary cultures of similar cells (bovine adrenals), release of adrenaline is enhanced

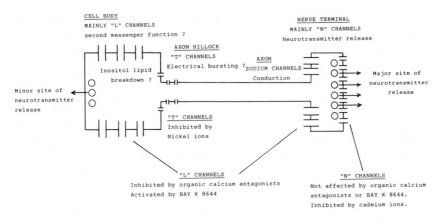

Fig. 2. *Different kinds of calcium channels on neurons.* The figure shows a schematic view of the site and roles of the various types of voltage-operated calcium channels on neurons. The nomenclature is that of Nowycky et al. (1985) and is based on the behavior of the channels under "patch-clamp" conditions. "L" channels are those with a *long* opentime, "T" channels open *transiently*, and "N" channels are *neither* (i.e., have an intermediate opentime). Much of the information on the likely site and function of these channels is reviewed in Miller (1987). "L" channels are the only type influenced in low concentrations by the organic "Ca^{2+} antagonists" such as verapamil, diltiazem, and dihydropyridines. Much of the evidence reviewed in this chapter suggests that central depressant drugs inhibit the opening of these Ca^{2+} channels (probably all types) and that an adaptive response involves an increase in numbers of the "L" subtype of Ca^{2+} channel.

at high external K^+ and Ca^{2+} concentrations from preparations grown in ethanol (Harper and Littleton, 1987). The increased Ca^{2+} flux and increased adrenaline release are sensitive to the dihydropyridine Ca^{2+} channel antagonists, confirming that "L" channels are probably responsible.

Evidence from adrenal cell cultures therefore suggests that chronic exposure to ethanol in vitro results in an increase in the number and function of "L" type Ca^{2+} channels in the cell membrane. Using similar techniques, it is now possible to assess whether the same changes occur in neuronal cell bodies in the central nervous system in vivo. Radioligand binding of dihydropyridine Ca^{2+} "antagonists" has been studied in the brains of animals treated chronically with ethanol by inhalation, a procedure that produced blood ethanol concentrations in excess of 50 mM. Such animals are severely intoxicated for several days, and demonstrate tolerance to the hypnotic effect of ethanol. Physical dependence can be demonstrated by a physical withdrawal syndrome on removal of the drug. This treatment increases the number of dihydropyridine binding sites on cerebral cortical membranes by 50% (Dolin et al., 1987b). This clearly suggests an increase in the number of "L" type Ca^{2+} channels on neuronal membranes associated with the development of ethanol dependence. Lucchi et al., 1985 also demonstrated that chronic ethanol treatment increased [^{3}H]-nitrendipine binding in rat brain, but decreased the sensitivity of this binding to calcium ions. In the above binding studies, tissues were obtained during intoxication, but the biochemical measurements were carried out after removal from chronic ethanol treatment.

Unlike the situation in adrenal cell cultures where "L" channels are unequivocally linked to adrenaline release, it is difficult to find evidence for functional alterations in "L" channels in the adult central nervous system. Release of neurotransmitters from neuronal cell bodies, where "L" channels predominate, almost certainly does occur (Suetake et al., 1981), but its physiological significance is uncertain and it is masked in brain slice preparations by the much greater release form nerve terminals. There are potentially two ways to circumvent these difficulties. One is to use an area of the brain in which cell bodies for a specific transmitter predominate and to study release of that transmitter. The second is to increase the contribution of "L" channels to depolarization-induced release by using a dihydropyridine activator of the "L" channels, such as BAY K 8644. We have employed both these strategies.

The dopaminergic nigrostriatal pathway has cell bodies in the substantia nigra and nerve terminals in the corpus striatum. The release of [^{3}H]-dopamine induced by K^+ stimulation from slices of substantia nigra and slices of corpus striatum of control rats and rats treated chronically with ethanol was compared for sensitivity to inhibition by a dihydropyridine Ca^{2+} "antagonist" (Pagonis and Littleton, 1987). In the terminal-rich slices of corpus striatum, the Ca^{2+} "antagonist" was without consistent effect on [^{3}H]-dopamine release in preparations from either treatment group. In the cell-body rich area (substantia nigra), the Ca^{2+} "antagonist" significantly inhibited release only in the preparations from ethanol-treated rodents, suggesting a greater contribution of "L" channels to the release process in these preparations.

Similarly, when release of either [³H]-noradrenaline (electrically induced release from cortical slices) or [³H]-dopamine (K^+- induced release from striatal slices) was potentiated by the "L" channel activator BAY K 8644, the potentiation was significantly greater in preparations from animals treated with ethanol by inhalation (*see above* for treatment details) (Pagonis and Littleton, 1987; Dolin et al., 1987a). This, too, suggests an increased number and/or functional activity of "L" type Ca^{2+} channels on neurons from ethanol-dependent animals.

As stated previously, the major function of "L" type Ca^{2+} channels on neurons may not be primarily concerned with transmitter release, and it has been suggested that they fulfill a "second-messenger" role related to inositol lipid breakdown (Baird and Nahorski, 1986). An increase in their function in this respect is strongly suggested by observations that the K^+ depolarization-induced breakdown of inositol lipids in rat cerebral cortical slices is greater in preparations from ethanol-dependent animals (Hudspith et al., 1987), and is also enhanced to a greater extent by BAY K 8644 (Dolin et al., 1987a) in these preparations.

Biochemically, therefore, considerable evidence supports the concept that neurons react to an inhibitory effect of ethanol on depolarization-induced Ca^{2+} entry by increasing the number of "L" type Ca^{2+} channels in the cell membrane. This increase in "L" Ca^{2+} channels seems likely to make the nerve cell body more responsive to excitatory stimuli. It can, therefore, be viewed as an appropriate response to the inhibitory effects of ethanol, both on neurotransmitter release (Clark et al., 1977) and on electrical activity on the soma itself (Eskuri and Pozos, 1987). Evidence that this "up-regulation" of neuronal Ca^{2+} channels may indeed underlie the development of tolerance and dependence on anesthetics and alcohols will be discussed later.

2.1.4. Effects on Cl⁻ Channels

The third potential mechanism for the inhibitory action of ethanol (and other alcohols and anesthetics) on neuronal excitability was suggested to be a potentiation of Cl⁻ flux through Cl⁻ channels in the neuronal membrane. The action of the inhibitory neurotransmitter, GABA, at the GABA/benzodiazepine receptor complex causes an increase in Cl^- flux across the neuronal membrane, which effectively clamps the membrane at the Cl^- reversal potential, making depolarizing stimuli relatively ineffective. As a consequence, the opening of voltage-operated channels (such as the Ca^{2+} channels described above) is inhibited, and neuronal excitability is reduced.

Electrophysiological experiments first demonstrated that ethanol potentiates the effects of GABA on neuronal excitability (Nestoros, 1980). Radioligand binding to GABA and benzodiazepine receptors, and to a third recognition site at which barbiturates and picrotoxin bind, suggested that ethanol interacted with the complex, decreasing binding at the TBPS/picrotoxin site and increasing binding to the low-affinity GABA site, but did not provide any clear indication as to functional consequences (Ticku, 1980; Davis and Ticku, 1981; Ramanjaneyulu and Ticku, 1984; Greenberg et al., 1984). Recent experiments on Cl^- flux into ''synaptoneurosomes'' do demonstrate a functional interaction between ethanol and GABA, with ethanol potentiating $^{36}Cl^-$ flux at concentrations as low as 20–30 mM (Suzdak et al., 1986). Similar experiments utilizing an experimental ligand for the benzodiazepine receptor provided evidence that this effect of ethanol on Cl^- flux is important in the acute intoxicating effects of the drug, as Ro 15-4513 selectively antagonized the intoxicating actions of ethanol, but not the lethal effects (Suzdak et al., 1986). However, Ro 15-4513 had the opposite action to ethanol, in that it decreased Cl^- flux, so the antagonism is not a simple interaction. These results suggest that one of the important effects of ethanol in low concentrations in the brain may be to reduce excitability of neurons that bear Cl^- channels. As stated previously, this will reduce the opening of voltage-operated ion channels in neuronal membranes.

As yet, few studies have been reported on the effects of chronic ethanol treatment, or treatment with any alcohols or anesthetics on the function of the GABA/benzodiazepine receptor complex in neuronal membranes, although there have been several studies on receptor binding. A decrease in low-affinity GABA binding has been reported during ethanol withdrawal (Ticku, 1980; Ticku and Burch, 1980), but neither TBPS binding nor the effect of ethanol at this site was altered after chronic ethanol treatment or during withdrawal (Liljequist et al., 1986; Ticku, 1987). Tolerance occurs, after chronic treatment, to the effect of ethanol in potentiating Cl^- flux (Allan and Harris, 1987). Durand and Carlen (1984) showed that, after withdrawal from chronic ethanol treatment, the amplitude of inhibitory potentials in the hippocampus was decreased. These potentials are likely to have been GABA mediated. There are some reports (e.g., Liljequist and Tabakoff, 1985; De Vries et al., 1987) that suggest modifications of the allosteric interactions between the subunits of the complex, but whether this has functional consequences is unknown. A reduction in Cl^- flux through these channels as an adaptive response to ethanol would not be expected to produce neuronal hyperexcitability *per se*, but might reduce inhibitory modulation. It is thus a potential mechanism for explaining ethanol tolerance and dependence, and should be investigated.

2.1.5. Conclusions

In general, the evidence that has been reviewed for ethanol favors an acute effect of the drug on synaptic transmission rather than nerve conduction. The action of ethanol on voltage-operated Ca^{2+} channels (which may in many cases be secondary to effects on Cl^- channels) is to reduce the release of transmitters from nerve terminals and to reduce postsynaptic excitatory impulses. Effects on Cl^- channels, which are observed at the lowest concentration of ethanol, potentiate inhibitory impulses. To date, most evidence for an adaptive response to ethanol favors an increase in neuronal excitability as a consequence of increased Ca^{2+} channel numbers and function. The Ca^{2+} channel that is "upregulated" may be primarily the "L" type present on cell bodies, and possibly modulating neuronal excitability via second messenger functions at this site.

2.2. Barbiturates

2.2.1. Introduction

The relation between potency and lipid solubility that characterizes anesthetics and alcohols also holds for the barbiturates (Hansch and Anderson, 1967), although the situation is complicated by the existence of barbiturate structures with excitatory actions (Downes et al., 1970). It is likely that the latter effects are caused by a different mechanism (Dunwiddie et al., 1986). In contrast to the actions of the nonspecific depressants discussed previously, there is little evidence that pharmacological effects of barbiturates are a consequence of inhibition of nerve conduction. The concentrations required to induce significant inhibition of nerve conduction are too high to be compatible with life. It is unlikely, therefore, that any effects of barbiturates on Na^+ channels are important to their actions in vivo. Most evidence favors depression of synaptic transmission as the site of barbiturate action. One important component of this effect probably occurs via the benzodiazepine/GABA receptor/Cl^- ionophore complex, although recent studies suggest that decreases in responses to excitatory amino acid transmitters may also play a role (Simmonds, 1988). To our knowledge, the effects of chronic barbiturate treatment on the actions of barbiturates on excitatory amino acid transmission have not been studied.

2.2.2. Effects on Cl^- Channels

The neuronal membrane protein that forms the Cl^- channel is a macromolecular complex containing the $GABA_A$ receptor (*see* Fig. 3). It has several high-affinity binding sites (Olsen, 1982), including one to which the barbiturates bind; this is distinct from the benzodiazepine binding site, which is also on the same complex (Olsen, 1982). The barbiturate site is near, or possibly identical to, the picrotoxin binding site, for which the ligand [^{3}H]-TBPS is often now used in binding studies (Squires et al., 1983).

The barbiturates potentiate the neuronal effects of the inhibitory transmitter GABA (Nicholl et al., 1975; Allan and Harris 1986), causing an increase in the duration of opening of chloride channels (Study and Barker, 1981). Barbiturates also open chloride channels of long duration in the absence of GABA (Mathers and Barker, 1980). The importance of these actions to the effects of the barbiturates seen in vivo is not clear. Evidence has been put forward that the general depressant actions, but not the anticonvulsant properties of these compounds, involve GABA potentiation (Allan and Harris, 1986), but it has also been suggested that the anticonvulsant actions of barbiturates correlate better with GABA potentiation, while their direct action on chloride channels may be

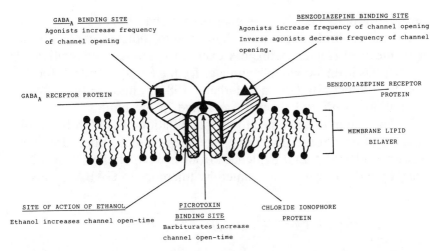

Fig. 3. *Interaction of central depressant drugs with the GABA$_A$ receptor: benzodiazepine receptor: chloride ionophore complex.* The figure shows a schematic cross-section through the macromolecular complex surrounding the chloride channel. Agonists at the GABA$_A$ receptor site increase the frequency of channel opening, and this is potentiated by benzodiazepine "agonists." "Inverse agonists" at the benzodiazepine binding site reduce the effectiveness of GABA$_A$ agonists in increasing the frequency of opening of the channel. Barbiturates bind to a separate site, closely related to the picrotoxin binding site on the chloride channel, and increase the duration of each opening (increase open time). Ethanol probably does not have a specific binding site, but alters the changes in the conformational states of the complex within the membrane bilayer, so that the open time of the channel is prolonged. Increased flux of Cl$^-$ through the channel stabilizes the membrane electrically, making opening of voltage-operated channels, including the various types of Ca^{2+} channels, less likely.

associated with general anesthetic activity (Macdonald and Barker, 1979). Evidence is therefore conflicting on this point (*see also* effects on Ca^{2+} channels, below), and Dunwiddie et al. (1986) have suggested that the difference between compounds with major depressant action, such as pentobarbitone, and more selective anticonvulsant compounds, such as phenobarbitone, may lie in the magnitude of the maximum potentiation of GABA that can be achieved. This is considerably lower for phenobarbitone than for pentobarbitone. The S(−) isomer of pentobarbitone is a more potent general anesthetic and a slightly more potent anticonvulsant than the (+)-isomer, with less excitatory action. Potentiation of GABA has been shown to be greater for the (−)isomer than for the (+)enantiomer (Barker and Mathers, 1981), although higher concentrations of the (+)isomer are found in the brain.

The extent of tolerance to the potentiation of GABA by barbiturates, after chronic treatment, has been studied little. Gray and Taberner (1985) found evidence, from in vivo studies, that tolerance to barbital was accompanied by tolerance to GABA agonists.

2.2.3. Effects on Ca^{2+} Channels

A large number of studies have demonstrated actions of barbiturates on Ca^{2+} channels. In 1975, Blaustein & Ector showed that barbiturates decreased the uptake of Ca^{2+} by synaptosomes at 400 μM, a concentration consistent with in vivo actions. Only the Ca^{2+} uptake stimulated by potassium depolarization, veratridine, or gramicidin was affected. This was later confirmed by others, who showed that $Ca2^{+}$ influx resulting from the ionophore A23187 was unaffected and Ca^{2+} uptake resulting from glutamate (which is not through voltage-dependent channels) was altered only by higher concentrations of barbiturates (Harris and Stokes, 1982). These authors did not consider that the actions of barbiturates were due to increased Na^{+}/Ca^{2+} exchange. Pentobarbitone (300 μM) decreased potassium-stimulated uptake of Ca^{2+} into synaptosomes from cerebral cortex, cerebellum, and brain stem, but not that from striatum, hypothalamus, or midbrain (Elrod and Leslie, 1980). The effective concentrations of pentobarbitone were found to be the same in several different species and strains (Friedman et al., 1980), and were within those found after in vivo administration causing hypnotic effects. Effects on Ca^{2+} uptake were suggested to cause the decreases in transmitter release seen with barbiturates, and to be involved in their sedative actions. Excitatory barbiturates had less effect (Harris and Stokes, 1982). These authors also found that the efflux of Ca^{2+} from synaptosomes was increased by pentobarbitone, though this is unaffected by ethanol.

ATP-dependent Ca^{2+} uptake by lysed synaptosomes (thought to represent intrasynaptosomal Ca^{2+} sequestration), and by mitochondria, was decreased by barbiturates. Higher concentrations (500 μM) were needed for this action than for effects of barbiturates on depolarization-induced uptake (Harris, 1981; Hood and Harris, 1980). It was suggested that this effect would raise intracellular Ca^{2+} concentrations and lead to increased resting transmitter release. Such an increase is found with high concentrations of barbiturates (Thomson and Turkanis, 1973).

Electrophysiological studies have also demonstrated actions of barbiturates on Ca^{2+} movements. Barbiturates decreased the Ca^{2+} currents of the action potentials of dorsal root ganglion cells; this effect was seen at 50–500 μM of pentobarbitone and 0.5–2 mM of phenobarbitone (Werz and Macdonald, 1985). These authors suggested that the barbiturates increase Ca^{2+} channel inactivation or cause open-channel block, as they increased the rate at which the currents decayed. An effect of several barbiturates on the decay phase of the Ca^{2+} current was also found by Nishi and Oyama (1983) using Helix neurons, at concentrations of 100 μM–1 mM. They suggested that the drugs increased the rate of voltage-dependent inactivation by causing conformational changes in the lipid surrounding the Ca^{2+} channel. The potential importance of the action of barbiturates on calcium ion movements was considered by Heyer and Macdonald (1982). These authors compared the concentrations of pentobarbitone and of phenobarbitone required to produce a variety of neuronal effects. Reductions in calcium action potentials were found at barbiturate concentrations that produced sedation and anesthesia in vivo. Anticonvulsant actions of barbiturates were said to correlate with concentrations causing decreases in excitatory synaptic transmission.

After chronic treatment with pentobarbitone, Harris and Stokes (1982) found tolerance to its effects on potassium-stimulated Ca^{2+} uptake; there was no change in baseline uptake or efflux. Rapid loss of tolerance to this action of pentobarbitone has been found during withdrawal (Harris et al., 1985; Leslie et al., 1980), and tolerance to the effect of pentobarbitone on potassium-stimulated Ca^{2+} uptake was demonstrated at a time when tolerance to the sedative actions of pentobarbitone was seen (Elrod and Leslie, 1980). Tolerance was also seen after chronic treatment to the action of pentobarbitone in decreasing ATP-dependent Ca^{2+} uptake into synaptosomes and that into mitochondria (Harris, 1981; Hood and Harris, 1980).

2.2.4. Conclusions

There are clear similarities between the behavioral effects and the likely neurochemical mechanisms for ethanol and the barbiturates. In the case of barbiturates, the evidence is clearly in favor of an action of these drugs on the Cl^- channel as being important in their CNS depressant action. There is, however, significant evidence of actions on Ca^{2+} flux, which may be via a direct effect on the voltage-operated Ca^{2+} channels or a consequence of decreased membrane excitability by action on the Cl^- channel. After chronic treatment with barbiturates, there is little evidence of adaptation within the GABA/benzodiazepine ionophore complex, but considerable evidence for alterations in the mechanisms that control Ca^{2+} homeostasis. The dynamic changes in "L" type Ca^{2+} channels reported in ethanol tolerance have not yet been sought in barbiturate tolerance, but given the similarities in the acute effects of the drugs, such changes may well play a role in the cross-tolerance and cross-dependence shown by these groups of drugs (*see below*).

2.3. Benzodiazepines

2.3.1. Introduction

Benzodiazepines have more specific pharmacological properties than barbiturates or other general depressant drugs; they have effects, such as the anticonvulsant, anxiolytic, and myorelaxant properties, at doses considerably lower than are required for other depressants. It is now well established that there are high-affinity binding sites in the CNS for the benzodiazepines (Squires and Braestrup, 1977), which are thought to mediate the pharmacological actions of these drugs, described above. These effects are antagonized by benzodiazepine receptor antagonists, such as Ro 15-1788 (Hunkeler et al., 1981) and ZK93426 (Jensen et al., 1984), which compete for the binding sites (*see* Fig. 3). However, the effects of high doses of benzodiazepines are probably not mediated through these receptors. Little and Bichard (1984) showed that the "general anesthetic" properties of high doses of several benzodiazepines were not antagonized by Ro 15-1788, but there was a correlation between their general anesthetic potency and the lipid solubility of the compounds. Another group of compounds with high affinity for benzodiazepine receptors are the benzodiazepine inverse agonists, which have pharmacological effects opposite to those of benzodiazepines, but which are thought to be mediated by the same receptors (Nutt et al., 1982). There is now compelling evidence that these receptors are, in the main, associated with the GABA$_A$ receptor Cl$^-$ ionophore complex in neuronal membranes.

2.3.2. Effects on Cl$^-$ Channels

Potentiation by benzodiazepines of the effects of GABA on Cl$^-$ channels is now well established (*see* Haefely et al., 1981 for review). GABA increases Cl$^-$ conductance by action at GABA$_A$ receptors. GABA action at these receptor sites results in either hyperpolarizing (Cl$^-$ influx) or depolarizing (Cl$^-$ efflux) synaptic potentials, according to the Cl$^-$ gradient across the a membrane. Both of these types of responses are potentiated by the benzodiazepines (Jahnsen and Laursen, 1981; Biscoe et al., 1983). GABA$_B$ receptors mediate other types of response that are not affected by benzodiazepines (Bowery et al., 1984). The potentiation by benzodiazepine agonists of the effects of GABA on Cl$^-$ channels is prevented by benzodiazepine receptor antagonists such as Ro 15-1788 (Hunkeler et al., 1981; Polc et al., 1981), which has no effect alone on the actions of GABA, except at very high concentrations (Little, 1984). Inverse agonists at the benzodiazepine receptor (such as β-CCE or FG7142) have opposite actions, decreasing Cl$^-$ conductance (Jensen and Lambert, 1983; 1984; Little, 1984). Although a direct action of benzodiazepines (at high concentrations) on Cl$^-$ channels has been reported (Barker and Owen, 1986), their main effect is thought to be due to their action on GABA, which occurs at micromolar concentrations of benzodiazepines. This order of concentration is achieved in the central nervous system after behaviorally effective doses of benzodiazepines. Study and Barker (1981) reported that diazepam increased the frequency of opening of Cl$^-$ channels, with some effect on the duration of channel opening, whereas pentobarbitone affected only the duration of channel opening. Parallel leftward shifts in dose–response curves for GABA and other GABA$_A$ receptor agonists are seen in the presence of benzodiazepines (Jensen and Lambert, 1984; Little, 1984). The Binding studies have shown that there are complex allosteric interactions between the three types of binding sites on the ionophore complex (Olsen, 1982), and these may influence the effects of drugs in vivo, but the exact mechanism of the potentiating action of benzodiazepines has yet to be fully elucidated. It does not seem to be accounted for by effects of benzodiazepines on GABA binding.

Tolerance occurs to all of the pharmacological properties of benzodiazepines, although it develops at different rates to the different effects, occurring rapidly for example to the sedative actions and more slowly to the anxiolytic effects (Brown et al., 1984). The mechanism is as yet unknown, although it does not appear to involve changes in benzodiazepine receptor binding. These have been reported only after extremely high levels of drug administration (e.g., Rosenberg and Chiu, 1979), and are not consistently found when tolerance is demonstrated (Braestrup et al., 1979). A decrease in postsynaptic sensitivity to GABA and in GABA stimulation of benzodiazepine binding has been reported after chronic benzodiazepine treatment (Gallager et al., 1984). However, GABA-stimulated benzodiazepine binding has also been reported to be increased after such chronic treatment (Stephens and Schneider, 1986). Only small and selective changes were found in GABA agonist actions after a chronic benzodiazepine treatment schedule, which caused tolerance to the sedative, anticonvulsant, and hypothermic effects of benzodiazepine agonists, although such treatment increased the actions of inverse agonists (Little, 1987).

2.3.3. Effects on Ca^{2+} Channels

Benzodiazepines do appear to have effects on Ca^{2+} fluxes, but the story is less clear than for the barbiturates or ethanol. Electrophysiological evidence (*see below*) suggests that Ca^{2+} channels may be more important in benzodiazepine action than generally supposed. In addition, effects on Ca^{2+} and Cl^- may be linked more closely than has been appreciated. A recent report showed that intracellular Ca^{2+} ions decrease the apparent affinity of the $GABA_A$ receptor (Inoue et al., 1986). In this study, on bullfrog sensory neurons, influx of Ca^{2+} was found to decrease subsequent GABA-activated

Cl^- fluxes. It is not yet clear whether this type of interaction occurs in the mammalian CNS.

Paul and coworkers found that benzodiazepines increased depolarization-induced Ca^{2+} uptake into synaptosomes, an effect prevented by CGS 8216, which is a weak inverse agonist at benzodiazepine receptors (Paul et al., 1982; Paul and Skolnick 1982). Mendelson et al. (1984a) showed that diazepam, at 1 μM, increased potassium-stimulated Ca^{2+} uptake by cortical synaptosomes. The Ca^{2+} channel "antagonists," nitrendipine and nifedipine (each at 1 μM), prevented this action of diazepam, in a "nonadditive" manner, but only nitrendipine decreased Ca^{2+} uptake when added alone. However, Peyton and Borowitz (1979) found decreased Ca^{2+} uptake into rat cortical synaptosomes after an ataxic dose (20 mg kg^{-1}, ip) of chlordiazepoxide. In addition, Leslie et al. (1980) showed that chlordiazepoxide (150 μM) inhibited potassium-stimulated Ca^{2+} uptake. This effect was also seen when chlordiazepoxide was given ip (20 mg kg^{-1}), suggesting that it may be involved in the behavioral actions of the benzodiazepine. These contradictory results are likely to have been the result of dose-related biphasic effects of benzodiazepines, since Ferendelli and Daniels-McQueen (1982) found depression of Ca^{2+} uptake into synaptosomes when diazepam was added at 10–100 μM, doses considerably higher than that used by Mendelson et al. (1984a). Such dose-related phenomena would be compatible with the actions of benzodiazepines on Ca^{2+} currents in electrophysiological studies.

Effects of midazolam have been demonstrated in the hippocampal slice preparation that appear to be due to effects on Ca^{2+} conductance (Carlen et al., 1983a). Low (nM) concentrations of midazolam caused hyperpolarizations and potentiated the after hyperpolarizations that followed spike trains. These effects were dependent on the presence of Ca^{2+}, but not on Cl^-. This, plus the observation that midazolam decreased the threshold and latency of Ca^{2+} action potentials, suggested that the hyperpolarizing effects were due to increases in Ca^{2+}-dependent potassium conductance. The effects of midazolam were blocked by Ro 14-7437, a benzodiazepine "antagonist," suggesting mediation via benzodiazepine receptors, but when given alone, Ro 14-7437 decreased Ca^{2+} spikes and Ca^{2+}-mediated potassium hyperpolarizations, so the interpretation of its antagonist action is difficult (Carlen et al., 1983b). Higher concentrations of midazolam were needed to potentiate the electrophysiological effects of GABA than to produce effects on Ca^{2+}-mediated conductances.

The situation is made more complex by the observation that, in addition to binding to the well-characterized receptor associated with the $GABA_A$ receptor/Cl^- ionophore complex, benzodiazepines also bind, with lower affinity, to another class of receptors, the so-called "peripheral" benzodiazepine sites, the function of which is, as yet, uncertain. These peripheral sites (which are also found in the CNS; Marangos et al., 1982) are thought to be associated with Ca^{2+} channels, although doubt has been cast on how close the association may be (Rampe and Triggle, 1986). Selective ligands for these sites, such as Ro 5-4864, block Ca^{2+} conductance in neurons (Skerritt et al., 1984); the peripheral receptor antagonist PK11195 prevented both this action and that of nitrendipine (Schramm et al., 1983). PK11195 antagonized the effects of BAY K 8644 on slow action potentials in guinea pig papillary muscle (Mestre et al., 1986). Doble et al. (1985) found no direct or allosteric interactions between dihydropyridine binding sites and peripheral benzodiazepine receptors in heart tissue, but concluded from the similarity of distribution that the two sites were in the same membrane compartment. Effects on voltage-sensitive Ca^{2+} channels in nerve terminals have been attributed to the "micromolar" benzodiazepine receptor, described by Taft and Delorenzo (1984), although the existence of this site is disputed (File et al., 1984). Little work seems to have been done on the "peripheral type" receptor after chronic benzodiazepine treatment. However, there is evidence in ethanol-dependent states for an increase in the number of these sites in rat brain (Schoemaker et al., 1983a; Tamborska and Marangos, 1986). The relation of this increase to the reported increase in Ca^{2+} channel binding sites in the same tissue in ethanol dependence (Dolin et al., 1987b) is not yet known.

Leslie et al. (1980) studied the effect of chronic treatment with chlordiazepoxide on the depression of synaptosomal uptake of Ca^{2+}, which they demonstrated to occur, both in vitro and in vivo (*see above*). Tolerance developed to this action in vitro after eight days of treatment with chlordiazepoxide in the diet, and this treatment also produced tolerance to the sedative effects of chlordiazepoxide. After the chronic treatment, the net potassium-induced increase in Ca^{2+} influx into synaptosomes, when no chlordiazepoxide was added in vitro, was not significantly different from control values. However, the total depolarized and nondepolarized accumulations were significantly decreased. The authors suggested that membrane binding of Ca^{2+} may have been reduced after the chronic treatment. The effects of chronic benzodiazepine treatment on the opposite action of diazepam, that of increasing Ca^{2+} influx into synaptosomes, shown by Mendelson et al., 1984a, have not been reported.

2.3.4. Conclusions

Once again, there are clear similarities between the behavioral and neurochemical effects of benzodiazepines, barbiturates, and ethanol. The benzodiazepines differ in that they appear to act via specific receptor binding sites, but they also affect Cl^- channels and Ca^{2+} channels. The evidence for any adaptation in the Cl^- channel system in response to benzodiazepines is at best equivocal, but there is evidence for adaptation in the Ca^{2+} system. As with barbiturates, no direct evidence for alterations in "L" type Ca^{2+} channels exists at present, but this is a possible explanation for cross-tolerance and cross-dependence between alcohol, barbiturates, and benzodiazepines (*see above*).

2.4. Opiates

2.4.1. Introduction

Although it cannot be denied that opiates cause depression of central nervous system activity, they do so in a way that differs considerably from that of drugs discussed previously. The opiates are "narcotic" in the sense that they cause a "cutting-off" from the environment, rather than being sedative or "hypnotic," i.e., producing a state resembling sleep. They also produce euphoria to a much greater extent than that shown by the central depressants discussed above. The molecular basis for the action of opiates is established as being via interaction with a family of receptors (μ, γ, κ, ϵ, δ) of which the μ-receptor is probably mainly responsible for the actions of morphine and the other abused opiate drugs. The mechanisms by which μ and other opiate receptors produce their cellular effects is not yet understood. It seems likely that they are coupled via GTP-binding proteins to intracellular enzymes such as adenyl cyclase or to ion channels (Hescheler et al., 1987). There is no evidence that these receptors modulate Cl^- channels at any site, but there is considerable evidence that neuronal Ca^{2+} channels can be modulated by the action of opiates on their receptors.

2.4.2. Effects on Ca^{2+} Channels

Opiates have been shown to modify depolarization-induced Ca^{2+} entry into neuronal preparations, producing a reduction in Ca^{2+} flux (Guerrero-Munoz et al., 1979) and a reduction in Ca^{2+} content (Ross et al., 1977). The interactions between opiates and Ca^{2+} ions have been reviewed by Chapman and Way (1980). These interactions suggest strongly that many of the acute effects of opiates can be explained by inhibition of Ca^{2+} entry into neurons. The mechanism may be via a direct effect of a GTP-binding protein, G_O with neuronal Ca^{2+} channels. The G_O protein is assumed to be regulated by an opiate receptor (Hescheler et al., 1987). The experimental protocol of these authors did not distinguish which Ca^{2+} channel type was affected. There is little doubt that opiates, via receptor activation, have other effects, e.g., modulation of adenyl cyclase activity, in neurons, and these may well be important in some actions of the drugs. However, the inhibitory effects on voltage-operated Ca^{2+} channels suggest a similar final mechanism of action between these highly selective drugs and those discussed earlier.

The rapid development of tolerance and physical dependence on opiates is a well-known therapeutic and social problem. The chronic administration of opiates to laboratory animals has been shown to produce an increase in Ca^{2+} uptake into various brain preparations (Ross et al., 1977; Harris et al., 1977; Guerrero-Munoz et al., 1979). These results can be interpreted as showing an increased Ca^{2+} flux through voltage-operated Ca^{2+} channels in neuronal membranes. As discussed earlier, this would be an appropriate adaptive response to the presence of a central depressant drug. It is likely that this phenomenon involves an increase in the number of "L" type Ca^{2+} channels in the brain, because the

[^{3}H]-DHP binding sites on central neuronal membranes of mice is increased about 60% by the institution of morphine tolerance (Ramakumar and El-Fakahany, 1984).

2.4.3. Conclusion

Although opiates have far more selective actions than many other central depressants, their mechanism for neuronal inhibition may turn out to be similar. Thus, by interaction with specific receptors on specific neuronal pathways, they reduce electrical excitability of the soma or terminals by inhibition of voltage-operated Ca^{2+} channels. There are possibly several adaptive mechanisms that could underlie opiate tolerance and dependence, including changes in cAMP and downregulation of opiate receptors. However, there is emerging evidence that an important mechanism involves the upregulation of the "L" subtype of Ca^{2+} channel. It is likely that, in the case of adaptation to opiates, such upregulation would be confined only to neurons in which opiate receptors are inhibitory to Ca^{2+} entry. This would clearly distinguish the end result of adaptation to morphine from that to, for example, ethanol where the relative nonspecificity of the mechanism of inhibition of Ca^{2+} entry should evoke an adaptive response in virtually all neurons (and possibly all excitable cells).

3. Acute Effects of Central Depressant Drugs — Summary

Taking a very simple property of neurons, their electrical excitability, we have suggested that central depressant drugs inhibit this property by interaction with voltage-operated cation channels and receptor-operated anion channels in the neuronal membrane. The balance between these actions depends on the group of drugs studied. The "nonspecific" depressant drugs, including alcohols and general anesthetics, are probably capable of affecting most ion channels in neuronal membranes at the high concentrations required for their anesthetic action. There is, however, increasing evidence that, of this group, ethanol may have unique actions on the Cl^- channels, which contribute to its intoxicating effect. Further refinement of this specificity for Cl^- channels is seen with the barbiturates, which bind to this channel and enhance GABA-mediated Cl^- flux. Barbiturates, however, probably also have inhibitory effects on Ca^{2+} channels that are not simply a consequence of their action on Cl^- channels. Benzodiazepines also are relatively selective for influencing the Cl^- channel via a receptor, but other ("peripheral" type) benzodiazepine receptors may be linked to Ca^{2+} channels, and a component of "classical" actions of benzodiazepines may be mediated through Ca^{2+} channels. In contrast, drugs such as morphine act selectively on receptors that may be coupled either to intracellular second-messenger systems or to membrane Ca^{2+} channels. In a molecular sense, these drugs are no more "specific" than benzodiazepines, but their lack of action on the ubiquitous GABA receptor/Cl^- ionophore system makes them more selective for specific neuronal pathways.

Stressing the similarities between these drugs, all are capable of modifying neuronal excitability via an action on voltage-operated Ca^{2+} channels. We have discussed at some length the possibility that ethanol evokes an adaptive response in neurons and that this involves an "upregulation" of "L" type Ca^{2+} channels. There is some evidence for a similar adaptive mechanism to barbiturates and benzodiazepines, and stronger evidence for this in morphine-induced adaptation.

If such a similarity in mechanisms and adaptation to these diverse drugs does exist, via their action on voltage-operated Ca^{2+} channels, then it has important consequences for acute and chronic behavioral effects of the drugs. These implications will be listed below, and the final section of the chapter considers to what extent the limited experimental evidence supports this hypothesis.

1. The acute effects of the CNS depressants should be potentiated by Ca^{2+} channel inhibitors, which cross the blood–brain barrier.
2. The acute effects of the CNS depressants should be inhibited by dihydropyridine Ca^{2+} channel activators, such as BAY K 8644.
3. The withdrawal syndrome from CNS depressants should be mimicked by administration of a Ca^{2+} channel activator such as BAY K 8644, and prevented by administration of Ca^{2+} channel inhibitors.

4. The development of tolerance (and dependence) to the CNS depressants should be prevented by agents that prevent upregulation of "L" type Ca^{2+} channels. This is not an essential prediction. Tolerance resulting from "decremental" adaptation need not be in any way affected.

4. Behavioral Interactions of Ca^{2+} Channel Inhibitors with CNS Depressants

The CNS contains high-affinity binding sites for dihydropyridines, which closely resemble the sites in the periphery (Gould et al., 1985). Calcium channel inhibitors, such as nitrendipine, have long been known to act on such sites in vascular smooth muscle, and are used in cardiovascular disorders. In addition, experimental evidence shows that Ca^{2+} channel activators, such as BAY K 8644, have the opposite effects, increasing Ca^{2+} conductance (Schramm et al., 1983). The calcium ionophore complex is thought to contain at least three binding sites, one of which has high affinity for dihydropyridines (Glossman et al., 1982). Until recently, however, these compounds appeared to have little action on the central nervous system, except in very high concentrations, despite the presence of many high-affinity binding sites. A major reason for this, as explained earlier, is that these compounds are selective for the "L" type of Ca^{2+} channel, which is not normally involved in processes such as transmitter release from nerve terminals.

In certain circumstances, neuronal effects of dihydropyridines can be demonstrated. Effects on cultured cells have been described in detail above. In addition, Turner and Goldin (1985) found that low concentrations of nitrendipine prevented the initial rapid phase of Ca^{2+} uptake into rat brain synaptosomes, but not the rest of the Ca^{2+} flux. Evidence for the existence of functional dihydropyridine-sensitive calcium channels, which varied in density in different areas of the CNS, was presented by Thayer et al., (1986). Another factor that may account for variations in the actions of dihydropyridines is that they have been suggested to bind to the inactivated state of the receptor/ionophore; the amount of time the complex spends in this state varies between tissues (Bean,

1984). The importance of the state of activation of the channel is illustrated by the actions of the dihydropyridines on the hippocampal slice. The action of these compounds on field potentials in this tissue was found to be limited, although small effects were seen at micromolar concentrations. However, when the neuronal activity was increased by the addition of the convulsant drug pentylenetetrazol, the paroxysmal activity induced by the convulsant disappeared after addition of 1 μM nifedipine (Louvel et al., 1986). Delorme and McGee (1986) showed that prolonged depolarization decreased the number of dihydropyridine binding sites in cultured neurons.

4.1. Interactions with Alcohols and Anesthetics

4.1.1. Potentiation by Ca^{2+} Channel Inhibitors

Dihydropyridine Ca^{2+} channel inhibitors, when given in vivo to rodents have no overt behavioral effects, except when near lethal doses are given. However, interactions with CNS depressants can be demonstrated at lower doses (*see* Table 1). Nimodipine potentiated the hypothermic and ataxic actions of ethanol, in mice, at doses that alone had no effect on these parameters (Isaacson et al., 1985). Dolin and Little (1986a,b) showed that the potencies of a range of general anesthetic agents, ethanol, pentobarbitone, argon, and nitrous oxide, measured by loss of righting reflex in mice, were considerably increased by intraperitoneal injection of nitrendipine of nimodipine (Fig. 4). The doses used gave central concentrations in the low micromolar range (Dolin et al., 1986a); loss of righting reflex was not seen with dihydropyridines alone even at much higher doses. We have found a similar effect on the actions of ethanol with the Ca^{2+} channel antagonist PN 200-110 (unpublished results).

The mechanism of such potentiating actions is not yet clear. While action on voltage-operated Ca^{2+} channels is likely to be an important factor in the effects of ethanol and other central depressants, it does not seem as if simple summation of actions is involved, as the Ca^{2+} inhibitors did not demonstrate general anesthesia, or even central depression, when given alone. It is possible that the binding of the dihydropyridine compounds alters the state of the Ca^{2+} ionophore, thereby altering the effects of the general anesthetics on the complex. Ethanol does not directly affect dihydropyridine binding, although pentobarbitone had some inhibitory action (Harris et al., 1985). Ethanol, and possibly the other anesthetics, may affect the "L" subtype of Ca^{2+} channel by an allosteric mechanism. It is possible that this changes the state of the receptor, allowing functional effects of the dihydropyridines. Alternately, it is possible that the actions of ethanol and other central depressants do not directly involve actions on the "L" Ca^{2+} channels, but that a decrease in the functions of these channels, such as occurs with the dihydropyridine calcium channel inhibitors, or an increase, such as we have demonstrated in ethanol tolerance (Dolin et al., 1986b; 1987b), have synergistic and antagonist effects, respectively, on the actions of central depressants. The latter possibility is compatible with the effects of the Ca^{2+} channel activator BAY K 8644 (*see below*).

The potentiation of the effects of ethanol by i.c.v. calcium (Erickson et al., 1978) is at first sight not compatible with the above results using Ca^{2+} antagonists, but increased extracellular calcium concentrations would decrease neuronal excitability, which would summate with the action of ethanol. This is a very different situation from the effects of blocking Ca^{2+} channels. The situation is further complicated by the fact that increased concentrations of calcium in vitro antagonized the action of ethanol on the guinea-pig ileum (Mayer et al., 1980). The actions of calcium channel antagonists, however, are likely to be considerably more selective than the effects of changing extracellular calcium concentrations, which could affect many neuronal processes. That the effects of these compounds, described above, were the result of actions on Ca^{2+} channels was suggested by the opposing actions of BAY K 8644 (described below). The stereospecificity of this action has not yet been studied, but in their effects on the ethanol withdrawal syndrome (*see below*), the stereoselectivity followed that demonstrated on calcium conductance. The potentiation of central depressant action by Ca^{2+} channel inhibitors, found in all the studies so far, is exactly as predicted by our hypothesis (*see above*).

4.1.2. Inhibition by Ca^{2+} Channel Activators

The Ca^{2+} channel activator, BAY K 8644, has been shown to increase Ca^{2+} conductance, both in the periphery (Schramm et al., 1983) and the CNS (Middlemiss and Spedding, 1985). In vivo it

Table 1[a]
Interactions *In Vivo* Between Central Depressant Drugs and Calcium Antagonists

Central depressant compound	Ca^{2+} antagonist	Action	Change	Reference
Ethanol	Nimodipine	Ataxia	Inc.	Isaacson et al., 1985
Ethanol	Nimodipine	Hypothermia	Inc.	Isaacson et al., 1985
Ethanol	Nitrendipine	G. anesthesia	Inc.	Dolin et al., (1986a)
Ethanol	Nimodipine	G. anesthesia	Inc.	Dolin et al., (1986a)
Ethanol	Flunarizine	G. anesthesia	Inc.	Dolin et al., (1986a)
Ethanol	Verapamil	G. anesthesia	Inc.	Dolin et al., (1986a)
Ethanol	PN 200–110	G. anesthesia	Inc.	Unpublished results
Ethanol	PN 200–110	Ataxia	Inc.	Unpublished results
Halothane	Verapamil	G. anesthesia	Inc.	Maze et al. (1983)
N$_2$O	Nitrendipine	G. anesthesia	Inc.	Dolin et al. (1986a)
Hexobar.	Nimodipine	G. anesthesia	Inc.	Hoffmeister et al. (1982)
Pentobarbitone	Nitrendipine	G. anesthesia	Inc.	Dolin et al. (1986a)
Pentobarbitone	Nimodipine	G. anesthesia	Inc.	Dolin et al. (1986a)
Pentobarbitone	Flunarizine	G. anesthesia	Inc.	Dolin et al. (1986a)
Argon	Nitrendipine	G. anesthesia	Inc.	Dolin et al. (1986a)
Argon	Nimodipine	G. anesthesia	Inc.	Dolin et al. (1986a)
Argon	Flunarizine	G. anesthesia	Inc.	Dolin et al. (1986a)

Drug	Calcium drug	Effect	Action	Reference
Flurazepam	Nifedipine (icv)	Sleep latency	Dec.	Mendelsen et al. (1984b)
Diazepam	Nimodipine	Hypothermia	Inc.	Draski et al. (1985)
Diazepam	Nimodipine	Ataxia	Unch.	Draski et al. (1985)
Midazolam	Nitrendipine	G. anesthesia	Inc.	Dolin et al. (1986a)
Flurazepam	PN 200–110	G. anesthesia	Inc.	Unpublished results
Flurazepam	PN 200–110	Ataxia	Unch.	Unpublished results
Flurazepam	PN 200–110	Sedation	Inc.	Unpublished results
Benzomorphans	D600 (icv)	Antinociception	Inc.	Ben-Sreti et al. (1983)
Fentanyl	Nimodipine	Antinociception	Inc.	Hoffmeister and Tettenborn (1986)
Morphine	Verapamil	Antinociception	Inc.	Benedek and Sziksay (1984)
Morphine	Diltiazem	Antinociception	Inc.	Benedek and Sziksay (1984)
Morphine	Verapamil	Hypothermia	Inc.	Benedek and Sziksay (1984)
Morphine	Diltiazem	Hypothermia	Inc.	Benedek and Sziksay (1984)
U-50, 488H	Nimodipine	Hypothermia	Inc.	Pillai and Ross (1986)
U-50, 488H	Diltiazem	Hypothermia	Inc.	Pillai and Ross (1986)
Morphine	Verapamil	Resp. depr.	Dec.	Sziksay et al. (1986)
Morphine	Nitrendipine	Resp. depr.	Dec.	Ventham et al. (1987)
Morphine	Nitrendipine	Antinociception	Unch.	Ventham et al. (1987)[b]

[a]Inc. = increased action, Dec. = decreased action, Unch. = action unchanged, by coadministration of Ca^{2+} antagonist. Hexobarb. = hexobarbitone, resp. depr. = respiratory depression. U-50,488H is a selective k-receptor agonist.

[b]Only one high dose of morphine tested, intended to cause tolerance, increased acute action may not have been detected.

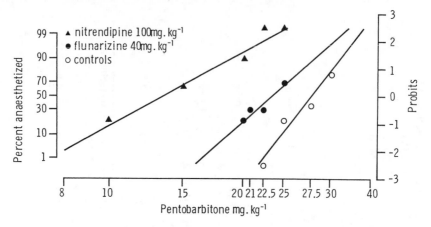

Fig. 4. *The effects of Ca^{2+} channel antagonists on the anesthetic potency of pentobarbitone in mice.* Mice were given either pentobarbitone plus vehicle (Tween 80, 0.5%) or pentobarbitone plus Ca^{2+} antagonist, ip. The dose of nitrendipine was 100 mg/kg^{-1} and that for flunarizine, 40 mg/kg^{-1}. The loss of righting reflex was assessed every 5 min until waking. Data presented are for 15 min after injections; each point represents results from 10 separate animals. Dose–response curves were constructed using Probit analysis; lines of best fit were compared using Chi-squared analysis. The dose–response curve for pentobarbitone anaesthesia was shifted significantly to the left by nitrendipine and by flunarizine. Open circles = controls, closed circles = flunarizine, closed triangles = nitrendipine. Groups were tested concurrently; body temperatures were maintained at 37 ± 1°C throughout. Data reproduced with permission from Dolin et al. (1986a).

produces a convulsive syndrome, which is prevented by intracerebral administration of nitrendipine (Bolger et al., 1985). At doses lower than those that cause this behavior, it antagonizes general anesthetic actions (Dolin and Little, 1986c; Dolin et al., 1987a). The dose–response curve for pentobarbitone anesthesia was shifted to the right in a parallel fashion, by BAY K 8644 at doses of 1, 5, and 10 mg/kg^{-1}. This pattern was also seen in ethanol anesthesia when 1 mg/kg^{-1} BAY K 8644 was used, but doses of 5 and 10 mg/kg^{-1} had the opposite effect, increasing the anesthetic action of ethanol. We do not yet know the reason for this difference, but the situation is complicated by the fact that the BAY K 8644 used was the racemic mixture; the ($+$) isomer has antagonist actions (Schramm et al., 1985). Studies using a pure "agonist" compound, such as ($-$)BAY K 8644 (Franckowiak et al., 1985) or ($-$)PN 200-110 (Hof et al., 1986), are needed to clarify these results. Another factor may be that ethanol and pentobarbitone differ in their effects on Ca^{2+} fluxes, as pentobarbitone increases efflux, in addition to decreasing influx, whereas ethanol affects only influx (Harris and Stokes, 1982). It is not clear how far these actions involve the "L" subtype of Ca^{2+} channels, on which dihydropyridines act.

Antagonism of central depressant actions by Ca^{2+} channel activators was predicted by our hypothesis (*see above*). With the exception of the interaction between high doses of BAY K 8644 and ethanol, this prediction has been verified by the experimental evidence. However, further studies are needed to examine the effects of drugs with more selective "agonist" action at the Ca^{2+} ionophore and to see whether the interactions demonstrated extend to properties of central depressants other than anesthesia, and to a wider range of compounds.

4.1.3. Prevention of Withdrawal Syndrome

Our hypothesis that an increase in number and function of the "L" subtype of Ca^{2+} channel underlies dependence on ethanol predicts that dihydropyridine Ca^{2+} channel inhibitors should prevent the ethanol withdrawal syndrome. As described above, these compounds have little effect on normal animals, in vivo. However, they had a clear and dramatic action in preventing the effects of ethanol withdrawal (Little et al., 1986). We showed that nitrendipine and nimodipine, at doses without behavioral actions in control animals, completely prevented the severe convulsions that were produced in rats by withdrawal from a liquid diet containing ethanol (Table 2). Other types of Ca^{2+} channel antagonists, verapamil and flunarizine, also afforded protection from the syndrome, but their actions were less clear than those of the dihydropyridines. A full behavioral rating of withdrawal signs other than convulsions, such as tremor, was not carried out in this study, but the animals receiving dihydropyridines appeared indistinguishable from normal rats, suggesting that all signs of withdrawal were affected. That Ca^{2+} channel inhibitors protect against ethanol withdrawal convulsions does not, of course, necessarily implicate Ca^{2+} channels in the ba-

Table 2[a]

The Incidence of Convulsions in a Period of 7½ H After Withdrawal
from Chronic Ethanol Treatment

Drug and dose (mg/gk^{-1})	No. rats showing spontaneous seizures	No. rats showing audiogenic seizures	Total no. animals having seizures	24 h mortality
Vehicle	5/9	4/5 +	8/9	8/9
Nitrendipine (100)	0/9[b]	0/9[d]	0/9[d]	0/9[d]
Vehicle	5/8	4/4[c]	8/8	8/8
Nimodipine (100)	0/8[b]	2/8	2/8[b]	1/8[c]
Vehicle	6/7	4/4	7/7	6/7
Diazepam (10)	2/8[b]	5/6	7/8	6/8

[a]Sprague-Dawley rats were given ethanol in a liquid diet for 5 wk. All drugs were suspended in Tween 80, 0.5%, and injected ip. The Ca^{2+} channel antagonists were given on withdrawal, and at 3 h intervals to 6 h. Diazepam was given 2 h after start of withdrawal. At 7½ h an electric bell was rung as an auditory stimulus. Statistical analysis was by Fisher's Exact Probability Test. Data reproduced with permission from Little et al. (1986).
[b]= P < 0.05.
[c]= P < 0.01.
[d]P < 0.001.
[b-d]Compared with concurrently tested, vehicle-treated, groups.
[e]Some animals did not survive the spontaneous seizures.

sis of this syndrome, but this anticonvulsant action of the dihydropyridines was selective, in that they are not effective in all types of convulsions. They had a slight protective effect against pentylenetetrazol seizures (Little et al., 1986; Dolin et al., 1986a) and prevented audiogenic convulsions in DBA/2 mice (Ascioti et al., 1986), but against other types of convulsions, such as those due to strychnine, they had no action, and the convulsive actions of N-methyl-d-aspartate were increased (Dolin et al., 1986a). They do, as expected, prevent the convulsive effects of BAY K 8644 (Bolger et al., 1985; Petersen, 1986).

We have also demonstrated the actions of nimodipine against the ethanol withdrawal syndrome in mice and have examined the stereoselectivity of the effect by using the isomers of the Ca^{2+} channel inhibitor PN 200-110 (Littleton and Little, 1987). The (+)isomer completely prevented handling-induced convulsions in mice undergoing ethanol withdrawal, whereas the (−)isomer was ineffective. This is the same pattern as that seen in the cardiovascular effects of the isomers on Ca^{2+} channel conductance (Hof et al., 1986) and their actions on inositol phospholipid hydrolysis (Kendall and Nahorski, 1985). The doses of Ca^{2+} channel antagonists required to affect ethanol withdrawal produced central concentrations in the low micromolar range (Dolin et al., 1986a) and were of the ranges that are required to displace radiolabeled dihydropyridines from central binding sites, when given systemically in vivo (Schoemaker et al., 1983b; Supavilai and Karobath, 1984). Nitrendipine was also found to be effective in preventing the withdrawal convulsions seen after mice have been allowed to breathe nitrous oxide for several days (Dolin and Little, 1986d).

Our hypothesis also predicted that Ca^{2+} channel activators would mimic the ethanol withdrawal syndrome. This was also found to be the case, since the convulsive effects of BAY K 8644, in rodents, resemble the ethanol abstinence syndrome, although high doses of BAY K 8644 appear to have an effect also on peripheral musculature, which is not seen after cessation of chronic ethanol treatment. When BAY K 8644 was given to mice undergoing a mild withdrawal syndrome, after inhalation of ethanol vapor, the withdrawal syndrome was intensified (Littleton and Little, 1987), and BAY K 8644 also prevented the ameliorative action of nimodipine in this withdrawal state. In this study, the behavioral ratings of the convulsive syndrome were in fact significantly higher after nimodipine plus BAY K 8644 than in untreated, withdrawn animals. This may indicate a more complex pattern than originally envisaged, but further investigation is necessary; again, use of a pure "agonist" rather than racemic BAY K 8644 might be informative.

4.1.4. Prevention of Tolerance

The next prediction from our hypothesis suggested that agents that prevent upregulation of Ca^{2+} channels should prevent the development of tolerance to, and dependence on, ethanol. We have evidence from a series of experiments that adaptation to ethanol can indeed be manipulated in this way (Little and Dolin, 1987). We treated rats for 10 d with either e0hanol alone, ethanol plus nitrendipine, nitrendipine alone, or vehicles. The ataxic effects of the compounds were tested using a rotorod apparatus. On first administration, the dose of nitrendipine used had only a very small ataxic action when given alone, but potentiated the effects of ethanol. In the tests at the end of the treatment, all animals received only ethanol, so that direct comparisons could be made. In these tests, there was significant tolerance to ethanol in the animals that had received chronic ethanol alone, but no tolerance in those given chronic ethanol plus chronic nitrendipine. Chronic treatment with nitrendipine alone did not alter the ataxic actions of ethanol. Measurement of brain nitrendipine concentrations, and in vivo [^{3}H]-nimodipine binding, excluded the possibility that residual nitrendipine contributed to the results. Binding studies showed that, after ethanol alone, central dihydropyridine binding was increased, as found previously (Dolin et al., 1986a; 1987a), while after ethanol plus nitrendipine, the number of binding sites was significantly decreased. A decrease in the B_{max} value for the dihydropyridine site after chronic dihydropyridine treatment in vivo was reported by Panza et al. (1985).

These results suggest that chronic administration of nitrendipine, when given with ethanol, prevents both functional adaptation (tolerance) to ethanol and the receptor changes that we have suggested may underlie such adaptation. It will be of great importance to determine whether or not this pattern extends to other situations. If this is the case, it may afford not only an interesting method of pharmacological manipulation, but also a potentially valuable therapeutic tool.

4.2. Interactions with Barbiturates and Benzodiazepines

There have been few studies of interactions between Ca^{2+} channel inhibitors and barbiturates. Hoffmeister et al. (1982) showed that nimodipine increased hexobarbitone anesthesia, as measured by duration of loss of righting reflex. We have reported that nitrendipine and nimodipine increased the anesthetic potency of pentobarbitone (Dolin and Little, 1986a), while the Ca^{2+} channel activator, BAY K 8644, antagonized this action of the barbiturate (Dolin et al., 1988). Possible interpretations of these results are discussed in section 4.1.1 above ("Interactions with Alcohols and Anesthetics. Potentiation by Ca^{2+} Channel Inhibitors").

Reports of interactions between calcium channel antagonists and benzodiazepines in vivo have been conflicting. These discrepancies may also be due to dosage, but the different routes of administration employed preclude direct comparison. However, in each case, the interaction seems to be confined to only some of the pharmacological effects of the benzodiazepines. In contrast, the interactions between calcium channel antagonists and ethanol occur whatever property of ethanol is measured, and were consistent in direction. Nifedipine, given i.c.v., has been reported to block the effect of flurazepam in decreasing sleep latency in the rat (Mendelson et al., 1984b). There was no effect of nifedipine alone on sleep latency and no effect on the action of flurazepam in the Vogel conflict test. Draski et al (1985) found that nimodipine potentiated the hypothermic effects of diazepam, but not the ataxic effects of the same doses, as measured on the rotorod. Both of these properties of benzodiazepines are blocked by benzodiazepine antagonists, suggesting mediation through benzodiazepine receptors. Nimodipine, in the dose used by Draski et al. (1985), 5 mg/kg ip, did not alter body temperature.

Interactions between dihydropyridine compounds affecting Ca^{2+} channels and benzodiazepines have been found when the general anesthetic actions of the latter were measured. When given to mice by the intraperitoneal route, the calcium channel activator BAY K 8644 decreased the general anesthetic potency of midazolam (Dolin et al., 1987b), whereas the calcium channel antagonist nitrendipine had the opposite action. Recent studies in our laboratory have shown that the dihydropyridine calcium channel antagonist PN 200-110 potentiated the general anesthetic, but not the ataxic, properties of flurazepam in mice (unpublished results).

From the studies so far it appears that, while all of the behavioral effects of ethanol are potentiated by Ca^{2+} channel inhibitors, the interactions of these compounds with benzodiazepines are more selective. This difference may provide important information concerning the mechanism of the acute effects of ethanol and the benzodiazepines. It is possible, for example, that the effects of benzodiazepines that are potentiated by Ca^{2+} channel inhibitors, such as general anesthesia and hypothermia (Dolin and Little,

1988; Draski et al., 1985), involve Ca^{2+} channels, while those that are unaffected, such as ataxia, have a different mechanism. The latter may be mediated through Cl^- rather than Ca^{2+} channels. We are currently investigating possible effects of dihydropyridines on benzodiazepine tolerance.

There is little information on the effects of dihydropyridine Ca^{2+} channel inhibitors on properties of barbiturates other than anesthesia. The syndromes produced by withdrawal from barbiturates and benzodiazepines were also decreased by the dihydropyridine calcium channel antagonists, when the latter were given chronically or acutely on withdrawal, suggesting an involvement of voltage-sensitive calcium channels (unpublished).

4.3. Interactions with Opiates

There have been several reports of interactions between Ca^{2+} antagonists and opiates. Calcium channel antagonists have been reported to increase the analgesic effects of fentanyl (v. Borman et al., 1985) in humans and the antinociceptive effects in rodents of benzomorphans (Ben-Sreti et al., 1983), of fentanyl (Hoffmeister and Tettenborn, 1986), and of morphine (Benedek and Sziksay, 1984). The respiratory depressant effects of morphine were, however, decreased by coadministration of verapamil (Sziksay et al., 1986). We have found that nitrendipine, when coadministered to mice with morphine, had no effect on the analgesic actions of the opiate (measured by hotplate reaction time), but decreased the respiratory depression (Ventham et al., 1987). Nitrendipine, when given alone, had a respiratory stimulant action. At the dose used in the latter studies, considerable potentiation of the general anesthetic effects of ethanol and of pentobarbitone was seen.

The organic calcium channel antagonists, verapamil and nimodipine, have been shown to decrease the withdrawal syndrome that is seen after chronic morphine administration, suggesting the adaptation in Ca^{2+} channels may be involved in morphine dependence (Bongianni et al., 1986). The number of [^{3}H]-nitrendipine binding sites in the CNS was increased by chronic morphine treatment (Ramkumar and El-Fakahany, 1984). This effect was localized to certain brain areas. We have investigated whether nitrendipine affected the development of opiate tolerance in mice (Ventham et al., 1987). Opiate tolerance has been suggested to be due to increases in intraneuronal calcium, which is an adaptation to the decreases in intraneuronal Ca^{2+} caused by acute administration of opiates. The abstinence syndrome would therefore be due to increased intracellular calcium. Presumably, this would occur in synapses that possess opiate receptors, but not necessarily in others.

4.4. Conclusions

The experimental evidence so far on the actions of the calcium channel modifiers, the dihydropyridines, shows that, in the case of ethanol, all the predictions from our hypothesis have been verified. This includes both the acute interactions with dihydropyridine compounds and the effects of the latter on the ethanol withdrawal syndrome and on tolerance development. The situation is not as clear for the benzodiazepines or for the opiates, and there is little information so far on the barbiturates.

We predicted that Ca^{2+} channel antagonists would potentiate the actions of central depressants. Ca^{2+} channel inhibitors potentiated the acute effects of ethanol, in every reported study, and also potentiated the general anesthetic actions of other compounds, including barbiturates and benzodiazepines. Of the other properties of benzodiazepines, potentiation was seen only in some cases; the ataxic action, for example, was unaffected. Potentiating effects of dihydropyridines on the acute actions of opiates have been reported. The prediction that Ca^{2+} channel activators, such as BAY K 8644, would antagonize central depressant action has been verified with respect to the general anesthetic properties of a variety of compounds, including pentobarbitone and benzodiazepines, although only the lower dose of BAY K 8644 antagonized ethanol anesthesia.

Prevention by Ca^{2+} channel antagonists of abstinence syndromes produced by central depressant drugs was our third prediction. This has been well substantiated for ethanol, for opiates, and for nitrous oxide, but corresponding data for barbiturates and benzodiazepines are not yet available. In addition, the Ca^{2+} channel activator, BAY K 8644, mimicked the ethanol withdrawal syndrome and prevented the ameliorative action of the antagonists.

Our final prediction was that drugs that prevent upregulation of the "L" subtype of Ca^{2+} channels may prevent the development of tolerance. Chronic treatment with the dihydropyridine

Ca^{2+} channel inhibitors caused downregulation of their high-affinity binding sites (Panza et al., 1985), suggesting downregulation of channels. Such treatment was found to prevent the development of tolerance to the ataxic effects of ethanol, exactly as predicted. In a single study on morphine tolerance, we did not find any effect of nitrendipine, but full investigation of the effects of the dihydropyridines on tolerance to compounds other than ethanol is needed. The type of tolerance involved in this single study may have involved a different mechanism.

5. Overall Conclusions

The evidence covered in this chapter has been selected to support the notion that similarities exist both in the mechanisms by which central depressant drugs produce their acute effects and in the adaptation that this evokes in the central nervous system. We accept that this is not the whole story, but the hypothesis that has evolved as a result is consistent with much of the evidence that has not been included. Put briefly, we suggest that central depressant drugs act on synaptic transmission either by inhibiting Ca^{2+} channel opening, by potentiating Cl^- channel opening, or both. The net result is a reduction in neuronal excitability. We further suggest that a basic adaptive mechanism that opposes this reduction in excitability involves an increase in a subtype of Ca^{2+} channel on neuronal cell bodies. This upregulation in Ca^{2+} channels restores the neuronal excitability towards normal levels in the presence of the depressant drug (tolerance) and, on its removal, causes neuronal hyperexcitability (withdrawal). Such a mechanism could have evolved as a basic means of controlling unicellular excitability in the face of fluctuations in external Ca^{2+} (or Cl^-) ions. It will be interesting to test this hypothesis in cultures of simple cells as well as mammalian neurons.

The happy chance that the subtype of Ca^{2+} channels that are upregulated is influenced by the dihydropyridine Ca^{2+} channel "antagonists" previously used in cardiovascular disease has important therapeutic implications. If depressant drug dependence is a consequence of this change, then the withdrawal syndrome should be prevented by treatment with Ca^{2+} "antagonists." Because these drugs are thought to have no abuse potential or sedative properties in normal individuals, they could represent a major advance in treatment. Other implications, too, should be obvious. Tolerance and dependence are major drawbacks in the therapeutic use of opiates as analgesics. Might concomitant administration of dihydropyridine Ca^{2+} antagonist prevent the development of tolerance as appears to be the case with ethanol? Might potentiation of depressant drugs with Ca^{2+} antagonists mean that much lower doses could be used with a corresponding margin of safety, for example in epilepsy?

This chapter has been deliberately controversial. The next few years of research will decide whether there is any merit in the hypothesis proposed. Even if it becomes evident that it is incorrect, we hope that it will have stimulated research along a logical and challenging path.

Acknowledgments

The work quoted in this chapter was supported by grants from the Wellcome Trust and the Brewers Society to J.M.L. and to H.J.L.

References

Allan A. M. and Harris R. A. (1986) Anaesthetic and convulsant barbiturates alter γ-aminobutyric acid-stimulated chloride flux across brain membrane. *J. Pharmacol. Exp. Ther.* **238**, 763–768.

Allan A. M. and Harris R. A. (1987) Acute and chronic ethanol treatments alter GABA receptor-operated chloride flux. *Pharmacol. Biochem. Behav.* **27**, 665–670.

Armstrong C. M. and Burstock G. (1964) The effects of several ethanols on the properties of the squid giant axon. *J. Gen. Physiol.* barbiturate48, 265–278.

Ascioti C., De Sarro G. B., Meldrum B. S., and Nistico G. (1986) Calcium entry blockers as anticonvulsant drugs in DBA/2 mice. *Br. J. Pharmacol.* **88**, 379P.

506   Littleton and Little

Baird J. G. and Nahorski S. R. (1986) Potassium depolarisation mark-
    edly   enhances   muscarinic   receptor   stimulated   inositol
    tetrakisphosphate accumulation in rat cerebral cortical slices.
    Biochem. Biophys. Res. Commun. 141, 1130–1137.
Barker J. L. and Mathers D. A. (1981) GABA receptors and the depress-
    ant action of pentobarbital. Trends in Neurosci. 4, 10–13.
Barker J. L. and Owen D. G. (1986) Electrophysiological pharmacology
    of GABA and diazepam in cultured CNS Neurones, in
    Benzodiazepine/GABA Receptors and Chloride Channels: Structure
    and Function. Alan R. Liss Inc., pp 135–165.
Bean B. P. (1984) Nitrendipine block of cardiac calcium channels: high
    affinity binding to the inactivated state. Proc. Natl. Acad. Sci. 81,
    6388–6392.
Benedek G. and Sziksay M. (1984) Potentiation of thermoregulatory and
    analgesic effects of morphine by calcium antagonists. Pharmacol.
    Res. Comm. 16, 1009–1018.
Ben-Sreti M. M., Gonzales J. P., and Sewell R. D. E. (1983) Effects of
    elevated calcium and calcium antagonists on 6,7-benzomorphan-
    induced analgesia. Eur. J. Pharmacol. 90, 385–391.
Bergmann M. C., Klee M. W., and Faber D. S. (1974) Different
    sensitivities to ethanol of early transient voltage clamp currents of
    Aplysia. Pfug. Arch. Physiol. 348 139–153.
Biscoe T. J., Duchen M. R., and Pascoe J. E. (1983) GABA/
    benzodiazepine interactions in the mouse hippocampal slice. J.
    Physiol. 341, 8P–9P.
Blaustein M. P. and Ector A. C. (1975) Barbiturate inhibition of calcium
    uptake by depolarized nerve terminals in vitro. Mol. Pharmacol. 11,
    369–378.
Blum K., Hamilton M. G., and Wallace, J. E. (1977) Alcohol and opi-
    ates: a review of common neurochemical and behavioral mecha-
    nisms, in Alcohol and Opiates — Neurochemical and Behavioral
    Mechanisms. Academic Press, New York, pp 203–236.
Boarder M. R., Marriott D., and Adams M. (1986) Stimulus-secretion
    coupling in cultured chromaffin cells. Dependency on external so-
    dium and on dihydropyridine-sensitive Ca^{2+} channels. Biochem.
    Pharmacol. 36, 163–167.
Bolger T. G., Weissman B. A., and Skolnick P. (1985) The behavioral
    effect of the calcium agonist BAY K 8644 in the mouse; antagonism
    by the calcium antagonist nifedipine. Naunyn-Schmiedebergs Arch.
    Pharmacol. 328, 373–377.

Bongianni F., Carla V., Moroni F., and Pellegrini-Giampieto D. E. (1986) Calcium channel inhibitors suppress the morphine withdrawal syndrome in rats. *Br. J. Pharmacol.* **88,** 561–567.

v.Bormann B., Boldt J., Sturm G., Kling D., Weidler B., Lohmann E., and Hemplemann G. (1985) Calciumantagonisten in der Analgesic. Additive Analgesie durch Nimodipin wahrend cardiochirurgischer Eingriffe. *Anaesthesist* **34,** 429–434.

Bowery N., Hill D. R., Hudson A. L., Doble A., Middlemiss D. N., Shaw J., and Turnbull M. (1984) (−)Baclofen decreases neurotransmitter release in the mammalian CNS at a novel GABA receptor. *Nature* **283,** 92–94.

Braestrup C., Nielsen M., and Squires R. F. (1979) No changes in benzodiazepine receptors after withdrawal from continuous treatment with lorazepam and diazepam. *Life Sci.* **24,** 347–350.

Brown C., Jones B., and Oakley N. R. (1984) Differential rate of tolerance development to the sedative, hypnotic and anticonvulsant effects of benzodiazepines. Proc. 14th CINP Meeting, Florence, c558.

Carlen P. L., Gurevich N., and Polc P. (1983a) Low-dose benzodiazepine neuronal inhibition: enhanced Ca^{2+}-mediated K^+-conductance. *Brain Res.* **271,** 358–364.

Carlen P. L., Gurevich N., and Polc P. (1983b) The excitatory effects of the specific benzodiazepine antagonist Ro 14-7434, measured intracellularly in hippocampal CA1 cells. *Brain Res.* **271,** 115–119.

Chapman D. and Way E. L. (1980) Metal ion interactions with opiates. *Ann. Rev. Pharmacol. Toxicol.* **20,** 553–579.

Clark J. W., Kalant H., and Carmichael F. J. (1977) Effect of ethanol tolerance on release of acetycholine and norepinephrine by rat cerebral cortical slices. *Can. J. Physiol. Pharmacol.* **55,** 758–768.

Collier H. O. J. (1965) A general theory in the genesis of drug dependence by induction of receptors. *Nature* **205,** 181–182.

Collier H. O. J. and Francis D. L. (1975) Morphine abstinence is associated with increased brain cyclic AMP. *Nature* **255,** 159–161.

Curtis D. R. and Johnston G. A. R. (1974) Amino acid transmitters in the mammalian nervous system. *Ergebn. Physiol.* **69,** 97–188.

Davis W. C. and Ticku M. J. (1981) Ethanol enhances [^{3}H]-diazepam binding at the benzodiazepine-GABA receptor-ionophore complex. *Mol. Pharmacol.* **20,** 287–294.

Delorme E. M. and McGee R. (1986) Regulation of voltage-dependent Ca^{2+} channels of neuronal cells by chronic changes in membrane potential. *Brain Res.* **397,** 189–192.

De Vries D. J., Ward L. C., Wilce P. A., Johnston G. A. R., and Shanley B. C. (1987) Effect of ethanol on the GABA-benzodiazepine receptor in brain, in *Advances in Biomedical Alcohol Research* Lindros, K. O., Ylikahri, R., and Kiianmaa, K. (eds.) Suppl. 1. *Alcohol & Alcoholism.* Pergamon Press, Oxford, New York, pp 657–662.

Doble A., Benavides J., Ferrir O., Bertand P., Menager J., Vaucher N., Burgevin M-C., Uzar A., Gueremy C., and Le Fur G. (1985) Dihydropyridine and peripheral type benzodiazepine binding site: subcellular distribution and molecular size. *Eur. J. Pharmacol.* **119,** 153–167.

Dolin S. J. and Little H. J. (1986a) Augmentation by calcium channels of general anaesthetic potency in mice. *Br. J. Pharmacol.* **88,** 909–914.

Dolin S. J. and Little H. J. (1986b) Effects of the calcium channel antagonist, nitrendipine, on nitrous oxide anaesthesia, tolerance and withdrawal. *Anesthesiology,* in press.

Dolin S. J. and Little H. J. (1986c) The effects of BAY K 8644 on the general anaesthetic potencies of ethanol and argon. *Br. J. Pharmacol.* **89,** 622P.

Dolin S. J. and Little H. J. (1986d) The dihydropyridine, nitrendipine, prevents nitrous oxide withdrawal seizures in mice. *Br. J. Addict.* **81,** 708.

Dolin S. J. and Little H. J. (1988) Differential interactions between benzodiazepines and dihydropyridines. *Br. J. Pharmacol.* **93,** 7P.

Dolin S. J., Grant A. J., Hunter A. B., and Little H. J. (1986a) Anticonvulsant profile and whole brain concentrations of nitrendipine and nimodipine. *Br. J. Pharmacol.* **89,** 866P.

Dolin S. J., Little H. J., Littleton J. M., and Pagonis, C. (1986b) Dihydropyridine-sensitive calcium channels are increased in ethanol physical dependence. *Br. J. Pharmacol.* **90,** 210P.

Dolin S. J., Halsey M. J., and Little H. J. (1987a) Antagonism of pentobarbitone and midazolam general anaesthesia by the calcium channel antagonist BAY K 8644. *J. Physiol.* **384,** 18P.

Dolin S. J., Little H. J., Hudspith M., Pagonis C., and Littleton J. (1987b). Increased dihydropyridine-sensitive calcium channels in rat brain may underlie ethanol physical dependence. *Neuropharmacology* **26,** 275–270.

Downes H., Perry R. S., Ostlund R. E., and Karler R. (1970) A study of the excitatory effects of barbiturates. *J. Pharmacol. Exp. Ther.* **175,** 692–699.

Draski L. J., Johnson J. E., and Isaacson R. L. (1985) Nimodipine's interactions with other drugs: II. Diazepam. *Life Sci.* **37,** 2123–2128.

Dunwiddie T. V., Worth T. S., and Olsen R. W. (1986) Facilitation of recurrent inhibition in rat hippocampus by barbiturate and related nonbarbiturate depressant drugs. *J. Pharmcol. Exp. Ther.* **238,** 564–575.

Durand D. and Carlen P. L. (1984) Decreased neuronal inhibition *in vitro* after long-term administration of ethanol. *Science* **224,** 1359–1361.

Elrod S. V. and Leslie S. W. (1980) Acute and chronic effects of barbiturates on depolarisation-induced calcium influx into synaptosomes from rat brain regions. *J. Pharmacol. Exp. Ther.* **212,** 131–136.

Erickson C. K., Tyler T. D., and Harris R. A. (1978) Ethanol: modification of acute intoxication by divalent cations. *Science* **199,** 1219–1221.

Eskuri S. A. and Pozos R. S. (1987) The effect of ethanol and temperature on calcium-dependent sensory neurone action potentials. *Alcohol & Drug Res.* **1,** 153–162.

Ferendelli J. A. and Daniels-McQueen S. (1982) Comparative actions of phenytoin and other anticonvulsant drugs on potassium- and veratridine-stimulated calcium uptake into synaptosomes. *J. Pharmacol. Exp. Ther.* **220,** 29–34.

File S. E., Grenn A. R., Nutt D. J., and Vincent N. D. (1984) On the convulsant action of Ro 5-4864 and the existence of a micromolar benzodiazepine binding site in rat brain. *Psychopharmacology* **82,** 199–202.

Franckowiak G., Bechem M., Schramm M., and Thomas G. (1985) The optical isomers of the 1,4-dihydropyridine BAY K 8644 show opposite effects on Ca^{2+} channels. *Eur. J. Pharmacol.* **114,** 223–226.

Friedman M. B., Erickson C. K., and Leslie S. W. (1980) Effects of acute and chronic ethanol administration on whole mouse brain synaptosomal calcium influx. *Biochem. Pharmacol.* **29,** 1903–1908.

Gallager D., Lakoski J. M., Gonsalves S. F., and Rauch S. L. (1984) Chronic benzodiazepine treatment decreases postsynaptic GABA sensitivity. *Nature* **290,** 74–77.

Glossman H., Ferry D. R., Lubbecke F., Mewes R., and Hofmann F. (1982) Calcium channels: direct identification with ligand binding sites. *Trends Pharmacol. Sci.* **3,** 431–433.

Goldstein D. B. (1972) Relationship of dose to intensity of withdrawal signs in mice. *J. Pharmacol. Exp. Ther.* **180,** 203–215.

Goldstein, D. B. and Goldstein, A. (1961) Possible role of enzyme inhibition and repression in drug tolerance and addiction. *Biochem. Pharmacol.* **8**, 48.

Goudie A. J. and Demellweek C. (1986) Conditioning factors in drug tolerance, in *Behavioral Analysis of Drug Dependence* (Goldberg S. R. and Stolerman I. P., eds.) Academic Press Inc., pp 225–285.

Gould R. J., Murphy K. M., and Snyder S. B. (1985) Autoradiographic localisation of calcium channel antagonist receptors in rat brain with [^{3}H]-nitrendipine. *Brain Res.* **330**, 217–223.

Gray P. and Taberner P. V. (1985) Evidence for GABA tolerance in barbiturate dependent and withdrawn mice. *Neuropharmacology* **24**, 437–444.

Greenberg D. A., Cooper E. C., Gordon A., and Diamond I. (1984) Ethanol and the γ-aminobutyric acid-benzodiazepine receptor complex. *J. Neurochem.* **42**, 1062–1068.

Guerrero-Munoz F., De Lourdes Guerrero M., and Leong-Way E. (1979) Effect of morphine on calcium uptake by lysed synaptosomes. *J. Pharmacol. Exp. Ther.* **211**, 370–374.

Haefely W., Pieri L., Polc P., and Schaffner R. (1981) General pharmacology and neuropharmacology of benzodiazepine derivatives, in *Handbook of Experimental Pharmacology*, Vol 55/II Hoffmeister F. and Stille G., eds. Springer-Verlag, Berlin, Heidelberg, New York, pp 13–262.

Hansch C. and Anderson S. M. (1967) The structure–activity relationship in barbiturates and its similarity to that in other narcotics. *J. Med. Chem.* **10**, 745–753.

Harper J. and Littleton J. M. (1987) Putative alcohol dependence in adrenal cell cultures. Relationship to Ca^{2+} channel activity. *Br. J. Pharmacol.* **92**, 661P.

Harris R. A. (1981) Ethanol and pentobarbital inhibition of intrasynaptosomal sequestration of calcium. *Biochem. Pharmacol.* **30**, 3209–3215.

Harris R. A. and Bruno P. (1985) Membrane disordering by anaesthetic drugs: relationship to synaptosomal sodium and calcium fluxes. *J. Neurochem.* **44**, 1274–1282.

Harris R. A. and Hood W. F. (1980) Inhibition of synaptosomal calcium uptake by ethanol. *J. Pharmcol. Exp. Ther.* **213**, 562–568.

Harris R. A. and Stokes J. A. (1982) Effects of a sedative and a convulsant barbiturate on synaptosomal calcium transport. *Brain Res.* **242**, 157–163.

Harris R. A., Jones S. B., Bruno P., and Bylund D. B. (1985) Effects of dihydropyridine derivatives and anticonvulsant drugs on [^{3}H]-nitrendipine binding and calcium and sodium fluxes in brain. *Biochem. Pharmacol.* **34,** 2187–2191.

Harris R. A., Yamamoto H., Loh H. H., and Way E. L. (1977) Discrete changes in brain calcium with morphine analgesia, tolerance-dependence, and abstinence. *Life Sci.* **20,** 501–506.

Hescheler J., Rosenthal W., Trautwein W., and Schulz G. (1987) The GTP-binding protein, G_O, regulates neuronal calcium channels. *Nature* **325,** 445–447.

Heyer E. J., and Macdonald R. L. (1982) Barbiturate reduction of calcium-dependent action potentials: correlation with anesthetic action. *Brain Res.* **236,** 157–171.

Himmelsbach C. K. (1942) Clinical studies of drug addiction. *Arch. Int. Med.* **69,** 766–772.

Hof R. P., Hof A., Ruegg U. T., Cook N. S., and Vogel A. (1986) Stereoselectivity at the calcium channel: different profiles of the hemodynamic activity of the enantiomers of the dihydropyridine derivative PN 200-110. *J. Cardivasc. Pharmacol.* **8,** 221–226.

Hoffmeister F. and Tettenborn D. (1986) Calcium agonists and antagonists of the dihydropyridine type: antinociceptive effects, interference with opiate-μ-receptor agonists and neuropharmacological actions in rodents. *Psychopharmacology* **90,** 299–307.

Hoffmeister F., Benz U., Heisse A., Krause H. P., and Neuser V. (1982) Behavioral effects of nimodipine in animals. *Arzneim.-Forsch./ Drug Res.* **32,** 347–360.

Hood W. F. and Harris A. R. (1980) Effects of depressant drugs and sulfhydryl reagents on the transport of calcium by isolated nerve endings. *Biochem. Pharmacol.* **29,** 957–959.

Hudspith M. J., Brennan C. H., Charles S., and Littleton J. M. (1987) Dihydropyridine-sensitive calcium channels and inositol phospholipid metabolism in ethanol physical dependence. *Ann. N. Y. Acad. Sci.* **492,** 156–170.

Hunkeler W., Mohler H., Pieri P., Polc P., Bonetti E. P., Cumin R., Schaffner R., and Haefely W. (1981) Selective antagonists of benzodiazepines. *Nature* **290,** 515–516.

Inoue M., Oomura Y., Yakushiji T., and Akaike N. (1986) Intracellular calcium ions decrease the affinity of the GABA receptor. *Nature* **234,** 156–158.

Isaacson R. L., Molina J. C., Draski L. J., and Johnston J. E. (1985)

Nimodipine's interactions with other drugs: 1. Ethanol. *Life Sci.* **36**, 2195–2199.

Jahsen H. and Laursen A. M. (1981) The effects of a benzodiazepine on the hyperpolarising and the depolarising responses of hippocampal cells to GABA. *Brain Res.* **207**, 214–217.

Jensen L. H., Petersen E. N., Braestrup C., Honore T., Kehr W., Stephens D. N., Schneider H. H., Siedelmann D., and Schmiechen R. (1984) Evaluation of the β-carboline ZK93426 as a benzodiazepine antagonist. *Psychopharmacology* **83**, 249–256.

Jensen M. S. and Lambert J. D. C. (1983) The interaction of the B-carboline derivative DMCM with inhibitory amino acid responses on cultured mouse neurones. *Neurosci. Lett.* **40**, 175–179.

Jensen M. S. and Lambert J. D. C. (1984) Modulation of the responses to the GABA-mimetics, THIP and piperidine-4-sulphonic acid, by agents which interact with benzodiazepine receptors. *Neuropharmacology* **23**, 1441–1450.

Kendall D. A. and Nahorski S. R. (1985) Dihydropyridine calcium channel activators and antagonists influence depolarisation-induced inositol phospholipid hydrolysis in brain. *Eur. J. Pharmacol.* **115**, 31–36.

Khanna J. M. and Meyer J. M. (1982) An analysis of cross-tolerance among ethanol, other general depressants and opioids. *Subst. Alc. Actions/Misuse* **3**, 243–257.

Khanna J. M., Le A. D., Kalant H., and Leblanc A. E. (1979) Cross-tolerance between ethanol and morphine with respect to their hypothermic effects. *Eur. J. Pharmacol.* **59**, 145–149.

Lê A. D., Poulos C. X., and Cappell H. D. (1979) Conditioned tolerance to the hypothermic effect of ethyl alcohol. *Science* **206**, 1109–1110.

Leslie S. W., Barr E., Judsen C., and Farrah R. P. (1983) Inhibition of fast and slow phase depolarisation dependent synaptosomal calcium uptake by ethanol. *J. Pharmacol. Exp. Ther.* **225**, 571–575.

Leslie S. W., Friedman M. B., Wilcox R. E., and Elrod S. V. (1980) Acute and chronic effects of barbiturates on depolarisation-induced calcium influx into rat synaptosomes. *Brain Res.* **185**, 409–417.

Liljequist S. and Tabakoff B. (1985) Binding characteristics of [^{3}H]-flunitrazepam and CL-218,872 in cerebellum and cortex of C57BL mice made tolerant to, and physically dependent on, ethanol. *Alcohol* **2**, 215–220.

Liljequist S., Culp S., and Tabakoff B. (1986) Effect of ethanol on the binding of [^{35}S]-t-butylbicyclophosphorothionate to mouse brain membranes. *Life Sci.* **38**, 1931–1939.

Little H. J. (1984) The effects of benzodiazepine agonists, inverse agonists and Ro 15-1788 on the responses of the superior cervical ganglion to GABA, *in vitro. Br. J. Pharmacol.* **83,** 57–68.

Little H. J. (1987) Chronic benzodiazepine treatment increases the effects of inverse agonists, in *Chloride Channels and Their Modulation by Neurotransmitters and Drugs. Adv. Biochem. Psychopharmacol.* Biggio G. and Costa E., (eds.) Raven, New York, in press.

Little, H. J. and Bichard A. R. (1984) Differential effects of the benzodiazepine antagonist Ro 15-1788 on the "general anaesthetic" effects of the benzodiazepines in mice. *Br. J. Anaesth.* **56,** 1153–1159.

Little H. J. and Dolin S. J. (1987) Lack of tolerance to ethanol after concurrent administration of nitrendipine. *Br. J. Pharmacol.,* in press.

Little H. J., Dolin S. J., and Halsey M. J. (1986) Calcium channel antagonists decrease the ethanol withdrawal syndrome. *Life Sci.* **39,** 2059–2065.

Littleton J. M. (1983) Tolerance and physical dependence on alcohol at the level of synaptic membranes: a review. *J. Roy. Soc. Med.* **76,** 593–601.

Littleton J. M., Harper J., Hudspith M., Pagonis C., Dolin S. J., and Little H. J. (1987) *Adaptation in Neuronal Calcium Channels May Cause Alcohol Physical Dependence.* British Association of Psychopharmacology Series, Oxford University Press, in press.

Louvel J., Abbes S. and Godfraind J. M. (1986) Effect of organic calcium channel blockers on calcium dependent processes. *Exp. Brain. Res.* **14,** 375–385.

Lucchi L., Govoni S., Battaini F., Passinetti G., and Trabucchi M. (1985) Ethanol administration *in vivo* alters calcium ions control in rat striatum. *Brain Res.* **332,** 376–379.

Lynch M. and Littleton J. M. (1983). Possible association of alcohol tolerance with increased synaptic calcium sensitivity. *Nature* **303,** 175–177.

Macdonald R. L. and Barker J. L. (1979) Anticonvulsant and anaesthetic barbiturates: different postsynaptic actions in cultured mammalian neurones. *Neurology* **29,** 432–447.

Mansfield J. G. and Cunningham C. L. (1980) Conditioning and extinction of tolerance to the hypothermic effects of ethanol in the rat. *J. Comp. Physiol. Psychol.* **94,** 962–969.

Marangos P. J., Patel J., Boulenger J-P., and Clerk-Rosenberg R. (1982) Characterisation of peripheral-type benzodiazepine binding sites in brain using [^{3}H]-Ro 5-4864. *Mol. Pharmacol.* **22,** 26–32.

Mathers D. A. and Barker J. L. (1980) (−)Pentobarbital opens ion channels of long duration in cultured mouse spinal neurones. *Science* **209**, 507–509.

Mayer J. M., Khanna J. M., and Kalant H. (1980) A role for calcium in the acute and chronic actions of ethanol *in vitro*. *Eur J. Pharmacol.* **68**, 223–227.

Mayer J. M., Khanna J. M., Kalant H., and Spero L. (1980) Cross-tolerance between ethanol and morphine in the isolated guinea-pig ileum myenteric plexus preparation. *Eur. J. Pharmacol.* **63**, 223–227.

Maze M., Mason D. M., and Kates R. E. (1983) Verapamil decreases MAC for halothane in dogs. *Anesthesiology* **59**, 327–329.

Mendelsen W. B., Skolnick P., Martin J. V., Luu M. D., Wagner R., and Paul, S. M. (1984a) Diazepam-stimulated increases in the synaptosomal uptake of $^{45}Ca^{2+}$: reversal by dihydropyridine calcium channel antagonists. *Eur. J. Pharmacol.* **104**, 181–183.

Mendelsen W. B., Owen C., Skolnick P., Paul S. M., Martin J. V., Ko G., and Wagner R. (1984b) Nifedipine blocks sleep induction by flurazepam in the rat. *Sleep* **7**, 64–68.

Messing R. O., Carpenter C. L., and Grennberg D. A. (1986) Ethanol regulates calcium channels in clonal neural cells. *Proc. Natl. Acad. Sci. U.S.A.* **83**, 6213–6215.

Mestre M., Carriot T., Belin C., Uzan A., Renault C., Dubroeucq M. C., Gueremy C., Doble A., and Le Fur G., (1986) Electrophysiological and pharmacological evidence that peripheral type benzodiazepine receptors are coupled to Ca^{2+} channels in the heart. *Life Sci.* **36**, 391–400.

Meyer H. H. (1901) Zw Theorie der Alkaholnarkose. *Arch. Exp. Pathol. Pharmakol.* **46**, 338.

Meyer J. M., Khanna J. M., Kalant H., and Sper L. (1980) Cross-tolerance between ethanol and morphine on the isolated guinea-pig ileum myenteric plexus preparation. *Eur. J. Pharmacol.* **63**, 223–227.

Middlemiss D. and Spedding M. (1985) A functional correlate for the dihydropyridine binding site in rat brain. *Nature* **314**, 94–96.

Miller R. J. (1987) Multiple calcium channels and neuronal function. *Science* **235**, 46–52.

Moore J. W., Ulbright W., and Takata M. (1964) Effect of ethanol on the sodium and potassium conductances of the squid giant axon membrane. *J. Gen. Physiol.* **48**, 279–288.

Mullin M. J. and Hunt W. A. (1985) Actions of ethanol on voltage-sensitive sodium channels: effects on neurotoxin-stimulated uptake in synaptosomes. *J. Pharmacol. Exp. Ther.* **232**, 413–420.

Nestoros J. N. (1980) Ethanol specifically potentiates GABA-mediated neurotransmission in feline cerebral coflex. *Science* **209**, 708–710.

Nicholl R. A., Eccles J. C., Oshima T., and Rubia F. (1975) Prolongation of hippocampal postsynaptic potentials by barbiturates. *Nature* **258**, 625–627.

Nishi K. and Oyama Y. (1983) Barbiturates increase the rate of voltage-dependent inactivation of the calcium current in snail neurones. *Br. J. Pharmacol.* **80**, 761–765.

Nowycky M. C., Fox A. P., and Tsien R. W. (1985) Three types of neuronal calcium channel with different calcium agonist sensitivity. *Nature* **316**, 440–443.

Nutt D. J., Cowen P. J., and Little H. J. (1982) Unusual interactions of benzodiazepine antagonists. *Nature* **295**, 436–439.

Olsen R. W. (1982) Drug interactions at the GABA receptor-ionophore complex. *Ann Rev. Pharmacol. Toxicol.* **22**, 245–277.

Overton, E. (1901) Studien uber die Narkose. Jena: Fischer.

Pagonis C. and Littleton J. M. (1987) The Ca^{2+} antagonist PN 200-110 inhibits [^{3}H]-dopamine release from nigral but not striatal slices from ethanol-dependent rats. *Br. J. Pharmacol.* in press.

Panza G., Grebb J. A., Sanna E., Wright A. G., and Hanbauer F. I. (1985) Evidence for down-regulation of [^{3}H]-nitrendipine recognition sites in mouse brain after long-term treatment with nifedipine or verapamil. *Neuropharmacology* **24**, 1113–1117.

Paul S. M. and Skolnick P. (1982) Comparative neuropharmacology of antianxiety drugs. *Pharmacol. Biochem. Behav.* **17**, (Suppl.), 37–40.

Paul S. M., Luu M. D., and Skolnick P. (1982) The effects of benzodiazepines on presynaptic calcium transport. In: Pharmacology of Benzodiazepines Usdin E., Skolnick P., Tallman J. F., Greenblatt D., and Paul S. M., eds. Macmillan, London, pp 87–92.

Petersen E. N. (1986) BAY K 8644 induces a reversible spasticity-like syndrome in rats. *Eur. J. Pharmacol.* **130**, 323–326.

Peyton J. C. and Borowitz J. L. (1979) Chlordiazepoxide and theophylline alter calcium levels in subcellular fractions of rat brain cortex. *Proc. Soc. Exp. Biol. Med.* **161**, 178–182.

Pillai N. P. and Ross D. H. (1986) Interaction of K-receptor agonists with Ca^{2+} channel antagonists in the modulation of hypothermia. *Eur. J. Pharmacol.* **132**, 237–244.

Polc P., Laurent J. P., Scherschlicht R., and Haefely W. (1981) Electrophysiological studies on the specific benzodiazepine antagonist Ro 15-1788. *Naunyn-Schmiedeberg's Arch. Pharmacol.* **316,** 317–325.

Ramajaneyulu R. and Ticku M. K. (1984) Binding characteristics and interactions of depressant drugs with [^{35}S]-butylbicyclophosphorothionate, a ligand that binds to the picrotoxin site. *J. Neurochem.* **42,** 221–229.

Ramkumar V. and El-Fakahany E. E. (1984) Increase in [^{3}H]-nitrendipine binding sites in the brain in morphine-tolerant mice. *Eur. J. Pharmacol.* **102,** 371–372.

Rampe D. and Triggle D. J. (1986) Benzodiazepine and calcium channel function. *Trends Pharmacol. Sci.* **7,** 461–463.

Ritzmann R. F. and Tabakoff B. (1976) Dissociation of alcohol tolerance and dependence. *Nature* **263,** 418–419.

Rosenberg H. C. and Chiu T. H. (1979) Decreased [^{3}H]-diazepam binding is a specific response to chronic benzodiazepine treatment. *Life Sci.* **24,** 803–808.

Ross D. H., Kibler B. and Cardenas H. L. (1977) Modification of glycoprotein residues as Ca^{2+} receptor sites after chronic ethanol exposure. *Drug & Alc. Dep.* **2,** 305–315.

Rottenberg H., Waring A., and Rubin E. (1981) Tolerance and crosstolerance in chronic alcoholics: reduced membrane binding of ethanol and other drugs. *Science* **213,** 583–585.

Schoemaker H., Smith T. L., and Yamamura H. L. (1983a) Effect of chronic ethanol consumption on central and peripheral type benzodiazepine receptors in the mouse brain. *Brain Res.* **258,** 347–350.

Schoemaker H., Lee R., Roeske W. R., and Yamamura H. I. (1983b) *In vivo* identification of calcium antagonist binding sites using [^{3}H]-nitrendipine. *Eur. J. Pharmacol.* **88,** 275–276.

Schon F. and Iversen L. L. (1974) The use of autoradiographic techniques for the identification and mapping of transmitter-specific neurones in the brain. *Life Sci.* **15,** 157–175.

Schramm M., Thomas G., Towert R., and Franckowiak G. (1983) Novel dihydropyridines with positive inotropic action through activation of Ca^{2+} channels. *Nature* **303,** 535–537.

Schramm M., Towert R., Lamp B., and Thomas G. (1985) Modulation of calcium ion influx by the 1,4-dihydropyridines, nifedipine and BAY K 8644. *J. Cardiovasc. Pharmacol.* **7,** 493–496.

Schwartzkroin P. A. and Slawsky M. (1977) Probable calcium spikes in hippocampal neurones. *Brain Res.* **135,** 157–161.

Siegel S. (1975) Evidence from rats that morphine tolerance is a learned response. *J. Comp. Physiol. Psychol.* **5**, 498–506.

Simmonds M. (1988) Barbiturates and excitatory amino acid interactions, in *Excitatory Amino Acids in Health and Disease* Lodge D., ed. John Wiley & Sons, Chichester, in press.

Sinclair J. D., Adkins J., and Walker S. (1973) Morphine-induced suppression of voluntary alcohol drinking in rats. *Nature* **246**, 425–427.

Skerritt J. H., Werz M. A., McLean M. J., and Macdonald R. L. (1984) Diazepam and its anomolous *p*-chloro-derivative Ro 5-4864: comparitive effects on mouse neurons in cell culture. *Brain Res.* **310**, 99–105.

Squires R. F. and Braestrup C. (1977) Benzodiazepine receptors in rat brain. *Nature* **266**, 732–734.

Squires R. F., Casida J. E., Richardson M., and Saederup E. (1983) [^{35}S]*t*-butylbicyclophosphoithionate binds with high affinity to brain specific sites coupled to γ-aminobutyric acid-A and ion recognition sites. *Mol. Pharmacol.* **23**, 326–336.

Stephens D. N. and Schneider H. H. (1986) Tolerance to the benzodiazepines in an animal model of anxiolytic activity. *Psychopharmacology* **87**, 322–327.

Study R. E. and Barker J. L. (1981) Diazepam and (−)pentobarbital: fluctuation analysis reveals different mechanisms for potentiation of γ-aminobutyric acid responses in cultured central neurones. *Proc. Natl. Acad. Sci.* **78**, 7180–7184.

Suzdak P., Glowa J. R., Crawley J. N., Schwartz R. D., Skolnick P., and Paul S. M., (1986) A selective imidazodiazepine antagonist of ethanol in the rat. *Science* **234**, 1243–1247.

Suetake K., Kojima H., Inanaga K., and Koketsu K. (1981) Catecholamine is released from non-synaptic cell-soma membrane: histochemical evidence in bullfrog sympathetic ganglion cells. *Brain Res.* **205**, 436–440.

Supervilai P. and Karobath M. (1984) The interaction of [^{3}H]-PY 108-068 and of [^{3}H]-PN 200-110 with calcium channel binding sites in the rat brain. *J. Neural. Trans.* **60**, 149–167.

Sziksay M., Snyder F. R., and London E. D. (1986) Interactions between verapamil and morphine on physiological parameters in rats. *J. Pharmacol. Exp. Ther.* **238**, 192–197.

Taft W. C. and de Lorenzo R. J. (1984) Micromolar-affinity benzodiazepine receptors regulate voltage-sensitive calcium channels in rat brain. *Proc. Natl. Acad. Sci. U.S.A.* **81**, 3118–3122.

Tambourska E. and Marangos P. J. (1986) Brain benzodiazepine binding sites in ethanol dependent and withdrawn states. *Life Sci.* **38,**. 465–472.

Thayer S. A., Murphy S. N., and Miller R. J. (1986) Widespread distribution of dihydropyridine-sensitive calcium channels in the central nervous system. *Mol. Pharmacol.* **30**, 505–509.

Thomson T. D. and Turkanis S. A. (1973) Barbiturate-induced release at a frog neuromuscular junction. *Br. J. Pharmacol.* **48**, 48–58.

Ticku M. K. (1980) The effects of acute and chronic ethanol administration and its withdrawal on γ-aminobutyric acid receptor binding in rat brain. *Br. J. Pharmacol.* **70**, 403–410.

Ticku M. K. (1987) Behavioural and functional studies indicate a role for GABA$_A$ergic transmission in the actions of ethanol, in *Advances in Biomedical Alcohol Research* Lindros K. O., Ylikahri R., and Kiianmaa, K. (eds.) Suppl. 1. *Alcohol & Alcoholism.* Pergamon Press, Oxford, New York, pp 657–662.

Ticku M. K. and Burch T. (1980) Alterations in γ-aminobutyric acid receptor sensitivity following acute and chronic ethanol treatments. *J. Neurochem.* **34**, 417–423.

Turner T. J. and Goldin S. M. (1985) Calcium channels in rat brain synaptosomes: identification and pharmacological characterisation. *J. Neurosci.* **5**, 841–849.

Ventham P., Dolin S. J., and Little H. J. (1987) Interactions between morphine and nitrendipine; analgesic and respiratory actions. *Br. J. Anaesth.*, in press.

Werz M. A. and Macdonald R. L. (1985) Barbiturates decrease voltage-dependent calcium conductance of mouse neurones in dissociated cell culture. *Mol. Pharmacol.* **28**, 269–277.

Part 3
Summary Chapters

Behavioral Tolerance and Sensitization

DEFINITIONS AND EXPLANATIONS

Derek E. Blackman

The interdisciplinary science of behavioral pharmacology continues to gather force, and the chapters of this book make an interesting contribution to this development. The authoritatively written reviews address various aspects of a clinically important and scientifically intriguing phenomenon, namely the waning or augmentation of a drug's effects with repeated administration. The present chapter is intended to focus on matters raised in the preceding reviews of different aspects of tolerance and sensitization that may appear to be of particular interest in the behavioral, as opposed to the pharmacological, domain, though it must be emphasized immediately that such a distinction should be approached cautiously, if indeed behavioral pharmacology is to be regarded as genuinely interdisciplinary.

In order to establish an initial position with respect to the material presented in the preceding chapters, some general and relatively uncontentious comments may be helpful. First, the impact of the detailed, empirical studies discussed here is impressive. Any empirical scientist would expect carefully designed and executed experiments to extend our appreciation of a natural phenomenon, and such is undoubtedly the case in the studies of those circumstances in which tolerance and sensitization occur or do not occur.

Second, and perhaps inevitably, the detailed discussion and the scope of the pharmacological and physiological analyses of the basic phenomena considered in some of the chapters are impressive to a behaviorally oriented reviewer. They illustrate the power of the generally accepted reductionistic explanatory systems favored in pharmacology, although it might also be said here that the dynamics and lability of tolerance and sensitization seem still to provide a considerable challenge to scientific analyses at this level. Third, and no doubt equally inevitably, a behaviorally oriented reviewer will feel pleased that there is here so substantial a body of detailed behavioral data to be incorporated into this area of behavioral pharmacology. Studies of the effects of drugs on behavior as a whole have long been characterized by more systematic manipulation of pharmacological parameters, such as dosage and route of administration, than of the variables of which behavior is known to be a function. The chapters of this book demonstrate unequivocally that the study of tolerance and sensitization to the effects of drugs has been significantly advanced, as well as diversified, by the inclusion of systematic studies of behavior.

In short, the substantive chapters of the present book provide a timely and important illustration of contemporary interests in one area of behavioral pharmacology. Furthermore, they are comprehensive, and they are cogently presented. There would seem, therefore, little to be gained by attempting here to supplement data that are presented, or even simply to pick out points of behavioral interest in the various chapters. Instead, attention is directed here to two broad questions, namely (a) the appropriate *definition* of tolerance and sensitization, and (b) the *explanation* of these phenomena. Both questions, of course, are addressed to a greater or lesser extent in each of the preceding chapters, and it should also be said that these two questions resist independent analysis because they are inevitably intimately interrelated.

1. The Definition of Tolerance and Sensitization

When faced with the excitement and impact of the cumulative growth of empirical knowledge about tolerance and sensitization, it may seem perverse to draw back to what might initially appear to be no more than conceptual pedantry. If such an approach is to be allowed, however, a natural starting point is to be found in the opening definitions offered by the editors of the present volume: "In current usage, tolerance is operationally defined as a shift to the right in the dose–response curve that occurs with chronic drug treatment; sensitization is correspondingly defined as a shift to the left in the dose–response curve" (Goudie and Emmett-Oglesby, this volume). The editors shortly thereafter give some emphasis to an additional feature of tolerance/sensitization, namely that such shifts in the dose–effect curve are "reversible." It is open to discussion whether reversibility constitutes an additional *definitional* criterion or "merely" an empirical generalization about the basic phenomenon, and it will be largely treated here as the latter, though its general relevance will be discussed again in this chapter.

Thus, two definitional criteria for tolerance/sensitization emerge: (a) a shift in a dose–response curve, which is (b) attributable to repeated drug administrations. A good empirical demonstration of tolerance expressed through measures of behavior is shown in Fig. 1, which displays data initially reported by Stolerman et al. (1974). These authors reported the effects of different doses of nicotine on two measures of activity exhibited by rats when they were tested in a maze 2 h after an injection of either saline or 0.75 mg/kg of nicotine. The behavioral measures were nominal entries into segments of the maze and the number of times the rats reared from the floor of the maze. The data are the means and standard errors for groups of nine rats exposed to the various doses of nicotine shown on the abscissa. It is instructive to discuss these data, which were obtained from a well-designed experiment and which showed clear empirical effects, in the light of the definitional criteria discussed above.

First it can be seen that a substantially increased dose of nicotine was required to produce a given degree of behavioral suppression with rats pretreated with the drug rather than with saline. Thus, the dose–response curve can indeed by said to be shifted to the right, and so one of the definitional criteria for tolerance can be said to be satisfied. These data are particularly interesting, however, because as Corfield-Sumner and Stolerman (1978) point out, the test dose of 0.5 mg/kg of nicotine exerted a facilitative, rather than a suppressive, effect on both measures of behavior after pretreatment with the drug. Data for this dose when considered alone might,

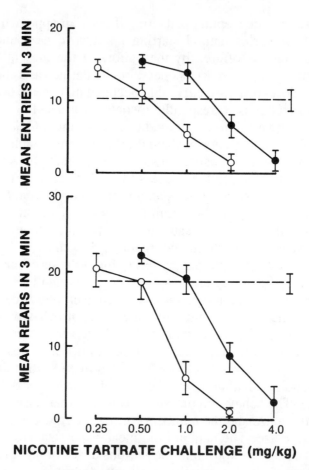

Fig. 1. Effects of pretreatment with nicotine (closed circles) or saline (open circles) on the activity of rats after further nicotine at doses
shown on abscissa. The vertical bars indicate standard errors from the
mean in groups of nine rats (from Stolerman et al., 1974).

therefore, suggest that the drug pretreatment led to sensitization
rather than to tolerance. Goudie (1989) has also emphasized the potential for "misleading" conclusions to emerge from studies with
single doses of a drug. However, the point might also be made here
that the potential confusion between tolerance and sensitization
illustrated by the data shown in Fig. 1 emerges from the *definition*
of these phenomena in terms of movements in dose–response
curves in specified directions, and would be avoided by an ap-

proach that focused simply on differences in drug effects attributable to prior drug experience. Again this point will be taken up later, but it should not be thought that an argument is here being made against exploring empirically a full range of effective doses of a drug, for such parametric analyses are an integral part of pharmacology and one mark of what is often a greater degree of pharmacological, as opposed to behavioral systematization in many experiments in behavioral pharmacology.

The study by Stolerman et al. (1974), in fact, illustrates this latter point, too: the independent variables that influence the mean number of entries by rats into segments of a maze and that influence the mean number of "rears" have not been well-defined experimentally and were certainly not manipulated in this study. In the light of many behavioral data reviewed in other chapters of this book (*see* especially the chapter by Barrett et al.), it is probable that different behavioral responses, or indeed even that similar responses measured in different conditions, would be affected differentially by drug treatments, even to the extent of being shifted in different directions, and this provides a further complication in the task of identifying in some global sense that repeated administrations of a psychoactive drug lead to behavioral tolerance or behavioral sensitization.

With respect to "shifts" in the behavioral effects of drugs, the data shown in Fig. 1 prompt a further comment. Here the shift takes the form of differences between different groups of rats, each of which was exposed to a single test condition. These intergroup differences are statistically significant and are, therefore, happily accepted by convention into the body of scientific knowledge, but it should be noted that this research design does not identify any shifts in a drug's effects within individual subjects. There is a sense then in which the term "shift" could be thought to be ambiguous, for it may be taken to imply movement rather than mere difference. The term "shift" appears to sit more easily within the clinical context in which the phenomena of drug tolerance and sensitization have much practical relevance. The tension here is illustrated by noting an earlier (but nevertheless still recent) alternative definition of tolerance and sensitization offered by Goudie (1989): "tolerance typically refers to the observation that during chronic drug treatment the effect(s) of some drugs may progressively reduce in magnitude"; "sensitization refers to the observation that during chronic treatment some drug effects actually increase in magnitude." Definitions such as these relate immediately of course to well-known clinical observations with individual patients (e.g., with opiates and antidepressants) and sometimes to individuals' personal experiences in nonclinical settings. As Goudie (1989) has noted, however, there is a paradox in attempting to consolidate the two approaches to the definition of tolerance and sensitization outlined above, for "in deriving dose/effect curves in within group experiments, the initial doses used in the assessment of the dose/effect curve may actually modify the response of animals to later doses of the drug." This is not the place to rehearse arguments about the relative advantages of within-group and between-group designs in behavioral pharmacology (*but see* Blackman, 1987 for discussion of this point). Enough has been said here, however, to indicate that the first definitional criterion of tolerance/sensitization is not entirely without tension.

The second definitional criterion offered above was that tolerance/sensitization should result from repeated or "chronic" drug treatment (*see* chapter by Goudie and Emmett-Oglesby). If this criterion is to be rigidly respected, the data of Stolerman et al. (1974) shown in Fig. 1 must be regarded as providing a limiting case, because in their study, a *single* pretreatment of nicotine proved sufficient to produce a rightwards "shift" in the dose–response curve to subsequent nicotine administration. However, once more, the definitional criterion encounters tensions, and these are also amply illustrated in the chapters of this book. For example, in his discussion of sensitization, Willner is quickly led to attempt to make a distinction between "true" and "pseudo" sensitization (*see* chapter by Willner, section 1.): if at the time of its second administration a first dose of a drug "has not yet been fully cleared from the body," then it would seem more appropriate to talk of cumulative dosing than of a sensitizing effect of the initial dose. This problem might be thought to be particularly pertinent to studies of sensitization rather than to studies of tolerance, and Willner is duly cautious in his review of the evidence for "true" behavioral sensitization, emphasizing that "it cannot be assumed merely from the fact that a response increases with repeated treatment." However, this inevitably raises the possibility that a distinction between "true" and "pseudo" *tolerance* in similar circumstances might be necessary. The idea that an *increase* in the net drug momentarily available in tissue might lead to a *decrease* in its effect is by no means implausible, as for example if an inverted-U shape dose–response curve is obtained in acute testing (*see* examples in the chapter by Barrett et al.). The second definitional criterion of prior drug administration may, therefore, appear to some to have been expressed too loosely in the definitions above.

Of even greater importance in the present context, however, is the issue of whether *nonpharmacological* variables can be said to lead to the development of tolerance. As noted above, by a strict application of the definitional criteria, shifts in a dose–response curve would be defined as tolerance or sensitization only if they resulted from prior drug administration. If such shifts result from nonpharmacological procedures, such as variations in food and water deprivation, or changes in diet or exercise, these effects might then be described as "simulations" of tolerance or sensitization. In practice, however, this distinction between "true" and such "simulated" tolerance/sensitization can prove difficult to sustain, as Goudie has discussed (1988). One potential link between "true" and "simulated" tolerance in this sense is to be found in the study of the dispositional mechanisms reviewed in the chapter by Lê and Khanna. If the acute effects of a drug vary because of differences in dispositional mechanisms, such as those affecting absorption and distribution, perhaps they should boldly be termed tolerance or sensitization, regardless of whether such differences result from prior drug administration or from nonpharmacological manipulations. This would of course be tantamount to defining tolerance/sensitization *in terms of* explanatory mechanisms, and this point will again become relevant later in this chapter.

However, it may be noted here that it can also be difficult to sustain a distinction between differences in dispositional mechanisms that would suffice for "true" tolerance resulting not from repeated drug administration, but nevertheless resulting from changes in dispositional mechanisms, and those that would not. For example, growth and aging may produce changes in dispositional mechanisms, but to accept their effects on dose–response curves as tolerance or sensitization would lead ever closer to identifying *any* differences in drug effects as tolerance/sensitization by definition. The concept of "innate tolerance" briefly mentioned by Lê and Khanna (*see* chapter by Lê and Khanna, Section 1.) provides the ultimate position in this respect, quite simply by in effect deleting the second definitional criterion for tolerance/sensitization and, thereby, allowing individual differences in dose–response curves to be incorporated as part of the study of tolerance/sensitization. It might be noted here that such ''innate tolerance'' refers to no more than the differences between individuals and can by definition, therefore, be demonstrated only by means of comparisons *between* individuals. Such tolerance/sensitization clearly cannot be reversed within subjects in any conventional sense by the manipulation of the relevant independent variable, although the manipulation of *other* variables, such as those that affect dispositional mechanisms, might produce reverses *in the differences* between individuals with respect to their dose–response curves.

The special relevance of behavioral manipulations should not be overlooked in this discussion of the definitional criteria for tolerance/sensitization. The phenomena of contingent tolerance and environmentally specific tolerance are fully reviewed in the chapters by Wolgin and by Siegel. Briefly, repeated administrations of a drug may lead to tolerance or sensitization only in certain circumstances in which well-understood behavioral principles of classical or operant conditioning are implicated. Thus, conditioning or learning seems to be an important mechanism in the development of tolerance/sensitization. These findings have greater impact than simply to introduce behavioral manipulations as possible influences on dispositional mechanisms, however. If they did no more than this, they would serve the same function as the relatively static differences resulting from changes in diet, and so on. In fact, the main challenges from the concept of contingent tolerance are more fundamental. First, contingent tolerance/sensitization suggests that, to adopt the language of pain research, dispositional mechanisms may be ''gated'' by behavioral manipulations: administrations of a drug may or may not produce a ''shift'' in a dose–response curve as a direct consequence of these behavioral manipulations, and thus although it may be insisted that chronic drug administration is necessary, it is often not sufficient for tolerance/sensitization to develop, even with drugs that are thought to be generally liable to lead to the development of tolerance/sensitization. Lê and Khanna point out that accompanying changes in pharmacokinetic parameters have not often been investigated in such situations, and so their occurrence cannot in principle be ruled out, but equally the challenge of identifying them remains.

The second and greater challenge from contingent tolerance is provided by the striking *within*-subject tolerance that may occur, dependent on the overall behavioral context within which drugs are administered. A notable recent example of this principle has been reported by Smith (1986) and is discussed by Wolgin (*see* Section 2.1.3. and Fig. 5 of Wolgin's chapter). Smith found first a *differential* tolerance within individual animals that were exposed to two alternating schedules of intermittent reinforcement: one pattern of behavior (A) showed tolerance to the effects of chronic *d*-amphetamine administration, whereas the other (B) did not. Smith then discontinued the schedule in which tolerance was exhibited (A), and found that tolerance *then* developed with the behavior that had previously been nontolerant (B). Finally, the more recently established tolerance of this second pattern of behavior (B) was obliterated by reintroducing intercurrent periods of the schedule that controlled behavior A once again, with *that* behavior showing tolerance to the drug once more. These dynamic, reversible, and strikingly context-specific contingent tolerance effects with an unchanging chronic drug regimen surely challenge accounts of tolerance/sensitization that are couched in terms of dispositional and functional mechanisms, perhaps to an extent as to raise doubts about the wisdom of *defining* tolerance/sensitization in such terms.

The behavioral data reviewed in the chapter by Barrett et al. present further problems for *definitions* of tolerance/sensitization. Barrett et al. review evidence that the acute effects of a drug on schedule-controlled operant behavior are subtle and depend on many environmental variables. For example, the size or even the direction of an effect produced by a given dose of a given drug will depend upon the behavior that is being investigated, and systematic knowledge is now accumulating about how the nature of the environmental contingencies that control the behavior contribute to the behavioral effects of a drug as well as the pharmacological nature of the drug. Barrett et al. also provide a detailed review of the literature that shows that the effects of a drug on behavior depend on the *context* in which the behavior occurs (i.e., on interactions between different schedule-controlled patterns of behavior exhibited by individual animals) and that an animal's past *history*, too, can determine not only the size of a drug's effect, but even its direction. Barrett et al. emphasize (*see* Barrett et al.'s chapter, section 1.2.) that such variables may "play as critical a role in determining the chronic effects of drugs as they do in determining effects when administered acutely." They support this claim by also citing Smith (1986), again to emphasize the context-specific tolerance effects of amphetamine discussed above. They also cite (*see* section 1.2. and Fig. 1 of Barrett et al.'s chapter) a study by Nader and Thompson (1987), in which the development of tolerance to chronic administration of methadone was shown to be a function "merely" (the present writer's term) of an animal's previous history of exposure to a different schedule of reinforcement. Examples of such subtle environmental determinants of tolerance to the behavioral effects of chronic drug administration serve, as noted above, to challenge attempts to define tolerance/sensitivity in terms of underlying dispositional and functional mechanisms. However, more generally, the work reviewed by Barrett et al. leads to the conclusion that nonpharmacological factors can have a major influence on the effects of drugs when they are administered chronically, as well as when they are given acutely. These data also, therefore, raise questions about whether the gross effects which nonpharmacological procedures can exert on dose–response curves could be classified as examples of tolerance/sensitization or merely as "simulations" of tolerance/sensitization. It is noteworthy that both Barrett et al. and the editors of this book (Goudie and Emmett-Oglesby) refer here to "tolerance-like" effects, but as Goudie (1989) has pointed out elsewhere, "the suggestion that tolerance can only be induced by pharmacological stimuli seems to assume *a priori* that pharmacological "causes" of tolerance are of greater importance than nonpharmacological ones."

It should also be emphasized that the work reviewed in the chapter by Barrett et al. presents problems for any suggestion that there is such a hypothetical entity as "*the* dose–response curve" for the behavioral effects of a drug. This conclusion is reminiscent of similar challenges to the concept of "*the* stimulus generalization gradient." As Nevin (1973) pointed out, the extent to which the control over behavior exerted by a stimulus generalizes to other stimuli is a function of the experimental conditions. Thus, both stimulus generalization gradients and dose–response curves depend crucially for their form of expression on the precise environmental circumstances in which subjects are trained and tested. A similar conclusion emerges from the review by Young and Sannerud of the development of tolerance to the discriminative stimulus properties of drugs. The conclusion is again prompted that adequate experimental analysis in behavioral pharmacology generally, and in the more specific study of tolerance/sensitization, requires systematic variations of behavioral as well as pharmacological parameters. Thus, the suggestion made by Nevin (1973, p. 148) that "any particular gradient (of stimulus generalization) is the result of the combined action of many factors, so that no underlying . . . process can be isolated" can be extended to any particular dose–response curve and, thus, shift in such a curve, which may be used to identify tolerance/sensitization.

The impact of these observations is to raise further uncertainties about the first definitional criterion for tolerance/sensitization discussed above, namely that there should be a "shift" from some control value. Any implication that there is some "true" effect of a drug on behavior that may be used as a comparator must be an oversimplification, if one is seeking to make generalizations about the development of tolerance/sensitization to the effects of a drug on "behavior" rather than on the quite specific patterns of behavior that are examined in specific conditions. Crucial determinants of changes in the effects of drugs on behavior cannot be eliminated simply by ignoring them, for as with the effects of schedules of reinforcement on operant behavior more generally, they will operate and thus influence behavioral outcomes whenever they can operate.

The dangers of asking what a word means are well known, and it may be thought that they have been illustrated only too well in the present discussion, which began with what appeared at least to be a clear definition of tolerance/sensitization. Interesting issues of varying types have been encountered, however, both with respect to the definitional criterion of a shift in dose–response curves and with the definition of tolerance/sensitization as exclusively a pharmacological effect. The latter could, in principle at least, be resolved simply by adhering to the criterion that tolerance/sensitization relates to chronic administration of a drug *tout court*, but this would remove its study from the study of very similar effects on dose–response curves that result from nonpharmacological manipulations. The former prompts consideration of whether tolerance/sensitization should be defined as a (reversible) *within-*

subject phenomenon, and to the relationship between data produced by between-group and within-subject methodologies.

The issues that have been raised cannot be resolved here, but perhaps the present discussion at least prompts the conclusion that tolerance/sensitization to the behavioral effects of drugs is neither a simple nor an isolated phenomenon. In particular, problems seem to emerge if the phenomenon is conceptualized in terms that are essentially static rather than dynamic. For the moment, it may be concluded that there is at least some ambiguity about whether tolerance/sensitization is to be conceptualized in terms of movement away from some hypothetical (or indeed empirically determined) point or in terms simply of differences in effects. Such ambiguities can extend even to the labeling of an effect as tolerance or as sensitization. Investigations of the effects of chronic drug administration on behavior reveal that there are difficulties in deciding to what extent the concept of tolerance/sensitization should be extended to nonpharmacological manipulations. The idea that the effects on behavior of repeated drug administration can somehow be evaluated free of "contaminating" behavioral/environmental variables, either present in an experiment or in a subject's past history, must be resisted. There is a need for systematic manipulations of behavioral as well as pharmacological parameters in the study of tolerance/sensitization, as in behavioral pharmacology as a whole.

2. The Explanation of Tolerance/Sensitization

When providing the operational definitions of tolerance and sensitization discussed above, Goudie and Emmett-Oglesby went on to comment: "Such definitions carry no implications concerning the mechanisms that may mediate these events." The present discussion now turns to consider how behavioral or contingent tolerance/sensitization may be explained or understood, i.e., to the mechanisms that might be thought to produce the phenomena.

As pointed out elsewhere by Goudie (1989), the term behavioral tolerance can prove to be a source of confusion. Goudie distinguished two usages of the term, first in a purely descriptive way (i.e., to describe changes in behavior with repeated drug administrations), and the second to identify "*only* those examples of tolerance which (are considered) to be mediated by behavioral or psychological processes." Goudie's first descriptive category is self-explanatory. His second category is illustrated by many of the studies reviewed in the chapter by Siegel, in which nonbehavioral data as well as behavioral data are presented, but in which the view is prompted by both that the development of tolerance is affected by behavioral principles incorporated in Pavlovian conditioning. There is indeed potential confusion in the two uses of the term behavioral tolerance, but as noted earlier, even the descriptive case of behavioral tolerance/sensitization focuses on data that have, in fact, provided an engaging challenge to any doctrine that tolerance/sensitization is ultimately to be understood in pharmacological, physiological, or dispositional terms.

Because of his limited pharmacological and physiological knowledge, it would certainly not be wise for the present writer to attempt to evaluate the adequacy of explanations of tolerance/sensitization couched in essentially reductionistic terms. However, Thompson (1984) has provided a brief analysis of the nature of such reductionistic enterprises in behavioral pharmacology. He distinguishes "explanatory" from "theoretical" reductionism, asserting that the former is based on the belief that an adequate explanatory account of the effects of drugs on behavior will be provided in terms of mechanisms acting at an essentially molecular (i.e., biophysical) level. There can surely be no objection to scientific attempts to identify such mechanisms involved in any example of tolerance/sensitization to the behavioral effects of drugs, although Thompson notes that behavior is integrated and functionally dynamic and that "as an analysis moves from one level of organization to another, the dynamic features at a given level are often nearly independent of the detailed structure of the various subsystems at a lower level of organization." (Thompson, 1984, p. 4). Thompson expresses greater reservations, however, about "theoretical reductionism," which he describes as the view that "the laws of one science can be shown to be special cases of theories or laws formulated in another branch of science" (p. 5). Such a view in behavioral pharmacology would reduce psychology to the position of a special case of physiology, with behavior conceptualized merely as an appendage to processes that occur at a different level. The necessity for such a view within psychology generally was, of course, challenged long ago by Skinner (1950), and it is notable that it is the development of his theoretical position that has led to

so much empirical work, enriching the science of behavior generally and, thus, behavioral pharmacology more specifically.

Some commentators might feel that the level of explanation
used in pharmacology/physiology is somehow more basic or fundamental than that available in the behavioral sciences. To put this
more crudely, it seems often to be assumed that behavioral phenomena may ultimately be explicable in terms of physiological
events, but not vice versa. Corfield-Sumner and Stolerman (1978)
point out that behavioral tolerance (in Goudie's descriptive sense)
may be accompanied by changes in neurotransmitter availability, in
the sensitivity of drug receptors, in the absorption, distribution, and
metabolism of drugs, "or in any other relevant mechanism," and
they go on to point out that ". . . because an environmental or
behavioral factor may influence the development of tolerance, it
does not follow that the tolerance is not mediated through physiological mechanisms" (Corfield-Sumner and Stolerman, 1978, p.
397). Even if reductionistic analyses are to be favored as somehow
the "real" explanations of behavioral tolerance/sensitization in
both the senses identified by Goudie (1989), the "elucidation of its
neuronal and molecular basis is clearly going to be difficult" (*see*
chapter by Littleton and Little, section 1.1.). There is, therefore, a
growing tendency to emphasize the importance of potential
"behavioral mechanisms" in the development of context-specific
tolerance and other behavioral aspects of tolerance/sensitization.
As Bignami et al. (1975) have pointed out: "The more one extends
the experimentation on a given tolerance phenomenon, the more
one finds evidence for the existence of multiple mechanisms."

Thus, the study of the development of tolerance/sensitization
to the behavioral effects of drugs and of the development of
context-specific tolerance more generally have emphasized the
need for explanations of these phenomena couched in terms of
behavioral as as well as pharmacological mechanisms. What
behavioral mechanisms might be considered here? Thompson
(1984, p. 3) has argued that a mechanism is "a description of a
given phenomenon in terms of more general principles." He argues
that the level of analysis that will lead to the identification of dis-
tinctively behavioral mechanisms in behavioral pharmacology is
not reductionistic in the senses discussed above, because that would
lead to an overemphasis on physiological or molecular mechanisms. Rather, it is to be found in terms of "a more general set of

environmental principles regulating behavior'' (1984, p. 5). The principles of Pavlovian and operant conditioning have been extensively studied in the experimental analysis of behavior, and one might therefore look to the integration of specific phenomena in the field of behavioral tolerance/sensitization and context-specific effects with these principles. The experiments reviewed in the chapter by Siegel, and of course his interpretation of their results, provide a striking exposition of explanatory behavioral mechanisms in this sense. The data from experiments on context-specific tolerance are, in turn, related elegantly to the general principles of sensory preconditioning (Siegel chapter, Section 3.1.3.), partial reinforcement (Siegel chapter, Section 3.2.1.), latent inhibition (Siegel chapter, Section 3.2.2.), extinction, (Siegel chapter, Section 3.3.1.), the effects of unpaired conditioned and unconditioned stimuli, (Siegel chapter, Section 3.3.2.), external inhibition (Siegel chapter, Section 3.3.3.), associative inhibition (Siegel chapter, Section 3.4.), "blocking" (Siegel chapter, Section 3.5.1.), and "overshadowing" (Siegel chapter, Section 3.5.2.). As a result, the phenomena of context-specific tolerance are elaborated, interpreted, and evaluated in relation to "a general set of environmental principles regulating behavior," in this case of course the principles of Pavlovian conditioning, and in this way, explanations of specific phenomena emerge in ways that aid our understanding of the phenomena. The extension of this explanatory account to less-controlled phenomena in the world outside the animal laboratory, as for example with "opiate overdose," represents an impressive extension of Siegel's explanatory schema. Wolgin provides a similar analysis, but with an emphasis on principles from operant or instrumental conditioning as an explanatory behavioral mechanism. The chapter by Barrett et al. also emphasizes behavioral mechanisms by drawing out the relevance or potential relevance to the development of behavioral tolerance of the principles of schedule-dependence, behavioral history, and context, thereby also placing specific data in an explanatory framework of a well-developed set of more general principles in behavioral pharmacology. Young and Sannerud conclude their chapter on tolerance to the discriminative effects of drugs by emphasizing that such behavioral as well as pharmacological principles are necessary to explain the factors that modulate the acquisition, maintenance, and generalization of drug stimulus control (Young and Sannerud's chapter, Section 6.).

There are, therefore, numerous and extensive examples in the chapters of this book of the kinds of "behavioral mechanisms" advocated by Thompson (1984), and these discussions contribute to greater behavioral sophistication in interdisciplinary study.

Just as explanations couched in terms of behavioral principles supplement, but cannot supplant, more reductionistic analyses that may identify molecular or neuronal mechanisms, it is becoming increasingly clear, as Dews (1978) has suggested, that more than one mechanism is often involved in the development of tolerance to a drug even in a single individual at a given moment. On the other hand, the task of identifying the set of behavioral principles that most adequately and parsimoniously provides an explanatory context for specific data can be pursued by testing predictions that arise from the different schemes. Evaluation of Siegel's interpretation of context-specific tolerance in terms of the classical conditioning of "compensatory" responses as opposed to interpreting them in terms of other behavioral principles derived from the study of habituation provides a good example of this (*see* Siegel's chapter Section 6.2.2.).

This last point draws attention to the fact that the more behaviorally orientated contributors to this book have sometimes gone beyond the strategy advocated by Thompson (1984) in their attempts to find appropriate behavioral mechanisms for the specific effects that they discuss. Siegel provides the most sustained example of this, consistently going beyond the environmental principles of Pavlovian conditioning in a systematic attempt to explain the data of context-specific tolerance in the specific terms of competing conditioned responses. Siegel admits that such compensatory conditioned responses can be elusive, but he notes the claim (made by others) that failures to demonstrate their existence "do not represent major challenges to the conditioning account of tolerance" (*see* chapter by Siegel, Section 6.1.6.). There is, therefore, perhaps some unresolved tension in Siegel's generalized conclusion that "it is clear that the effects of a drug are importantly modulated by drug-anticipatory responses." There would be less tension about an alternative formulation that asserted that context-specific tolerance can be understood in terms of the *principles* of Pavlovian conditioning. Siegel's emphasis on "anticipation," couched at one point in the admittedly casual suggestion that "the rat that *expects* the barbiturate is hyperresponsive (to cocaine)" (Siegel's chapter, Section 1., emphasis added) provides a further extension to explanations in terms of the types of behavioral mechanisms advocated by Thompson (1984), although to be sure this apparently more cognitive account does of course make contact with some animal learning theorists who currently favor cognitive explanations of Pavlovian (or "associative") conditioning (e.g., Dickinson, 1980).

Wolgin is drawn into similar diversions in his chapter. When reviewing the important study by Smith (1986) discussed above, he moves from the impeccable summary statement that "behavioral tolerance is influenced by the 'global' density of reinforcement" to the more contentious suggestion that "subjects are keenly *aware* of the rate at which they receive reinforcement" (Wolgin chapter, Section 2.1.3., emphasis added). As with Siegel, this lapse into apparent cognitivism is an isolated event. However, again like Siegel, Wolgin's attempts to show the role of "instrumental learning in the development of tolerance" leads him to try to identify what is conditioned in behavioral tolerance, i.e., "what the subject has learned to do in order to overcome the initial effect of the drug" (Wolgin's chapter, Section 4.). As with Siegel, therefore, a concern for explaining the effects of behavioral tolerance *in terms of* (or *as*) a conditioning process is subserved to attempts to identify often unobserved and therefore "circumstantial" (Wolgin's chapter, Section 4.) responses that might mediate tolerance. Finally, it may be noted that even that normally scrupulous functional analyst Barrett and his colleagues entertain a similar account in their chapter when they suggest (Barrett et al. chapter, section 1.2.) that "exposure to a (specified) schedule . . . may provide a behavioral repertoire that then facilitates . . . the development of tolerance," a statement that is undoubtedly true, but that is not necessarily well directed in the most parsimonious explanation of the phenomena they review.

It may seem unreasonable to have devoted such space in this short review to tracing those occasions on which contributors have ventured beyond Thompson's "behavioral mechanisms," usually by appealing to the existence of competing or facilitating responses, but occasionally even by what some might describe as lapses into an apparent cognitivism that adds little by way of explanatory power. However, a most important point is to be found here, central to the whole question of how behavioral and context-specific tolerance/sensitization are to be explained. As Poling (1986, p. 63) has pointed out, "The notion that the description of functional relations provides an adequate explanation of behavior, or any other phenomena, may not be intuitively obvious. However, in behavioral psychology as in science in general, it is held that something is 'explained' when we can specify the events that cause it." For exponents of functional analysis, therefore, a description of a demonstrably functional set of environmental principles provides an explanation of the phenomena of behavioral tolerance/ sensitization, without the need for assumed "underlying" processes or responses that might be thought to mediate such phenomena. From the perspective of functional analysis, therefore, behavioral tolerance and sensitization emerge not as an appendage to some other ("underlying"?) events, but instead as a phenomenon of scientific interest in its own right that is explicable, at least in part so far, in terms of environmental principles acting upon behaving organisms.

In summary, issues concerning the *explanation* of tolerance/ sensitization provide a forum for debate about the nature of explanation in behavioral pharmacology generally. There is no doubt that the phenomena of tolerance/sensitization are multiply determined (Dews, 1978). Some chapters of this book indicate the progress that has been made in identifying physiological or molecular mechanisms in tolerance. Studies of tolerance to the behavioral effects of drugs, and of the phenomena of context-specific tolerance generally, emphasize however that these effects can be fully understood only by incorporating environmental influences that provide the context for the effects of drugs on intact biological organisms. These two levels of explanation are mutually supportive within an interdisciplinary science, and it should not be expected that either can be replaced by the other. The behavioral mechanisms can be adequately described in functional terms, "in terms of more a general set of environmental principles regulating behavior" (Thompson, 1984). If behavioral tolerance and context-specific tolerance can be explained in this way, functional analysts will wish to take their cue from Poling (1986) and resist claims that it is necessary to "explain" the phenomena further in terms of competing or facilitating responses that are often elusive, or in terms of assumed cognitive processes that may be no more than "translations" of the environmental contingencies that are acting on behavior as it interacts with the pharmacological effects of a drug.

3. Concluding Comments

It is the purpose of this brief interpretative review to identify for discussion general issues that emerge from the other chapters of this book. It is certainly not its purpose, however, to be anything other than fully appreciative of those more substantive contributions. As noted at the outset, the chapters that are more data-oriented provide clear and authoritative reviews that amply illustrate the excitement that is to be found in contemporary experimental investigations of the important phenomena of drug tolerance/sensitization. There can be no doubt that the study of these phenomena provides an area in which behavioral and pharmacological analyses are being brought into fruitful interplay, nor can there be any doubt that the accumulation of empirical data has made a significant contribution to our understanding of the phenomena of tolerance and sensitization.

Goudie (1989) has pointed out that "though it may be relatively easy to demonstrate that tolerance develops to a specific drug effect in a behavioral procedure, it is *much* more difficult to determine what mechanism(s) are involved." Furthermore, since behavioral tolerance/sensitization focuses on the effects of drugs on behavior, it is nothing less than inevitable that both behavioral and pharmacological mechanisms must be involved. To this extent, we must agree with the comment made by Bignami et al. (1975) that "Tolerance to the effect of any given agent is a complex phenomenon and it is arbitrary to draw a qualitative limit between 'true pharmacological tolerance' and other tolerance phenomena."

In an evaluative review of behavioral pharmacology in general, Branch (1984) has suggested that most research in this field is designed to explore pharmacological rather than behavioral questions, to such an extent that behavioral pharmacology is largely "a subdiscipline within pharmacology wherein drug similarity is examined with intact organisms" (p. 511). Branch cites, in support of his view, the fact that the study of the discriminative stimulus properties of drugs has become in effect a standardized behavioral assay, with minimal manipulation of behavioral parameters that are in fact known to be important in determining the extent to which the discriminative properties of a given stimulus (which might be a drug) generalize to other stimuli. The behavioral data that emerge from such studies are increasingly interpreted or "explained" in terms of putative drug mechanisms at drug-receptor sites, i.e., in pharmacological terms. Branch believes that these tendencies result from the fact that "pharmacology is a more well-developed science than is psychology," and whereas "its principles provide an organized, logical framework from which experimental ideas can be interpreted, behavioral principles, by contrast, are not as widely accepted and they tend to provide a weaker interpretive base" (p. 518).

It is inevitable that the study of tolerance and sensitization should be marked by the imbalance that Branch noted and, thus, for the pharmacological parameters of its behavioral phenomena sometimes to be regarded as somehow more important or even "real" than the behavioral. At the same time, it is perhaps this field of behavioral pharmacology that currently offers most potential for the true impact of an interdisciplinary science to be recognized. There are a number of reasons for this, of which two will be mentioned here. Both arise from the observation that the very definition of the phenomenon of tolerance/sensitization finds within it a tension as to the extent to which it can be limited merely to changes in drug effects resulting from the manipulation of pharmacological parameters (i.e., repeated drug administration).

First, the effects reviewed extensively in chapters of this book demonstrate that whether or not repeated administrations of a drug will lead to the development of tolerance depends critically on the behavioral context in which the drug is administered. Thus context-specific tolerance (which is not of course limited to strictly behavioral data) shows that tolerance/sensitization cannot be conceptualized *solely* in terms of the pharmacological effects of the drug.

Second, again as explored in detail in preceding chapters, shifts in a dose–response curve can result as readily from the manipulation of behavioral parameters as of pharmacological, and this raises the question of whether such effects of nonpharmacological procedures should be regarded as tolerance/sensitization or merely as simulations of such effects.

In both these cases, any simple dichotomy between "real" explanations of tolerance/sensitization in terms of pharmacological principles and somehow "less" real accounts couched in behavioral terms breaks down. Thus, studies of contingent effects and of behavioral tolerance/sensitization both suggest that the phenomena should be or even must be characterized as being interdisciplinary in nature.

The unavoidable necessity to attempt some form of integration of the pharmacological and behavioral parameters of these phenomena also leads to a forceful recognition in this field of behavioral pharmacology that it is essentially a *dynamic* system that forms the focus of study. This must surely be the case in all areas of behavioral pharmacology, but as Branch (1984) has pointed out in some other fields, the relative importance of behavioral and pharmacological parameters can be distorted, and thus their dynamic interaction can fail to be recognized. In the context of studies of tolerance/sensitization, dynamism is illustrated at every level and can also be appreciated in the form of intrinsic interactions between different levels of analysis.

The pharmacological aspects of tolerance/sensitization are dynamic in many ways, not least of course in the important dimension of the pharmacodynamics of dispositional factors. It was noted earlier in the context of discussion of the data reproduced from Stolerman et al. (1974) that there is always a need for systematic variations in the pharmacological parameter of the dose, if a study is to identify whether pretreatment with a drug leads to a subsequent potentiation or reduction in its effect. This point may be extended here: in some situations, it is possible that a drug that would be described in general as producing tolerance as a result of its repeated administration (because it produces a rightward shift in the *overall* dose–response curve) may nevertheless produce an effect at a single dose that does not reflect this general outcome or that even appears to be contradictory to it. This is essentially a pharmacological effect, and, although such a specific outcome may not seem intuitively obvious, it is not difficult to appreciate that it might occur, nor is it difficult to recognize its potential significance within the dynamic context of pharmacological analysis.

Exactly similar points may be made with respect to behavioral parameters in the study of tolerance/sensitization, however. Behavior is a dynamic function of environmental contingencies that provide the current context or the historical context in which the behavior is studied. A manipulation of some aspect of these environmental contingencies may be crucially important in determining whether repeated administrations of a drug produce a rightward shift in the dose–response curve (contingent tolerance). Systematic manipulation of these behavioral parameters along a continuum (as for example by manipulating the frequency of reinforcement maintaining the behavior or the degree of specific discriminative control exerted by a stimulus that controls the behavior differentially) may produce an effect at a single value that does not reflect this general outcome or that even appears to be contradictory to it. This is essentially a behavioral effect, and, as with the pharmacological case above, its importance and orderliness can be understood in a systematic study even if it may not seem intuitively obvious.

It is to such potential subtle effects that the added dynamism of *interaction* between pharmacological and behavioral parameters must be added. The first of the two preceding paragraphs has been expressed as if, in this pharmacological case, behavioral variables are somehow irrelevant and can be ignored. Similarly, in the behavioral case outlined in the immediately preceding paragraph, pharmacological variables have been disregarded as if they were not relevant. In fact, of course, *both* pharmacological and behavioral variables must be important in the development of a drug's effects on behavior, and so neither can be ignored in this way.

At the end of their chapter, Barrett et al. (Section 5.) emphasize that ''behavior is not a passive transducer of drug action but appears to actively impart direction and magnitude to drug effects,'' and this role is of course dependent on the variables that control that behavior. Similarly, it is well understood that a drug is not a passive influence on behavior and can impart direction as well as magnitude to behavioral change, a role dependent on pharmacological variables such as dispositional factors. The emergence of tolerance or of sensitization when drugs are repeatedly administered and when their effects are measured in behavioral terms can be properly understood only when the real importance of both pharmacological and behavioral systems is recognized, so that their mutual interactions can begin to be appreciated. This is of course true of any phenomenon within behavioral pharmacology, but as noted above, it has not always been fully recognized in some areas of the interdisciplinary science. However, it is a requirement that is so prominent and inevitable in the study of tolerance and sensitization that there is a more pressing need to ensure that pharmacological mechanisms are not overemphasized at the cost of the less widely appreciated behavioral mechanisms. This is indeed the case in the contributions to this book taken as a whole, for they illustrate well the impact of this general theoretical point in the best way possible — by presenting rich and challenging experimental data that show both how much has been achieved and how much more there is to do in the study of sensitization and tolerance.

Nevertheless, there is perhaps a paradox at the conceptual heart of this whole field. Behavioral tolerance/sensitization is increasingly recognized to be the result of dynamic interactions between the pharmacological and the behavioral determinants of drug effects, each of which is in turn a dynamic system. The basic phenomena have even been defined in essentially dynamic terms, such as "shifts" in a drug's effects, although there may be some ambiguity about the meaning of this term. Yet, it is too easy to fall back on the idea that there is some value that represents a baseline from which these shifts diverge, and this idea introduces a false and more static conception of what the field of study is about. In fact, on *any* occasion that a drug is given, its effects on behavior are crucially dependent on the values of the specific behavioral and pharmacological parameters that characterize both the present circumstances of the investigation and the past history of the experimental subject. The very notion of "tolerance/sensitization" may paradoxically serve to suggest that there is here an identifiable and separate area of behavioral pharmacology, namely the study of the waxing and waning of a drug's effects, but the thrust of the present review is to suggest that the study of these phenomena in the behavioral domain may be difficult to distinguish from the study of any other phenomena in behavioral pharmacology.

To end with a provocative suggestion, perhaps the terms tolerance and sensitization might even be dispensed with here. The terms have been taken from the discipline of pharmacology, which has developed largely through the study of drugs on quite specific mechanisms in vitro. Perhaps their use gives undue emphasis to the pharmacological aspects of the dynamic effects of drugs on behavior, and thereby inject an unwarranted and unnecessary degree of stasis to an essentially dynamic subject of study, which by its very nature can be carried out only in vivo.

References

Bignami G., Rosic N., Michalek H., Milosevic M., and Gatti G. L. (1975) Behavioral toxicity of anticholinesterase agents: Methodological, neurochemical and neuropsychological aspects, in *Behavioral Toxicology* (Weiss B. and Laties V. G., eds.), Plenum Press, New York, pp. 155–215.

Blackman D. E. (1987) Experimental psychopharmacology: Past, present, and future, in *Experimental Psychopharmacology* (Greenshaw A. J. and Dourish C. T., eds.), Humana Press, Clifton, New Jersey pp. 1–25.

Branch M. N. (1984) Rate dependency, behavioral mechanisms, and behavioral pharmacology. *J. Exper. Anal. Behav.* **42,** 511–522.

Corfield-Sumner P. K. and Stolerman I. P. (1978) Behavioral tolerance, in *Contemporary Research in Behavioral Pharmacology* (Blackman D. E. and Sanger D. J., eds.), Plenum, New York, pp. 391–448.

Dews P. B. (1978) Behavioral tolerance, in *Behavioral Tolerance: Research and Treatment* (Krasnegor N., ed.), N.I.D.A. Research Monograph No. 18, U.S. Government Printing Office, Washington D.C., pp. 18–26.

Dickinson A. (1980) *Contemporary Animal Learning Theory*. Cambridge University Press, Cambridge.

Goudie A. J. (1989) Behavioral techniques for assessing drug tolerance and sensitization, in *Neuromethods Vol. 13: Psychopharmacology* (Boulton, A. A., Baker G. B., and Greenshaw A. G., eds.), Humana Press, Clifton, New Jersey, in press.

Nader M. A. and Thompson T. (1987) Interaction of methadone, reinforcement history and variable-interval performance. *J. Exper. Anal. Behav.,***48,** 303–315.

Nevin J. A. (1973) Stimulus control, in *The Study of Behavior: Learning, Motivation, Emotion, and Instinct* (Nevin J. A., ed.), Scott Foresman and Co., Glenview, Illinois, pp. 114–152.

Poling A. (1986) *A Primer of Human Behavioral Pharmacology*. Plenum, New York.

Skinner B. F. (1950) Are theories of learning necessary? *Pschol. Rev.*, **57,** 193–216.

Smith J. B. (1986) Effects of chronically administered *d*-amphetamine on spaced responding maintained under multiple and single-component schedules. *Psychopharmacol.* **88,** 296–300.

Stolerman I. P., Fink R., and Jarvik M. E. (1974) Nicotine tolerance in rats: Role of dose and dose interval. *Psychopharmacol.* **34,** 317–324.

Thompson T. (1984) Behavioral mechanisms of drug dependence, in *Advances in Behavioral Pharmacology, Vol. 4* (Thompson T., Dews P. B, and Barrett J. E., eds.), Academic, New York, pp. 1–45.

Drug Tolerance and Sensitization:

A PHARMACOLOGICAL OVERVIEW

Harold Kalant

Any reader of this book who is familiar with the reviews on alcohol and drug tolerance published during the 1970s cannot help but be struck by the marked elaboration of knowledge and concepts concerning tolerance that have taken place in the past 10–15 ys. This is particularly true of the behavioral aspects of the subject. The older simplistic picture of tolerance as a homeostatic reaction to the presence of a drug in a living system has been replaced by a much more complex picture that justifies its prominent place in the literature of behavioral pharmacology. The editors and most of the contributors to this book have stressed the interaction of drug, behavior, environment, and individual history as a complex, dynamic system that affects the result in each individual case.

Yet the classical pharmacologist would be justified in reminding the reader that this is not really unique with respect to behavioral pharmacology. Every interaction between a drug and a biological system above the level of isolated organelles or proteins is also a complex dynamic system. For example, the Langendorff heart preparation, or the isolated segment of guinea pig ileum, also responds differently to the same concentration of the same drug, depending upon the degree of stretch imposed upon the muscle fibers, the temperature at which the preparation is kept, the composition of the perfusion fluid or incubation medium, and the nutritional and other antecedent history of the animal from which the organ was obtained. It should, therefore, be no surprise that the same considerations apply, with special importance in the case of a phenomenon so complex as behavior.

Nevertheless, the difficulty is certainly compounded in the case of behavioral pharmacology by the fact that the types of experimental approach and conceptualization derived from its two parent disciplines have not yet fused satisfactorily. Four of the chapters in this book can be labeled as essentially behavioral in approach, and the other four as essentially pharmacological. The difference is reflected in a fundamental contrast in the nature of what is meant by "explanation." According to The Shorter Oxford English Dictionary, "explain," derived from the Latin root **planus** or flat, has the various meanings of to make smooth, to spread out flat, to unfold, to make plain or intelligible, to interpret, and to account for. The last three meanings are all used in science, but in somewhat different senses by different disciplines. The statistician "explains" variance in a set of results by parceling it out to different factors, so as to account for it quantitatively by showing how much of the total variance is "explained" by each factor. Behavioral researchers, especially those engaged in the study of operant behavior, consider that an analysis of the functional relations between different factors explains the behavior in the sense that it permits prediction of the effect that a given manipulation is likely to have on the behavior of a given individual or group in a given set of circumstances. To the biological scientist, however, an explanation must be an elucidation of the mechanisms by which these factors have produced their end results.

In the latter sense, explanation is comparable to an onion from which successive layers are peeled off, to reveal yet other layers underneath. For the biologist, it is an article of faith that no behavior occurs in any organism unless there is a system for transducing a stimulus applied to the organism into an effector response, that in higher organisms that transducing system is the central nervous system, that the nervous system functions through the interconnections of neurons (and glia?), that the interconnection occurs through the release of neurotransmitters and neuromodulators, which act upon receptors on other neurons, and that both the neurotransmitter release and the postsynaptic response to these transmitters depend upon a complex series of molecular events in all the neurons involved. In this view, behavioral or environmental modification of a response to a drug, including drug effects on behavior, must ultimately depend upon the occurrence of some molecular event or events. In accordance with that view, the explanation is not complete until the last details of the molecular events have been identified, and "explained" or laid out flat (to make clear in *which* sense they are "explained"). For the biologist, therefore, the question is not whether a functional analysis of the relationships between factors affecting behavior explains that behavior, but rather, at what level it explains it.

The broad gap that remains between the behavioral and pharmacological components of behavioral pharmacology is clearly visible in the two sections of this book. The behavioral chapters do not venture across the border into the territory of cellular events, much less of molecular events. In contrast, the pharmacological chapters, with rare exceptions, do not examine the impact of behavioral or environmental manipulations upon cellular and molecular events. It is necessary in experimental research to isolate the component parts of a complex system in order to study them individually, but the objective should always be to synthesize a pattern of interrelations that approximates as closely as possible the properties of the living system. A failure to remain constantly aware of the need for synthesis can only work against the objective set out by the editors, "to relate the results of fundamental research to the real world and establish the 'ecological validity' of the various mechanisms for the induction of tolerance/sensitization that are studied in the experimental laboratory."

1. Definitions of Tolerance and Sensitization

In their introductory chapter, the editors have adopted definitions of tolerance and sensitization that are based on shifts, to the right or to the left, respectively, in the dose–response (DR) curve. It must be noted that the term "tolerance," when defined in this manner, is a convenient bit of shorthand for "acquired change in tolerance." Without such qualification, tolerance means simply the ability to tolerate or withstand something, and in the present context, the *degree* of ability to withstand the action of a drug. This is often referred to, in the drug tolerance literature, as initial tolerance (i.e., on first exposure to the drug), and reflects both genetic factors (the "innate tolerance" mentioned by Lê and Khanna) and experiential factors. Most authors, however, use the word "tolerance" in the shorthand sense mentioned above, and it is important to remember this, in order to avoid confusion. The editors' definitions omit cases of tolerance that are attributable not to a rightward shift of the DR curve, but to a "flattening" or reduction in the maximum effect obtainable, regardless of how much the dose is increased. More will be said about this "flattening" phenomenon below.

Nevertheless, from a pharmacological viewpoint, the editors are essentially correct in choosing to define tolerance and sensitization in terms of DR curve shifts. Many drugs are known to have biphasic DR curves for their effects on a variety of different phenomena at different levels of organization, from whole-animal behavior to transmembrane potentials in individual neurons. Simple categorization of drugs tends at times to make one overlook such biphasic effects. For example, Littleton and Little (Littleton and Little's chapter, section 1.2.) state that "central depressant drugs all, by definition, depress electrical activity in the neurons of the central nervous system." It is true that most studies of the effects of ethanol, barbiturates, benzodiazepines, and other "central depressant drugs," especially in isolated cell or tissue preparations, are carried out typically in high concentration ranges that are indeed depressant. Nevertheless, many of the effects experienced by humans and by intact animals of other species at lower dose or concentration ranges indicate phenomena of excitation. For example, the release of acetylcholine by isolated cerebral cortex slices exposed to depolarizing stimuli (Fig. 1) is increased by low concentrations of pentobarbital and decreased only by much higher ones (Kalant and Grose, 1967). In vivo, relatively low doses of ethanol or barbiturates produce behavioral excitation rather than depression, accompanied by signs of electroencephalographic arousal, which has been referred to as the "beta buzz" (Murphree 1973).

Dews (1969) has argued eloquently against the use of such terms as "central depressant" and "central stimulant," because so many drugs can produce changes in either direction, depending on the dose and circumstances. Under in vivo conditions, only *some*, and not "*the*," neurons will show decreased electrical activity as a

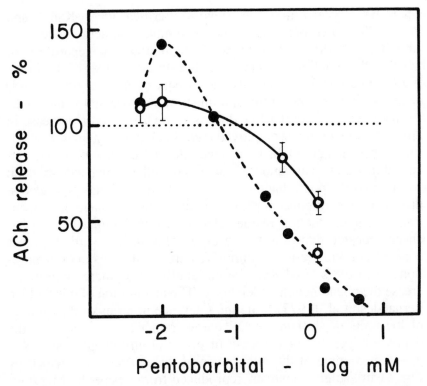

Fig. 1. Effect of pentobarbital on release of acetylcholine from guinea-pig cerebral cortex slices in vitro: ● data from McLennan and Elliott (1951); ○ data from Kalant and Grose, (1967); —incubation in medium containing 5 mM K$^+$; ----incubation in medium containing 27 mM K$^+$. 100% = release from corresponding control slices. Vertical bars represent SEM.

result of the effects of central depressant drugs, whereas others will show no change or actual enhancement of electrical activity, depending on the dose, manner of administration, and circumstances (Kalant and Woo, 1981).

Since biphasic responses can be shown even in isolated cell or tissue preparations, it seems probable that the biphasicity is inherent in the response to the most basic molecular action of the drug. For example, an effect of ethanol that results in a progressive gradual decrease in the transmembrane resting potential will at first increase the excitability of the affected cell, and then decrease it when the depolarization goes beyond a critical level (Kalant and Woo, 1981). Yet, presumably the same molecular action is responsible for the different degrees of depolarization, regardless of whether the functional effect is apparently excitatory or inhibitory. If the experimenter examines the effects of only a single dose, changes produced by chronic administration may result in an apparent increase in effect if a low dose is being used, and a decrease in effect if a higher one is employed (*see* Blackman's chapter, Fig. 1). The experimenter might then conclude that sensitization had occurred in the one case and tolerance in the other. Such a conclusion would miss the point that the whole spectrum of effects produced at different doses or concentrations requires a higher range of concentrations for production of the effects after the animal has experienced repeated exposure to the drug. This shift towards a higher concentration range is a common feature at all doses or concentrations, regardless of whether the initial effect is to increase or to decrease the phenomenon under study. The recognition of this fact led Fernandes et al. (1977) to define tolerance in terms of the number of log units of shift in the DR curve, rather than in terms of the degree of increase or decrease of effect at any given dose.

An example of the importance of this concept is provided by Fig. 11 in Wolgin's chapter, reproduced from a paper by Margules and Stein (1968). These investigators trained rats on a typical approach–avoidance behavior in which bar-pressing to obtain food reward was inhibited by electric footshock, which was also delivered by some lever presses ("punished responding"). During alternating periods of punished and unpunished responding by the rats, oxazepam markedly increased the rate of punished responses ("anxiolytic effect"), but initially reduced the rate of unpunished responding markedly. Tolerance to this effect on unpunished responding developed over a period of 4 or 5 d, but the rate of response rose not only to the control level, but 100% above it. A possible explanation for this finding is that a typical low-dose effect of oxazepam might be stimulation of such behavior, but that the dose employed in these studies was initially high enough to produce behavioral depression. The development of tolerance might then cause this higher dose to now have an excitatory effect on behavior that would be normally characteristic of a lower dose in the drug-naive animal. An examination of the whole DR curve, before and after chronic treatment, is necessary in order to test the validity of this explanation.

A further problem arises from the interrelationship of competing phenomena produced by the same drug action. For example, Bläsig et al. (1973) found that the various manifestations of the morphine withdrawal reaction constitute a response hierarchy, such that when one sign is present, it may prevent the expression of another one. Therefore, depending upon the intensity of the withdrawal reaction, which in turn depends upon the size of the previous chronic treatment dose and the length of time after the last administration of the drug, some signs may appear early, then be suppressed by other signs appearing as the reaction intensifies, and then return when the dominant sign has again diminished. A similar example is provided in the case of morphine by the competing hypothermic and hyperthermic responses produced by the drug. Initially, the hypothermic effect predominates and the body temperature falls during the early period after injection of a moderately large dose of morphine. As this effect wears off, the body temperature returns to and then above the baseline, so that a period of hyperthermia ensues. With chronic treatment with morphine, the hypothermic phase becomes progressively less marked and this is clearly recognized as tolerance. As a result, however, the hyperthermic phase appears earlier and is of greater maximum intensity; this has been regarded by some authors as sensitization of the hyperthermic effect (Gunne, 1960). However, it is possible to prevent the intital hypothermic reaction by raising the ambient temperature; under these circumstances, the entire hyperthermic phase is unmasked. When this is done, chronic treatment is shown to produce tolerance, rather than sensitization, to the hyperthermic response (Mucha et al., 1987).

Pharmacology is essentially a statistical science; the DR curve itself is a statistical statement of the probability of occurrence of a given event at a particular dose of the drug. Therefore, despite the variability of drug effect that is seen in different individuals with different heredity, age, nutritional status, and environmental and behavioral history, it is still perfectly valid to define statistically the pattern or degree of response that most subjects in a given population will show under the prevailing conditions. Such a DR curve constitutes a legitimate statistical baseline, against which it is equally legitimate to define tolerance and sensitization in terms of shifts of the DR curve, either within a population or in a given individual. The population change defines the general case, whereas individual differences that do not match the general case help to analyze the mechanisms by calling attention to differences in response that set those individuals apart from the others.

2. Major Identified Influences on Tolerance and Sensitization

It is self-evident that, in analyzing the influences that affect the development of tolerance and sensitization, we can deal only with those factors that we currently recognize. The history of the study of tolerance has shown a progressive change in concept. It was originally seen as a molecular adaptation to the presence of the drug. The landmark report by Schuster et al. (1966) showed that this could not be a sufficient explanation, because it failed to account for the influence of changes in reinforcement density (*see* chapters by Wolgin and by Barrett et al., in this volume). Some years later, the importance of practice under the influence of a drug was demonstrated (Chen 1968), and this in turn was followed by further analysis that gave rise to the hypothesis of behavioral augmentation of tolerance (LeBlanc et al., 1973). In the late 1970s and all of the present decade, a growing body of evidence (reviewed by Siegel in this volume) has attested to the importance of Pavlovian conditioning as a factor in the intensity or speed of onset of tolerance. One cannot assert that these are all of the factors that play a role in the development of tolerance, and past experience suggests that more will yet be found.

The present book, then, reflects the current state of knowledge. The chapters by Lê and Khanna, and by Willner deal with the traditional pharmacological differentiation of pharmacokinetic from pharmacodynamic factors in the development of tolerance and sensitization. Three chapters deal with various aspects of the role of instrumental behavior and conditioning in tolerance: that by Wolgin deals with the role of current instrumental behaviors, that by Barrett et al. considers the impact of past behavioral history on current drug reponses, and the chapter by Young and Sannerud examines the importance of tolerance to drug discriminative stimuli as determinants of behavioral responses. Siegel, who has contributed the most to the study of Pavlovian conditioning in drug tolerance, reviews the large and rapidly growing body of experimental findings on this topic. Finally, the chapters by Greenshaw et al., Willner, and Littleton and Little deal with selected aspects of neurotransmitter–receptor interaction and membrane transduction mechanisms, which may be responsible in cellular terms for the tolerance and sensitization that are observed at behavioral levels.

However, the question remains unanswered, and indeed largely unasked, whether these are all separate processes of tolerance, or separate influences that modulate a common process. The question can also be asked in a different way, if we define tolerance, as the editors have done, as a change in the DR curve for a given effect of a drug. If that definition is accepted, then, also by definition, all of the factors mentioned above would have to be considered modulatory influences and the question then becomes one of how far upstream from the end result these different factors act or interact, and by what cellular or molecular means. The following sections of this overview will not answer that question, but will attempt to point out some considerations and subsidiary questions that have to be dealt with in attempting to answer it.

3. Pharmacokinetic Factors

A fundamental question in the analysis of tolerance and sensitization is whether the observed change in response magnitude is the result of (a) a change in the amount or concentration of drug acting at the target site, or (b) a change in the responsiveness of the target site to the same drug concentration. Obviously, these two alternatives have very different implications with respect to mechanisms. A change in the concentration of a drug acting at its target site could result from alterations in absorption, distribution, biotransformation, or elimination of the drug as a result of previous exposure to it. The evidence concerning these classical pharmacological considerations of drug disposition is thoroughly reviewed by Lê and Khanna in relation to different categories of drugs, and most elegantly and lucidly by Willner in relation to sensitization to antidepressants.

In recent years, this concept has been carried even to the subcellular level. Littleton and Little refer to the work of Rottenberg et al. (1981), who reported that membrane fragments obtained from alcohol-tolerant subjects took up less ethanol from the incubation medium when the distribution of alcohol at equilibrium was studied in vitro. Littleton and Little (section 1.1.) suggest that this is a form of dispositional tolerance or, to use their term, decremental tolerance. However, it is not at all clear that decreased partitioning of alcohol from the medium into the membranes, or by implication from the blood into the cell membranes in the living subject, is the cause of tolerance rather than a manifestation of it. As Rottenberg et al. (1981) suggest, the decreased partitioning of alcohol into the membranes is probably a consequence of well-recognized changes in the lipid composition of the membranes in the tolerant subject (Goldstein et al., 1980; Littleton, 1980a). This would presumably change the relative solubility of ethanol in the membranes in the same way that its solubility or partition coefficient is different if the nonpolar phase is olive oil than if it is *n*-hexane or some other organic solvent. However, the change in lipid composition may *per se* be the cause of tolerance (Chin et al., 1978) by altering the pharmacodynamic interaction of ethanol with other components of the cell membrane, such as the ion channels discussed by Littleton and Little, receptors, or membrane-bound enzymes.

On the other hand, a number of as-yet-unexplained anomalies may conceivably have a dispositional basis. As Littleton and Little point out, there is a clear similarity between the actions of ethanol and barbiturates, which has long been recognized by their common classification as central depressants. The same authors, however, also point out differences among the barbiturates: for example, pentobarbital is more potent than phenobarbital in potentiating the effect of gamma-aminobutyric acid (GABA) on chloride ion channels, and this difference correlates with the greater anesthetic effect of pentobarbital than of phenobarbital at doses that have the same anticonvulsant effect. No explanation is provided for this difference, but it is noteworthy that pentobarbital has a higher lipid/water partition coefficient than phenobarbital or ethanol. On a variety of test procedures in vivo, cross-tolerance is easily demonstrated between ethanol and barbital (Gougos et al., 1986), but only with considerable difficulty and under limited behavioral circumstances between ethanol and pentobarbital (Khanna and Mayer, 1982). Conceivably, these facts reflect the influence of differences in drug disposition, on either the macro or micro scale, between pentobarbital on the one hand and ethanol and barbital on the other. If true, this would imply that at least some aspect of tolerance is dependent upon the precise locus within the cell at which each drug produces its initial effect.

The importance of induction or inhibition of drug-biotransforming enzymes as mechanisms of drug/drug interaction has been well recognized in all fields of pharmacology, and not only in behavioral pharmacology (Conney, 1986). For example, phenobarbital has long been known to produce sensitization (in Willner's terminology, pseudosensitization) to a wide variety of other drugs that are inactivated by the hepatic cytochrome P-450 system, because phenobarbital binds to the cytochrome P-450 with high affinity and competitively inhibits the biotransformation of the other drugs. Chronically, however, phenobarbital is an excellent inducer of the hepatic cytochrome P-450 system, and causes the other drugs that are substrates for that enzyme system to be metabolized much more rapidly, so that they produce less effect than they otherwise would. This is an instance of dispositional cross-tolerance. Recent developments have refined the picture through the use of monoclonal antibodies to specific cytochromes of the P-450 series (Gelboin and Friedman, 1985). This work has shown, for example, that chronic alcohol administration induces a very large increase in the amount of a specific $P-450_{3a}$, which normally makes up only a very small fraction of the total cytochrome P-450 in the liver cell (Koop and Coon, 1985). This induction may account for some of the dispositional tolerance to ethanol and cross-tolerance to other drugs that follow chronic alcohol administration.

Of much greater potential interest to behavioral pharmacologists, however, is the recent finding that at least some of the behavioral and environmental influences on tolerance may occur through alterations in drug disposition. This evidence has been reviewed by Lê and Khanna (section 6.1.). Several groups have now shown that, in animals in which environmental conditioning has contributed to the development of tolerance, the early absorption of the drug in question is significantly reduced, and the apparent volume of distribution of the drug in the body may be altered. One possible explanation for this phenomenon could be that the response to the conditional stimuli includes altered autonomic influences on gastrointestinal motility and splanchnic blood flow, both of which would affect the rate of absorption of drug from the intestine, and possibly the rate of biotransformation on first pass through the liver. Similar autonomic nervous system influences on blood vessels in skeletal muscle could contribute to changes in the apparent early volume of distribution. These findings, which have so far been examined by very few investigators, underscore an important point made by Lê and Khanna, i.e., that very little work on behavioral factors in drug tolerance has included detailed examination of drug kinetics. Therefore, one cannot conclude yet whether these factors alter the sensitivity of the central nervous system or merely the concentration of drug to which it is exposed. There is a real need for much more extensive study of pharmacokinetics in different models of tolerance.

4. Instrumental Learning

The evidence reviewed extensively and critically in the chapters by Wolgin and by Barrett el al., and by Demellweek and Goudie (1983) elsewhere, makes clear that practice of a task under the influence of a drug, especially when the drug effect causes a loss of reinforcement that would normally be obtained from that performance, is a very important factor in the development of tolerance, in both single-dose (acute) and multiple-dose (chronic) paradigms. The classic example that is commonly seen as the point of departure for research of this type was the study by Schuster et al. (1966). These investigators trained rats to obtain food reward by lever-pressing on two schedules, DRL and FI, which alternated through each experimental session. *d*-Amphetamine initially increased the response rate on both components, but tolerance developed rapidly on the DRL component in which the initial effect of amphetamine had caused loss of reward, whereas no tolerance developed on the FI component in which the increased rate had no effect on the number of reinforcements obtained. As Wolgin points out, since this differential reinforcement on the two components of the schedule was seen within the same animal during the same test session, it is most improbable that pharmacokinetic factors were responsible.

The phenomenon of stereotypy can be produced by a single large dose of amphetamine or other central nervous system stimulants. It can also appear gradually with repeated administration of smaller doses, and has frequently been interpreted as evidence of sensitization of dopaminergic axon terminals (e.g., in the nigrostriatal pathway) to the effects of the stimulants. We have observed that this development of stereotypy can also be related to reinforcement density (Kalant H. and Woo N., unpublished). Using a combination schedule involving DRL and FR components, we observed that, as expected, amphetamine initially increased response rate on both components, and that tolerance to the rate-increasing effect developed on the DRL component but not on the FR. However, with continued administration of amphetamine, stereotypy appeared during both components. In the FR component, stereotypy resulted in loss of rewards, and tolerance to the stereotypy soon appeared. In contrast, in the DRL component the stereotypy helped the animal to space out its responses to meet the requirements of the schedule, and stereotypy not only showed no tolerance, but was actually enhanced, i.e., showed "sensitization."

Wolgin (section 3.3.) points out that tolerance to amphetamine-induced stereotypy occurs only when stereotypy interferes with feeding, as in the example cited above. He suggests that, if another drug, such as fenfluramine, which is believed to act through serotonergic rather than catecholaminergic pathways, produces "anorexia" by means other than stereotypy, then no cross-tolerance between amphetamine and the second drug would be expected to occur. The fact that reinforcement deprivation by itself is insufficient to cause the cross-tolerance carries an important implication, i.e., that response contingency can be only a modifying factor rather than a primary mechanism of tolerance development. Wolgin also suggests that tolerance resulting from changes in neurotransmitter or receptor function cannot account for the observation that tolerance to stereotypy occurs during the feeding period, but not before or after it. This is not necessarily a valid conclusion. Unlike pharmacokinetic changes, which are relatively slow, changes in the level of neuronal activity can occur within milliseconds. It is quite conceivable that the stimuli associated with feeding might alter activity in neuronal pathways mediating associated behaviors, in such a way as to alter their sensitivity to the effect of the drug.

Despite the large mass of evidence supporting the reinforcement density hypothesis and, by extension, the role of instrumental learning in tolerance and sensitization, there are observations that do not appear to be consistent with it. One example is provided by the work on discriminative stimulus properties of drugs. The evidence reviewed by Young and Sannerud (this volume) demonstrates that drug discrimination by the trained subject remains remarkably constant over long periods of time, showing no change in discrimination threshold or generalization gradient, even when tolerance has developed clearly to other effects of the drug. Discrimination training typically involves differential reward of correct lever responses in the presence of the subjective stimuli produced by the drug. Any tolerance to the discriminative stimuli would cause errors in response that would give rise to a loss of reinforcements. Therefore, the prediction of the reinforcement density hypothesis should indeed be that no tolerance would occur. However, when the training dose is gradually increased, tolerance does develop, and lower doses lose their ability to sustain drug discrimination.

Similarly, if discrimination training is interrupted while chronic treatment with a higher drug dose is carried out, retesting at the end of the treatment period reveals that the discrimination threshold and the generalization gradient have shifted upwards. In these instances, the development of tolerance appears to be at odds with the reinforcement density hypothesis.

Other examples of findings inconsistent with the reinforcement density hypothesis are provided by both Wolgin and Barrett et al. (this volume), who point out problems of several kinds. They cite instances in which no tolerance developed despite drug-induced loss of reinforcement. In other cases that they reviewed, tolerance developed despite no loss of reinforcement caused by the drug. For example, in the study by Branch (1979), amphetamine increased responding for food or shock avoidance on an FI 5-min schedule, which would have no consequences with respect to the frequency of reinforcement, yet chronic administration of amphetamine shifted the DR curve to the right.

Another problem cited in both chapters is that raised by the observations by Smith (1986). Rats were trained on a food-reinforced lever-pressing task on a multiple schedule containing alternating random-ratio (RR) and DRL components. Amphetamine reduced the initially high response rate on the RR component, but increased the rate on the DRL. In both components, these effects resulted in a loss of reinforcements. Tolerance developed to the effects on the RR component, but not on the DRL. When the RR component was removed, tolerance appeared on the DRL schedule; when the RR component was reinstated, tolerance on the DRL component disappeared. The conclusion was drawn that tolerance was more influenced by global changes in the consequences of responding during the entire experimental session, rather than by the changes during individual components of the schedule. Nevertheless, it leaves unanswered, and indeed unasked, the question as to why these results differed from those obtained by Schuster et al. (1966). Why can the rat maintain appropriate responding on both components of one multiple schedule, whereas on another multiple schedule, the effects on one component override those on another? If the analysis of these differences is confined to consideration of schedule effects and their interrelationships, without attempting to understand the neural mechanisms responsible for the differences, the neurobiologist is likely to see this as mere restatement of the observations rather than true explanation.

Despite these problems, it is clear that there is a very large body of evidence supporting the role of instrumental learning in tolerance to central stimulants. In relation to ethanol tolerance, the terminology tends to be somewhat different, in that reference is made more frequently to the importance of intoxicated practice than to changes in reinforcement density. Nevertheless, many of the tests used for ethanol tolerance do involve reinforcement contingencies. For example, the circular maze test devised by Chen (1968) involved food reinforcement; the moving belt test devised by Gibbins et al. (1968) involved negative reinforcement by shock avoidance; the rotarod test (Dunham and Miya, 1957) involved avoidance of either footshock or falling from a considerable height. In these and other tests that have been used, the more rapid or more marked development of tolerance under conditions of intoxicated practice is compatible with the loss-of-reinforcement hypothesis. The conclusion by Wenger et al. (1981) that *no* tolerance develops in the absence of intoxicated practice is an even stronger statement of the same principle.

However, a similar phenomenon has been observed with other tests in which the presence or absence of reinforcement is less clear, for example, the hypothermia test. The development of tolerance to the hypothermic effect of ethanol has been shown to be retarded or prevented if the hypothermic effect of the training doses of alcohol is prevented either by whole body diathermy (Hjeresen et al., 1986) or by giving the training doses when the rat is in an environment at a temperature above that of thermoneutrality (Lê et al., 1986a). What is the reinforcement involved in these experiments? One possibility is that hypothermia is an aversive stimulus. Pohorecky and Rizek (1981) observed that ethanol-treated rats, which showed biochemical signs of stress concurrent with hypothermia when kept at an ambient temperature (T_A) of 22°C, showed much less evidence of stress when the hypothermia was prevented by exposure to a T_A of 35°C. Moreover, rats experiencing mild hypothermia, resulting from an ethanol withdrawal reaction, spent much more time in the arm of a T-maze in which they could obtain radiant heat to restore their body temperature to normal, in preference to the arms kept at normal or low temperature (Brick and Pohorecky, 1977). This would suggest that negative reinforcement is involved in "learning" to prevent the hypothermic effect of ethanol. Again, however, the biologist would like to know what the rat actually learns in order to accomplish an elevation of body temperature, and how it can bring about the maintenance of body temperature in spite of the presence of a drug that normally impairs thermoregulation.

The problem is complicated when the concept of learning is broadened to include intracerebral or peripheral mechanisms. For example, in the studies by Barondes' group (Barondes et al., 1979; Traynor et al., 1980), the isolated abdominal ganglion of *Aplysia californica*, when exposed continuously to ethanol in an incubation bath, developed tolerance to the effects of ethanol on synaptic transmission if the ganglion was stimulated electrically throughout the period of ethanol exposure, but not if it was exposed to the same ethanol concentration without electrical stimulation. Similarly, Jorgensen et al. (1984, 1986) found that, in the rat with complete transection of the spinal cord above the level of the segments mediating the tailflick spinal reflex, repeated exposure to ethanol resulted in tolerance to the inhibitory effects on the tailflick reflex if the reflex was repeatedly elicited during the periods of ethanol exposure, but not if the training doses were not accompanied by elicitation of the reflex.

In these examples, in which the nature of the preparation itself precluded any cognitive component, what does "learning" mean, and how does it differ from neuronal adaptation? A neurobiological explanation would be that the activated synapse is more plastic than the resting synapse, and that activation of the synapse therefore permits more opportunity for adaptation to the effects of the drug. This interpretation would be compatible with the observation that ethanol tolerance elicited by both multiple-dose ("chronic") and two-dose ("rapid") regimens with repeated trials under drug was prevented by concurrent administration of the cerebral protein synthesis inhibitors cycloheximide (LeBlanc et al., 1976) and anisomycin (Speisky and Kalant 1987). In this formulation, tolerance produced by instrumental learning is simply one instance of behaviorally augmented tolerance (LeBlanc et al., 1973). The capability for adaptation to sustained stimuli or altered environment is inherent in every living cell. According to the concept of behavioral augmentation of tolerance, response requirements during the period of exposure to a drug simply provide or enhance the stimulus for calling into play this inherent adaptive capability.

Further support for such a view of tolerance is provided by the observations on dose-dependence reviewed by Wolgin. Several investigators (Jorgensen et al., 1986; Lê et al., 1986b; LeBlanc et al., 1973) have observed that animals treated with relatively low doses of alcohol require behavioral experience under the drug in order to develop tolerance, but that those treated with high doses can develop tolerance without such practice or other behavioral experience. Moreover, repeated behavioral experience under the drug is not always sufficient in itself to produce tolerance. For example, Fig. 10 in Wolgin's chapter, reproduced from LeBlanc et al. (1973), shows that rats trained in a food-reinforced circular maze task under the influence of saline, but tested every fourth day under ethanol, did not develop tolerance at all, even though they had the same density of intoxicated practice on these test days as rats that received ethanol after the daily practice sessions between test days, and that did develop tolerance. In other words, the extra alcohol received by the "after" group on the intervening days was necessary for the development of tolerance, which otherwise was not produced by the periodic behavioral experience under the test doses.

Similarly, Wolgin surveys other evidence indicating that behavioral experience under the drug is not necessary for the development of tolerance to opiates, benzodiazepines, or tetrahydrocannabinol. With all of these drugs, behavioral experience under the influence of the drug could facilitate tolerance, but was not essential for it. As in the case of ethanol, it would appear that behavioral experience can enhance the stimulus to tolerance development when the training doses are small or the drug is one with a very short half-life, but that higher doses, more frequent administration, or the use of drugs with longer half-life, can provide enough pharmacological stimulus to the development of tolerance without the need for added contribution from intoxicated practice. The problem therefore remains whether the tolerance produced under these differing conditions is of different types, or is of a single type to which the additive effects of separate stimuli have contributed.

5. Pavlovian Conditioning

The same question must be asked with respect to the classical conditioning models of tolerance reviewed by Siegel (this volume) and by Goudie and Demellweek (1986). Does conditioning *produce* tolerance or simply *accelerate* an inherent biological adaptive response? Siegel's formulation appears to imply that classical conditioning merely accelerates an innate response. In answer to the question as to why a conditional response (CR) to drug-associated cues is opposite to the typical drug effect rather than similar to it, Siegel draws on the proposal by Eikelboom and Stewart (1982) that the drug effect is not a UCR but a UCS, and the UCR is the acute adaptive response that is opposite in direction to the drug effect. This proposal does not account for the fact that the CR to morphine-paired stimuli was found by Krylov (*see* Chapter by Siegel, Section 2.1.) to be morphine-like rather than morphine-opposite. Nevertheless, the hypothesis advanced by Eikelboom and Stewart implies that acute tolerance, i.e., tolerance developing within the duration of exposure to a single dose of a drug, is an intrinsic biological phenomenon resembling any other adaptive response to a sustained stimulus of any type. This adaptive UCR becomes a CR through repeated pairings with a CS derived from the drug-linked environment, and acts as an anticipatory compensation that diminishes the drug effect that follows. In the case of saline injection, i.e., when the drug does not follow, the CR then constitutes the basis of the withdrawal reaction, which is also typically drug-opposite rather than drug-like.

Genetic studies provide strong support for the idea that an innate biological adaptive process is essential for the development of tolerance. Interspecies comparisons, as well as comparisons among various genetically selected strains within the same species, have shown clearly the existence of important biologically determined differences in initial sensitivity to ethanol and other drugs, and in the ability to acquire tolerance, both acute (within-session) and chronic (Crabbe and Belknap, 1980; Deitrich and Spuhler, 1984). Especially the genetic differences in acute tolerance, which are independent of previous pairings of the drug with behavioral or environmental factors, are indicative of the existence of innate adaptive mechanisms.

If an innate adaptive response (acute tolerance) is the UCR that forms the basis of environment-contingent or conditional tolerance, this raises another question still current in the alcohol research literature as to the relationship between acute tolerance and chronic tolerance (Kalant et al., 1971; Littleton, 1980b). The most frequently used paradigms for the production of chronic tolerance involve the administration of repeated doses of a drug and repeated testing, and are therefore ideal for generating drug-administration cues that can serve as a CS. With such a model, the possibility clearly arises that chronic tolerance, measured as a gradual decrease in response produced at a fixed time after each succeeding dose of drug, could very well be the same as the acute tolerance constituting the UCR and the CR. In contrast, other models of sustained drug administration, such as the pellet-implantation technique (Gibson and Tingstad 1970) used extensively in the study of tolerance to opiates and barbiturates, or the vapor inhalation model used for the study of alcohol tolerance and physical dependence (Goldstein and Pal 1971), really constitute one single greatly prolonged dose. In such cases, it is possible that chronic tolerance is in fact the one episode of acute tolerance sustained indefinitely.

If these conjectures are correct, it should be possible to test them experimentally by the use of extinction trials. The single-episode tolerance developed by pellet implantation or vapor exposure should disappear rapidly after removal of the drug, without any need for extinction trials, in the same way that the innate acute tolerance disappears within hours when the period of action of a single dose has ended. In contrast, the multidose "chronic" tolerance should disappear more rapidly if extinction procedures are used, and evidence reviewed by Siegel (section 3.3.1.) indicates that this is indeed true. On such grounds, one might conclude that the conditioning process does not produce the tolerance, but elicits it.

Nevertheless, at least two different biological mechanisms are contributing. Environment-conditional tolerance to ethanol, pentobarbital, and probably other drugs is associated with a decreased early rate of drug absorption from its site of administration, and possibly an alteration in the apparent volume of distribution, which could account for a rapid decrease in observed drug effect at early times after administration of the dose (*see* Lê and Khanna chapter, section 6.1.1.). Possible autonomic nervous mechanisms for these changes have been suggested above. At the same time, such an explanation cannot account for the drug-opposite counterresponse that follows the injection of saline. This must be related to central nervous system adaptation, as discussed above in relationship to acute tolerance. The existence of these two separate mechanisms could explain the fact that tolerance can occur without conditioning, but that it is more rapid and more marked if conditioning does occur.

Again, as with instrumental learning, conditioning is not required for the production of tolerance when high treatment doses are employed. Siegel deals with this situation by reference to the work of several groups on the role of drug-generated interoceptive cues as CSs that can elicit tolerance as a CR to the later effects of larger doses of drug. However, this formulation fails to account for the development of tolerance in the pellet-implantation or vapor-inhalation models, in which there is in fact only one single prolonged exposure to a drug, with no opportunities for repeated pairing of early drug cues with later drug effects.

6. Task-Specificity of Tolerance

All of the hypotheses discussed above, except for that of acute tolerance as an innate adaptive response, offer some basis for task specificity of tolerance. Behavioral augmentation of tolerance, instrumental learning, and classical conditioning all involve some link between the drug effects and a particular test performance, regardless of whether that link is through enhanced behavioral demand, reinforcement loss, or drug-anticipatory cues. In all of these cases, it would be predicted on either a behavioral or a neurophysiological basis that those functions that are specifically exercised in the presence of the drug should be the most likely to show adaptive changes to the drug effects. In their chapter, Young and Sannerud illustrate this specificity with the examples in which tolerance developed to some effects of drugs, but did not develop to stimulus discrimination of the same drugs.

However, other influences could also contribute to task-specificity of tolerance. The evidence is reviewed by Barrett et al., concerning the effects of past history (whether pharmacological, behavioral, environmental, or physical history) on the acute effects of a drug on specific tasks. If past history influences the acute effects of a drug, it should therefore also affect the stimulus to adaptation. This is illustrated, for example, in the study by Lê et al. (1986a) of the effect of ambient temperature on tolerance to the hypothermic effect of ethanol. Groups of animals were exposed repeatedly to the same dose of ethanol at three different environmental temperatures, 36, 22, and 4°C. The same impairment of thermoregulation caused no change in body temperature at 36°C, some moderate hypothermia at 22°C, and marked hypothermia at 4°C. As predicted, the development of tolerance to the hypothermic effect (tested at 22°C in all groups) was much greater in the animals trained at 4°C than in the other two groups.

Another example is provided by the demonstration of carry-over effects from one cycle of development and loss of tolerance to another (Kalant et al., 1978; de Souza Moreira, et al. 1981). In those experiments, it was notable that the facilitation of tolerance development by a previous history of having been tolerant was independent of any response contingencies. In other words, the history of having acquired tolerance under conditions of repeated intoxicated practice on the test facilitated the development of tolerance in a subsequent cycle without intoxicated practice, and vice versa. On the other hand, Young and Sannerud point out an instance in which the past history is of no importance, i.e., the development of tolerance to drug discriminative stimuli, in which only the current training dose matters.

7. Possible Biological Mechanisms Underlying Pharmacodynamic Tolerance

All of the foregoing considerations have failed to answer clearly the question as to whether we are dealing with multiple separate mechanisms of tolerance, or multiple factors facilitating or retarding the operation of a single biological adaptive capability. One might hope to answer this question by examining the biological changes in drug effects on cells within the central nervous system, on such phenomena as cell excitability, neurotransmitter turnover, receptor function, and membrane transducer systems, in order to see whether the alterations are the same or different in the various models of tolerance. Unfortunately, this has not yet come to pass.

The changes in dopamine receptor function reviewed by Greenshaw et al. (this volume) have been in all possible directions. Changes in receptor binding, and in receptor-linked neurochemical responses, after chronic administration of amphetamine, various dopaminergic agonists, and cocaine were found by different investigators to include increases, decreases, and no change, without any apparent correlation with behavioral tolerance or sensitization. The same lack of consistency has been found with respect to changes in dopamine, norepinephrine, acetylcholine, serotonin, and other receptors in animals treated chronically with ethanol (Tabakoff and Hoffman 1980). It has also been evident in the studies reviewed by Willner (this volume) involving the chronic administration of antidepressant drugs, and Willner concludes that the neurochemical evidence to date provides no real explanation of sensitization to these drugs.

A further complication arises from comparison of tolerance to different classes of drugs acting by different cellular mechanisms.

As pointed out above, tolerance is not always characterized by a parallel rightward shift of the DR curve, but sometimes by a flattening of the curve. This has been shown clearly in experiments involving high-dose treatment with morphine (Mucha and Kalant 1980) and benzodiazepines (Lê et al., 1986c), and has been interpreted as evidence for a receptor desensitization or downregulation. Obviously, in the case of ethanol and general anesthetics, which do not act through stereospecific receptors, such a mechanism cannot apply and flattening of the DR curve has not been observed. In the case of opiates and benzodiazepines, on the other hand, multiple mechanisms can exist and indeed tolerance to these drugs has been shown to involve at first a parallel rightward shift of the DR curve and then a subsequent flattening.

These complexities in relation to neurotransmitter turnover and receptor properties have led to a recent shift of emphasis to the postreceptor transduction systems such as the adenylate cyclase system, the inositol phosphate cycle, voltage-dependent calcium channels, and GABA-activated chloride channels reviewed by Littleton and Little (this volume). In recent years, there has been a rapid advance in the development of very sensitive and elegant techniques for studying these processes, such as the patch-clamp method of studying the drug effects on a single macromolecule, the protein constituting an ion channel, in the neuronal membrane. These techniques permit the study of tolerance in the respective subcellular processes in preparations obtained from animals treated chronically in vivo with various drugs. Yet, remarkable as these techniques are, they have failed so far to identify clearly the basis of tolerance or sensitization to any of the drugs studied. The reasons for this failure are multiple.

The first problem is that most of the processes that have been studied in single cells or subcellular preparations require very high drug concentrations to produce the acute drug effects. For example, the calcium channels discussed by Littleton and Little typically require ethanol concentrations of 50 mM or higher to produce significant changes in ion flux during acute exposure to the drug. This concentration in plasma water would correspond to a blood alcohol level of nearly 290 mg/dL, which is profoundly intoxicating in humans and quite intoxicating even in the rat. It is therefore difficult to visualize such changes as a primary mechanism of alcohol intoxication at lower concentrations. Nevertheless, preparations obtained from animals made tolerant to chronic administration of lower doses or concentrations in vivo do show increased resistance to the effects of ethanol on calcium channels in vitro. Therefore, one must consider the possibility that these in vitro changes observed in the preparation from a tolerant animal are secondary to alcohol effects of another kind, produced at lower alcohol concentrations, elsewhere in the neuron or in the nervous system. When the tolerance must be developed in response to local changes within the same neurons, as in the cell culture studies reviewed by Littleton and Little, increases in the number of "L" channels occurred after 6 d of exposure to 200 mM ethanol, a concentration that would be lethal in the intact animal.

The second problem is that all of the changes of this type that have been studied provide at best only correlational evidence. Animals are made tolerant by drug exposure in vivo, and then the effects of drug upon some neurochemical or biophysical process are compared in preparations from tolerant and nontolerant groups. Under those circumstances, it is impossible to know whether the observed changes are *mechanisms* of tolerance, *manifestations* of tolerance, or entirely coincidental. Ideally, one would wish to be able to produce the same neurochemical or biophysical changes by other means than drug administration, and then see whether those changes produced in other ways confer increased resistance to the effects of the drug on first exposure. Unfortunately, there are no means known at present for reliably producing such changes by nondrug influences.

A third problem is that there are many inconsistencies and gaps in the evidence, which make it difficult to draw any firm conclusions. For example, Littleton and Little review evidence showing that nitrendipine, a dihydropyridine (DHP) type of calcium "L" channel blocker, increased the effect of ethanol on rotarod performance, presumably by synergistic action with ethanol to increase the degree of blockade of calcium entry into neurons. Yet, chronic nitrendipine treatment blocked the development of ethanol tolerance and blocked the increase in B_{max} for DHP-type calcium channel blockers, i.e., the chronic nitrendipine treatment blocked the ethanol-induced increase in the number of "L" channels. This is difficult to understand. One might have predicted that chronic nitrendipine treatment should increase the B_{max} rather than prevent an ethanol-induced increase. For example, the evidence reviewed by Greenshaw et al. (this volume) indicates that chronic administration of dopamine receptor blockers resulted in an increased B_{max} for both D_1 and D_2 types of dopamine receptors. A similar compensatory upregulation might well have been expected with chronic nitrendipine treatment, and that should have increased the degree of ethanol tolerance or produced cross-tolerance to ethanol when the latter was tested for the first time. A related difficulty is that the DHP-type calcium channel blockers enhance the effect of acute administration of ethanol, pentobarbital, nitrous oxide, and general anesthetic in rats tested for ataxia and loss of righting reflex, yet the blockers do not produce these effects by themselves. How, then, is it possible to explain the ethanol effects as being the result of blockade of such channels when specific channel blockers do not produce the same behavioral effects?

Another problem, pointed out by Littleton and Little, is that the "L" channels occur principally on the neuronal cell bodies, and not in the axon terminals where neurotransmitter release occurs. Rather, the "N" type calcium channels are apparently more closely related to neurotransmitter release, yet chronic ethanol administration produces an increase in the numbers of "L" channels and not of "N" channels. This is difficult to reconcile with the conclusion reached by many investigators that the main in vivo effect of ethanol and of other similar drugs is on synaptic transmission. If that is so, why would increases in the number of "L" channels on the cell body compensate for the effects of ethanol on synaptic transmission? Moreover, an increase in the number of "L" channels on the neuronal cell body would be expected to increase the degree of after-hyperpolarization, and thus to increase the duration of the refractory period. Therefore, the end result should be a decrease in the number of impulses initiated at the cell body, and a consequent decrease in release of neurotransmitter at the axon terminal. If anything, this should add to the effect of ethanol rather than provide a basis for tolerance to it. These and other related questions can be answered only by a careful concurrent study of the effects of ethanol, channel blockers, and electrical manipulation of neuronal membrane function, so as to map in detail all of the drug effects on all aspects of neuronal function in the same cells under the same conditions, at the same range of concentrations in vitro and after the same chronic drug treatment in vivo. Unfortunately, such studies have not yet been carried out, and the calcium channel hypothesis is an attractive but unproven one.

8. Conclusion

Throughout this overview, a number of themes and questions have been raised repeatedly, some explicitly and some implicitly. From the biologist's point of view, a purely behavioral explanation of drug tolerance or sensitization is a description, rather than an explanation, unless it attempts to link the behavioral influences with cellular mechanisms. Equally, a cellular explanation is not an explanation if it does not take into account the very important influences that behavioral and environmental factors can exert upon the development of tolerance or sensitization. Unless both approaches are employed in an integrated manner, it will be impossible to answer the question as to whether the various behavioral and pharmacological factors elicit the same type of tolerance or act through entirely different mechanisms.

One of the main difficulties in resolving this problem is the degree of methodological specialization that sophisticated modern techniques almost inevitably impose upon their users. As Lê and Khanna point out, very few researchers who have studied behavioral or environmental influences on tolerance have examined the basic pharmacokinetic issues. Equally, most investigators of cellular mechanisms of tolerance have failed to carry out complete dose–response studies, and almost none have tested the applicability of their postulated mechanisms to tolerance induced by behavioral or environmental manipulation of the drug response. Even within the broadly defined biological approach to the subject, as reviewed by Greenshaw et al., Willner, and Littleton and Little, remarkably little correlation has been attempted by the actual experimenters using various research techniques. For example, in the study of neurochemical effects of drugs and possible neuronal mechanisms of drug tolerance, studies on receptor properties or on membrane transducer systems are characteristically carried out on homogenates of whole brain or, at best, of gross anatomical subdivisions of it. The electrophysiologist using the patch-clamp technique works with isolated neurons in culture, or with subcellular fragments such as synaptoneurosomes formed by reclosure of the membrane of the axon terminal after the tissue has been disrupted. Yet whole organism studies have shown clearly that there are differences between the responses of different pathways, regions, and structures in the brain to the same dose of the same drug. It is incumbent upon the investigator who works with specific processes in small, isolated portions of the nervous system to attempt to retain some perspective on the features of drug actions, tolerance, and sensitization in the integrated nervous system and the whole organism in its normal environment.

Real understanding of the nature of tolerance and sensitization to drug effects is unlikely to come about until behavioral pharmacology has matured to the point of producing investigators who feel equally at home when studying behavioral, neurophysiological, and neurochemical aspects of the subject. This may be an unrealistic goal for the near future, but it must surely be the goal towards which we all work.

References

Barondes S. H., Traynor M. E., Schlapfer W. R., and Woodson P. B. J. (1979) Rapid adaptation to neuronal membrane effects of ethanol and low temperature: some speculations on mechanism. *Drug Alc. Dep.* **4,** 155–166.

Bläsig J., Herz A., Reinhold K., and Zieglgänsberger S. (1973) Development of physical dependence on morphine in respect to time and dosage and quantification of the precipitated withdrawal syndrome in rats. *Psychopharmacologia (Berl.)* **33,** 19–38.

Branch M. N. (1979) Consequent events as determinants of drug effects on schedule-controlled behavior: modification of effects of cocaine and *d*-amphetamine following chronic amphetamine administration. *J. Pharmacol. Exp. Ther.* **210,** 354–360.

Brick J. and Pohorecky L. A. (1977) Ethanol withdrawal: altered ambient temperature selection in rats. *Alcholism: Clin. Exp. Res.* **1,** 207–211.

Chen C. -S. (1968) A study of the alcohol-tolerance effect and an introduction of a new behavioral technique. *Psychopharmacologia (Berl.)* **12,** 433–440.

Chin J. H., Parsons L. M., and Goldstein D. B. (1978) Increased cholesterol content of erythrocytes and brain membranes in ethanol-tolerant mice. *Biochim. Biophys. Acta* **513,** 358–363.

Conney A. H. (1986) Induction of microsomal cytochrome P-450 enzymes. *Life Sci.* **39**, 2493–2518.

Crabbe J. C. and Belknap J. K. (1980) Pharmacogenetic tools in the study of drug tolerance and dependence. *Subst. Alcohol Actions Misuse* **1**, 385–413.

Deitrich R. A. and Spuhler K. (1984) Genetics of alcoholism and alcohol actions; in *Research Advances in Alcohol and Drug Problems*, Vol. 8 (Smart R. G., Cappell H. D., Glaser F. B., Israel Y., Kalant H., Popham R. E., Schmidt W., and Sellers E. M., eds.) pp. 47–98, Plenum Press, New York.

Demellweek C. and Goudie A. J. (1983) Behavioral tolerance to amphetamine and other psychostimulants: the case for considering behavioral mechanisms. *Psychopharmacology* **80**, 287–307.

De Souza Moreira L. F., Capriglione M. J., and Masur J. (1981) Development and reacquisition of tolerance to ethanol administered pre- and post-trial to rats. *Psychopharmacology* **73**, 165–167.

Dews P. B. (1969) General discussion, in *Experimental Approaches to the Study of Drug Dependence* (Kalant H. and Hawkins R. D., eds.) p. 150 and p. 168, University of Toronto Press, Toronto.

Dunham N. W. and Miya T. S. (1957) A note on a simple apparatus for detecting neurological deficit in rats and mice. *J. Am. Pharm. Assoc.* **46**, 208–209.

Eikelboom R. and Stewart J. (1982) Conditioning of drug-induced physiological responses. *Psychol. Rev.* **89**, 507–528.

Fernandes M., Kluwe S., and Coper H. (1977) Quantitative assessment of tolerance to and dependence on morphine in mice. *Naunyn-Schmiedeberg's Arch. Pharmacol.* **297**, 53–60.

Gelboin H. V. and Friedman F. K. (1985) Monoclonal antibodies for studies on xenobiotic and endobiotic metabolism. *Biochem. Pharmacol.* **34**, 2225–2234.

Gibbins R. J., Kalant H., and LeBlanc A. E. (1968) A technique for accurate measurement of moderate degrees of alcohol intoxication in small animals. *J. Pharmacol. Exp. Ther.* **159**, 236–242.

Gibson R. D. and Tingstad J. E. (1970) Formulation of a morphine implantation pellet suitable for tolerance-physical dependence studies in mice. *J. Pharm. Sci.* **59**, 426–427.

Goldstein D. B. and Pal N. (1971) Alcohol dependence produced in mice by inhalation of ethanol: grading the withdrawal reaction. *Science* **172**, 288–290.

Gouldstein D. B., Chin J. H., McComb J. A., and Parsons L. M. (1980)

Chronic effects of alcohols on mouse biomembranes. *Adv. Exp. Med. Biol.* **126,** 1–5.

Goudie A. J. and Demellweek C. (1986) Conditioning factors in drug tolerance, in *Behavioral Analysis of Drug Dependence* (Goldberg S. R. and Stolerman I. P., eds.) pp. 225–285, Academic Press, New York.

Gougos A., Khanna J. M., Lê A. D., and Kalant H. (1986) Tolerance to ethanol and cross-tolerance to pentobarbital and barbital. *Pharmacol. Biochem. Behav.* **24,** 801–807.

Gunne L. M. (1960) The temperature response in rats during acute and chronic morphine administration, a study of morphine tolerance. *Arch. Int. Pharmacodyn. Ther.* **129,** 416–428.

Hjeresen D. L., Reed D. R., and Woods S. C. (1986) Tolerance to hypothermia induced by ethanol depends on specific drug effects. *Psychopharmacology* **89,** 45–51.

Jorgensen H. A. and Hole K. (1984) Learned tolerance to ethanol in the spinal cord. *Pharmacol. Biochem. Behav.* **20,** 789–792.

Jorgensen H. A., Fasmer O. B., and Hole K. (1986) Learned and pharmacologically-induced tolerance to ethanol and cross-tolerance to morphine and clonidine. *Pharmacol. Biochem. Behav.* **24,** 1083–1088.

Kalant H. and Grose W. (1967) Effects of ethanol and pentobarbital on release of acetylcholine from cerebral cortex slices. *J. Pharmacol. Exp. Ther.* **158,** 386–393.

Kalant H., LeBlanc A. E., and Gibbins R. J. (1971) Tolerance to, and dependence on, some non-opiate psychotropic drugs. *Pharmacol. Rev.* **23,** 135–191.

Kalant H. and Woo N. (1981) Electrophysiological effects of ethanol on the nervous system. *Pharmacol. Ther.* **14,** 431–457.

Kalant H., LeBlanc A. E., Gibbins R. J., and Wilson A. (1978) Accelerated development of tolerance during repeated cycles of ethanol exposure. *Psychopharmacology* **60,** 59–65.

Khanna J. M. and Mayer J. M. (1982) An analysis of cross-tolerance among ethanol, other general depressants and opioids. *Subst. Alc. Actions Misuse* **3,** 243–257.

Koop D. R. and Coon M. J. (1985) Role of alcohol P-450-oxygenase (APO) in microsomal ethanol oxidation. *Alcohol* **2,** 23–26.

Lê A. D., Kalant H., and Khanna J. M. (1986a) The influence of ambient temperature on the development and maintenance of tolerance to ethanol-induced hypothermia. *Pharmacol. Biochem. Behav.* **25,** 667–672.

Lê A. D., Kalant H., and Khanna J. M. (1986b) Effects of treatment dose and intoxicated practice on the development of tolerance to ethanol-induced motor impairment. *Alcohol & Alcoholism*, **Suppl. 1,** 435–439.

Lê A. D., Khanna J. M., Kalant H., and Grossi F. (1986c) Tolerance to and cross-tolerance among ethanol, pentobarbital and chlordiazepoxide. *Pharmacol. Biochem. Behav.* **24,** 93–98.

LeBlanc A. E., Gibbins R. J., and Kalant H. (1973) Behavioral augmentation of tolerance to ethanol in the rat. *Psychopharmacologia (Berl.)* **30,** 117–122.

LeBlanc A. E., Matsunaga M., and Kalant H. (1976) Effects of frontal polar cortical ablation and cycloheximide on ethanol tolerance in rats. *Pharmacol. Biochem. Behav.* **4,** 175–179.

Littleton J. M. (1980a) The effects of alcohol on the cell membrane: a possible basis for tolerance and dependence, in *Addiction and Brain Damage* (Richter D., ed.) pp. 46–74, University Park Press, Baltimore.

Littleton J. M. (1980b) The assessment of rapid tolerance to ethanol, in *Alcohol Tolerance and Dependence* (Rigter H. and Crabbe J. C., Jr., eds.) pp. 53–79, Elsevier/North Holland, Amsterdam.

Margules D. L. and Stein L. (1968) Increase of "antianxiety" activity and tolerance of behavioral depression during chronic administration of oxazepam. *Psychopharmacologia (Berl.)* **13,** 74–80.

McLennan H. and Elliott K. A. C. (1951) Effects of convulsant and narcotic drugs on acetylcholine synthesis. *J. Pharmacol. Exp. Ther.* **103,** 35–43.

Mucha R. F. and Kalant H. (1980) Log dose/response curve flattening in rats after daily injection of opiates. *Psychopharmacology* **71,** 51–61.

Mucha R. F., Kalant H., and Kim C. (1987) Tolerance to hyperthermia produced by morphine in the rat. *Psychopharmacology* **92,** 452–458.

Murphree H. B. (1973) Electroencephalographic and other evidence for mixed depressant and stimulant actions of alcoholic beverages. *Ann. N. Y. Acad. Sci.* **215,** 325–331.

Pohorecky, L. A. and Rizek A. E. (1981) Biochemical and behavioral effects of acute ethanol in rats at different environmental temperatures. *Psychopharmacolgy* **72,** 205–209.

Rottenberg H., Waring A., and Rubin E. (1981) Tolerance and cross-tolerance in chronic alcoholics: reduced membrane binding of ethanol and other drugs. *Science* **213,** 583–585.

Schuster C. R., Dockens W. S., and Woods J. H. (1966) Behavioral variables affecting the development of amphetamine tolerance. *Psychopharmacologia (Berl.)* **9,** 170–182.

Smith J. B. (1986) Effects of chronically administered *d*-amphetamine on spaced responding maintained under multiple and single-component schedules. *Psychopharmacology* **88,** 296–300.

Speisky M. B. and Kalant H. (1987) Effects of anisomycin, PLG and DGAVP on rapid tolerance to ethanol in the rat. *Abst. Xth Internat. Congr. Pharmacol.*, Sydney, Australia. Abst. P781.

Tabakoff B. and Hoffman P. L. (1980) Alcohol and neurotransmitters, in *Alcohol Tolerance and Dependence* (Rigter H. and Crabbe J. C., Jr., eds.) pp. 201–226, Elsevier/North Holland, Amsterdam.

Traynor M. E., Schlapfer W. T., and Barondes S. H. (1980) Stimulation is necessary for the development of tolerance to a neuronal effect of ethanol. *J. Neurobiol.* **11,** 633–637.

Wenger J. R., Tiffany T. M., Bombardier C., Nicholls K., and Woods, S. C. (1981) Ethanol tolerance in the rat is learned. *Science* **213,** 575–577.

Contributing Authors

Glen B. Baker is Professor and Associate Director of the Neurochemical Research Unit, Department of Psychiatry, University of Alberta. He is also an Honorary Professor in the Faculty of Pharmacy and Pharmaceutical Sciences at the University of Alberta. Following completion of his Ph.D. in Biological Psychiatry at the University of Saskatchewan, Canada, he worked as a Postdoctoral Fellow at the Medical Research Council Neuropharmacology Unit, University of Birmingham, U.K. His principal research interests are in the mechanisms of action of antidepressant and neuroleptic drugs and the biochemical bases of psychiatric disorders, with particular reference to depression and attention deficit disorder. He has published numerous refereed articles, and is currently coeditor of the extensive *Neuromethods* series, and Vice-President of the Canadian College of Neuropsychopharmacology.

James E. Barrett is Professor in the Departments of Psychiatry, Pharmacology and Medical Psychology at the F. Edward Hebert School of Medicine, Uniformed Services University of the Health Sciences, Bethesda, Maryland, USA. His main research interests are in behavioral pharmacology and neurochemistry. He is coeditor, with T. Thompson and P. B. Dews, of the influential series of review chapters published as *Advances in Behavioural Pharmacology*.

Derek Blackman is Professor and Head of the School of Psychology at the University of Wales, College of Cardiff, UK. He obtained his Ph.D. from the Queen's University of Belfast, and subsequently researched in the Universities of Nottingham and Birmingham. His research interests are broadly those of a radical behaviorist, with specific interests in the experimental analysis of behavior and in behavioral pharmacology. He has served as a member of the editorial board of the *Journal of the Experimental Analysis of Behaviour*. He wrote *Operant Conditioning* (1974), and coedited (with D. J. Sanger) *Contemporary Research in Behavioural Pharmacology* (1978) and *Aspects of Psychopharmacology*

579

(1984). Professor Blackman has served as President of the British Psychological Society and as the editor of the *British Journal of Psychology* (1983–89).

Michael Emmett-Oglesby is Professor of Pharmacology at Texas College of Osteopathic Medicine, Fort Worth, Texas, USA. He views tolerance and dependence from both behavioral and pharmacological perspectives. These views are natural continuations of undergraduate training in Experimental Psychology at the University of Chicago (1969), and Ph.D. training in Pharmacology at The State University of New York at Buffalo (1973). Professor Emmett-Oglesby first became interested in tolerance when in the laboratory of Lewis Seiden in Chicago, where he was a post-doctoral fellow from 1973–1975. Seiden was studying tolerance to psychostimulants, and Emmett-Oglesby was intrigued by the biochemical basis of behavioral tolerance—he is still looking for this! Recently, he has been using drug discrimination procedures to explore animal models of human subjective responses to drugs, with particular reference to tolerance and withdrawal, and he has published numerous articles on these topics. In 1986, he was the recipient of a Humbolt Fellowship and spent a sabbatical year in the laboratory of Professor Albert Herz, at the Max-Planck Institute of Psychiatry in Munich.

John R. Glowa received his Ph.D from the University of Maryland, where he studied the behavioral effects of alcohol and chronic amphetamine administration, and the aversive properties of drugs. He was then appointed to the faculty of Harvard Medical School, where he studied the behavioral effects of volatile solvents and worked on the historical determinants of behavioral effects of drugs. As a member of the Clinical Neuroscience and Neuroendocrinology Branches of the National Institute of Mental Health, Dr. Glowa has recently extended his work on the behavioral effects of solvents, and developed interests in the relationships between neuroendocrine systems and the behavioral effects of drugs.

Andrew J. Goudie is currently Senior Lecturer in Psychology at Liverpool University. He graduated from Oxford University in 1971 with 1st class honors in Psychology and Physiology, and completed his doctoral research at Liverpool University in 1979. His main research interests are in the role of learning processes in the development of drug tolerance and dependence, with more specific interests in the actions of drugs as aversive and discriminative stimuli. He is a member of the editorial boards of the journals *Psychopharmacology* and *Neuropharmacology*. He has published over 50 papers in refereed journals, as well as a number of chapters for edited books.

Andrew J. Greenshaw is a Scholar of the Alberta Heritage Foundation for Medical Research, and Assistant Professor of Psychiatry at the University of Alberta in Edmonton, Canada. Following completion of his Ph.D. research at University College, Cardiff UK and at the Institute of Physiology of the Czech Academy of Sciences in Prague, Dr. Greenshaw worked at the Psychiatric Research Division at the University of Saskatchewan, Canada. Currently his research interests focus on the mechanisms of psychotherapeutic drug action with particular interest in the effects of chronic drug treatment. Dr. Greenshaw has coedited *Experimental Psychopharmacology: Concepts and Methods* and *Neuromethods: Psychopharmacology* and has published numerous research articles.

Harold Kalant, M.D. Ph.D, has been director of The Biobehavioural Research Section of The Addiction Research Foundation of Toronto, Ontario, Canada since 1979. He has published over 250 scientific articles and book chapters, and is a member of the editorial boards of a number of scientific journals, including: *The Journal of Studies on Alcohol; Drug and Alcohol Dependence; Psychopharmacology; Pharmacology, Biochemistry and Behavior;* and *Neuroscience and Biobehavioural Reviews.* He has received a number of awards for his research work, including the Jellinek Memorial Award for Research on Alcoholism; the Upjohn Award from the Pharmacological Society of Canada, and the Nathan B. Eddy Award from the Committee on Problems on Drug Dependence, USA. He has also acted as scientific adviser to a range of international bodies, including the World Health Organization, the Committee on Problems on Drug Dependence, and the National Institute on Alcoholism and Alcohol Abuse, USA.

Jatinder M. Khanna obtained his Ph.D. at the University of Connecticut, USA in 1964. Since 1976, he has been Professor of Pharmacology at the University of Toronto. He is a Senior Scientist and Section Head in the Behavioural Pharmacology and Drug Analysis Sections of the Alcoholism and Drug Addiction Research Foundation of Toronto, Ontario, Canada. His current major research interests are into the mechanisms of tolerance and cross-tolerance to alcohol and other psychotropic drugs and into the relationship between drug tolerance and self-administration.

A. D. Lê completed his M.Sc (1979) and Ph.D. (1982) in Pharmacology at the University of Toronto. He is currently a Research Scientist in the Biobehavioral Research Department of the Addiction Research Foundation in Toronto, Canada. His main research interests have focused on the role of behavioral (operant and classical conditioning) and pharmacological factors in the regulation of drug tolerance and cross-tolerance. He has published many papers on various aspects of tolerance, including studies on the role of genetic factors in tolerance

Hilary J. Little is currently Wellcome Trust Lecturer in Mental Health in the Pharmacology Department, Bristol University, England, having worked previously in the Pharmacology Department of Oxford University after obtaining M.Sc. and Ph.D. degrees at The University of Manchester. Her major research interests have always been in drug dependence, having worked initially on opiate and then on benzodiazepine dependence. Her current research is devoted to dependence on ethanol and other sedative/hypnotics. She has published numerous papers in these areas, including influential papers on the benzodiazepine receptor and its response to chronic drug treatment.

John M. Littleton is Professor of Pharmacology in the Department of Pharmacology, Kings College, London, England. He has a long-standing interest in the cellular basis of ethanol tolerance and dependence, on which he has published a number of influential papers.

Michael A. Nader received his Bachelor of Science degree, with honors, in Psychology at Wayne State University in 1982. His graduate training was with Dr. Travis Thompson at the University of Minnesota, USA, where he received a Ph.D. in 1985. He was a Postdoctoral Fellow and Research Associate in Professor James E. Barrett's laboratory at Uniformed Services University of the Health Sciences. Currently, he is a Research Associate in the Department of Psychiatry at the University of Chicago. His research interests include examination of the ways in which environmental variables influence the effects of drugs on schedule-controlled behavior, and the role of behavioral and pharmacological variables in the maintenance of drug self-administration.

Christine A. Sannerud is an Instructor in Psychiatry and Behavioural Sciences at The Johns Hopkins University School of Medicine, Baltimore, USA. She obtained her Ph.D. in 1985 at Wayne State University, Detroit, Michigan, and was subsequently (1985–1988) a Postdoctoral Fellow in Neuroscience and Psychiatry at The Johns Hopkins University School of Medicine. Her main research interests center on tolerance to, and dependence on, benzodiazepines.

Shepard Siegel is Professor of Psychology at McMaster University in Hamilton, Ontario, Canada. He has published extensively in the areas of learning and pharmacology, and has served on the editorial boards of several journals. He is a Fellow of the American Psychological Association, of the Canadian Psychological Association, and of the Academy of Behavioural Medicine Research. In 1985, he was the recipient of the Distinguished Psychopharmacologist Award of the Canadian Psychological Association.

Paul Willner currently holds a Personal Chair in Psychology at the City of London Polytechnic. He graduated from Oxford in 1969 with 1st class honors in Psychology and Physiology, and completed his doctoral research at Oxford University in 1974. He was a founder member of the European Behavioural Pharmacology Society. His main research interests are in the psychobiology of depression and in the psychopharmacology of feeding and reward. He has published over 60 full papers in journals, and in 1985, he wrote the major text *Depression: A Psychobiological Synthesis.* He is currently editing a book on *Behavioural Models in Psychopharmacology,* which is due to be published in 1989.

Thomas B. Wishart is Professor of Psychology, University of Saskatchewan, Canada. Following completion of his Ph.D. at the University of Western Ontario, he engaged in postdoctoral work at the University of Paris investigating the neuropsychological control of movement. His current research interests are in the neural and neurochemical control of behavior, into animal models of human psychiatric disorders and brain self-stimulation. Dr. Wishart has spent sabbatical years at the Institute of Neurology in London, UK and at the Department of Psychiatry at the University of Alberta. He is the author of numerous research articles and is a member of the editorial board of the *Canadian Journal of Psychology.*

David L. Wolgin's interest in drug tolerance developed when, as a doctoral student at Rutgers University USA, he collaborated with Peter Carlton on a project that led to the important discovery of "contingent tolerance" to the anorexigenic effect of amphetamine. Following a research position at the University of Illinois, he joined the Department of Psychology at Florida Atlantic University in 1975, where he is currently Professor of Psychology and Director of the Institute for the Study of Alcohol and Drug Dependence.

Alice M. Young is Associate Professor of Psychology at Wayne State University, Detroit, Michigan, USA. She received her Ph.D. in experimental psychology under the direction of Travis Thompson at the University of Minnesota, and postdoctoral training with James H. Woods at the University of Michigan. Her major research interests include the role of behavioral processes in opioid tolerance and opioid discriminative stimulus control. She is a recipient of a NIDA Research Scientist Development Award.

Index